T0281216

Introduction to LINEAR OPTIMIZATION

Introduction to LINEAR OPTIMIZATION

Arkadi Nemirovski
Georgia Institute of Technology, USA

NEW JERSEY • LONDON • SINGAPORE • BEIJING • SHANGHAI • HONG KONG • TAIPEI • CHENNAI • TOKYO

Published by

World Scientific Publishing Co. Pte. Ltd.
5 Toh Tuck Link, Singapore 596224
USA office: 27 Warren Street, Suite 401-402, Hackensack, NJ 07601
UK office: 57 Shelton Street, Covent Garden, London WC2H 9HE

Library of Congress Cataloging-in-Publication Data
Names: Nemirovskiĭ, A. S. (Arkadiĭ Semenovich), author.
Title: Introduction to linear optimization / Arkadi Nemirovski,
Georgia Institute of Technology, USA.
Description: New Jersey : World Scientific, [2024] | Includes bibliographical references and index.
Identifiers: LCCN 2023016700 | ISBN 9789811277900 (hardcover) |
ISBN 9789811278730 (paperback) | ISBN 9789811277917 (ebook for institutions) |
ISBN 9789811277924 (ebook for individuals)
Subjects: LCSH: Linear programming--Textbooks. | Mathematical optimization--Textbooks.
Classification: LCC QA402.5 .N4538 2024 | DDC 519.7/2--dc23/eng/20231012
LC record available at https://lccn.loc.gov/2023016700

British Library Cataloguing-in-Publication Data
A catalogue record for this book is available from the British Library.

For any available supplementary material, please visit
https://www.worldscientific.com/worldscibooks/10.1142/13457#t=suppl

Desk Editors: Soundararajan Raghuraman/Nijia Liu

Typeset by Stallion Press
Email: enquiries@stallionpress.com

Printed in Singapore

Preface

This graduate-level textbook, based upon lecture notes created by the author during his 15+ years of teaching the semester-long graduate course *Introduction to Linear Optimization* (ISyE 6661) at the Georgia Institute of Technology School of Industrial and Systems Engineering, addresses the basics of modeling, theory, and algorithmic tools of linear optimization (LO).

The Contents

The topics we cover are as follows:

A. *Descriptive theory of LO*:

  (a) *LO Modeling*: Here, aside from presenting illustrative examples (several traditional decision-making problems like diet, production planning, transportation, and applications in engineering — compressive sensing and synthesis of linear controllers), we focus on how to recognize optimization problems that can be posed as LO programs; scientifically speaking, this is the calculus of polyhedrally representable sets and functions.

  (b) Geometry of polyhedral sets.

  (c) Theory of systems of linear inequalities (general theorem on alternative) and linear programming duality, including the geometry of primal–dual pairs of LO programs and applications, e.g., in robust LO.

We do our best to represent the above topics rigorously and in full detail.

B. *Classical pivoting algorithms of LO — Primal and dual simplex method and network simplex method*:

While presentation here is still rigorous and self-contained, it is more sketchy than in **A** — with all due respect to pivoting algorithms that, historically, played a crucial role in the dawning of LO and in the first 40+ years of its development as a mathematical theory with a broad spectrum of real-life applications, today their role in LO *per se* is relatively moderate. Besides this, we believe that the majority of our potential readers in the future will *apply* LO in various subject areas rather than *develop* LO algorithms; with this in mind, we do *not* think that algorithms, traditional and modern alike, should constitute a significant part in an introductory graduate textbook on LO.[1] As a consequence of our attitude, we do not touch at all some subjects, such as revised simplex algorithm, Dantzig–Wolfe and Benders decompositions, etc., which are usually included in more traditional graduate LO textbooks like the highly popular, and rightly so, work of (Bertsimas and Tsitsiklis, 1997).

C. *Ellipsoid Algorithm and polynomial time solvability of LO with rational data (Khachiyan, 1979).*

D. *Beyond LO: Conic programming and interior point path-following methods*, including the following:

(a) conic programming and "magic" conic problems: linear, Second-Order Conic, and Semidefinite Optimization;

(b) conic programming duality theorem and geometry of primal–dual pairs of conic problems;

[1]As a "common life" analogy to our attitude: A driver in a country with developed car service does not need to be an expert in car engines; of course, it makes no harm to have some general idea of the engine, e.g., to know that 150 horsepowers are not the same as 150 horses under the hood... 30 years ago, pivoting algorithms formed the entire toolbox of LO; today, the default setting in all known to us commercial LO solvers is interior point computation; who knows what will be LO "working horses" in 20 years from now? In contrast, we believe that the descriptive theory of LO is essentially complete and will stay "as is" for decades to come; as a result, this theory should, to the best of our understanding, be the backbone of an introductory LO textbook.

(c) conic representations of sets and functions and expressive abilities of second-order and semidefinite optimization;
(d) path-following interior point polynomial methods (IPMs) for LO and SDO.

While presentation here, as everywhere in the textbook, is rigorous and (essentially) self-contained, it, same as in B, is sketchy, but by reasons quite different from those in B — somehow, "full' presentation of the topic, including details on conic quadratic and semidefinite representations of functions and sets or on long-step path-following methods, or interior point methods for second-order conic programming, not speaking about IPMs for general conic problems, would require incomparably larger space and, in our opinion, goes far beyond the scope of an introductory textbook on LO. On the other hand, *some* essential information on this subject seems to be necessary — calculus of conic representations and conic duality are natural extensions of their "LO counterparts," and IPM-related material provides readers with a basic knowledge of the current state of LO and SDO algorithmic toolbox.

Audience and prerequisites: Our intended audience is graduate (and perhaps senior undergraduate) math and engineering students interested in the rigorous presentation of the mathematics of modeling, theoretical, and algorithmic components of linear optimization. As for the prerequisites, they primarily reduce to the most elementary linear algebra; the concluding "Conic Programming and IPM" part of the book requires minimal acquaintance with real analysis, calculus, and linear algebra basics on symmetric matrices. All these prerequisites are summarized, without proofs, in the corresponding appendices. The most important prerequisite, however, is not the acquaintance with the outlined subject fields, it is the general mathematical culture — interest not just in facts, but in the "driving forces" beyond these facts, and the ability to comprehend the related mathematical reasoning.

Organization of material is fully traditional. The titles of some sections start with an asterisk $*$, indicating that the section is devoted to specific, usually somehow advanced, material and can be skipped in the first reading. All "LO proper" chapters are augmented by exercises, some of them, like "mark by **P** polyhedral sets in the following list," allowing the reader to check how well he/she is acquainted with the corresponding material, and some offering to investigate less straightforward questions and thus to "put the acquired knowledge in motion." Solutions to selected exercises can be found at the end of the textbook.

Acknowledgments: I am extremely grateful to the numerous colleagues, including first and foremost Aharon Ben-Tal, Dimitris Bertsimas, Stephen Boyd, Anatoli Juditsky, Yuri Nesterov, Boris Polyak, Alex Shapiro and Craig Tovey who over the years influenced my understanding of our profession in general and of the subject of this textbook in particular. My separate gratitude goes to my younger colleagues and former Ph.D. students from whom I learned a great deal — Alexander Goldenshluger, Fatma Kilinc-Karzan and George Lan. Last, but not least, I am extremely thankful to Ms. Rochelle Kronzek, Executive Editor at World Scientific Publishing Company, for encouraging me to convert the lecture notes posted for many years on my websites into a printed textbook. Needless to say, the responsibility for all kinds of drawbacks lies solely on me.

Arkadi Nemirovski
Atlanta, USA

About the Author

Arkadi Nemirovski is John Hunter Chair and Professor at H. Milton Stewart School of Industrial and System Engineering, Georgia Institute of Technology. He got his Ph.D. (Math, 1974) from Moscow State University, Russia. His research areas are convex optimization and nonparametric statistics. He has been awarded Fulkerson Prize (MPS and AMS, joint with L. Khachiyan and D. Yudin), Dantzig Prize (MPS and SIAM, joint with M. Groetschel), John von Neumann Theory Prize (INFORMS, joint with M. Todd) and Norbert Wiener Prize in Applied Mathematics (AMS and SIAM, joint with M. Berger). He is a member of the American Academy of Arts and Sciences and the US National Academies of Science and of Engineering.

Contents

Main Notational Conventions

Vectors and matrices: By default, all vectors in this book are column vectors. Usually, we utilize "MATLAB notation:" a vector with coordinates $x_1, \ldots, x_n$ is written down as $x = [x_1; \ldots; x_n]$. More generally, if $A_1, \ldots, A_m$ are matrices with the same number of columns, we write $[A_1; \ldots; A_m]$ to denote the matrix which is obtained when writing A_2 beneath A_1, A_3 beneath A_2, and so on. If $A_1, \ldots, A_m$ are matrices with the same number of rows, then $[A_1, \ldots, A_m]$ stands for the matrix which is obtained when writing A_2 to the right of A_1, A_3 to the right of A_2, and so on. For example,

$$[1, 2, 3, 4] = [1; 2; 3; 4]^T,$$

$$[[1, 2; 3, 4], [5, 6; 7, 8]] = [1, 2, 5, 6; 3, 4, 7, 8] = \begin{bmatrix} 1 & 2 & 5 & 6 \\ 3 & 4 & 7 & 8 \end{bmatrix}.$$

$\mathrm{Diag}\{A_1, A_2, \ldots, A_m\}$ denotes the block-diagonal matrix with the diagonal blocks $A_1, \ldots, A_m$. For example,

$$\mathrm{Diag}\{[1, 2], 3, [4, 5; 6, 7]\} = \left[\begin{array}{cc|c|cc} 1 & 2 & & & \\ \hline & & 3 & & \\ \hline & & & 4 & 5 \\ & & & 6 & 7 \end{array}\right],$$

where blank spaces are filled with zeros.

For a square nonsingular matrix A, A^{-T} means $[A^{-1}]^T$.

The zero vectors and matrices are, as always, denoted by 0; if we have reasons to point out the sizes of a zero vector/matrix, we write something like $0_{3\times 4}$. The unit $m \times m$ matrix is denoted by I or I_m.

We write $A \succeq B$ (or, which is the same, $B \preceq A$) to express the fact that A, B are symmetric matrices of the same size such that $A - B$ is positive semidefinite; $A \succ B$ (or, which is the same, $B \prec A$) means that A, B are symmetric matrices of the same size such that $A - B$ is positive definite. We abbreviate "if and only if" to "iff."

Chapter 1

Introduction to LO: Examples of LO Models

In this chapter, we define the main entity we are interested in our book — a *Linear Optimization (LO) problem*, provide a number of instructive examples and address the question of when an optimization problem can be posed as an LO one.

1.1 LO program: Definition

1.1.1 *An LO program*

A *Linear Optimization problem*, or *program* (also called *Linear Programming (LP)* problem/program) is the problem of optimizing a *linear function* $c^T x$ of an n-dimensional vector x under *finitely many linear* equality and nonstrict inequality constraints. For example, the following mathematical programming problems

$$\min_x \left\{ x_1 : \left\{ \begin{array}{c} x_1 + x_2 \leq 20 \\ x_1 - x_2 = 5 \\ x_1, x_2 \geq 0 \end{array} \right\} \right. \tag{1.1}$$

and

$$\max_x \left\{ x_1 + x_2 : \left\{ \begin{array}{c} 2x_1 \geq 20 - x_2 \\ x_1 - x_2 = 5 \\ x_1 \geq 0 \\ x_2 \leq 0 \end{array} \right. \right\} \tag{1.2}$$

are LO programs. In contrast to this, the optimization problems

$$\min_x \left\{ \exp\{x_1\} : \left\{ \begin{array}{c} x_1 + x_2 \leq 20 \\ x_1 - x_2 = 5 \\ x_1, x_2 \geq 0 \end{array} \right. \right\} \tag{1.3}$$

and

$$\max_x \left\{ x_1 + x_2 : \left\{ \begin{array}{l} \quad ix_1 \geq 20 - x_2, i = 2, 3, ... \\ x_1 - x_2 = 5 \\ \quad x_1 \geq 0 \\ \quad x_2 \leq 0 \end{array} \right. \right\} \tag{1.4}$$

are *not* LO programs: (1.3) has a nonlinear objective, and (1.4) has infinitely many constraints.

> A careful reader could say that (1.3) is "the same" as (1.1) (since the exponent is monotone, it is the same what to minimize, x_1 or $\exp\{x_1\}$). Similarly, (1.4) is "the same" as (1.2), since for $x_1 \geq 0$, the infinitely many constraints $ix_1 + x_2 \geq 20$, $i = 2, 3, ...$ are equivalent to the single constraint $2x_1 + x_2 \geq 20$. Note, however, that we classify optimization problems according to *how they are presented*, and not according to *what they can be equivalent/reduced to.*

Now, we can somehow "standardize" the format in which an LO program is presented. Specifically, we have the following:

- every linear equality/inequality can be equivalently rewritten in the form where the left-hand side is a weighted sum $\sum_{j=1}^n a_j x_j$ of variables x_j with coefficients, and the right-hand side is a real constant; e.g., $2x_1 \geq 20 - x_2$ is equivalent to $2x_1 + x_2 \geq 20$;
- the sign of a nonstrict linear inequality always can be made "$\leq$," since the inequality $\sum_j a_j x_j \geq b$ is equivalent to $\sum_j [-a_j] x_j \leq [-b]$;
- a linear equality constraint $\sum_j a_j x_j = b$ can be represented equivalently by the pair of opposite inequalities $\sum_j a_j x_j \leq b$, $\sum_j [-a_j] x_j \leq [-b]$;
- minimizing a linear function $\sum_j c_j x_j$ is exactly the same as maximizing the linear function $\sum_j [-c_j] x_j$.

Canonical form of an LO program: In view of the above observations, *every* LO program can be equivalently written down as a problem of maximizing a linear objective under finitely many nonstrict linear inequality constraints of the "$\leq$"-type, i.e., in the *canonical* form

$$\begin{aligned}
&\mathrm{Opt} = \max_x \left\{ \sum_{j=1}^n c_j x_j : \sum_{j=1}^n a_{ij} x_j \leq b_i,\ 1 \leq i \leq m \right\} \\
&\qquad\qquad [\text{“term-wise” notation}] \\
&\Leftrightarrow \mathrm{Opt} = \max_x \left\{ c^T x : a_i^T x \leq b_i,\ 1 \leq i \leq m \right\} \qquad (1.5) \\
&\qquad\qquad [\text{“constraint-wise” notation}] \\
&\Leftrightarrow \mathrm{Opt} = \max_x \left\{ c^T x : Ax \leq b \right\} \\
&\qquad\qquad [\text{“matrix-vector” notation}]
\end{aligned}$$

where $c = [c_1; \ldots; c_n]$, $a_i = [a_{i1}; \ldots; a_{in}]$, $A = [a_1^T; a_2^T; \ldots; a_m^T]$, $b = [b_1; \ldots; b_m]$.

Standard form of an LO program: An LO program in the *standard* form reads

$$\begin{aligned}
&\mathrm{Opt} = \max_x \left\{ \sum_{j=1}^n c_j x_j : \begin{array}{l} \sum_{j=1}^n a_{ij} x_j = b_i,\ 1 \leq i \leq m \\ x_j \geq 0,\ j = 1, \ldots, n \end{array} \right\} \\
&\qquad\qquad [\text{“term-wise” notation}] \\
&\Leftrightarrow \mathrm{Opt} = \max_x \left\{ c^T x : \begin{array}{l} a_i^T x = b_i,\ 1 \leq i \leq m \\ x_j \geq 0,\ 1 \leq j \leq n \end{array} \right\} \qquad (1.6) \\
&\qquad\qquad [\text{“constraint-wise” notation}] \\
&\Leftrightarrow \mathrm{Opt} = \max_x \left\{ c^T x : Ax = b, x \geq 0 \right\} \\
&\qquad\qquad [\text{“matrix-vector” notation}],
\end{aligned}$$

where $c = [c_1; \ldots; c_n]$, $a_i = [a_{i1}; ...; a_{in}]$, $A = [a_1^T; a_2^T; ...; a_m^T]$, $b = [b_1; ...; b_m]$. As compared with (1.5), in the standard form of an LO all "general" linear constraints are equalities, and the inequality constraints are *sign constraints*, specifically, the restrictions of nonnegativity, imposed on all variables. The standard form is as "universal" as the canonical one.

Observation 1.1. Every LO program can be straightforwardly converted into an equivalent program in the standard form.

Proof. We lose nothing by assuming that the original form of the program is the canonical one, specifically,

$$\max_y \left\{e^T y : Py \leq p\right\} \tag{!}$$

(note change in the notation). Now, to say that $Py \leq p$ is exactly the same as saying that $Py + u = p$ for certain *nonnegative* vector u; in addition, every real vector y can be represented as the difference of two nonnegative vectors: $y = v - w$. It follows that (!) is equivalent to the problem

$$\max_{x=[u;v;w]} \left\{c^T x := e^T[v-w] : Ax := P(v-w) + u = b := p, x \geq 0\right\}$$

which is an LO in the standard form. □

In the sequel, when investigating the "geometry" of LO, it will be more convenient to use the canonical form of an LO program; the standard form is preferable when presenting traditional LO algorithms.

1.1.2 *LO terminology*

We are about to present the most basic "vocabulary" of LO. For the sake of definiteness, in our presentation we refer to the canonical format of an LO program (1.5), leaving the "translation" to the case of a program in the standard form to the reader. The vocabulary is as follows:

- The variable vector x in (1.5) is called the *decision vector* of the program; its entries x_j are called *decision variables.* The linear function $c^T x$ is called the *objective function* (or just *objective*) of the problem, and the inequalities $a_i^T x \leq b_i$ are called the *constraints*. Sometimes, with slight abuse of wording, we refer to the vector c itself as to the objective.

- The *structure* of (1.5), given the way we are writing the problem down, reduces to the *sizes* m (number of constraints) and n (number of variables). The *data* of an LO program is the collection of numerical values of the coefficients in the *cost vector* (or simply *objective*) c, in the *right-hand side vector* b and in the *constraint matrix* A.
- A *solution* to (1.5) is an arbitrary value of the decision vector. A solution x is called *feasible* if it satisfies the constraints: $Ax \leq b$. The set of all feasible solutions is called the *feasible set* of the program; the program is called *feasible* if the feasible set is nonempty, and is called *infeasible* otherwise.
- Given a program (1.5), there are three possibilities:
  - The program is infeasible. In this case, its optimal value Opt, by definition, is $-\infty$ (this convention is logical, since in the case in question one cannot point out a feasible solution with the value of the objective $> -\infty$).
  - The program is feasible, and the objective is *not* bounded from above on the feasible set, meaning that for every real a one can point out a feasible solution x such that $c^T x > a$. In this case, the program is called *unbounded*, and its optimal value Opt is, by definition, $+\infty$.

    The program which is not unbounded is called *bounded*; a program is bounded iff its objective is bounded from above on the feasible set (e.g., due to the fact that the latter is empty).
  - The program is feasible, and the objective is bounded from above on the feasible set: there exists a real a such that $c^T x \leq a$ for all feasible solutions x. In this case, the optimal value Opt is the supremum, over the feasible solutions, of the values of the objective at a solution.

    Thus, to say that Opt $= 5$ means to say two things: first, there does not exist a feasible solution with the value of the objective > 5, and, second, for every $\epsilon > 0$ there exists a feasible solution with the value of the objective $\geq 5 - \epsilon$.

- A solution to the program is called *optimal* if it is feasible, and the value of the objective at the solution equals to Opt. A program is called *solvable* if it admits an optimal solution.

Remarks. A. The above terminology is aimed at the *maximization* LO in the canonical form. The terminology in the case of a *minimization* problem "mirrors" the one we have described, specifically,

- the optimal value of an infeasible program is $+\infty$;
- the optimal value of a feasible and *unbounded* program (unboundedness now means that the objective to be minimized is not bounded *from below* on the feasible set) is $-\infty$, while the optimal value of a *bounded and feasible* LO is the *infimum* of values of the objective at feasible solutions to the program.

B. The notions of feasibility, boundedness, solvability and optimality can be straightforwardly extended from LO programs to arbitrary MP ones. With this extension, a solvable problem definitely is feasible and bounded (why?), while the inverse is not necessarily true, as illustrated by the program

$$\mathrm{Opt} = \max_x \{-\exp\{-x\} : x \geq 0\},$$

where the optimal value — the supremum of the values taken by the objective at the points of the feasible set — clearly is 0; this value, however, is not achieved — there is no feasible solution where the objective is equal to $0 = \mathrm{Opt}$, and, as a result, the program is unsolvable. Thus, in general, the facts that an optimization program has a "legitimate" — real, and not $\pm\infty$ — optimal value, is *strictly weaker* than the fact that the program is solvable (i.e., has an optimal solution). In LO the situation is much better; eventually we shall prove that *an LO program is solvable iff it is feasible and bounded.*

1.2 Examples of LO models

Here, we present a short series of (mostly) standard examples of LO problems. In every one of them, we start with a certain semi-verbal story and then "translate" this story into an LO program; this is called *modeling* — building a mathematical model, in our case, of the LO type, of a "practical" situation. It should be stressed that in applications of Optimization, modeling plays a crucial role: on the one hand, we need to end up with a model which is not "oversimplified," that is, captures all important relations and dependencies between the entities involved for the application in question, and, on the other hand, is not too complicated, so that we can specify all the relevant data and process the resulting problem numerically at a reasonable computational cost. A proper balance between these two conflicting goals requires both deep understanding of the subject area to which the application belongs and good knowledge of optimization theory and algorithms. This being said, note that modeling *per se*, being a "go-between for reality and Mathematics," is beyond the scope of our book.

1.2.1 *Examples of LO models in Operations Research*

1.2.1.1 *Diet problem*

> *There are n types of products and m types of nutrition elements. A unit of product # j contains p_{ij} grams of nutrition element # i and costs c_j. The daily consumption of a nutrition element # i should be at least a given quantity $\underline{b}_i$ and at most a given quantity $\overline{b}_i$. Find the cheapest possible "diet" — mixture of products — which provides appropriate daily amounts of every one of the nutrition elements.*

Denoting x_j the amount of jth product in a diet, the LO model of the problem reads[1]

$$
\begin{array}{lll}
\min\limits_x & \sum\limits_{j=1}^n c_j x_j & \text{[Diet's cost to be minimized]},\\
\text{subject to} & & \\
& \left.\begin{array}{l}\sum\limits_{j=1}^n p_{ij}x_j \geq \underline{b}_i\\ \sum\limits_{j=1}^n p_{ij}x_j \leq \overline{b}_i\end{array}\right\}, 1 \leq i \leq m & \begin{bmatrix}\text{bounds on the contents}\\ \text{of nutrition elements}\end{bmatrix},\\
& x_j \geq 0, 1 \leq j \leq n & \begin{bmatrix}\text{one cannot use negative}\\ \text{amounts of products}\end{bmatrix}.
\end{array}
\tag{1.7}
$$

[1]Here and in the subsequent examples, we do not bother to convert the model into a specific form, e.g., the canonical one, since this (completely straightforward and "mechanical") process would only obscure the construction of a model. Note that existing LO solvers also do not require from a user to input the problem in certain particular form and use preprocessors to convert an LO program into the format directly accessible for the solver. A "standard" format is convenient when investigating LO's as "mathematical beasts," not when building LO models!

The Diet problem is one of the first LO models, and today it is routinely used in many areas, e.g., in mass-production of poultry. With regard to nourishment of human beings, the model is of not much use, since it completely ignores factors like food's taste, food diversity requirements, etc.

Here is the optimal daily human diet as computed by the software at https://neos-guide.org/case-studies/om/the-diet-problem/ (when solving the problem, I allowed to use all 64 kinds of food offered by the code):

Food	Serving	Cost
Raw Carrots	0.12 cups shredded	0.02
Peanut Butter	7.20 Tbsp	0.25
Popcorn, Air-Popped	4.82 Oz	0.19
Potatoes, Baked	1.77 cups	0.21
Skim Milk	2.17 C	0.28

Note: Daily cost $ 0.96.

1.2.1.2 *Production planning*

A factory consumes R types of resources (electricity, raw materials of various kinds, various sorts of manpower, processing times at different devices, etc.) *and produces P types of products. There are n possible production processes, jth of them can be used with "intensity" x_j* (you may think of these intensities as fractions of the planning period (say, 1 month) during which a particular production process is used). *Used at unit intensity, production process # j consumes A_{rj} units of resource r, $1 \leq r \leq R$, and yields C_{pj} units of product p, $1 \leq p \leq P$. The profit of selling a unit of product p is c_p. Given upper bounds $b_1, ..., b_R$ on the amounts of various resources available during the planning period, and lower bounds $d_1, ..., d_P$ on the amount of products to be produced, find a production plan which maximizes the profit under the resource and the demand restrictions.*

Denoting by x_j the intensity at which production process j is used, the LO model reads as follows:

$$\max_x \sum_{j=1}^{n} \left(\sum_{p=1}^{P} c_p C_{pj} \right) x_j \quad \text{[profit to be maximized]},$$

subject to

$$\sum_{j=1}^{n} A_{rj} x_j \leq b_r,\ 1 \leq r \leq R \quad \begin{bmatrix} \text{upper bounds on consumed} \\ \text{resources should be met} \end{bmatrix},$$

$$\sum_{j=1}^{n} C_{pj}x_j \geq d_p,\ 1 \leq p \leq P \qquad \begin{bmatrix}\text{lower bounds on products'}\\ \text{yield should be met}\end{bmatrix},$$

$$\left.\begin{array}{l}\sum_{j=1}^{n} x_j \leq 1\\ x_j \geq 0,\ 1 \leq j \leq n\end{array}\right\} \qquad \begin{bmatrix}\text{total intensity should be} \leq 1 \text{ and}\\ \text{intensities must be nonnegative}\end{bmatrix}. \tag{1.8}$$

Note that the simple model we have presented tacitly assumes that all what is produced can be sold, that there are no setup costs when switching from one production process to another one, that products are infinitely divisible (we produce needles rather than Boeings, so to speak), and makes a lot of other implicit assumptions.

1.2.1.3 *Inventory*

An inventory operates over the time horizon $1, ..., T$ (say, T days) and handles K types of products:

- *Products share common warehouse with storage space C. The space required to store a unit of product k in the warehouse is $c_k \geq 0$, and the holding cost (the per-day cost of storing a unit of product k in the warehouse) is h_k.*
- *The inventory is replenished via ordering from a supplier; a replenishment order sent in the beginning of day t is executed immediately, and ordering a unit of product k costs o_k.*
- *The inventory is affected by external demand which amounts to d_{tk} units of product k in day t. While backlogged demand is allowed, a day-long delay in supplying a customer by unit of product k costs p_k.*

Given the initial amounts s_{0k}, $k = 1, ..., K$, of products in the warehouse, all the cost coefficients (which are nonnegative) and the demands d_{tk},[2] *we want to specify the replenishment orders v_{tk} (*v_{tk} is the amount of product k which is ordered from the supplier at the beginning of period t*) in such a way that at the end of period T there is no backlogged demand, and we want to meet this requirement at total inventory management costs that are as small as possible.*

[2]The latter assumption — that the demands are known in advance — more often than not is unrealistic. This issue will be addressed in the mean time, when speaking about LO problems with uncertain data.

In order to convert this story into an LO model, we first introduce the *state variables* s_{tk} representing the amount of product k in the warehouse at the end of period t (or, which is the same, at the beginning of period $t+1$); we allow these state variables to be negative as well as positive, with a negative value of s_{tk} interpreted as "at the end of period t, the inventory owes the customers $|s_{tk}|$ units of product k." With this convention, our problem can be modeled as the optimization program

$$
\begin{array}{ll}
\min\limits_{U,v,s} & U \\
\text{subject to} & \\
& U = \sum_{k=1}^{K}\sum_{t=1}^{T}[o_k v_{tk} + h_k \max[s_{tk},0] + p_k \max[-s_{tk},0]] \quad (a)\\
& s_{tk} = s_{t-1,k} + v_{tk} - d_{tk}, 1 \le t \le T, 1 \le k \le K \quad (b)\\
& \sum_{k=1}^{K} c_k \max[s_{tk},0] \le C,\ 1 \le t \le T \quad (c)\\
& s_{Tk} \ge 0, 1 \le k \le K \quad (d)\\
& v_{tk} \ge 0, 1 \le k \le K, 1 \le t \le T \quad (e)
\end{array}
\tag{1.9}
$$

(the s-variables s_{tk} have $t \ge 1$, s_{0k} being part of problem's data). In this model, we have the following:

- The variable U is the overall inventory management cost which we want to minimize.
- Constraint (a) expresses the fact that U indeed is the overall inventory management cost — the total, over the K products and the T days, ordering cost $(o_k v_{tk})$, holding cost $(h_k \max[s_{tk}, 0])$ and penalty for backlogged demand $(p_k \max[-s_{tk}, 0])$ associated with product k and period t. Further, constraints (b) express the evolution of states, and constraints (c) express the restrictions on the space available for storing the products.

  The implicit assumptions underlying the latter claims are as follows: the replenishment orders v_{tk} are issued in the beginning of day t and are executed immediately. As a result, the amount of product k in the inventory in the beginning of day t jumps from $s_{t-1,k}$ to $s_{t-1,k} + v_{tk}$.

Immediately after this, the demands d_{tk} of day t become known and the products are shipped to the customers, which reduces — again immediately! — the inventory level by d_{tk}, so that the resulting level $s_{t-1,k} + v_{tk} - d_{tk}$, if nonnegative, is the amount of product k stored in the inventory during day t, otherwise the modulus of this level is the backlogged demand on product k during day t.

From the story we have just told we see, first, that the states s_{tk} of the inventory, as defined above, evolve according to (b). Second, our expenses, associated with product k, in day t include the ordering cost $o_k v_{tk}$, and on top of it, *either* the holding cost $h_k s_{tk}$ (this is so if s_{tk} is nonnegative), *or* the penalty $p_k[-s_{tk}]$ for backlogged demand (when s_{tk} is negative). We see that (a) correctly represents the expenses. Further, we see that the "physical" amount of product k stored in the warehouse during day t is $\max[s_{tk}, 0]$, so that (c) correctly represents the restriction on the space available for storage of products.

- Constraint (d) expresses equivalently the requirement that at the end of the planning period (i.e., at the end of day t) there is no backlogged demand.
- Finally, constraints (c) express the implicit assumption that we can only order from the supplier, while return of products to the supplier is forbidden.

Note that we lose nothing when replacing the *equality* constraint (1.9(a)) with the *inequality* constraint

$$U \geq \sum_{k=1}^{K} \sum_{t=1}^{T} [o_k v_{tk} + h_k \max[s_{tk}, 0] + p_k \max[-s_{tk}, 0]] \qquad (a')$$

which corresponds to minimizing an *upper bound* U on the actual inventory management cost; since nothing prevents us from setting this bound to be equal to the right-hand side in (1.9(a)) (which in any case will be enforced by minimization), the problem modified in this way is equivalent to the original one. In the remaining discussion, we assume that (1.9(a)) is replaced with (a$'$).

We have modeled our verbal problem by (a slightly modified version of) (1.9); note, however, that the resulting model is *not* an LO program, due to the presence of *nonlinear* in our design variables terms $\max[\pm s_{tk}, 0]$. We are about to demonstrate (pay maximal attention to this construction!) that *we can handle these types of nonlinearities via LO*. Specifically, assume that we have a constraint of the form

$$\alpha_1 \mathrm{Term}_1(y) + \cdots + \alpha_M \mathrm{Term}_M(y) \leq b, \qquad (!)$$

where α_ℓ are *nonnegative* constant coefficients, b is a constant, and every term $\mathrm{Term}_\ell(y)$ is either a linear function of our design variables y (let it be so for $L < \ell \leq M$), or is a *piecewise linear* function of the form

$$\mathrm{Term}_\ell(y) = \max\left[a_{1\ell}^T y + b_{1\ell}, a_{2\ell}^T y + b_{2\ell}, \ldots, a_{p\ell}^T y + b_p\right] \qquad (*)$$

(the latter is the case for $1 \leq \ell \leq L$).

Note that *converted equivalently to the form of* (!), *all constraints* (a'), (1.9(b–e)) *are of this form*, which is an immediate corollary of the nonnegativity of the cost coefficients h_k, p_k and the space coefficients c_k.

Now, let us replace every piecewise linear term Term_ℓ in (!) with a new decision variable w_ℓ ("slack variable" in the slang of LO) and augment this action by imposing the constraints

$$w_\ell \geq a_{\nu\ell}^T y + b_{\nu\ell}, \quad 1 \leq \nu \leq p$$

on the original and the slack variables. As a result, the constraint (!) will be replaced by the system

$$\begin{aligned} &\sum_{\ell=1}^{L} \alpha_\ell w_\ell + \sum_{\ell=L+1}^{M} \alpha_\ell \mathrm{Term}_\ell(y) \leq b \\ &w_\ell \geq a_{\nu\ell}^T y + b_{\nu\ell}, \ 1 \leq \nu \leq p, 1 \leq \ell \leq L \end{aligned}$$

of *linear* inequalities in the variables $y, w_1, \ldots, w_L$. Taking into account that $\alpha_1, \ldots, \alpha_L$ are nonnegative, it is clear that *this system says about our original variables y exactly the same as the constraint* (!), meaning that y can be extended, by properly chosen values of slack variables w_ℓ, $1 \leq \ell \leq L$, to a feasible solution of the system iff y satisfies (!). If now (!) is a constraint in certain optimization problems, then, augmenting the variables of the problem by slack variables w_ℓ and replacing the constraint in question with the aforementioned system, we arrive at an *equivalent* problem where the nonlinear constraint in question is replaced with a system of linear constraints. If all constraints with nonlinearities in an optimization problem admit the outlined treatment, we can apply the outlined procedure "constraint by constraint" and end up with an LO which is equivalent to the original problem.

Let us apply this recipe to problem (1.9) (with constraint (a) replaced with (a′), which, as we remember, keeps the problem intact up to equivalence). Specifically, we introduce the slack variables (upper bounds) y_{tk} for the quantities $\max[s_{tk}, 0]$ and z_{tk} for the quantities $\max[-s_{tk}, 0]$ and replace the nonlinearities with these upper bounds, augmenting the resulting system of constraints with linear constraints (constraints (f), (g) below) expressing equivalently the fact that y_{tk}, z_{tk} indeed are upper bounds on the corresponding nonlinearities. The resulting program reads

$$\begin{array}{lll}
\min\limits_{U,v,s,y,z} & U & \\
\text{subject to} & & \\
& U \geq \sum_{k=1}^{K}\sum_{t=1}^{T}[o_k v_{tk} + h_k y_{tk} + p_k z_{tk}] & \text{(a)} \\
& s_{tk} = s_{t-1,k} + v_{tk} - d_{tk}, 1 \leq t \leq T, 1 \leq k \leq K & \text{(b)} \\
& \sum_{k=1}^{K} c_k y_{tk} \leq C,\ 1 \leq t \leq T & \text{(c)} \\
& s_{Tk} \geq 0, 1 \leq k \leq K & \text{(d)} \\
& v_{tk} \geq 0, 1 \leq k \leq K, 1 \leq t \leq T & \text{(e)} \\
& y_{tk} \geq s_{tk},\ y_{tk} \geq 0, 1 \leq k \leq K, 1 \leq t \leq T & \text{(f)} \\
& z_{tk} \geq -s_{tk},\ z_{tk} \geq 0, 1 \leq k \leq K, 1 \leq t \leq T & \text{(g)}
\end{array} \tag{1.10}$$

and is an LO program which is equivalent to (1.9) and thus models our inventory problem.

Warning: The outlined "eliminating nonlinearities" heavily exploits the facts that

(1) (!) is a constraint with piecewise linear nonlinearities which are *maxima* of linear forms,
(2) all the nonlinearities are to the left of "≤"-sign, and
(3) the coefficients at these nonlinearities are nonnegative.

Indeed, assume that we are given a constraint with the terms which are either linear functions of the variables, or piecewise linear functions "maximum of linear terms" multiplied by constant coefficients. We always can rewrite this constraint in the form of (!), but the coefficients at nonlinearities in this constraint should not necessarily be nonnegative. Of course, we can "move the coefficients into the nonlinearities," noting that

$$c \max\left[a_1^T y + b_1, \ldots, a_p^T y + b_p\right]$$

is either $\max[ca_1^T y + cb_1, \ldots, ca_p^T y + b_p]$ when $c \geq 0$, or $\min[ca_1^T y + cb_1, \ldots, ca_p^T y + b_p]$ when $c < 0$. Now, all nonlinearities have coefficients 1 and are to the left of "$\leq$," but there, in general, are two types of them: maxima of linear forms and minima of linear forms. The above construction shows that

> If *all the nonlinearities in the constraint*
>
> $$\text{Term}_1(x) + \cdots + \text{Term}_M(x) \leq b \qquad (!)$$
>
> *are maxima of linear forms, we can eliminate them at the cost of introducing slack variables and converting* (!) *into a system of linear inequalities. The number of slack variables we need is equal to the number of nonlinearities we are eliminating, and the number of linear inequalities we end up with is by one greater, than the total number of linear forms participating in the nonlinearities.*

The situation changes dramatically when among the nonlinearities are minima of linear forms. Given such a nonlinearity, say, $\min[a_1^T y + b_1, \ldots, a_p^T y + b_p]$, we can, of course, replace it with a slack variable w at the cost of augmenting the list of constraints by the constraint $w \geq \min[a_1^T y + b_1, \ldots, a_p^T y + b_p]$ ("isolating" the nonlinearity, so to speak). The difficulty is that now this additional constraint cannot be immediately reduced to a system of linear inequalities: instead of expressing the fact that w is $\geq$ the maximum of $a_i^T y + b_i$ over i, that is, that $w \geq a_1^T y + b_1$ AND $w \geq a_2^T y + b_2$ AND ... AND $w \geq a_p^T y + b_p$ (which is just a system of p linear inequalities on w and y), we need to express the fact that w is $\geq$ the minimum of $a_i^T y + b_i$ over i, that is, that $w \geq a_1^T y + b_1$ OR $w \geq a_2^T y + b_2$ OR ... OR $w \geq a_p^T y + b_p$, which is *not* a system of linear inequalities on w, y. Of course, it is possible to eliminate nonlinearities of the min-type by "branching" on them: to eliminate nonlinearity $w \geq \min[a_1^T y + b_1, \ldots, a_p^T y + b_p]$, we build p uncoupled problems where this nonlinearity is substituted subsequently by every one of the linear inequalities $w \geq a_i^T y + b_i$, $i = 1, \ldots, p$. However, if we have several "bad" — requiring branching — nonlinearities in an optimization problem, when eliminating all of them, we need to consider separately all combinations of the above substitutions across the bad nonlinearities. As a result, if in the original problem we have K "bad" piecewise linear nonlinearities and kth of them involves p_k linear functions, their elimination results in the necessity to consider separately $N = p_1 p_2 \ldots p_K$ "LO variants" of the original problem. Since the number of variants grows exponentially fast with the number of bad nonlinearities, this approach, at least in its outlined straightforward form, can be used only when K and p_i, $1 \leq i \leq K$, are small.

1.2.1.4 *Transportation and network flows*

There are I warehouses, ith of them storing s_i units of product, and J customers, jth of them demanding d_j units of product. Shipping a unit of product from warehouse i to customer j costs c_{ij}. Given the supplies s_i, the demands d_j and the costs C_{ij}, we want to decide on the amounts of product to be shipped from every warehouse to every customer. Our restrictions are that we cannot take from a warehouse more product than it has, and that all the demands should be satisfied; under these restrictions, we want to minimize the total transportation cost.

Introducing decision variables x_{ij}, with x_{ij} being the amount of product to be shipped from warehouse i to customer j, the story can be modeled by the LO program

$$
\begin{array}{ll}
\min_x \sum_{i,j} c_{ij} x_{ij} & \text{[transportation cost to be minimized]}, \\
\text{subject to} & \\
\sum_{j=1}^{J} x_{ij} \leq s_i,\ 1 \leq i \leq I & \left[\begin{array}{l}\text{we should respect capacities} \\ \text{of the warehouses}\end{array}\right], \\
\sum_{i=1}^{I} x_{ij} = d_j,\ j = 1, \ldots, J & \text{[we should satisfy the demands]}, \\
x_{ij} \geq 0, 1 \leq i \leq I, 1 \leq j \leq J & \left[\begin{array}{l}\text{you cannot ship negative} \\ \text{amount of a product}\end{array}\right].
\end{array}
\tag{1.11}
$$

We end up with what is called a *transportation problem.* Note that when building the model, we have assumed implicitly that the product is infinitely divisible.

A far-reaching generalization of the transportation problem is the *multicommodity network flow problem* as follows. We are given a *network* (an oriented graph) — a finite set of *nodes* $1, \ldots, n$ along with a finite set Γ of *arcs* — ordered pairs $\gamma = (i, j)$ of distinct $(i \neq j)$ nodes. We say that an arc $\gamma = (i, j)$ *starts* at node i, *ends* at node j and *links* node i to node j. As an example, you may think of a road network, where the nodes are road junctions, and the arcs are the one-way segments of roads "from a junction to a neighboring one;" a two-way road segment can be modeled by two opposite arcs. (Of course, many other interpretations of a network are possible). Now imagine that there are N types of "commodities" moving along the network, and let s_{ki} be the "external supply" of kth commodity at node i. This supply can be positive (meaning that the node "pumps" into the network s_{ki} units of commodity k), negative (the node "drains" from the network $|s_{ki}|$ units of commodity k) and zero. You may think of kth commodity as about the stream of cars originating within a time unit (say, one hour) at a particular node (say, at GaTech campus) and moving to a particular destination (say, Northside Hospital); the corresponding s_{ki} are zeros for all nodes i except for the origin and the destination ones. For the origin node i, s_{ki} is the per hour amount c of cars leaving the node for the destination in question, while for the destination node i, $s_{ki} = -c$, so that $|s_{ki}|$ is the per hour amount of cars arriving at the destination from a given origin.[3] Now, the propagation of commodity k through the network can be represented by a vector f^k with entries f^k_γ indexed by the arcs of the network; f^k_γ is the amount of the commodity moving through the arc γ. Such a vector is called a *feasible flow* if it is nonnegative and meets the *conservation law* as follows: *for every node i in the network, the total amount of the commodity k arriving at the node plus the supply s_{ki} of the commodity k at the node equals the total amount of the commodity k leaving the node*:

$$\sum_{p \in P(i)} f^k_{pi} + s_{ki} = \sum_{q \in Q(i)} f^k_{iq},$$

[3]In our static "traffic illustration" we assume implicitly that the traffic is in steady state.

where $P(i)$ is the set of all nodes p such that (p, i) is an arc in the network, and $Q(i)$ is the set of all nodes q such that (i, q) is an arc in the network.

The *multicommodity flow* problem reads as follows: *Given*

- *a network with n nodes $1, \ldots, n$ and a set Γ of arcs,*
- *a number K of commodities along with supplies s_{ki} of nodes $i = 1, \ldots, n$ to the flow of commodity k, $k = 1, \ldots, K$,*
- *the per unit cost $c_{k\gamma}$ of transporting commodity k through arc γ,*
- *the capacities h_γ of the arcs,*

find the flows $f^1, \ldots, f^K$ of the commodities which are nonnegative, respect the conservation law and the capacity restrictions (that is, the total, over the commodities, flow through an arc does not exceed the capacity of the arc) *and minimize, under these restrictions, the total, over the arcs and the commodities transportation cost.*

In our "traffic illustration," you may think about a shipping cost $c_{k\gamma}$ as about the time required for a car to travel through arc γ (with this interpretation, $c_{k\gamma}$ should be independent of k), in which case the optimization problem becomes the problem of finding *social optimum* — the routing of cars in which the *total*, over all cars, traveling time of a car is as small as possible.

To write down an LO model of the problem, let us define the *incidence matrix* $P = [P_{i\gamma}]$ of a network as the matrix with rows indexed by the nodes $i = 1, \ldots, n$ of the network and columns indexed by the arcs $\gamma \in \Gamma$ of the network, with the entry $P_{i\gamma}$

- equal to 1 when the arc γ starts at node i,
- equal to -1 when the arc γ ends at node i,
- equal to 0 in all remaining cases.

For example, the incidence matrix of the graph

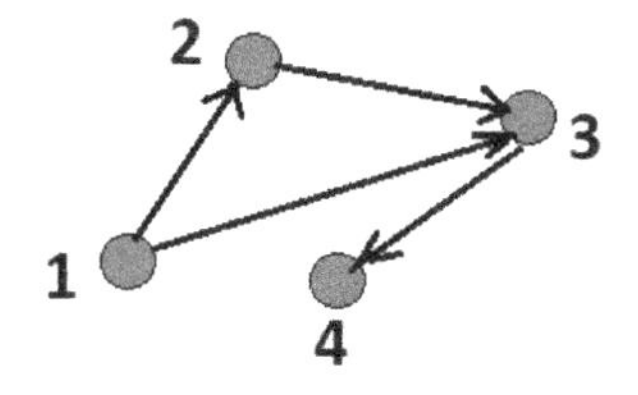

is

$$P = \begin{bmatrix} 1 & 1 & 0 & 0 \\ -1 & 0 & 1 & 0 \\ 0 & -1 & -1 & 1 \\ 0 & 0 & 0 & -1 \end{bmatrix}$$

$$\left[\begin{array}{c} \text{indexes of rows, top–bottom: nodes 1,2,3,4} \\ \text{indexes of columns, left–right: arcs (1,2),(1,3),(2,3),(3,4)} \end{array}\right].$$

With this notation, the conservation law for a flow f, the supplies being $s_1, \ldots, s_n$, reads (check it!)

$$\sum_\gamma P_{i\gamma} f_\gamma = s_i, \quad i = 1, \ldots, n.$$

Now we can write down an LO program modeling the multicommodity flow problem as follows:

$$\begin{array}{ll}
\min_{f^1,\ldots,f^K} \sum_{k=1}^K \sum_{\gamma\in\Gamma} c_{k\gamma} f^k_\gamma & [\text{total transportation cost}], \\
\text{subject to} & \\
Pf^k = s^k := [s_{k1}; \ldots; s_{kn}], \quad k = 1, \ldots, K & \left[\begin{array}{l}\text{flow conservation law for the} \\ \text{flow of every commodity}\end{array}\right], \\
f^k_\gamma \geq 0,\ 1 \leq k \leq K, \gamma \in \Gamma & [\text{flows must be nonnegative}], \\
\sum_{k=1}^K f^k_\gamma \leq h_\gamma, \gamma \in \Gamma & \left[\begin{array}{l}\text{we should respect bounds} \\ \text{on capacities of the arcs}\end{array}\right].
\end{array} \tag{1.12}$$

Note that the transportation problem is a very specific case of the multicommodity flow problem. Indeed, in the situation of the transportation problem, let us start with $I + J$-nodal graph with I red nodes representing warehouses and J green nodes representing the customers, and IJ arcs leading from every warehouse to every customer; the arc from ith warehouse to jth customer has infinite capacity and transportation cost c_{ij}. Further, let us add to this graph one extra node, called *source*, and I arcs linking the source and the warehouses. The arc "source-warehouse #i" is assigned with zero transportation cost and capacity s_i (the amount of product in

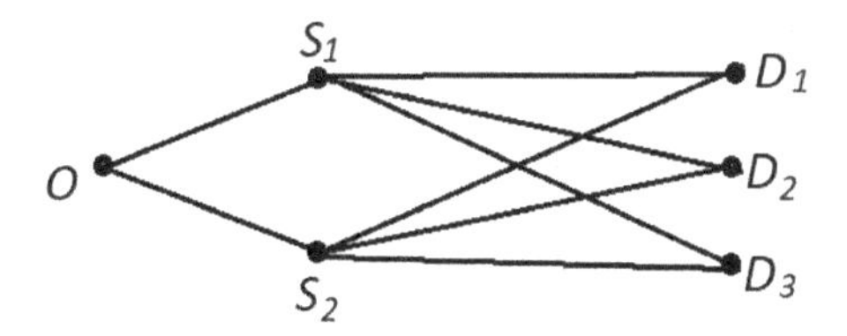

Fig. 1.1. Network Flow reformulation of the transportation problem $\min_x\{\sum_{i\le 2,j\le 3} c_{ij}x_{ij} : \sum_{j\le 3} x_{ij} \le s_i, i \le 2, \sum_{i\le 2} x_{ij} = d_j, j \le 3, x_{ij} \ge 0, i \le 2, j \le 3\}$. Arc capacities: s_i in arcs (O, S_i), $i \le 2$, $+\infty$ in arcs (S_i, D_j); External supplies: $d_1 + d_2 + d_3$ at O, $-d_j$ at D_j, $j \le 3$, zeros at S_i, $i \le 2$; Transportation costs: zeros in (O, S_i), $i \le 2$, c_{ij} in (S_i, D_j), $i \le 2$, $j \le 3$.

warehouse i). Finally, let there be a single commodity, with external supply equal to $D = \sum_{j=1}^{J} d_j$ at the source node and equal to $-d_j$ at the green node # j (the node representing jth customer); see Fig. 1.1. Clearly, the resulting single-commodity version of the multicommodity flow problem (1.12) is nothing but the transportation problem (1.11).

LO programs on networks form a special, extremely nice part of LO. Here is one of the most beautiful problems of this type — the *Maximal Flow* problem as follows: *We are given a network with arcs γ assigned with nonnegative capacities h_γ. One of the nodes is designated as source, another one — as sink. We are looking at the maximal flow from source to sink, that is, for the largest s such that the external supply "s at the source, $-s$ at the sink, zero at all other nodes" corresponds to certain feasible flow respecting the arc capacity bounds.*

The LO model of this problem reads:

$$\max_{f,s} s \qquad \text{[total flow from source to sink to be maximized]},$$

subject to

$$\begin{aligned}
&\sum_\gamma P_{i\gamma} f_\gamma = \begin{cases} s, & i \text{ is the source node,} \\ -s, & i \text{ is the sink node} \\ 0 & \text{for all other nodes,} \end{cases} \qquad \text{[flow conservation law]},\\
&f_\gamma \ge 0,\ \gamma \in \Gamma \quad \text{[flows in the arcs should be nonnegative]},\\
&f_\gamma \le h_\gamma, \gamma \in \Gamma \quad \text{[we should respect arc capacities]}.
\end{aligned} \tag{1.13}$$

The beauty of network flow problems stems from the fact that one can utilize additional and very specific structures coming from the associated network; as a result, numerous network flow LOs admit specialized highly efficient solution algorithms which *within their scope* by far outperform "general purpose" LO methods.

1.2.2 *Engineering examples*

Traditionally, LO models and algorithms were considered as part of Operations Research and as such were primarily associated with decision-making applications. Power of LO in engineering applications was realized essentially later, and "penetrating" of LO in these areas seems to be still in progress. Applications of this type include synthesis of linear controllers for discrete time linear dynamical systems, and various applications in Data Mining and Signal Processing. Here we present just two illustrations, one of them teaching us important "modeling tricks," and the other one selected due to its crucial role in sparsity-oriented signal processing. Application to synthesis of linear controllers relies on LO duality and will be considered in Chapter 3 (Section 3.3.7).

1.2.2.1 *Fitting parameters in linear regression models*

Imagine that we have observed m pairs of input $a_i \in \mathbf{R}^n$ and output $y_i \in \mathbf{R}$ to and from a "black box," respectively. Sometimes we have reasons to believe that this output is a corrupted-by-noise version of the "existing in principle, but unobservable, ideal output" $y_i^* = a_i^T x^*$, which is just a linear function of the inputs (this is called "linear regression model"). Our goal is to convert actual observations a_i, y_i, $1 \leq i \leq m$, into estimates of the *unknown* vector of parameters x^*. This problem would be easy if there were no observation errors (y_i were exactly equal to $a_i^T x^*$) and we were possessing a rich enough set of observations, so that among the vectors $a_1, \ldots, a_m$ ("regressors" in the terminology of linear regression) there were $n = \dim x^*$ linearly independent. In this case the "true" vector of unknown parameters would be a solution to the system of linear equations $y_i = a_i^T x$, $1 \leq i \leq m$, in variables x, the solution to this system being unique (since among the vectors $a_1, \ldots, a_m$ there are n linearly independent); it remains to note that finding the unique solution to a solvable system of linear equations is a simple Linear Algebra problem.

The situation changes dramatically when there are observation noises and/or the number n of "degrees of freedom" of the regression model — the dimension of the vector of parameters, or, which is the same, of the regressor vectors a_i — is larger than the number m of observations. Because of observation noises, the system

$$a_i^T x = y_i, \quad i = 1, \ldots, m \tag{$*$}$$

in variables x can become infeasible (this will be typically the case when $m > n$) and even when feasible, it is unclear what is the relation of its solution(s) to the true value of the parameter vector (which now is *not* a solution to the system). Likewise, with a nonunique solution (this will be typically the case when $m < n$), it is unclear which one of the solutions to the system to take — and this is so even if there are no observation errors, that is, when we know in advance that the true vector of parameters is among the solutions to the system.

There exists a wide (and constantly extending) spectrum of various techniques for parameter estimation in a linear regression, differing from each other primarily in what is our *a priori* information on the nature of the observation errors, the structure of the true vector of parameters, etc.; some of these techniques heavily utilize LO. For example, we can choose a simple "discrepancy measure" — a kind of distance between the vector of outputs $Ax = [a_1^T x; \ldots; a_m^T x]$ of our hypothetical model (here A is the $m \times n$ matrix with the rows $a_1^T, \ldots, a_m^T$) and the vector of observed outputs $y = [y_1; \ldots; y_m]$, and look for the vector of parameters which minimizes this discrepancy. This amounts to the necessity to solve the optimization problem

$$\min_x \phi(Ax, y), \tag{$*$}$$

where $\phi(u, v)$ is the discrepancy between vectors u and v. Note that this approach does not make much sense when the number of observations m is less than the number n of unknown parameters (think why); it is used, if at all, when $m \gg n$.

There are two simple cases when the outlined problem reduces to LO. The first is the case when we are interested in the *uniform fit*:

$$\phi(u,v) = \|u-v\|_\infty := \max_{1\le i\le \dim u} |u_i - v_i|.$$

The second case corresponds to the ℓ_1 *fit*

$$\phi(u,v) = \|u-v\|_1 := \sum_{i=1}^{\dim u} |u_i - v_i|.$$

- With the uniform fit, $(*)$ reads

$$\min_x \max_{1\le i\le m} |a_i^T x - y_i|;$$

while literally this is not an LO program, we can easily convert it to the LO form by introducing slack variable τ which should be an upper bound on all the quantities $|a_i^T x - y_i|$ and minimizing this bound. The resulting problem reads

$$\min_{\tau,x} \left\{\tau : a_i^T x - y_i \le \tau, y_i - a_i^T x \le \tau, 1 \le i \le m\right\},$$

which is an LO program.

- With the ℓ_1-fit, $(*)$ reads

$$\min_x \sum_{i=1}^m |a_i^T x - y_i|, \tag{1.14}$$

which again is *not* an LO program. There are two ways to convert it into LO — a good and a bad one. The good way is to note that $|r| = \max[r, -r]$, that is, $|a_i^T x - y_i|$ is the maximum of two linear forms of x, and to use the trick we remember from processing the inventory problem; the resulting LO equivalent to the problem of interest reads

$$\min_{x,w} \left\{\sum_{i=1}^m w_i : a_i^T x - y_i \le w_i, y_i - a_i^T x \le w_i, 1 \le i \le m\right\}.$$

A bad way is to note that $\sum_{i=1}^{m}|r_i| = \max_{\epsilon_1=\pm1,\epsilon_2=\pm1,\dots,\epsilon_m=\pm1} \sum_{i=1}^{m}\epsilon_i r_i$, which allows to write the problem of interest down as an LO solely in the original variables x, augmented by a *single* slack variable τ, specifically, as

$$\min_{\tau,x}\left\{\tau : \tau \geq \sum_{i=1}^{m}\epsilon_i[a_i^T x - y_i]\ \forall \epsilon_1 = \pm1, \epsilon_2 = \pm1, \dots, \epsilon_m = \pm1\right\}. \tag{1.15}$$

While being legitimate, this conversion is indeed bad, since we end up with an LO with 2^m linear constraints; numerical handling of the resulting LO program will be completely impractical already for $m = 10$, and will be impossible for $m = 30$.

1.2.2.2 *Sparsity-oriented signal processing and ℓ_1 minimization*

Compressed Sensing addresses, essentially, the same linear regression problem as above, but in the case opposite to the one we have just considered, specifically, when the number of observations m is much less than the number n of unknown parameters. Thus, we are in the situation when the m-dimensional vector of observations is obtained from an *unknown* n-dimensional vector of parameters x^* according to

$$y = Ax^*, \tag{!}$$

(for the time being, there is no observation error), and A is a given $m \times n$ *sensing matrix*. Our goal is to recover x^* given y, and the Compressed Sensing situation is the one where $m \ll n$. At first glance, our goal is unreachable: when $m \ll n$, (!), treated as a system of linear equations in variables x, is heavily underdetermined: if solvable, it has infinitely many solutions, including those which are very far from each other, since the solution set of (!), if nonempty, is unbounded (why?). It follows that we have no chances to recover the true solution, unless we augment the observations with certain additional information. In Compressed Sensing, this additional information is the one of *sparsity* of x^*, specifically, the a priory knowledge of an upper bound $s \ll m$ on the number of nonzero entries in x^*.

Note that in many applications we indeed can be sure that the true vector of parameters x^* is sparse. Consider, e.g., the following story about signal detection (from the applied viewpoint, this story is *not* of "linear regression" flavor, but the mathematical model looks exactly the same).

> *There are n locations where signal transmitters could be placed, and m locations with the receivers. The contribution of a signal of unit magnitude originating in location j to the signal measured by receiver i is a known quantity A_{ij}, and signals originating in different locations merely sum up in the receivers; thus, if x^* is the n-dimensional vector with entries x_j^* representing the magnitudes of signals transmitted in locations $j = 1, 2, \ldots, n$, then the m-dimensional vector y of measurements of the m receivers is $y = Ax^*$. Given this vector, we intend to recover x^*.*

Now, if the receivers are hydrophones registering noises emitted by submarines in a certain part of the Atlantic, tentative positions of submarines being discretized with resolution 500 m, the dimension of the vector x^* (the number of points in the discretization grid) will be in the range of tens of thousands, if not tens of millions. At the same time, the total number of submarines (i.e., nonzero entries in x^*) can be safely upper-bounded by 50, if not by 20.

It should be added that typical images and audio signals, *when represented by their coefficients in properly selected bases*[4] admit tight sparse approximations, making sparsity-oriented signal recovery an extremely promising, literally revolutionary, technique in, e.g., Medical Imaging, where it allows to reduce by significant factor the acquisition time in procedures like MRI without sacrificing quality of the resulting images.[5]

In view of the just outlined "Signal Processing" interpretation of the situation we are in, in the sequel we use the words "true signal" as an equivalent to the words "the true vector of parameters."

---

[4]Specifically, wavelet bases, whatever this means, in the case of images, and in the Fourier basis in the case of audio signals.

[5]Excellent outline of the role of Compressed Sensing in Medical Imaging can be found in the Gauss Prize lecture of one of the founders of Compressed Sensing, David Donoho, https://www.bing.com/videos/search?q=donoho+Gauss+lecture&view=detail&mid=42235C42695AFBBEAA4E42235C42695AFBBEAA4E&FORM=VIRE.

Given in advance that x^* has at most $s \ll m$ nonzero entries, the possibility of *exact* recovery of x^* from observations y becomes quite natural. Indeed, let us try to recover x^* by the following "brute force" search: we inspect, one by one, all subsets I of the index set $\{1, \ldots, n\}$ — first the empty set, then n singletons $\{1\}, \ldots, \{n\}$, then $\frac{n(n-1)}{2}$ 2-element subsets, etc., and each time try to solve the system of linear equations

$$y = Ax, \quad x_j = 0 \text{ when } j \notin I;$$

when arriving for the first time at a solvable system, we terminate and claim that its solution is the true vector x^*. It is clear that we will terminate before all sets I of cardinality $\leq s$ are inspected. It is also easy to show (do it!) that if every $2s$ distinct columns in A are linearly independent (when $m \geq 2s$, this indeed is the case for a matrix A in a "general position"[6]), then the procedure is correct — it indeed recovers the true vector x^*.

A bad news is that the outlined procedure becomes completely impractical already for "small" values of s and n because of the astronomically large number of linear systems we need to process.[7] A partial remedy is as follows. The outlined approach is, essentially,

[6]Here and in the sequel, the phrase "in general position" means the following. We consider a family of objects, with a particular object — an instance of the family — identified by a vector of real parameters (you may think about the family of $n \times n$ square matrices; the vector of parameters in this case is the matrix itself). We say that an instance of the family possesses certain property *in general position*, if the set of values of the parameter vector for which the associated instance does *not* possess the property is of measure 0. Equivalently, randomly perturbing the parameter vector of an instance, the perturbation being uniformly distributed in a (whatever small) box, we with probability 1 get an instance possessing the property in question, e.g., a square matrix "in general position" is nonsingular.

[7]When $s = 5$ and $n = 100$, this number is $\approx 7.53\text{e}7$ — much, but perhaps doable. When $n = 200$ and $s = 20$, the number of systems to be processed jumps to $\approx 1.61\text{e}27$, which is by many orders of magnitude beyond our "computational grasp"; we would be unable to carry out that many computations even if the fate of the mankind were dependent on them. And from the perspective of Compressed Sensing, $n = 200$ still is a completely toy size, by 3–4 orders of magnitude less than we would like to handle.

a particular way to solve the optimization problem

$$\min\{\text{nnz}(x) : Ax = y\}, \tag{$*$}$$

where $\text{nnz}(x)$ is the number of nonzero entries of a vector x. At the present level of our knowledge, this problem looks completely intractable (in fact, we do not know algorithms solving the problem essentially faster than the brute force search), and there are strong reasons to believe that it indeed is intractable. Well, if we do not know how to minimize under linear constraints the "bad" objective $\text{nnz}(x)$, let us "approximate" this objective with one which we do know how to minimize. The true objective is separable: $\text{nnz}(x) = \sum_{i=1}^{n} \xi(x_j)$, where $\xi(s)$ is the function on the axis equal to 0 at the origin and equal to 1 otherwise. As a matter of fact, the separable functions which we do know how to minimize under linear constraints are sums of *convex* functions of $x_1, \ldots, x_n$.[8] The most natural candidate to the role of *convex* approximation of $\xi(s)$ is $|s|$; with this approximation, $(*)$ converts into the ℓ_1*-minimization problem*

$$\min_x \left\{ \|x\|_1 := \sum_{j=1}^{n} |x_j| : Ax = y \right\}, \tag{1.16}$$

which, as we know, is equivalent to the LO program

$$\min_{x,w} \left\{ \sum_{j=1}^{n} w_j : Ax = y, x_j \leq w_j, -x_j \leq w_j, 1 \leq j \leq n \right\}.$$

For the time being, we were focusing on the (unrealistic!) case of noiseless observations. A realistic model is that the observation contains noise ξ:

$$y = Ax^* + \xi$$

[8] A real-valued function $f(s)$ on the real axis is called *convex*, if its graph, between every pair of its points, is below the chord linking these points, or, equivalently, if $f(x + \lambda(y - x)) \leq f(x) + \lambda(f(y) - f(x))$ for every $x, y \in \mathbf{R}$ and every $\lambda \in [0, 1]$. For example, maxima of (finitely many) affine functions $a_i x + b_i$ on the axis are convex. For more detailed treatment of convexity, see Chapter 2 (Section 2.1.3).

and we know an upper bound δ on the "magnitude" $\|\xi\|$ of the noise. In this case, ℓ_1 minimization becomes

$$\min_x \{\|x\|_1 : \|Ax - y\| \leq \delta\}. \tag{1.17}$$

When $\|\cdot\|$ is either $\|\cdot\|_\infty$, or $\|\cdot\|_1$, the latter problem again reduces to LO, specifically, to the LO program

$$\min_{x,w} \left\{ \sum_{j=1}^n w_j : \left\{ \begin{array}{l} -\delta \leq [Ax - y]_i \leq \delta,\ 1 \leq i \leq m \\ -w_j \leq x_j \leq w_j,\ 1 \leq j \leq n \end{array} \right\} \right.$$

when $\|\cdot\| = \|\cdot\|_\infty$, and to the LO program

$$\min_{x,w,z} \left\{ \sum_{j=1}^n w_j : \left\{ \begin{array}{l} -z_i \leq [Ax - y]_i \leq z_i,\ 1 \leq i \leq m \\ \sum_{i=1}^m z_i \leq \delta \\ -w_j \leq x_j \leq w_j,\ 1 \leq j \leq n \end{array} \right\} \right.$$

when $\|\cdot\| = \|\cdot\|_1$.

1.2.2.3 **How good is ℓ_1 minimization in the compressed sensing context?*

s-goodness and nullspace property: Let us say that a sensing matrix A is *s-good* if in *the noiseless case* ℓ_1 minimization (1.16) recovers correctly all s-sparse signals x. It is easy to say when this is the case: the necessary and sufficient condition for A to be s-good is the following *nullspace property:*

$$\forall (z \in \mathbf{R}^n : Az = 0, z \neq 0, I \subset \{1, \ldots, n\}, \mathrm{Card}(I) \leq s) : \sum_{i \in I} |z_i| < \frac{1}{2}\|z\|_1. \tag{1.18}$$

In other words, for every nonzero vector $z \in \mathrm{Ker}\, A$, the sum $\|z\|_{s,1}$ of the s largest magnitudes of entries in z should be strictly less than half of the sum of magnitudes of all entries.

The necessity and sufficiency of the nullspace property for s-goodness of A can be derived "from scratch" — from the fact that s-goodness means that every s-sparse signal x should be the unique optimal solution to the associated LP $\min_w\{\|w\|_1 : Aw = Ax\}$ combined with the LP optimality conditions. Another option, which we use here, is to *guess* the condition and then to prove that it is indeed necessary and sufficient for s-goodness of A. The necessity is evident: if the nullspace property does *not* take place, then there exists $0 \neq z \in \operatorname{Ker} A$ and s-element subset I of the index set $\{1,\ldots,n\}$ such that if J is the complement of I in $\{1,\ldots,n\}$, then the vector z_I obtained from z by zeroing out all entries with indexes not in I along with the vector z_J obtained from z by zeroing out all entries with indexes not in J satisfy the relation $\|z_I\|_1 \geq \frac{1}{2}\|z\|_1 = \frac{1}{2}[\|z_I\|_1 + \|z_J\|_1]$, that is,

$$\|z_I\|_1 \geq \|z_J\|_1.$$

Since $Az = 0$, we have $Az_I = A[-z_J]$, and we conclude that the s-sparse vector z_I is *not* the unique optimal solution to the LP $\min_w\{\|w\|_1 : Aw = Az_I\}$, since $-z_J$ is a feasible solution to the program with the value of the objective at least as good as the one at z_I, on the one hand, and the solution $-z_J$ is different from z_I (since otherwise we should have $z_I = z_J = 0$, whence $z = 0$, which is not the case) on the other hand.

To prove that the nullspace property is *sufficient* for A to be s-good is equally easy: indeed, assume that this property does take place, and let x be s-sparse signal, so that the indexes of nonzero entries in x are contained in an s-element subset I of $\{1,\ldots,n\}$, and let us prove that if $\widehat{x}$ is an optimal solution to the LP (1.16), then $\widehat{x} = x$. Indeed, denoting by J the complement of I, setting $z = \widehat{x} - x$ and assuming that $z \neq 0$, we have $Az = 0$. Further, in the same notation as above we have

$$\|x_I\|_1 - \|\widehat{x}_I\|_1 \leq \|z_I\|_1 < \|z_J\|_1 = \|\widehat{x}_J\|_1$$

(the first inequality is due to the Triangle inequality, the second — due to the nullspace property, the equality is due to $x_J = 0$, that is, $z_J = \widehat{x}_J$), whence $\|x\|_1 = \|x_I\|_1 < \|\widehat{x}_I\|_1 + \|\widehat{x}_J\| = \|\widehat{x}\|_1$, which contradicts the origin of $\widehat{x}$.

From nullspace property to error bounds for imperfect ℓ_1 recovery: The nullspace property establishes necessary and sufficient condition for the validity of ℓ_1 recovery in the noiseless case, whatever be the s-sparse true signal. We are about to show that after appropriate quantification, this property implies meaningful error bounds in the case of *imperfect* recovery (presence of observation noise, near, but not exact, s-sparsity of the true signal, approximate minimization in (1.17)).

Let us associate with an $m \times n$ sensing matrix A and positive integer s the quantity

$$\gamma_s(A) = \min_{\gamma \geq 0} \{\gamma : \|z\|_{s,1} \leq \gamma \|z\|_1 \,\forall z \in \operatorname{Ker} A\}. \tag{1.19}$$

Nullspace property says that A is s-good if and only if

$$\gamma_s(A) < 1/2. \tag{1.20}$$

We claim that for a given A, s and norm $\|\cdot\|$ on $\mathbf{R}^m$ there exists $\beta < \infty$ such that

$$\|z\|_{s,1} \leq \beta \|Az\| + \gamma_s(A)\|z\|_1 \quad \forall z \in \mathbf{R}^n. \tag{1.21}$$

Indeed, let P be an orthogonal projector on $\operatorname{Ker} A$. For some $\alpha < \infty$ and all z, we have $\|(I-P)z\|_1 \leq \alpha \|A(I-P)z\|$ and $A(I-P)z = Az$, whence

$$\begin{aligned}
\|z\|_{s,1} &\leq \|(I-P)z\|_{s,1} + \|Pz\|_{s,1} \\
&\leq \|(I-P)z\|_1 + \gamma_s(A)\|Pz\|_1 \quad [\text{due to } Pz \in \operatorname{Ker} A] \\
&\leq \|(I-P)z\|_1 + \gamma_s(A)[\|z\|_1 + \|(I-P)z\|_1] \\
&\leq (1+\gamma_s(A))\|(I-P)z\|_1 + \gamma_s(A)\|z\|_1 \\
&\leq \alpha(1+\gamma_s(A))\|\underbrace{A(I-P)z}_{=Az}\| + \gamma_s(A)\|z\|_1 \\
&= \underbrace{\alpha(1+\gamma_s(A))}_{\beta}\|Az\| + \gamma_s(A)\|z\|_1.
\end{aligned}$$

From now on, we fix some norm $\|\cdot\|$ on $\mathbf{R}^m$ and denote by $\beta_s(A) = \beta_{s,\|\cdot\|}(A)$ the smallest β satisfying (1.21).

Now, consider imperfect ℓ_1 recovery $x \mapsto y \mapsto \widehat{x}$, where

(1) $x \in \mathbf{R}^n$ can be approximated within some accuracy ρ, measured in the ℓ_1 norm, by an s-sparse signal, or, which is the same,

$$\|x - x^s\|_1 \leq \rho,$$

where x^s is the best s-sparse approximation of x (to get this approximation, one zeros out all but the s largest in magnitude entries in x, the ties, if any, being resolved arbitrarily);

(2) y is a noisy observation of x:

$$y = Ax + \eta, \quad \|\eta\| \leq \delta;$$

(3) $\widehat{x}$ is a μ-suboptimal and ϵ-feasible solution to (1.17), specifically,

$$\|\widehat{x}\|_1 \leq \mu + \min_w \{\|w\|_1 : \|Aw - y\| \leq \delta\} \;\&\; \|A\widehat{x} - y\| \leq \epsilon.$$

Theorem 1.1. *Let A, s be given, and let the relation*

$$\forall z : \|z\|_{s,1} \leq \beta\|Az\| + \gamma\|z\|_1 \tag{1.22}$$

hold true with some parameters $\gamma < 1/2$ and $\beta < \infty$ (as definitely is the case when A is s-good, $\gamma = \gamma_s(A)$ and $\beta = \beta_{s,\|\cdot\|}(A)$). Then for the outlined imperfect ℓ_1 recovery the following error bound holds true:

$$\|\widehat{x} - x\|_1 \leq \frac{2\beta(\delta + \epsilon) + \mu + 2\rho}{1 - 2\gamma}, \tag{1.23}$$

i.e., the recovery error is of the order of the maximum of the "imperfections" mentioned in (1)–(3).

Proof. Let I be the set of indexes of the s largest in magnitude entries in x, J be the complement of I, and $z = \widehat{x} - x$. Observing that x is feasible for (1.17), we have $\min_w \{\|w\|_1 : \|Aw - y\| \leq \delta\} \leq \|x\|_1$, whence

$$\|\widehat{x}\|_1 \leq \mu + \|x\|_1,$$

or, in the same notation as above,

$$\underbrace{\|x_I\|_1 - \|\widehat{x}_I\|_1}_{\leq \|z_I\|_1} \geq \underbrace{\|\widehat{x}_J\|_1 - \|x_J\|_1}_{\geq \|z_J\|_1 - 2\|x_J\|_1} - \mu,$$

whence

$$\|z_J\|_1 \leq \mu + \|z_I\|_1 + 2\|x_J\|_1,$$

so that

$$\|z\|_1 \leq \mu + 2\|z_I\|_1 + 2\|x_J\|_1. \tag{a}$$

We further have

$$\|z_I\|_1 \leq \beta\|Az\| + \gamma\|z\|_1,$$

which combines with (a) to imply that

$$\|z_I\|_1 \leq \beta\|Az\| + \gamma[\mu + 2\|z_I\|_1 + 2\|x_J\|_1],$$

whence, in view of $\gamma < 1/2$ and due to $\|x_J\|_1 \leq \rho$,

$$\|z_I\|_1 \leq \frac{1}{1-2\gamma}\,[\beta\|Az\| + \gamma[\mu + 2\rho]].$$

Combining this bound with (a), we get

$$\|z\|_1 \leq \mu + 2\rho + \frac{2}{1-2\gamma}[\beta\|Az\| + \gamma[\mu + 2\rho]].$$

Recalling that $z = \widehat{x} - x$ and that therefore $\|Az\| \leq \|Ax - y\| + \|A\widehat{x} - y\| \leq \delta + \epsilon$, we finally get

$$\|\widehat{x} - x\|_1 \leq \mu + 2\rho + \frac{2}{1-2\gamma}[\beta[\delta + \epsilon] + \gamma[\mu + 2\rho]]. \qquad \square$$

Compressed sensing: Limits of performance: The compressed sensing theory demonstrates that

(1) For given m, n with $m \ll n$ (say, $m/n \leq 1/2$), there exist $m \times n$ sensing matrices which are s-good for the values of s "nearly as large as m," specifically, for $s \leq O(1)\frac{m}{\ln(n/m)}$.[9] Moreover, there are natural families of matrices where this level of goodness "is a rule." For example, when drawing an $m \times n$ matrix at random from the Gaussian or the ± 1 distributions (i.e., filling the matrix with independent realizations of a random variable which is either Gaussian (zero mean, variance $1/m$), or takes values $\pm 1/\sqrt{m}$ with probabilities 0.5[10]), the result will be s-good, for the outlined value of s, with probability approaching 1 as m and n grow. Moreover, for the indicated values of s and randomly selected matrices A, one has $\beta_{s,\|\cdot\|_2}(A) \leq O(1)\sqrt{s}$ with probability approaching one when m, n grow.

[9] From now on, $O(1)$'s denote positive *absolute constants* — appropriately chosen numbers like 0.5, or 1, or perhaps 100,000. We could, in principle, replace all $O(1)$'s by specific numbers; following the standard mathematical practice, we do not do it, partly from laziness, partly because the particular values of these numbers in our context are irrelevant.

[10] Entries "of order of $1/\sqrt{m}$" make the Euclidean norms of columns in $m \times n$ matrix A nearly one, which is the most convenient for Compressed Sensing normalization of A.

(2) The aforementioned results can be considered as good news. The bad news is that we do *not* know how to check efficiently, given an s and a sensing matrix A, that the matrix is s-good. Indeed, we know that a necessary and sufficient condition for s-goodness of A is the nullspace property $\gamma_s(A) < 1/2$; this, however, does not help, since the quantity $\gamma_s(A)$ is difficult to compute: computing it via its definition requires, on a close inspection, solving $N = 2^s\binom{n}{s}$ LO programs, which is an astronomic number already for moderate n unless s is really small, like 1 or 2. And no alternative efficient way to compute $\gamma_s(A)$ is known.

As a matter of fact, not only do we not know how to check s-goodness efficiently; there is still no efficient recipe allowing to build, given m, an $m \times 2m$ matrix A which is provably s-good for s larger than $O(1)\sqrt{m}$ — a much smaller "level of goodness" than the one ($s = O(1)m$) promised by theory for typical randomly generated matrices.[11] The "common life" analogy of this pitiful situation would be as follows: you know that with probability at least 0.9, a brick in your wall is made of gold, and at the same time, you do not know how to tell a golden brick from a usual one.[12]

---

[11]Note that the naive algorithm "generate $m \times 2m$ matrices at random until an s-good, with s promised by the theory, matrix is generated" is *not* an efficient recipe, since we do not know how to check s-goodness efficiently.

[12]This phenomenon is met in many other situations. For example, in 1938 Claude Shannon (1916–2001), "the father of Information Theory," made (in his M.Sc. Thesis!) a fundamental discovery, as follows. Consider a Boolean function of n Boolean variables (i.e., both the function and the variables take values 0 and 1 only); as it is easily seen there are 2^{2^n} functions of this type, and every one of them can be computed by a dedicated circuit comprising "switches" implementing just three basic operations AND, OR and NOT (like computing a polynomial can be carried out on a circuit with nodes implementing just two basic operations: addition of reals and their multiplication). The discovery of Shannon was that every Boolean function of n variables can be computed on a circuit with no more than $Cn^{-1}2^n$ switches, where C is an appropriate absolute constant. Moreover, Shannon proved that "nearly all" Boolean functions of n variables require circuits with *at least* $cn^{-1}2^n$ switches, c being another absolute constant; "nearly all" in this context means that the fraction of "easy to compute" functions (i.e., those computable by circuits with less than $cn^{-1}2^n$ switches) among all Boolean functions of n variables goes to 0 as n goes to ∞. Now, computing Boolean functions by circuits comprising switches was an important technical task already in 1938;

Verifiable sufficient conditions for s-goodness: As it was already mentioned, we do not know efficient ways to check s-goodness of a given sensing matrix in the case when s is not really small. The difficulty here is standard: to certify s-goodness, we should verify (1.20), and the most natural way to do it, based on computing $\gamma_s(A)$, is blocked: by definition,

$$\gamma_s(A) = \max_z \{\|z\|_{s,1} : Az = 0, \|z\|_1 \leq 1\}, \tag{1.24}$$

that is, $\gamma_s(A)$ is the *maximum* of a convex function $\|z\|_{s,1}$ over the convex set $\{z : Az = 0, \|z\|_1 \leq 1\}$. Although both the function and the set are simple, maximizing *convex* function over a convex set typically is difficult. The only notable exception here is the case of maximizing a convex function f over a convex set X given as the convex hull of a finite set:

$$X = \mathrm{Conv}\{v^1, \ldots, v^N\} := \left\{ \sum_{i=1}^N \lambda_i v^i : \lambda \geq 0, \sum_i \lambda_i = 1 \right\}.$$

In this case, a maximizer of f on the finite set $\{v^1, \ldots, v^N\}$ (this maximizer can be found by brute force computation of the values of f at v^i) is the maximizer of f over the entire X (check it yourself or see Section 2.1.3).

Given that the nullspace property "as it is" is difficult to check, we can look for "the second best thing" — efficiently computable *upper and lower bounds* on the "goodness" $s_*(A)$ of A (i.e., on the largest s for which A is s-good).

Let us start with efficient lower bounding of $s_*(A)$, that is, with efficiently verifiable *sufficient conditions* for s-goodness. One way to derive such a condition is to specify an efficiently computable *upper bound* $\widehat{\gamma}_s(A)$ on $\gamma_s(A)$. With such a bound at our disposal, the efficiently verifiable condition $\widehat{\gamma}_s(A) < 1/2$ will clearly be a sufficient condition for the validity of (1.20).

---

its role in our life today can hardly be overestimated — the outlined computation is nothing but what is going on in a computer. Given this observation, it is not surprising that the Shannon discovery of 1938 was the subject of countless refinements, extensions, modifications, etc. What is still missing, is a *single individual example* of a "difficult to compute" Boolean function: as a matter of fact, all multivariate Boolean functions $f(x_1, \ldots, x_n)$ people managed to describe explicitly are computable by circuits with just *linear* in n number of switches!

The question is, how to find an efficiently computable upper bound on $\gamma_s(A)$, and here is one of the options:

$$\begin{aligned}
\gamma_s(A) &= \max_z \{\|z\|_{s,1} : Az = 0, \|z\|_1 \leq 1\} \\
\Rightarrow \forall H \in \mathbf{R}^{m\times n} : \gamma_s(A) &= \max_z \left\{\|[1 - H^T A]z\|_{s,1} : Az = 0, \|z\|_1 \leq 1\right\} \\
&\leq \max_z \left\{\|[1 - H^T A]z\|_{s,1} : \|z\|_1 \leq 1\right\} \\
&= \max_{z\in Z} \|[I - H^T A]z\|_{s,1}, \ Z = \{z : \|z\|_1 \leq 1\}.
\end{aligned}$$

We see that whatever be "design parameter" $H \in \mathbf{R}^{m\times n}$, the quantity $\gamma_s(A)$ does not exceed the maximum of a convex function $\|[I - H^T A]z\|_{s,1}$ of z over the unit ℓ_1-ball Z. But the latter set is perfectly well suited for maximizing convex functions: it is the convex hull of a small (just $2n$ points, $\pm$ basic orths) set. We end up with

$$\begin{aligned}
\forall H \in \mathbf{R}^{m\times n} : \gamma_s(A) &\leq \max_{z\in Z} \|[I - H^T A]z\|_{s,1} \\
&= \max_{1\leq j\leq n} \|\mathrm{Col}_j[I - H^T A]\|_{s,1},
\end{aligned}$$

where $\mathrm{Col}_j(B)$ denotes jth column of a matrix B. We conclude that

$$\gamma_s(A) \leq \widehat{\gamma}_s(A) := \min_H \underbrace{\max_j \|\mathrm{Col}_j[I - H^T A]\|_{s,1}}_{\Psi(H)} . \tag{1.25}$$

The function $\Psi(H)$ is efficiently computable and convex, this is why its minimization can be carried out efficiently. Thus, $\widehat{\gamma}_s(A)$ is an *efficiently computable* upper bound on $\gamma_s(A)$.

Some instructive remarks are in order:

(1) *The trick* which led us to $\widehat{\gamma}_s(A)$ is applicable to bounding from above the maximum of a convex function f over the set X of the form $\{x \in \mathrm{Conv}\{v^1, \ldots, v^N\} : Ax = 0\}$ (i.e., over the intersection of an "easy for convex maximization" domain and a linear subspace). The trick is merely to note that if A is $m \times n$, then

for every $H \in \mathbf{R}^{m\times n}$ one has

$$\max_x \left\{ f(x) : x \in \mathrm{Conv}\{v^1, \ldots, v^N\}, Ax = 0 \right\} \\ \leq \max_{1\leq i\leq N} f([I - H^T Ax]v^i) \qquad (!)$$

Indeed, a feasible solution x to the left-hand side of the optimization problem can be represented as a convex combination $\sum_i \lambda_i v^i$, and since $Ax = 0$, we have also $x = \sum_i \lambda_i [I - H^T A]v^i$; since f is convex, we have therefore $f(x) \leq \max_i f([I - H^T A]v^i)$, and (!) follows. Since (!) takes place for every H, we arrive at

$$\max_x \left\{ f(x) : x \in \mathrm{Conv}\{v^1, \ldots, v^N\}, Ax = 0 \right\} \\ \leq \widehat{\gamma} := \min_H \max_{1\leq i\leq N} f([I - H^T A]v^i),$$

and, same as above, $\widehat{\gamma}$ is efficiently computable, provided that f is an efficiently computable convex function.

(2) The efficiently computable upper bound $\widehat{\gamma}_s(A)$ is *polyhedrally representable* — it is the optimal value in an explicit LP program. To derive this problem, we start with polyhedral representation of the function $\|z\|_{s,1}$, which is important by itself.

Lemma 1.1. *For every $z \in \mathbf{R}^n$ and integer $s \leq n$, we have*

$$\|z\|_{s,1} = \min_{w,t} \left\{ st + \sum_{i=1}^n w_i : |z_i| \leq t + w_i,\ 1 \leq i \leq n, w \geq 0 \right\}. \tag{1.26}$$

Proof. Indeed, if (w, t) is feasible for (1.26), then $|z_i| \leq w_i + t$, whence the sum of the s largest magnitudes of entries in z does not exceed st plus the sum of the corresponding s entries in w, and thus — since w is nonnegative — does not exceed $st + \sum_i w_i$. Thus, the right-hand side in (1.26) is $\geq$ the left-hand side. On the other hand, let $|z_{i_1}| \geq |z_{i_2}| \geq \ldots \geq |z_{i_s}|$ be the s largest magnitudes of entries in z (so that $i_1, \ldots, i_s$ are distinct from each other), and let $t = |z_{i_s}|$, $w_i = \max[|z_i| - t, 0]$. It is immediately seen that (t, w) is feasible for the right-hand side problem in (1.26) and that $st + \sum_i w_i = \sum_{j=1}^s |z_{i_j}| = \|z\|_{s,1}$. Thus, the right-hand side in (1.26) is $\leq$ the left-hand side. $\square$

Lemma 1.1 straightforwardly leads to the following polyhedral representation of $\widehat{\gamma}_s(A)$:

$$\widehat{\gamma}_s(A) := \min_H \max_j \|\mathrm{Col}_j[I - H^T A]\|_{s,1}$$
$$= \min_{H,w^j,t_j,\tau} \left\{ \tau : \begin{array}{c} -w_i^j - t_j \leq [I - H^T A]_{ij} \leq w_i^j + t_j \,\forall i,j \\ w^j \geq 0 \,\forall j, st_j + \sum_i w_i^j \leq \tau \,\forall j \end{array} \right\}.$$

(3) The quantity $\widehat{\gamma}_1(A)$ is *exactly equal* to $\gamma_1(A)$ rather than being an upper bound on the latter quantity. This fact can be easily verified via what is called LO Duality (it will be our subject in Chapter 3). Right now the reader should just take the claim for granted.

Observe that an optimal solution H to the problem

$$\widehat{\gamma}_1(A) = \min_H \max_{i,j} |[I_n - H^T A]_{i,j}|$$

can be found column by column, with jth column h^j of H being an optimal solution to the LP

$$\min_h \|e_j - A^T h\|_\infty,$$

where e_j is the jth standard basic orth in $\mathbf{R}^n$. This is in nice contrast with computing $\widehat{\gamma}_s(A)$ for $s > 1$, where we should solve a single LP with $O(n^2)$ variables and constraints, which is typically much more time-consuming than solving $O(n)$ LP's with $O(n)$ variables and constraints each, as it is the case when computing $\widehat{\gamma}_1(A)$.

Observe also that if p, q are positive integers, then for every vector z one has $\|z\|_{pq,1} \leq q\|z\|_{p,1}$, and in particular $\|z\|_{s,1} \leq s\|z\|_{1,1} = s\|z\|_\infty$. It follows that if H is such that $\widehat{\gamma}_p(A) = \max_j \|\mathrm{Col}_j[I - H^T A]\|_{p,1}$, then $\widehat{\gamma}_{pq}(A) \leq q \max_j \|\mathrm{Col}_j[I - H^T A]\|_{p,1} \leq q\widehat{\gamma}_p(A)$. In particular,

$$\widehat{\gamma}_s(A) \leq s\widehat{\gamma}_1(A),$$

meaning that the easy-to-verify condition

$$\widehat{\gamma}_1(A) < \frac{1}{2s}$$

is sufficient for the validity of the condition

$$\widehat{\gamma}_s(A) < 1/2$$

and is thus sufficient for s-goodness of A.

(4) Assume that A and s are such that s-goodness of A can be certified via our verifiable sufficient condition, that is, we can point out an $m \times n$ matrix H such that

$$\gamma := \max_j \|\mathrm{Col}_j[I - H^T A]\|_{s,1} < 1/2.$$

Now, for every $m \times n$ matrix B, any norm $\|\cdot\|$ on $\mathbf{R}^m$ and every vector $z \in \mathbf{R}^n$ we clearly have

$$\|Bz\| \leq \left[\max_j \|\mathrm{Col}_j[B]\|\right] \|z\|_1$$

(Why?) Therefore, from the definition of γ, for every vector z we have $\|[I - H^T A]z\|_{s,1} \leq \gamma\|z\|_1$, so that

$$\begin{aligned} \|z\|_{s,1} &\leq \|H^T Az\|_{s,1} + \|[I - H^T A]z\|_{s,1} \\ &\leq \left[s \max_j \|\mathrm{Col}_j[H]\|_*\right] \|Az\| + \gamma\|z\|_1, \\ \text{where } \|h\|_* &= \max_{y:\|y\|\leq 1} h^T y \end{aligned}$$

meaning that H certifies not only the s-goodness of A, but also an inequality of the form (1.22) and thus — the associated error bound (1.23) for imperfect ℓ_1 recovery.

1.2.2.4 **Supervised binary machine learning via LP support vector machines*

Imagine that we have a source of *feature vectors* — collections x of n measurements representing, e.g., the results of n medical tests taken from patients, and a patient can be affected, or not affected, by a particular illness. "In reality," these feature vectors x go along with *labels* y taking values ± 1; in our example, the label -1 says that the patient whose test results are recorded in the feature vector x does *not* have the illness in question, while the label $+1$ means that the patient is ill.

We assume that there is certain dependence between the feature vectors and the labels, and our goal is to predict, given a feature vector alone, the value of the label. What we have in our disposal is a *training sample* (x^i, y^i), $1 \leq i \leq N$, of *examples* (x^i, y^i) where we know both the feature vector and the label; given this sample, we want to build a *classifier* — a function $f(x)$ on the space of feature vectors x taking values ± 1 — which we intend to use to predict, given the value of a new feature vector, the value of the corresponding label. In our example this setup reads: we are given medical records containing both the results of medical tests and the diagnoses of N patients; given this data, we want to learn how to predict the diagnosis given the results of the tests taken from a new patient.

The simplest predictors we can think about are just the "linear" ones looking as follows. We fix an affine form $w^T x + b$ of a feature vector, choose a positive threshold γ and say that if the value of the form at a feature vector x is "well positive" — is $\geq \gamma$ — then the proposed label for x is $+1$; similarly, if the value of the form at x is "well negative" — is $\leq -\gamma$, then the proposed label will be -1. In the "gray area" $-\gamma < w^T x + b < \gamma$ we decline to classify. Noting that the actual value of the threshold is of no importance (to compensate a change in the threshold by certain factor, it suffices to multiply by this factor both w and b, without affecting the resulting classification); from now on, we normalize the situation by setting the threshold to the value 1.

Now, we have explained how a linear classifier works, but from where do we take it? An intuitively appealing idea is to use the training sample in order to "train" our potential classifier — to choose w and b in a way which ensures correct classification of the examples in the sample. This amounts to solving the system of linear inequalities

$$w^T x^i + b \geq 1 \quad \forall (i \leq N : y^i = +1) \;\&\; w^T x^i + b \leq -1 \quad \forall (i : y^i = -1),$$

in variables w and b, which can be written equivalently as

$$y^i (w^T x^i + b) \geq 1 \quad \forall i = 1, \ldots, N.$$

Geometrically speaking, we want to find a "stripe"

$$-1 < w^T x + b < 1 \tag{$*$}$$

between two parallel hyperplanes $\{x : w^T x + b = -1\}$ and $\{x : w^T x + b = 1\}$ such that all "positive examples" (those with the label $+1$) from the training sample are on one side of this stripe, while all negative (the label -1) examples from the sample are on the other side of the stripe. With this approach, it is natural to look for the "thickest" stripe separating the positive and the negative examples.

Since the geometric width of the stripe is $\frac{2}{\sqrt{w^T w}}$ (why?), this amounts to solving the optimization program

$$\min_{w,b} \left\{ \|w\|_2 := \sqrt{w^T w} : y^i(w^T x^i + b) \geq 1, \, 1 \leq i \leq N \right\}. \tag{1.27}$$

The latter problem, of course, is not necessarily feasible: it well can happen that it is impossible to separate the positive and the negative examples in the training sample by a stripe between two parallel hyperplanes. To handle this possibility, we can allow for classification errors and minimize a weighted sum of $\|w\|_2$ and total penalty for these errors. Since the absence of classification penalty at an example (x^i, y^i) in our context is equivalent to the validity of the inequality $y^i(w^T x^i + b) \geq 1$, the most natural penalty for misclassification of the example is $\max[1 - y^i(w^T x^i + b), 0]$. With this in mind, the problem of building "the best on the training sample" classifier becomes the optimization problem

$$\min_{w,b} \left\{ \|w\|_2 + \lambda \sum_{i=1}^{N} \max[1 - y^i(w^T x^i + b), 0] \right\}, \tag{1.28}$$

where $\lambda > 0$ is responsible for the "compromise" between the width of the stripe $(*)$ and the "separation quality" of this stripe; how to choose the value of this parameter, this is an additional story we do not touch here. Note that the outlined approach to building classifiers is the most basic and the most simplistic version of what in Machine Learning is called "Support Vector Machines."

Now, (1.28) is *not* an LO program: we know how to get rid of nonlinearities $\max[1 - y^i(w^T x^i + b), 0]$ by adding slack variables and linear constraints, but we cannot get rid of the nonlinearity brought by the term $\|w\|_2$. Well, there are situations in Machine Learning where it makes sense to get rid of this term by "brute force," specifically, by replacing the $\|\cdot\|_2$ with $\|\cdot\|_1$. The rationale behind this "brute force" action is as follows. The dimension n of the feature vectors can be large. In our medical example, it could be in the range of tens, which perhaps is "not large;" but think about digitalized images of handwritten letters, where we want to distinguish between handwritten letters "A" and "B;" here the dimension of x can well be in the range of thousands, if not millions. Now, it would be highly desirable to design a good classifier with *sparse* vector of weights w, and there are several reasons for this desire. First, intuition says that a good-on-the-training-sample classifier which takes into account just three of the features should be more "robust" than a classifier which ensures equally good classification of the training examples but uses 10,000 features for this purpose; we have all reasons to believe that the first classifier indeed "goes to the point," while the second one adjusts itself to random, irrelevant-for the-"true classification" properties of the training sample. Second, having a good classifier which uses small number of features is definitely better than to have an equally good classifier which uses a large number of them (in our medical example: the "predictive power" being equal, we definitely would prefer predicting diagnosis via the results of three tests to predicting via the results of 20 tests). Finally, if it is possible to classify well via a small number of features, we hopefully have good chances *to understand the mechanism of the dependencies between these measured features and the feature whose presence/absence we intend to predict* — it usually is much easier to understand interactions between 2–3 features than between 2,000–3,000 of them. Now, the SVMs (1.27), (1.28) are not well suited for carrying out the outlined *feature selection* task, since minimizing $\|w\|_2$ norm under constraints on w (this is what explicitly goes on in (1.27) and implicitly goes on in (1.28)[13]) typically results in "spread" optimal solution, with many small nonzero components. In view of our "Compressed Sensing" discussion, we could expect that minimizing the ℓ_1-norm of w will result in "better concentrated" optimal solutions, which leads us to what is called "LO Support Vector Machine." Here the classifier is given by the solution of the $\|\cdot\|_1$-analogy of (1.28), specifically, the optimization problem

$$\min_{w,b}\left\{\|w\|_1+\lambda\sum_{i=1}^{N}\max\left[1-y^i\left(w^Tx^i+b\right),0\right]\right\}. \tag{1.29}$$

[13]To understand the latter claim, take an optimal solution (w_*,b_*) to (1.28), set $\Lambda=\sum_{i=1}^N\max[1-y^i(w_*^Tx^i+b_*),0]$ and note that (w_*,b_*) solves the optimization problem

$$\min_{w,b}\left\{\|w\|_2:\sum_{i=1}^{N}\max\left[1-y^i\left(w^Tx^i+b\right),0\right]\le\Lambda\right\}$$

(why?).

This problem clearly reduces to the LO program

$$\min_{w,b,v,\xi}\left\{\sum_{j=1}^{n} v_j + \lambda\sum_{i=1}^{N}\xi_i : -v_j \le w_j \le v_j,\ 1 \le j \le n,\right.$$
$$\left.\xi_i \ge 0,\ \xi_i \ge 1 - y^i(w^T x^i + b),\ 1 \le i \le N\right\}. \qquad (1.30)$$

Concluding remarks: A reader could ask, what is the purpose of training the classifier on the training set of examples, where we know the labels of all the examples from the very beginning? Why should a classifier which classifies well on the training set be good at new examples? Well, intuition says that if a simple rule with a relatively small number of "tuning parameters" (as it is the case with a sparse linear classifier) recovers well the labels in examples from a large enough sample, this classifier should have learned something essential about the dependency between feature vectors and labels, and thus should be able to classify well new examples. Machine Learning theory offers a solid probabilistic framework in which "our intuition is right," so that under assumptions (not too restrictive) imposed by this framework it is possible to establish quantitative links between the size of the training sample, the behavior of the classifier on this sample (quantified by the $\|\cdot\|_2$ or $\|\cdot\|_1$ norm of the resulting w and the value of the penalty for misclassification), and the *predictive power* of the classifier, quantified by the probability of misclassification of a new example; roughly speaking, good behavior of a linear classifier achieved at a large training sample ensures low probability of misclassifying a new example.

1.3 What can be reduced to LO?

Looking at the collection of LO models we have presented, we see that mathematical models which can finally be formulated as LO programs are *not always* "born" in this form; we became acquainted with several tricks which, with luck, allow to convert a nonLO optimization problem into an equivalent LO program. This section is devoted to "in-depth" investigating of these tricks.

1.3.1 *Preliminaries*

We start with several useful, although "philosophical," remarks. What we are interested in our book is *mathematics* of Linear Optimization, so that the main entities to be considered are specific functions and sets. Say, an LO program is the program of maximizing a linear *function* $c^T x$ over a *polyhedral subset* X of $\mathbf{R}^n$, that is, over the solution set $\{x \in \mathbf{R}^n : Ax \leq b\}$ of a finite system of nonstrict linear inequalities in n variables x. Now, sets and functions are abstract "mathematical beasts;" the concept of a set is *the* basic mathematical concept which we do *not* define in terms of simpler concepts[14]; the concept of a function is a derivative of the concept of a set,[15] and both these concepts have nothing to do with particular *representations* of these entities; representations are by far not the same as the entities being described. For example, the segment $[-1, 1]$ of the real line is a set, and this set admits various representations, e.g.,

- the representation as a solution set of the system of two linear inequalities $x \geq -1$, $x \leq 1$ in real variable x,
- the representation as the set of all values taken by the function $\sin(x)$ on the real axis,

and a countless variety of other representations. Similarly, a linear function $f(x) = x$ on the real axis can be represented as $f(x) = x$, or $f(x) = x + \sin^2(x) + \cos^2(x) - 1$, and in a countless variety of other forms. Thus, we should distinguish between sets/functions as abstract "objects of our perceptions and our thoughts" and their *concrete representations*, keeping in mind that a particular "object of our

[14]Whether one believes that the concept of a set is an abstract "derivative" of our experience in thinking of/handling various collections of "physical" entities, or, following Plato, thinks that this concept is a shadow of a certain "idea" existing in some ideal sense, no one offers a formal definition of this fundamental concept, just *illustrates* it. Perhaps, the best illustration is the famous citation from George Cantor, the founder of Set Theory: "By a "set" we mean any collection M into a whole of definite, distinct objects m (which are called the "elements" of M) of our perception [Anschauung] or of our thought."

[15]A function f defined on a set X and taking values in a set Y can be identified with its *graph*, which is the subset of $X \times Y := \{(x, y) : x \in X, y \in Y\}$ comprising pairs $(x, f(x))$; a subset F of $X \times Y$ indeed represents a function if every $x \in X$ is the first component of *exactly one* pair from F.

thoughts" admits many different representations. We should distinguish well between properties of an object and properties of its particular representation. For example, the nonemptiness is a property of the *set* $[-1, 1]$, while the number of linear inequalities (namely, 2) in its representation as the solution set of the system $x \geq -1, x \leq 1$ in real variable x clearly is a property of the representation in question, not of the set, since the same set can be represented as the solution set of a system of, say, 10 linear inequalities (add to the previous system inequalities $x \leq 3, x \leq 4, \ldots, x \leq 10$). In a sense, nearly all we intend to do in our book (or, more widely, what a significant part of Mathematics is about), is to understand how to derive conclusions on properties of the "abstract beasts" — sets and functions — from representations of these beasts in certain concrete formats. This is a highly challenging and highly nontrivial task, even when speaking about such a simple, at first glance, property as emptiness.[16]

Now, the abstract form of an optimization problem with n real decision variables is minimizing a given real-valued function $f(x)$ over a given feasible set $X \subset \mathbf{R}^n$; LO deals with this abstract problem in the particular case when f is linear, and X is polyhedral, and even in this particular case it does not deal with this problem *per se*, but with particular *representations* of the entities involved: f as $c^T x$, X as $\{x : Ax \leq b\}$, with explicitly — just by listing the values of the coefficients — given data c, A, b. As it was already mentioned, the "maiden" representation of a problem, the one in which the problem "is born," is not always the one required by LO; most typically, the initial representation is in the Mathematical Programming form

$$\max_x \{f_0(x) : x \in X = \{x : f_i(x) \leq 0, \, i = 1, \ldots, m\}\} \qquad \text{(MP)}$$

with explicitly given (analytically or algorithmically) functions f_0, $f_1, \ldots, f_m$. Thus, we need tools allowing (a) to recognize the possibility of translating a representation of the form (MP) in an LO representation, and (b) to implement the translation when its possibility is recognized. Our goal in the rest of this section is to develop a toolbox of this type.

[16] To illustrate the point: the Great Fermat Theorem merely states that the set with extremely simple representation (quadruples of positive integers x, y, z, p with $p > 2$ satisfying the equation $x^p + y^p = z^p$) possesses an extremely simple property of being empty.

A reader might ask what is the purpose for all this "scholastics" about the difference between optimization problems and their representations, and why we intend to operate with representations of mathematical entities rather than to work directly with these entities. The answer is very simple: an algorithm (and at the end of the day we want the problem to be processed and thus need algorithms) by its nature cannot work with abstract mathematical entities, only with their representations; to some extent, the same is true for human beings, as can be witnessed by everybody with even a minimal experience in solving mathematical problems, no matter which ones, building proofs or crunching numbers.

1.3.2 *Polyhedral representations of sets and functions: Definitions and Fourier–Motzkin elimination*

When converting an optimization problem (MP) with explicitly given objective and constraints into an equivalent LO program, our goal is twofold: (a) to end up with a linear objective represented as $c^T y$, and (b) to end up with a feasible set represented as $\{y : Ay \leq b\}$ (we write y instead of x, keeping in mind the possibility of augmenting the original decision variables with slack ones). It is easy to achieve the first goal: to this end it suffices to add a slack variable t and to rewrite (MP) equivalently as

$$\max_{x,t} \{t : t - f_0(x) \leq 0, f_1(x) \leq 0, \ldots, f_m(x) \leq 0\};$$

the objective in the resulting problem is linear in the new design vector $[x; t]$, and the constraints are "as explicitly given" as those in the original problem. To save notation, assume that this transformation is done in advance, so that the problem we intend to convert into an LO program from the very beginning is of the form

$$\min_{x \in X} c^T x, X = \{x \in \mathbf{R}^n : f_i(x) \leq 0, 1 \leq i \leq m\}. \tag{1.31}$$

This assumption "costs nothing" and allows us to focus solely on the constraints and on the feasible set X they define.

Now, our experience with slack variables suggests a good formalization of the informal task "to end up with a feasible set represented as $\{y : Ay \leq b\}$," specifically, as follows.

Definition 1.1 (Polyhedral representation of a set). A polyhedral representation (p.r.) of a set $X \subset \mathbf{R}^n$ is a representation of the form

$$X = \{x \in \mathbf{R}^n : \exists w \in \mathbf{R}^s : Px + Qw \le r\}, \tag{1.32}$$

i.e., it is a finite system of nonstrict linear inequalities in variables x, w such that $x \in X$ iff x can be extended, by properly chosen w, to a feasible solution of the system.

Geometrically, polyhedral representation of X means the following: We take a set, given by an explicit system $Px+Qw \le r$ of linear inequalities, in the space of x, w-variables and project this set onto the subspace of x-variables; the system $Px + Qw \le r$ polyhedrally represents X iff the projection is exactly X.

The role of polyhedral representability in our context stems from the following evident fact.

Observation 1.2. Given a polyhedral representation (1.32) of a set $X \subset \mathbf{R}^n$, we can immediately and straightforwardly convert problem (1.31) into an LO program, specifically, into the program

$$\max_{x,w} \{c^T x : Px + Qw \le r\}.$$

Example: Let us look at the linear regression problem with ℓ_1-fit (problem (1.14)) which we now rewrite as a problem with linear objective

$$\min_{\tau,x} \left\{ \tau : \sum_{i=1}^m |a_i^T x - y_i| \le \tau \right\}.$$

The feasible set of this problem admits an immediate polyhedral representation:

$$\begin{aligned} &\left\{ [x;\tau] : \sum_{i=1}^m |x^T a_i - y_i| \le \tau \right\} \\ &\quad = \left\{ [x;\tau] : \exists w : -w_i \le a_i^T x - y_i \le w_i \, i = 1, \dots, m, \sum_{i=1}^m w_i \le \tau \right\} \end{aligned} \tag{1.33}$$

which allows to rewrite the problem equivalently as the LO program

$$\min_{x,\tau,w} \left\{ \tau : -w_i \le a_i^T x - y_i \le w_i \, i = 1, \dots, m, \sum_{i=1}^m w_i \le \tau \right\}. \tag{1.34}$$

This is exactly what we did with the problem of interest in the previous section.

1.3.2.1 *Fourier–Motzkin elimination*

We have seen that all we need in order to convert an optimization program with linear objective into an LO program is a polyhedral representation of the feasible set X of the problem. This need is easy to satisfy if X is a polyhedral set represented as $\{x : Ax \leq b\}$. A polyhedral representation of a set is something more flexible — now we do not want to represent X as the solution set of a system of linear inequalities, only as a projection of such a solution set onto the space where X lives. At this point, it is unclear whether the second type of representation indeed is more flexible than the first one, that is, we do not know whether the projection of a polyhedral set in certain $\mathbf{R}^n$ onto a linear subspace is or is not polyhedral. The answer is positive:

Theorem 1.2. *Every polyhedrally representable set is polyhedral.*

This important theorem can be obtained as a byproduct of the conceptually simple *Fourier–Motzkin elimination scheme.*

Fourier–Motzkin elimination scheme. Let $X = \{x : \exists w : Px + Qw \leq r\}$, that is, X is the projection on the space of x-variables of the polyhedral set $Q = \{[x; w] : Px + Qw \leq r\}$ in the space of x, w-variables. We want to prove that X can be represented as the solution set of a finite system of linear inequalities solely in variables x. Let $w = [w_1; ...; w_k]$. We start with eliminating from the polyhedral description of X the variable w_k. To this end, let us set $z = [x; w_1; ...; w_{k-1}]$, so that the system of linear inequalities $Px + Qw \leq r$ can be rewritten in the form

$$a_i^T z + b_i w_k \leq c_i, \quad 1 \leq i \leq m. \tag{S}$$

Let us "color" an inequality of the system in red, if $b_i > 0$, in green if $b_i < 0$, and in white, if $b_i = 0$, and let I_r, I_g and I_w be the sets of indices of red, green and white inequalities, respectively. Every red inequality can be rewritten equivalently as $w_k \leq c_i/b_i - a_i^T z/b_i =: e_i^T z + f_i$, and every green inequality can be rewritten equivalently as $w_k \geq c_i/b_i - a_i^T z/b_i =: e_i^T z + f_i$. It is clear that z can be extended, by a properly chosen w_k, to a feasible solution of (S) if and only if, first, z satisfies every white inequality and, second, every "red" quantity $e_i^T z + f_i$ (which should be an upper bound on w_k) is $\geq$ every "green" quantity $e_i^T z + f_i$ (which should be a lower bound on w_k). In other words, z can be extended to a feasible solution of (S) if and only if z satisfies the system of linear inequalities

$$a_i^T z \leq b_i \quad \forall i \in I_w; e_i^T z + f_i \geq e_j^T z + f_j \quad \forall (i \in I_r, j \in I_g). \qquad (S')$$

We see that the projection of Q on the space of the variables $x, w_1, \ldots, w_{k-1}$ is the solution set Q' of a finite system of linear inequalities in these variables; note that X is the projection of Q' on the space of x-variables, that is, we have built a polyhedral representation of X using $k-1$ slack variables $w_1, \ldots, w_{k-1}$. Proceeding in the same fashion, we can eliminate all slack variables one by one, thus ending up with a desired polyhedral representation of X "free of slack variables."

Note that the Fourier–Motzkin elimination is an algorithm, and we can easily convert this algorithm into a finite algorithm for solving LO programs. Indeed, given an LO program $\max_x\{c^T x : Ax \leq b\}$ with n variables $x_1, \ldots, x_n$ and augmenting these variables by a new variable τ, we can rewrite the program equivalently as

$$\max_{y=[\tau;x]} \{\tau : Ax \leq b, \tau - c^T x \leq 0\}. \qquad (P)$$

The set of feasible values of τ — those which can be extended by properly chosen x to feasible solutions of (P) — is the projection of the feasible set of (P) on the τ-axis; applying the above elimination scheme, we can represent this set as the set of solutions of a finite system $\mathcal{S}$ of nonstrict linear inequalities *in variable τ alone*. It is immediately seen that the solution set of such a system

(a) either is empty,
(b) or is a ray of the form $\tau \leq \beta$,
(c) or is a nonempty segment $\alpha \leq \tau \leq \beta$,
(d) or is a ray of the form $\tau \geq \alpha$,
(e) or is the entire τ-axis.

Given $\mathcal{S}$, it is easy to recognize which one of these cases actually takes place, and what are the corresponding α and β. In the case of (a), (P) is infeasible, in the cases (d,e) (P) is feasible and unbounded, in the cases (b,c) it is feasible and bounded, β is the optimal value in (P), and $\tau = \beta$ is a feasible solution to $\mathcal{S}$. Starting with this solution and using the elimination scheme in a backward fashion, we can augment $\tau = \beta$ by values of the variables $x_1, \ldots, x_n$, one at a time, in such a way that $[\tau = \beta; x_1; ...; x_n]$ will be feasible (and then optimal) for (P). Thus, we can identify in finite time the "feasibility status" (infeasible/feasible and unbounded/feasible and bounded) of (P) and point out, also in finite time, an optimal solution, provided that the problem is feasible and bounded.

Note that as a byproduct of our reasoning, we see that our former claim that *a feasible and bounded LO program admits an optimal solution* is indeed true.

A bad news is that the outlined finite algorithm for solving LO programs is of purely academic value; as a practical tool, it can handle extremely small problems only, with few (like 2–3) variables and perhaps few tens of constraints. The reason is that every step of the elimination scheme can increase dramatically the number of linear constraints we should handle. Indeed, if the original system (S) has m inequalities, half of them red and half of them green, after eliminating the first slack variable we will get a system of $m_1 = m^2/4$ inequalities, at the second step we can get as many as $m_1^2/4 = m^4/64$ inequalities, and so on; now take $m = 16$ and look what happens after five steps of the recurrence $m := m^2/4$.

The fact that a polyhedrally representable set is polyhedral and thus can be represented by a system of linear inequalities not involving slack variables in no sense diminishes the importance of slack variables and polyhedral representations involving these variables. Indeed, the possibility to represent the set of interest as the solution set of a finite system of linear inequalities is not all we are looking for when building LO models; we definitely do not want to handle astronomically many inequalities. In this latter respect, adding slack variables (i.e., passing to general-type polyhedral representations) can result in dramatic reduction in the number of linear inequalities we need to handle as compared to the case when no slack variables are used. For example, when speaking about linear regression with ℓ_1 fit, we have seen that the problem can indeed be rewritten equivalently as an LO in the original variables x, τ of (1.33), specifically, as the LO (1.15). Note, however, that the latter LO has $n+1$ variables and as many as 2^m constraints, which is astronomically large already for moderate m. In contrast to this, the LO program (1.34), while involving "slightly more" variables $(n+m+1)$, has just $2m+1$ constraints.

What is ahead? Observation 1.2 suggests that a good way to understand what can be reduced to LO is to understand how to recognize that a given set is polyhedral and if it is the case, how to point out a polyhedral representation of the set. It does not make sense to pose this question as a formal mathematical problem — we could recognize polyhedrality only by working with certain initial description of the set; we have assumed that this is a description by m "explicitly given" constraints $f_i(x) \leq 0$, but the words "explicitly given" are too vague to allow for well-defined constructions and provable results. Instead, we are about to develop a kind of "calculus" of polyhedral representations, specifically, to indicate basic examples of p.r.'s, augmented by calculus rules which say that such and such operations with polyhedral sets result in a polyhedral set, and a p.r. of this set can be built in such and such fashion from the p.r.'s of the operands. As a result of these developments, we will be able to conclude that a set which is obtained by such and such sequence of operations from such and such "raw materials" is polyhedral, with such and such p.r.[17]

One last remark before passing to calculus of polyhedral representability. In optimization, feasible sets usually are given by finite systems of constraints $f_i(x) \leq 0$, that is, as the intersection of *sublevel sets* of several functions.[18] In order to catch this phenomenon, it makes sense to introduce the notion of a *polyhedrally representable function* (p.r.f. for short). This notion is a kind of "derivative" of the

[17]The outlined course of actions is very typical for Mathematics. We know what a differentiable function is — there is a formal definition of differentiability expressed in terms of a function as a "mathematical beast," without reference to any particular representation of this beast. This definition, however, does not allow to recognize differentiability even when a function is given by an analytic formula of a simple structure (since the formula can contain nonsmooth components which in fact cancel each other, but this cancellation is very difficult to discover), let alone when the function is given in a more complicated fashion. What we routinely use to establish differentiability and to compute derivatives is the usual calculus, where we start with "raw materials" — elementary functions like ln, sin, exp, etc., where we check differentiability and compute the derivatives "by bare hands," by working with the definition of the derivative. We then augment the "raw materials" by calculus rules which explain to us when an operation with functions, like multiplication, addition, taking superposition, etc., preserves differentiability and how to express the derivatives of the result via derivatives of the operands, thus getting a key to differentiating a huge spectrum of functions, including quite complicated ones.

[18]A sublevel, or a Lebesgue, set of a function f is, by definition, a set of the form $\{x : f(x) \leq a\}$, where a is a real number.

notion of a polyhedral set, and the corresponding definitions are as follows.

Definition 1.2. Let $f(x)$ be a function on $\mathbf{R}^n$ taking real values and the value $+\infty$.

(i) The domain $\mathrm{Dom}\, f$ of f is the set of all x where $f(x)$ is finite;
(ii) The epigraph of f is the set $\mathrm{Epi}\{f\} = \{[x;\tau] \in \mathbf{R}^n \times \mathbf{R} : x \in \mathrm{Dom}\, f, \tau \geq f(x)\}$;
(iii) f is called polyhedrally representable if its epigraph $\mathrm{Epi}\{f\}$ is a polyhedral set, so that for appropriate matrices P, Q and vectors p, r it holds

$$\begin{aligned}\{[x;\tau] : x \in \mathrm{Dom}\, f, \tau \geq f(x)\}\\ = \{[x;\tau] : \exists w : Px + \tau p + Qw \leq r\}.\end{aligned} \tag{1.35}$$

We refer to a polyhedral representation of the epigraph of f as a polyhedral representation (p.r.) of f itself.

Observation 1.3. A sublevel set $\{x : f(x) \leq a\}$ of a p.r.f. (polyhedrally representable function) f is a polyhedral set, and a p.r. of this set is readily given by a p.r. of the function, specifically, (1.35) implies that

$$\begin{aligned}\{x : f(x) \leq a\} &= \{x : [x;a] \in \mathrm{Epi}\{f\}\}\\ &= \{x : \exists w : Px + Qw \leq r - ap\}.\end{aligned}$$

Example: Consider the function $f(x) = \|x\|_1 := \sum_{i=1}^n |x_i| : \mathbf{R}^n \to \mathbf{R}$. This function is polyhedrally representable, e.g., by the p.r.

$$\begin{aligned}\mathrm{Epi}\{f)\} &:= \left\{[x;\tau] : \tau \geq \sum_{i=1}^n |x_i|\right\}\\ &= \left\{[x;\tau] : \exists w : -w_i \leq x_i \leq w_i,\ 1 \leq i \leq n, \sum_{i=1}^n w_i \leq \tau\right\}.\end{aligned}$$

Remarks. Some remarks are in order.

A. **Partially defined functions:** Normally, a scalar function f of n variables is specified by indicating its *domain* — the set where the function is well defined, and by the description of f as a real-valued function in the domain. It is highly convenient to combine both components of such a description in a single description by allowing the function to take "a fictional" value $+\infty$ outside of the actual domain. A reader should not look for something "mystical" in this approach: this is just a convention allowing to save a lot of words. In order to allow for basic operations with partially defined functions, like their addition or comparison, we augment our convention with the following agreements on the arithmetics of the "extended real axis" $\mathbf{R} \cup \{+\infty\}$.

- *Addition*: for a real a, $a + (+\infty) = (+\infty) + (+\infty) = +\infty$.
- *Multiplication by a nonnegative real* λ: $\lambda \cdot (+\infty) = +\infty$ when $\lambda > 0$, and $0 \cdot (+\infty) = 0$.
- *Comparison*: for a real a, $a < +\infty$ (and thus $a \leq +\infty$ as well), and of course $+\infty \leq +\infty$.

As far as operations with $+\infty$ are concerned, our arithmetic is severely incomplete — operations like $(+\infty) - (+\infty)$ and $(-1) \cdot (+\infty)$ remain undefined. Well, we can live with it.

B. **Convexity of polyhedral sets and polyhedrally representable functions:** A set $X \in \mathbf{R}^n$ is called convex if whenever two points x, y belong to X, the entire segment $\{x + \lambda(y - x) = (1 - \lambda)x + \lambda y : 0 \leq \lambda \leq 1\}$ belongs to X.

To understand the link between the informal — verbal — and the formal — algebraic — parts of this definition, note that when x, y are two distinct points, then all points $(1 - \lambda)x + \lambda y$ form, geometrically, the line passing through x and y, and the part of this line corresponding to the range $0 \leq \lambda \leq 1$ of λ "starts" at x ($\lambda = 0$) and "ends" at y ($\lambda = 1$) and thus is exactly what is natural to call "the segment linking x, y." When $x = y$, the above line, same as its part corresponding to the range $0 \leq \lambda \leq 1$ of values of λ, collapses to the singleton $\{x\}$ which again is the only natural candidate to the role of "segment linking x and x."

A function $f : \mathbf{R}^n \to \mathbf{R} \cup \{+\infty\}$ is called convex iff its epigraph is convex, or, which is the same (*check why),

$$\forall (x, y \in \mathbf{R}^n, \lambda \in [0, 1]) : f((1 - \lambda)x + \lambda y) \leq (1 - \lambda)f(x) + \lambda f(y).$$

To complete the terminology, a function f taking values in $\mathbf{R} \cup \{-\infty\}$ is called *concave* if $-f$ is convex (in this description, and everywhere else, $-(-\infty) = \infty$). In view of this definition, handling concave functions reduces to handling convex ones, and we prefer to stick to this possibility in the sequel. Note that there is no notion "concave set."

One can immediately verify (do it!) that a polyhedral set is convex, whence a polyhedrally representable function also is convex. It follows that

- lack of convexity makes impossible polyhedral representation of a set/function,
- consequently, operations with functions/sets allowed by "calculus of polyhedral representability" we intend to develop should be convexity-preserving operations.

To illustrate the latter point: taking intersection of two sets and taking maximum or sum of two functions are convexity-preserving operations, and indeed we shall see them in our calculus. In contrast to this, taking union of two sets and taking minima or difference of two functions does not necessarily preserve convexity, and we shall not see these operations in our calculus.

C. **Structure of a proper polyhedrally representable function:** A function $f : \mathbf{R}^n \to \mathbf{R} \cup \{+\infty\}$ is called *proper* if $\mathrm{Dom}\, f \neq \emptyset$, i.e., the function is finite at least at one point.

Proposition 1.1. *A proper function $f(x) : \mathbf{R}^n \to \mathbf{R} \cup \{+\infty\}$ is polyhedrally representable if and only if the domain of the function is a nonempty polyhedral set, and in this domain f is piecewise linear — it is the maximum of finitely many affine functions.*

Proof. In one direction: if $D = \{x \in \mathbf{R}^n : Ax \leq b\}$ is a nonempty polyhedral set and $f_i(x) = a_i^T x + b_i$, $1 \leq i \leq I$, is a nonempty finite collection of affine functions, then the function

$$f(x) = \begin{cases} +\infty, & x \notin D \\ \max_i f_i(x), & x \in D \end{cases}$$

is p.r.:

$$\{[x;\tau] : \tau \geq f(x)\} = \{[x;\tau] : Ax \leq b, \tau \geq f_i(x), i \leq I\}.$$

In the opposite direction: if f is proper p.r.f., then $\mathrm{Epi}\{f\}$ is polyhedrally representable and therefore is polyhedral, Thus,

$$\mathrm{Epi}\{f\} = \{[x;\tau] : a_i^T x + b_i \tau \leq c_i, i \leq I\}.$$

We claim that all b_i are nonpositive. Indeed, f is proper, that is, for some x the solution set of system of inequalities $b_i \tau \leq c_i - a_i^T x$,

$i \leq I$, in variable τ should be a ray of the form $[w, \infty)$ with some real w; this, of course, is impossible when some of b_i are positive. Let $J = \{i : b_i < 0\}$ and $J^o = \{i : b_i = 0\}$. We claim that $J \neq \emptyset$. Indeed, otherwise the set $\{\tau : a_i^T x + b_i \tau \leq c_i, i \leq I\}$ for every x would be, depending on x, either empty, or the entire real line, which definitely is not the case for intersections of the epigraph of a proper function with the "vertical lines" $x = \text{const}$. The bottom line of our observations is that

$$\text{Epi}\{f\} = \left\{[x;\tau] : a_i^T x \leq c_i, i \in J^o, \tau \geq \max_{i \in J} |b_i|^{-1} \left[a_i^T x - c_i\right]\right\},$$

that is, the domain of f is a nonempty polyhedral set, and on this domain f is piecewise linear. □

The fact that a polyhedrally representable function is just a piecewise linear function restricted onto a polyhedral domain does not nullify the usefulness of the notion of a p.r.f. Indeed, similar to the case of polyhedral sets, a p.r. function $f(x)$ admitting a "compact" p.r. can require astronomically many "pieces" $a_i^T x + b_i$ in a piecewise linear representation $f(x) = \max_i [a_i^T x + b_i]$ (think of $f(x) = \sum_i |x_i|$).

1.3.3 *Polyhedral representations of sets and functions: Calculus*

The "raw materials" in our calculus are really simple:

- "elementary" polyhedral sets are those represented as $X = \{x \in \mathbf{R}^n : a^T x \leq b\}$ (when $a \neq 0$, or, which is the same, when the set is nonempty and differs from the entire space, such a set is called *half-space*);
- "elementary" polyhedrally representable functions are just affine functions represented in the standard form

$$f(x) = a^T x + b.$$

An affine function indeed is a p.r.f., since its epigraph

$$\{[x;\tau] : \tau \geq f(x)\}$$

is a solution set of a linear inequality in variables x, τ and thus is polyhedral.

Calculus of polyhedral representability: Sets: The basic rules here are as follows:

S.1. **Taking finite intersections:** *If the sets $X_i \subset \mathbf{R}^n$, $1 \le i \le k$, are polyhedral, so is their intersection, and a p.r. of the intersection is readily given by p.r.'s of the operands.*

Indeed, if

$$X_i = \{x \in \mathbf{R}^n : \exists w^i : P_i x + Q_i w^i \le r_i\},\ i = 1, \ldots, k,$$

then

$$\bigcap_{i=1}^{k} X_i = \{x : \exists w = [w^1; \ldots; w^k] : P_i x + Q_i w^i \le r_i,\ 1 \le i \le k\},$$

which is a polyhedral representation of $\bigcap_i X_i$.

S.2. **Taking direct products.** Given k sets $X_i \subset \mathbf{R}^{n_i}$, their *direct product* $X_1 \times \cdots \times X_k$ is the set in $\mathbf{R}^{n_1+\cdots+n_k}$ comprising all block-vectors $x = [x^1; ...; x^k]$ with blocks x^i belonging to X_i, $i = 1, \ldots, k$. For example, the direct product of k segments $[-1, 1]$ on the axis is the unit k-dimensional box $\{x \in \mathbf{R}^k : -1 \le x_i \le 1,\ i = 1, \ldots, k\}$. The corresponding calculus rule is as follows:

If the sets $X_i \subset \mathbf{R}^{n_i}$, $1 \le i \le k$, are polyhedral, so is their direct product, and a p.r. of the product is readily given by p.r.'s of the operands. Indeed, if

$$X_i = \{x^i \in \mathbf{R}^{n_i} : \exists w^i : P_i x^i + Q_i w^i \le r_i\},\ i = 1, \ldots, k,$$

then

$$\begin{aligned} &X_1 \times \cdots \times X_k \\ &\quad = \{x = [x^1; \ldots; x^k] : \exists w = [w^1; \ldots; w^k] : P_i x^i + Q_i w^i \le r_i, \\ &\qquad 1 \le i \le k\}. \end{aligned}$$

S.3. **Taking affine image:** *If $X \subset \mathbf{R}^n$ is a polyhedral set and $y = Ax + b : \mathbf{R}^n \to \mathbf{R}^m$ is an affine mapping, then the set $Y =$*

$AX + b := \{y = Ax + b : x \in X\} \subset \mathbf{R}^m$ is polyhedral, with p.r. readily given by the mapping and a p.r. of X.

Indeed, if $X = \{x : \exists w : Px + Qw \leq r\}$, then

$$\begin{aligned} Y &= \{y : \exists[x; w] : Px + Qw \leq r, y = Ax + b\} \\ &= \{y : \exists[x; w] : Px + Qw \leq r, y - Ax \leq b, Ax - y \leq -b\}. \end{aligned}$$

Since Y admits a p.r., Y is polyhedral (Theorem 1.2).

Note: This is the point where we see the importance of slack variables (i.e., the advantage of general-type polyhedral representations $X = \{x : \exists w : Px + Qw \leq r\}$ as compared to straightforward ones $X = \{x : Ax \leq b\}$). When taking intersections and direct products of "straightforwardly represented" polyhedral sets, building a straightforward representation of the result is easy; when taking affine image of the set as simple as the k-dimensional unit box, a straightforward representation of the result exists, but is, in general, intractable, since it can require an *exponential in k* number of linear inequalities.

S.4. **Taking inverse affine image:** *If $X \subset \mathbf{R}^n$ is polyhedral, and $x = Ay + b : \mathbf{R}^m \to \mathbf{R}^n$ is an affine mapping, then the set $Y = \{y \in \mathbf{R}^m : Ay + b \in X\} \subset \mathbf{R}^m$ is polyhedral, with p.r. readily given by the mapping and a p.r. of X.*

Indeed, if $X = \{x : \exists w : Px + Qw \leq r\}$, then

$$\begin{aligned} Y &= \{y : \exists w : P[Ay + b] + Qw \leq r\} \\ &= \{y : \exists w : [PA]y + Qw \leq r - Pb\}. \end{aligned}$$

S.5. **Taking arithmetic sum:** *If the sets $X_i \subset \mathbf{R}^n$, $1 \leq i \leq k$, are polyhedral, so is their arithmetic sum $X_1 + \cdots + X_k := \{x = x_1 + \cdots + x_k : x_i \in X_i, 1 \leq i \leq k\}$, and a p.r. of the sum is readily given by p.r.'s of the operands.*

Indeed, the arithmetic sum of the sets $X_1, \ldots, X_k$ is the image of their direct product under linear mapping $[x^1; \ldots; x^k] \mapsto x^1 + \cdots + x^k$, and both operations preserve polyhedrality. Here is an explicit p.r. for the sum: if $X_i = \{x : \exists w^i : P_i x + Q_i w^i \leq r_i\}$,

$1 \leq i \leq k$, then

$$X_1 + \cdots + X_k$$
$$= \left\{ x : \exists x^1, \ldots, x^k, w^1, \ldots, w^k : P_i x^i + Q_i w^i \leq r_i, \right.$$
$$\left. 1 \leq i \leq k, x - \sum_{i=1}^{k} x^i \right\},$$

and it remains to replace the vector equality in the right-hand side by a system of two opposite vector inequalities.

Calculus of p.r. functions: Here the basic calculus rules read as follows:

F.1. **Taking linear combinations with positive coefficients:** *If $f_i : \mathbf{R}^n \to \mathbf{R} \cup \{+\infty\}$ are p.r.f.'s and $\lambda_i > 0$, $1 \leq i \leq k$, then $f(x) = \sum_{i=1}^k \lambda_i f_i(x)$ is a p.r.f., with a p.r. readily given by those of the operands.*

Indeed, if $\{[x;\tau] : \tau \geq f_i(x)\} = \{[x;\tau] : \exists w^i : P_i x + \tau p_i + Q_i w^i \leq r_i\}$, $1 \leq i \leq k$, are p.r.'s of $f_1, \ldots, f_k$, then

$$\left\{ [x;\tau] : \tau \geq \sum_{i=1}^{k} \lambda_i f_i(x) \right\}$$
$$= \left\{ [x;\tau] : \exists t_1, \ldots, t_k : t_i \geq f_i(x), 1 \leq i \leq k, \quad \sum_i \lambda_i t_i \leq \tau \right\}$$
$$= \left\{ [x;\tau] : \exists t_1, \ldots, t_k, w^1, \ldots, w^k : \right.$$
$$\left. P_i x + t_i p_i + Q_i w^i \leq r_i, 1 \leq i \leq k, \sum_i \lambda_i t_i \leq \tau \right\}.$$

F.2. **Direct summation.** *If $f_i : \mathbf{R}^{n_i} \to \mathbf{R} \cup \{+\infty\}$, $1 \leq i \leq k$, are p.r.f.'s, then so is their direct sum*

$$f([x^1; \ldots; x^k]) = \sum_{i=1}^{k} f_i(x^i) : \mathbf{R}^{n_1 + \cdots + n_k} \to \mathbf{R} \cup \{+\infty\}$$

and a p.r. for this function is readily given by p.r.'s of the operands.

Indeed, if $\{[x^i;\tau] : \tau \geq f_i(x^i)\} = \{[x^i;\tau] : \exists w^i : P_i x^i + \tau p_i + Q_i w^i \leq r_i\}$, $1 \leq i \leq k$, are p.r.'s of f_i, then

$$\left\{[x^1;\ldots;x^k;\tau] : \tau \geq \sum_{i=1}^{k} f_i(x^i)\right\}$$

$$= \left\{[x^1;\ldots;x^k;\tau] : \exists t_1,\ldots,t_k : t_i \geq f_i(x^k), 1 \leq i \leq k,\right.$$

$$\left.\sum_i t_i \leq \tau\right\}$$

$$= \left\{[x^1;\ldots;x^k;\tau] : \exists t_1,\ldots,t_k, w^1,\ldots,w^k :\right.$$

$$\left. P_i x^i + t_i p_i + Q_i w^i \leq r_i, 1 \leq i \leq k, \sum_i t_i \leq \tau\right\}.$$

F.3. **Taking maximum:** *If $f_i : \mathbf{R}^n \to \mathbf{R} \cup \{+\infty\}$ are p.r.f.'s, so is their maximum $f(x) = \max[f_1(x),\ldots,f_k(x)]$, with a p.r. readily given by those of the operands.* In particular, *a piecewise linear function $\max[a_1^T x + b_1,\ldots,a_m^T x + b_m]$ is a p.r.f.*

Indeed, if $\{[x;\tau] : \tau \geq f_i(x)\} = \{[x;\tau] : \exists w^i : P_i x + \tau p_i + Q_i w^i \leq r_i\}$, $1 \leq i \leq k$, then

$$\left\{[x;\tau] : \tau \geq \max_i f_i(x)\right\}$$
$$= \{[x;\tau] : \exists w^1,\ldots,w^k : P_i x + \tau p_i + Q_i w^i \leq r_i, 1 \leq i \leq k\}.$$

Note that this rule mirrors the rule on p.r. of the intersection of polyhedral sets, due to $\mathrm{Epi}\{\max_i f_i\} = \bigcap_i \mathrm{Epi}\{f_i\}$.

F.4. **Affine substitution of argument:** *If a function $f(x) : \mathbf{R}^n \to \mathbf{R} \cup \{+\infty\}$ is a p.r.f. and $x = Ay + b : \mathbf{R}^m \to \mathbf{R}^n$ is an affine mapping, then the function $g(y) = f(Ay+b) : \mathbf{R}^m \to \mathbf{R} \cup \{+\infty\}$ is a p.r.f., with a p.r. readily given by the mapping and a p.r. of f.*

Indeed, if $\{[x;\tau]:\tau\geq f(x)\}=\{[x;\tau]:\exists w:Px+\tau p+Qw\leq r\}$ is a p.r. of f, then

$$\begin{aligned}\{[y;\tau]:\tau\geq f(Ay+b)&=\{[y;\tau]:\exists w:P[Ay+b]+\tau p\\&\qquad+Qw\leq r\}\\&=\{[y;\tau]:\exists w:[PA]y+\tau p\\&\qquad+Qw\leq r-Pb\}.\end{aligned}$$

Note that this rule mirrors the rule on p.r. of the inverse affine image of a polyhedral set, since Epi$\{f(Ay+b)\}$ is the inverse image of Epi$\{f\}$ under the affine mapping $[y;\tau]\mapsto[Ay+b;\tau]$.

F.5. **Theorem on superposition:** *Let $f_i(x):\mathbf{R}^n\to\mathbf{R}\cup\{+\infty\}$, $i\leq m$, be p.r.f.'s, and let $F(y):\mathbf{R}^m\to\mathbf{R}\cup\{+\infty\}$ be a p.r.f. which is nondecreasing w.r.t. every one of the variables $y_1,\ldots,y_m$. Then the superposition*

$$g(x)=\begin{cases}F(f_1(x),\ldots,f_m(x)), & f_i(x)<+\infty,1\leq i\leq m\\+\infty, & \textit{otherwise}\end{cases}$$

of F and $f_1,\ldots,f_m$ is a p.r.f., with a p.r. readily given by those of f_i and F.

Indeed, let

$$\begin{aligned}\{[x;\tau]:\tau\geq f_i(x)\}&=\{[x;\tau]:\exists w^i:P_ix+\tau p_i+Q_iw^i\leq r_i\},\\\{[y;\tau]:\tau\geq F(y)\}&=\{[y;\tau]:\exists w:Py+\tau p+Qw\leq r\}\end{aligned}$$

be p.r.'s of f_i and F. Then

$$\begin{aligned}\{[x;\tau]:\tau\geq g(x)\}&\overbrace{=}^{(*)}\{[x;\tau]:\exists y_1,\ldots,y_m:y_i\geq f_i(x),\\&\qquad 1\leq i\leq m,\quad F(y_1,\ldots,y_m)\leq\tau\}\\&=\{[x;\tau]:\exists y,w^1,\ldots,w^m,w:P_ix+y_ip_i+Q_iw^i\leq r_i,\\&\qquad 1\leq i\leq m,\quad Py+\tau p+Qw\leq r\},\end{aligned}$$

where $(*)$ is due to the monotonicity of F.

Note that if some of f_i, say, $f_1,\ldots,f_k$, are affine, then the superposition theorem remains valid when we require the monotonicity of F w.r.t. the variables $y_{k+1}, \ldots, y_m$ only; a p.r. of the superposition in this case reads

$$
\begin{aligned}
&\{[x;\tau] : \tau \geq g(x)\} \\
&\quad = \{[x;\tau] : \exists y_{k+1} \ldots, y_m : y_i \geq f_i(x),\ k+1 \leq i \leq m, \\
&\qquad\qquad\qquad\qquad F(f_1(x), \ldots, f_k(x), y_{k+1}, \ldots, y_m) \leq \tau\} \\
&\quad = \{[x;\tau] : \exists y_1, \ldots, y_m, w^{k+1}, \ldots, w^m, w : y_i = f_i(x), \\
&\qquad\qquad\qquad\qquad 1 \leq i \leq k, \\
&\qquad P_i x + y_i p_i + Q_i w^i \leq r_i, k+1 \leq i \leq m, Py + \tau p + Qw \leq r\},
\end{aligned}
$$

and the linear equalities $y_i = f_i(x)$, $1 \leq i \leq k$ can be replaced by pairs of opposite linear inequalities.

Note that when taking superposition, some monotonicity requirements on the outer function are natural, since otherwise this operation does *not* preserve convexity (think of superposition of $f(x) = \max[x_1, x_2]$ and $F(y) = -y$).

1.4 *Fast polyhedral approximation of the second-order cone

We have seen that taking projection onto a subspace can convert a polyhedral set $X = \{x \in \mathbf{R}^n : Ax \leq b\}$ which is "simple" — is defined by a moderate number of linear inequalities — into a polyhedral set Y which is "complex" — its representation in the form $\{y : By \leq b\}$ requires a much larger number of linear inequalities. An example, already known to us, is

$$
\begin{aligned}
Y &:= \left\{ y \in \mathbf{R}^k : \sum_{i=1}^{k} |y_i| \leq 1 \right\} \\
&= \left\{ y \in \mathbf{R}^k : \exists w : -w_i \leq y_i \leq w_i\ 1 \leq i \leq k, \sum_{i=1}^{k} w_i \leq 1 \right\}.
\end{aligned}
$$

Here the left-hand side set $Y \subset \mathbf{R}^k$ is represented as the projection onto the y-plane of the set

$$
X = \left\{ [y;w] : -w_i \leq y_i \leq w_i\ 1 \leq i \leq k, \sum_{i=1}^{k} w_i \leq 1 \right\}
$$

which "lives" in $\mathbf{R}^{2k}$ and is given by $2k+1$ linear inequalities; it can be proved that *every* representation of Y in the form $\{y : Cy \leq c\}$ requires *at least* 2^k linear inequalities.

Given this observation, a natural question is whether it is possible to approximate well a *nonpolyhedral* set Y by the projection $\widehat{X}$ of a "simple" polyhedral set X in higher dimension. The motivation here might be (and, as far as the construction we intend to present, actually was) the desire to approximate well the problems of optimizing a linear objective over Y (which is not an LO program, since Y is nonpolyhedral) by the problem of minimizing the same objective over the close-to-Y set $\widehat{X}$; the latter problem reduces to an LO program with the simple polyhedral feasible set X and thus is within the grasp of LO algorithms.

The answer to our question depends, of course, on what is the set Y we want to approximate. We intend to demonstrate that when Y is a n-dimensional ball, the answer to the above answer is affirmative. Specifically, we intend to prove the following.

Theorem 1.3 (Fast polyhedral approximation of the second-order cone). *Let*

$$\mathbf{L}^{n+1} = \{[x;\tau] \in \mathbf{R}^n \times \mathbf{R} : \tau \geq \|x\|_2 := \sqrt{x^T x}\}$$

(this set is called the $(n+1)$-dimensional second-order (a.k.a. Lorentz, or Ice-Cream) cone). For every n and every $\epsilon \in (0, 1/2)$ one can explicitly point out a system

$$Px + \tau p + Qw \leq 0 \tag{1.36}$$

of homogeneous linear inequalities in the original variables x, τ and slack variables w such that

- *the number $I(n,\epsilon)$ of inequalities in the system is $\leq O(1)n\ln(1/\epsilon)$,*
- *the number $V(n,\epsilon) = \dim w$ of slack variables in the system is $\leq O(1)n\ln(1/\epsilon)$,*

and the projection

$$\widehat{L}^{n+1} = \{[x;\tau] : \exists w : Px + \tau p + Qw \leq 0\} \tag{1.37}$$

of the solution set of this system on the space of (x,τ)-variables is in between the second-order cone and its "$(1+\epsilon)$-extension:"

$$\begin{aligned}\mathbf{L}^{n+1} &:= \{[x;\tau] \in \mathbf{R}^n \times \mathbf{R} : \tau \geq \|x\|_2\}\\ &\subset \widehat{L}^{n+1} \subset \mathbf{L}^{n+1}_\epsilon := \{[x;\tau] \in \mathbf{R}^n \times \mathbf{R} : (1+\epsilon)\tau \geq \|x\|_2\}.\end{aligned} \tag{1.38}$$

To get an impression of the constant factors in the theorem, look at a sample of values of $I(n,\epsilon)$, $V(n,\epsilon)$:

	$\epsilon = 10^{-1}$		$\epsilon = 10^{-6}$		$\epsilon = 10^{-14}$	
n	$V(n,\epsilon)$	$I(N,\epsilon)$	$V(n,\epsilon)$	$I(n,\epsilon)$	$V(n,\epsilon)$	$I(n,\epsilon)$
4	6	17	31	69	70	148
16	30	83	159	345	361	745
64	133	363	677	1458	1,520	3,153
256	543	1,486	2,711	5,916	6,169	12,710
1,024	2,203	6,006	10,899	23,758	24,773	51,050

Note: You can see that $V(n,\epsilon) \approx 0.7n \ln \frac{1}{\epsilon}$, $I(n,\epsilon) \approx 2n \ln \frac{1}{\epsilon}$.

Several comments are in order.

A. When $\epsilon = 1.\text{e} - 17$ or something like, a usual computer cannot distinguish between 1 and $1+\epsilon$, so that with such an ϵ, $\widehat{L}^{n+1}$ "for all practical purposes" is the same as the Lorentz cone $\mathbf{L}^{n+1}$. On the other hand, with $\epsilon = 1.\text{e-}17$ the numbers of inequality constraints and slack variables in the polyhedral representation of $\widehat{L}$ by the system (1.36) are moderate multiples of n (indeed, while 10^{-17} is "really small," $\ln(10^{17}) \approx 39.1439$ is quite a moderate number).

B. After we know how to build a fast polyhedral approximation of the second-order cone, we know how to build such an approximation for a Euclidean ball $B^R_n = \{x \in \mathbf{R}^n : \|x\|_2 \leq R\}$. Indeed, from (1.38), it follows that the projection $\widehat{B}^R_n$ onto the x-plane of the polyhedral set

$$\{[x;w] : Px + Qw \leq -Rp\}$$

in the space of (x,w)-variables is in between B^R_n and $B^{(1+\epsilon)R}_n$:

$$B^R_\tau \subset \widehat{B}^R_n \subset B^{(1+\epsilon)R}_n. \tag{1.39}$$

C. In principle, there is nothing strange in the fact that a "good" nonpolyhedral set Y in $\mathbf{R}^n$ can be approximated, within a whatever accuracy, by a polyhedral set. Such a possibility definitely exists when Y is a closed and bounded convex set, as it is the case with the ball B_n^R. Let us focus on this case; to simplify notation, w.l.o.g. let us set $R = 1$ and $B_n = B_n^1$. It is intuitively clear (and indeed is true) that given an $\epsilon > 0$ and taking a dense enough finite "grid" $x_1, \ldots, x_N$ on the boundary of B_n, that is, on the unit sphere $S_{n-1} = \{x \in \mathbf{R}^n : \|x\|_2 = 1\}$, the polytope bounded by the tangent to S_{n-1} at x_i hyperplanes $\{x : x_i^T(x - x_i) = 0\}$, $1 \leq i \leq N$, will contain B_n and be contained in $B_n^{1+\epsilon}$. The problem, however, is how many hyperplanes should we take for this purpose, and the answer is as follows: *For every polyhedral set $\widehat{B} \subset \mathbf{R}^n$ such that $B_n \subset \widehat{B} \subset B_n^{1+\epsilon}$, the number N of linear inequalities in a "straightforward" polyhedral representation $\widehat{B} = \{x \in \mathbf{R}^n : Cx \leq c\}$ is at least* $\exp\{O(1)n\ln(1/\epsilon)\}$, provided $\epsilon \leq 0.1$. We see that a "straightforward" approximation of B_n within a fixed accuracy ϵ, say $\epsilon = 0.1$, by a solution set of a system of linear inequalities requires an exponential in n, and thus astronomically large already for moderate n, number of inequalities. In contrast to this, to approximate B_n within the same accuracy by the *projection* onto the space where B_n lives of a solution set of a system of linear inequalities requires just linear in n numbers of inequalities and slack variables.

It is highly instructive (and not difficult) to understand where the exponential in n lower bound on a number N of linear inequalities in a system in variables $x \in \mathbf{R}^n$ with the solution set well approximating B_n comes from. Assume that a polyhedral set $\widehat{B} = \{x \in \mathbf{R}^n : a_i^T x \leq b_i, 1 \leq i \leq N\}$ is in between B_n and $(1+\epsilon)B_n = B_n^{1+\epsilon}$, where ϵ is not too large, say, $\epsilon \leq 0.1$. We can assume w.l.o.g. that $a_i \neq 0$ and then normalize a_i to ensure that $\|a_i\|_2 = 1$ for all i, which we assume from now on. Now, b_i should be ≥ 1, since otherwise the constraint $a_i^T x \leq b_i$ would cut off B_n, a nonempty set (due to $\max_{x \in B_n} a_i^T x = \|a_i\|_2 = 1$) and thus the inclusion $B_n \subset \widehat{B}_n$ would be impossible. By the same token, if $b_i > 1$, then, replacing it with 1, we do not affect the validity of the inclusion $B_n \subset \widehat{B}$ and only decrease $\widehat{B}$, thus preserving the validity of the inclusion $\widehat{B} \subset B_n^{1+\epsilon}$. The bottom line is that we lose nothing by assuming that $\widehat{B} = \{x \in \mathbf{R}^n : a_i^T x \leq 1, 1 \leq i \leq N\}$, with all a_i being unit vectors.

Now, saying that $\widehat{B} := \{x : a_i^T x \leq 1, i \leq N\} \subset B_n^{1+\epsilon}$ is exactly the same as saying that $\widetilde{B} := \{x : a_i^T x \leq (1-\epsilon), i \leq N\} \subset B_n$. Thus, we are in the situation when

$$\widetilde{B} \subset B_n \subset \widehat{B},$$

so that the boundary S_{n-1} of B_n should be contained in the set $\widehat{B}\backslash(\text{int }\widetilde{B})$. The latter set is contained in the union, over $i \leq N$, of the "stripes" $P_i = \{x \in \mathbf{R}^n : (1-\epsilon) \leq a_i^T x \leq 1\}$, whence also

$$S_{n-1} \subset \bigcup_{i=1}^{N} \overbrace{[P_i \cap S_{n-1}]}^{H_i}. \tag{$*$}$$

Geometrically, H_i is a "spherical hat," and all these hats are congruent to the set

$$H_\epsilon = \{x \in \mathbf{R}^n : x^T x = 1, x_n \geq 1-\epsilon\}.$$

Denoting α_ϵ the $(n-1)$-dimensional "spherical area" of H_ϵ and by α the area of the entire S_{n-1} and taking into account that $H_1, \ldots, H_N$ cover S_{n-1}, we get

$$N\alpha_\epsilon \geq \alpha \quad \Rightarrow \quad N \geq \frac{\alpha}{\alpha_\epsilon}. \tag{!}$$

It remains to bound the ratio α/α_ϵ from below. Let us start with bounding α_ϵ from above. The projection of the spherical hat H_ϵ onto the plane $x_n = 0$ is a $(n-1)$-dimensional ball of radius $r = \sqrt{1-(1-\epsilon)^2} = \sqrt{2\epsilon - \epsilon^2} \leq \sqrt{2\epsilon}$. When ϵ is ≤ 0.1, this projection, up to a factor $O(1)$, preserves the $(n-1)$-dimensional volume, so that $\alpha_\epsilon \leq O(1)r^{n-1}\beta_{n-1}$, where β_{n-1} is the $(n-1)$-dimensional volume of B_{n-1}^1. Now, the projection of the "northern part" $\{x : x^T x = 1, x_n \geq 0\}$ of S_{n-1} onto the plane $x_n = 0$ is B_{n-1}^1, and this projection clearly reduces the $(n-1)$-dimensional volume; thus, $\alpha \geq 2\beta_{n-1}$. We end up with $N \geq O(1)r^{-(n-1)} \geq O(1)(2\epsilon)^{-(n-1)/2} \geq \exp\{O(1)n\ln(1/\epsilon)\}$, as claimed.

A rigorous reasoning goes as follows. Assume that $n \geq 3$. Let $\pi/2 - \phi$ be the "altitude angle" of a point e on S_{n-1}, that is, ϕ is the angle between e and the direction of nth coordinate axis. Then H_ϵ is the set of all points on the unit sphere S_{n-1} for which $\phi \geq \phi_0 = \text{asin}(r)$. Denoting by γ_{n-1} the $(n-2)$-dimensional volume of S_{n-2}, we clearly have $\alpha_\epsilon = \gamma_{n-1}\int_0^{\phi_0} \sin^{n-2}(\phi)d\phi$ and $\alpha = 2\gamma_{n-1}\int_0^{\pi/2} \sin^{n-2}(\phi)d\phi$. It follows that

$$\begin{aligned}\alpha_\epsilon &\leq \gamma_{n-1}\int_0^{\phi_0} \sin^{n-2}(\phi)\frac{\cos(\phi)}{\cos(\phi_0)}d\phi = \gamma_{n-1}(n-1)^{-1}\cos^{-1}(\phi(0))\sin^{n-1}(\phi_0)\\ &= \gamma_{n-1}(n-1)^{-1}\cos^{-1}(\phi_0)r^{n-1} = \gamma_{n-1}(n-1)^{-1}(1-\epsilon)^{-1}r^{n-1}.\end{aligned}$$

At the same time, it can be easily checked numerically that $\cos(x) \geq 0.99\exp\{-x^2/2\}$ when $0 \leq x \leq 0.5$, whence

$$\begin{aligned}
\alpha &= 2\gamma_{n-1}\int_0^{\pi/2}\sin^{n-2}(\phi)d\phi = 2\gamma_{n-1}\int_0^{\pi/2}\cos^{n-2}(\phi)d\phi \\
&\geq 1.98\gamma_{n-1}\int_0^{1/2}\exp\{-\phi^2(n-2)/2\}d\phi \\
&= 1.98\gamma_{n-1}(n-2)^{-1/2}\int_0^{\frac{1}{2}\sqrt{n-2}}\exp\{-s^2/2\}ds \\
&= 1.98\gamma_{n-1}(n-2)^{-1/2}(1+\epsilon_n)\int_0^{\infty}\exp\{-s^2/2\}ds \\
&= 0.99\sqrt{2\pi}\gamma_{n-1}(n-2)^{-1/2}(1+\epsilon_n),
\end{aligned}$$

where $\epsilon_n \to 0$ as $n \to \infty$. Thus,

$$\begin{aligned}
N &\geq \alpha/\alpha_\epsilon \geq 0.99(1+\epsilon_n)\sqrt{2\pi}(n-2)^{-1/2}(n-1)(1-\epsilon)r^{-(n-1)} \\
&\geq 0.99\sqrt{2\pi}(1+\delta_n)\sqrt{n}(2\epsilon)^{-(n-1)/2}
\end{aligned}$$

with $\delta_n \to 0$ as $n \to \infty$. We see that for large n (on a closest inspection, "large" here means $n \geq 10$) it definitely holds $N \geq 2n^{1/2}(2\epsilon)^{-(n+1)/2}$, provided $\epsilon \leq 0.1$. When $n = 100$ and $\epsilon = 0.1$, this lower bound on N is as large as 7.9e35.

D. In fact, it is well known that the Euclidean ball B_n can be easily approximated by the projection of a very simple polyhedral set — just a box — "living" in higher-dimensional space. Specifically, it is not difficult to prove that if $D_N = \{x \in \mathbf{R}^N : \|x\|_\infty \leq 1\}$ is the unit box in $\mathbf{R}^N$, $N > n$, then a "random projection" of D_N on $\mathbf{R}^n$ is, typically, close to a Euclidean ball in $\mathbf{R}^n$, with the discrepancy going to 0 as N/n grows. The precise statement is as follows: if $A_{n,N}$ is a random matrix with independent standard (zero mean, unit variance) Gaussian entries and n, N, ϵ are linked by the relation $\epsilon \geq O(1)\sqrt{\frac{n}{N}\ln\left(\frac{N}{n}\right)}$, then, with probability approaching 1 as n grows, the set $A_{n,N}D_N \subset \mathbf{R}^n$ is in between two concentric Euclidean balls with the ratio of radii $\leq 1+\epsilon$.

The advantages of the result stated in Theorem 1.3 as compared to this well-known fact are twofold: first, the numbers $N-n$ of "slack variables" and $2N$ of inequality constraints in the "ϵ-accurate" polyhedral approximation of B_n are "nearly" of order of $n\epsilon^{-2}$ (in fact, by a logarithmic in $1/\epsilon$ factor worse), while Theorem 1.3 speaks about "much more compact" ϵ-approximate representation of an n-dimensional ball, with the numbers of slack variables and inequality constraints of the order of $n\ln(1/\epsilon)$. Second, the construction underlying Theorem 1.3 is explicit and thus possesses not only academic but also a practical value (see what follows). In contrast to this, in spite of the provable fact that a random projection of a box, under the assumptions of the "random projection" statement, with high probability is close to a ball, no individual examples of projections with this property are known (cf. the story about sensing matrices with high level of goodness). This being said, it should be mentioned that in the "random projection" approximation of a ball, the polyhedral set we are projecting is centrally symmetric, which is important in certain academic applications. In contrast to this, the construction underlying Theorem 1.3 approximates a ball by the projection of a highly asymmetric polyhedral set; whether results similar to those stated in Theorem 1.3 can be obtained when projecting a centrally symmetric polyhedral set is an interesting open academic problem.

E. Theorem 1.3, essentially, says that "for all practical purposes," a *conic quadratic* optimization problem, that is, an optimization problem P with linear objective and (finitely many) *conic quadratic* constraints, that is, constraints of the generic form

$$\|Ax + b\|_2 \leq c^T x + d \qquad (*)$$

is an LO program (a rigorous statement reads "problem P can be approximated *in a polynomial time fashion* by an LO program;" eventually, we shall understand what "polynomial time fashion" means). Indeed, Theorem says that "for all practical purposes" the set $\|u\|_2 \leq \tau$ is polyhedrally representable with an explicit (and "short") polyhedral representation. But then, by calculus of polyhedral representability, the same is true for the feasible set of $(*)$ (rule on inverse affine image) and thus for the feasible set of P (rule on taking finite intersections), so that P reduces to an LO program.

Now, "expressive abilities" of conic quadratic problems are surprisingly rich. For example, a quadratic constraint of the form

$$x^T A^T A x \leq b^T x + c$$

is equivalent (check it!) to the conic quadratic constraint

$$\| [2Ax; 1 - b^T x - c] \|_2 \leq 1 + b^T x + c;$$

some constraints which do not look quadratic at all, e.g., the constraints in variables x, t

$$\prod_{i=1}^{n} x_i^{-\pi_i} \leq \tau, x > 0,$$

where $\pi_i > 0$ are rational numbers, or

$$\prod_{i=1}^{n} x_i^{\pi_i} \geq \tau, \; x \geq 0,$$

where $\pi_i > 0$ are rational numbers and $\sum_i \pi_i \leq 1$ can be reduced, *in a systematic way*, to a system of conic quadratic constraints in the original variables x, τ and additional slack variables. Theorem 1.3 states that "for all practical purposes" (or, scientifically, up to a polynomial time approximation) the rich expressive abilities of conic quadratic programs are shared by the usual LO's.

As a more striking example, consider the exponential function $\exp\{x\}$. The exponent which "lives" in a computer is somehow different from the exponent which lives in our mind: the latter is well-defined and nonzero on the entire real axis. The former makes sense in a moderate segment of values of x: if you ask MATLAB what is $\exp\{-800\}$, the answer will be 0, and if you ask what is $\exp\{800\}$, the answer will be $+\infty$. Thus, "for all practical purposes," the "real life" exponent $\exp\{x\}$ can be considered as the restriction of the "ideal" exponent which lives in our mind on a finite, and even not very large, segment $-T \leq x \leq T$. Now, for the "ideal" exponent we have

$$\exp\{x\} = (\exp\{2^{-k} x\})^{2^k},$$

and when $|x| \leq T$ and $2^k \gg T$, $\exp\{2^{-k} x\}$ is pretty close to $1 + 2^{-k} x$. It follows that

$$|x| \leq T \quad \Rightarrow \quad \exp\{x\} \approx (1 + 2^{-k} x)^{2^k}.$$

Note that the right-hand side function, restricted onto the segment $[-T, T]$ with $T \leq 2^k$, can be expressed via a "short series" of linear and quadratic inequalities and a single linear equation:

$$\begin{aligned}&\{[x;\tau] \in \mathbf{R}^2 : -T \le x \le T, \tau \ge (1+2^{-k}x)^{2^k}\}\\ &= \left\{[x;\tau] \in \mathbf{R}^2 : \exists u_0, u_1, \ldots, u_k : \begin{array}{r} -T \le x \le T, u_0 = 1 + 2^{-k}x, \\ u_0^2 \le u_1, u_1^2 \le u_2, \ldots, u_{k-1}^2 \\ \le u_k, u_k \le \tau \end{array}\right\}.\end{aligned}$$

Now, the quadratic inequality $u^2 \le v$ in variables u, v, as we already know, can be represented by a conic quadratic constraint and thus, "for all practical purposes," can be represented by a short system of *linear* constraints in u, v and additional variables. As a result, *the exponent which "lives in a computer" is, for all practical purposes, a polyhedrally representable function.* The precise statement is as follows:

Given $T > 0$ and $\epsilon \in (0,1)$, one can point out an explicit system of linear inequalities in scalar variables x, τ and additional variables w such that

- *the number of inequalities in the system, same as the dimension of the vector w, are bounded by polynomials in T and $\ln(1/\epsilon)$;*
- *the projection of the solution set of the system on the 2D plane of x, τ-variables is the epigraph* Epi$\{f\}$ *of a function $f(x)$ (which thus is polyhedrally representable) such that* $\mathrm{Dom}\, f = [-T, T]$ *and*

$$|x| \le T \Rightarrow \exp\{x\} \le f(x) \le (1+\epsilon)\exp\{x\}.$$

Remark 1.1. The result we have just formulated is correct, but to interpret it as the justification of the claim "that the exponent which lives in the computer for all practical purposes, whatever that means, is a p.r.f." involves some cheating. Indeed, when explaining why the exponent which lives in the computer, let us call it Exp(s), is a function with bounded (and not too large) domain, we referred to the fact that floating point operations are imprecise, and with their standard implementation we do have $\exp\{800\} = \infty$ and $\exp\{-800\} = 0$. However, when speaking about high accuracy p.r. approximation of Exp$(\cdot)$, we acted as if the operations involved into building this approximation were operations of precise real arithmetics. In fact, these operations involve close to 1 number $1+2^{-k}x$, and for k like few tens and x, say, of the order of one the standard floating point arithmetics will handle such a number with significant loss of accuracy. Thus, *in principle* the same argument which justifies our interest in Exp(s) as a "real life substitution of $\exp\{s\}$" can be used against the conclusion that Exp$(\cdot)$ is, "for all practical purposes," polyhedrally representable. Luckily, this potential threat is indeed just potential; it can be shown (Ben-Tal and Nemirovski, 2022, Section 2.3.6) that the outlined construction in slightly refined form allows to build p.r.f. which, *when computed via the standard floating point arithmetics,* approximates $\exp\{s\}$ in the range $|s| \le 700$ with relative error $\le 2\mathrm{e}{-11}$.

It follows that when solving a Geometric Programming problem — an optimization problem with linear objective and finitely many constraints of the form

$$\sum_{i=1}^{k} \exp\{a_i^T x + b_i\} \le 1 \tag{!}$$

(perhaps augmented by linear constraints), we can "regularize" the problem by replacing every constraint (!) with the system of constraints

$$0 \ge w_i \ge a_i^T x + b_i, w_i \ge -T, \exp\{w_i\} \le u_i, 1 \le i \le k, \sum_{i=1}^{k} u_i \le 1, \tag{!!}$$

where u_i, w_i are slack variables and T is a once-for-ever-fixed moderate constant (say, $T = 60$). As far as actual computations are concerned, this regularization does not alter the constraint.

Indeed, whenever x can be extended by properly chosen u_i, w_i, to a feasible solution of (!!), x is clearly feasible for (!), and "nearly vice versa." Specifically, assuming that x is feasible for (!) and setting $w_i = \max\{-T, a_i^T x + b_i\}$, $u_i = \exp\{w_i\}$, we get a solution to (!!) which satisfies all constraints in (!!) except, perhaps, the last of them, $\sum_i u_i \le 1$; instead of the latter constraint, its "slightly weakened" version $\sum_i u_i \le 1 + k\exp\{-T\}$ is satisfied. With $T = 60$ and k as large as 10^9, one has $k\exp\{-T\} \le 1.\mathrm{e}{-17}$, so that the computer cannot tell the difference between validity of the actual and the weakened versions of the last constraint in (!!).

We have seen that "numerically speaking," the regularized problem is the same as the original Geometric Programming program. On the other hand, the only nonlinear constraints in the regularized problem are $\exp\{w_i\} \le u_i$, and w_i is restricted to sit in $[-T, 0]$. It follows that we can approximate every one of the nonlinear constraints, in a polynomial time fashion, by a "short" system of linear inequalities and thus reduce the original Geometric Programming problem to an LO one.

We do not claim that the best way to solve real-world conic quadratic and Geometric Programming problems is to approximate them by LO's; both these classes of problems admit dedicated theoretically (and to some extent, also practically) efficient solution algorithms which can solve practical problems faster than LO algorithms as applied to the LO approximations of these problems. What is good and what is bad in practice, it depends on the available software. Two decades ago, there were no efficient and reliable software for medium-size (few thousands of variables) conic quadratic problems, and Theorem 1.3 was a byproduct of attempts to handle these problems with the existing software. Today there exists efficient and reliable software for solving conic quadratic and Geometric Programming problems, and there is basically no necessity to reduce these problems to LO. This being said, note that the outlined "fast polyhedral approximation" results are of definite theoretical interest.

Proof of Theorem 1.3. Let $\epsilon > 0$ and a positive integer n be given. We intend to build a polyhedral ϵ-approximation of the second-order cone $\mathbf{L}^{n+1}$. We may assume w.l.o.g. (think why) that n is an integer power of 2: $n = 2^\kappa$, $\kappa \in \mathbf{N}$.

1^0 The key: fast polyhedral approximation of $\mathbf{L}^3$: Consider the system of linear inequalities in variables y_1, y_2, y_3 and additional variables ξ^j, η^j, $0 \leq j \leq \nu$ (ν is a parameter of the construction) as follows:

$$
\begin{array}{ll}
(a) & \left\{\begin{array}{l} \xi^0 \geq |y_1|, \\ \eta^0 \geq |y_2|, \end{array}\right. \\
(b) & \left\{\begin{array}{l} \xi^j = \cos\left(\frac{\pi}{2^{j+1}}\right)\xi^{j-1} + \sin\left(\frac{\pi}{2^{j+1}}\right)\eta^{j-1}, \\ \eta^j \geq \left|-\sin\left(\frac{\pi}{2^{j+1}}\right)\xi^{j-1} + \cos\left(\frac{\pi}{2^{j+1}}\right)\eta^{j-1}\right| \end{array}\right., \quad j = 1, \dots, \nu, \\
(c) & \left\{\begin{array}{l} \xi^\nu \leq y_3, \\ \eta^\nu \leq \tan\left(\frac{\pi}{2^{\nu+1}}\right)\xi^\nu \cdot \end{array}\right.
\end{array}
\tag{1.40}
$$

Note that (1.40) can be straightforwardly rewritten equivalently as a system of linear homogeneous inequalities

$$\Pi^{(\nu)}[y; w] \leq 0 \tag{S_ν}$$

in variables $y \in \mathbf{R}^3$ and additional variables w, specifically, as follows. (1.), we add slack variables and associated linear constraints to eliminate the nonlinearities $|\cdot|$ which are present in (a) and in (b); (2.) we use the equality constraints in (b) to eliminate the variables ξ^j, $1 \leq j \leq \nu$. The numbers of variables and linear inequalities in the resulting system (S_ν) clearly are $\leq O(1)\nu$. (S_ν) is a p.r. of the polyhedral set

$$Y^\nu = \{y \in \mathbf{R}^3 : \exists w : \Pi^{(\nu)}[y; w] \leq 0\}.$$

Lemma 1.2. *Y^ν is a polyhedral $\chi(\nu)$-approximation of $\mathbf{L}^3 = \{y \in \mathbf{R}^3 : y_3 \geq \sqrt{y_1^2 + y_2^2}\}$:*

$$\mathbf{L}^3 \subset Y^\nu \subset \mathbf{L}^3_{\delta(\nu)} := \left\{y \in \mathbf{R}^3 : (1 + \chi(\nu))y_3 \geq \sqrt{y_1^2 + y_2^2}\right\} \tag{1.41}$$

with

$$\chi(\nu) = \frac{1}{\cos\left(\frac{\pi}{2^{\nu+1}}\right)} - 1. \tag{1.42}$$

Proof. We should prove the following:

(i) If $y \in \mathbf{L}^3$, then y can be extended to a solution to (1.40).
(ii) If $y \in \mathbf{R}^3$ can be extended to a solution to (1.40), then $\|[y_1; y_2]\|_2 \leq (1 + \chi(\nu))y_3$.

Let $P_j = [\xi^j; \eta^j]$, $0 \leq j \leq \nu$. System (1.40) can be considered as a set of geometric constraints on the points $[y_1; y_2]$, $P_0, \ldots, P_\nu$ on the 2D plane, namely, as follows (in the following story, we treat x_1 as the abscissa, and x_2 as the ordinate on the 2D plane where the points live):

(a) says that P_0 should belong to the first quadrant, and its coordinates should be $\geq$ the magnitudes of the respective coordinates in $[y_1; y_2]$.
(b) says that the link between P_j and P_{j+1} should be as follows (see Fig. 1.2):

— given $P = P_j$, we rotate the vector P_j clockwise by the angle $\pi/2^{j+2}$ and then reflect the resulting point w.r.t. the x_1-axis,

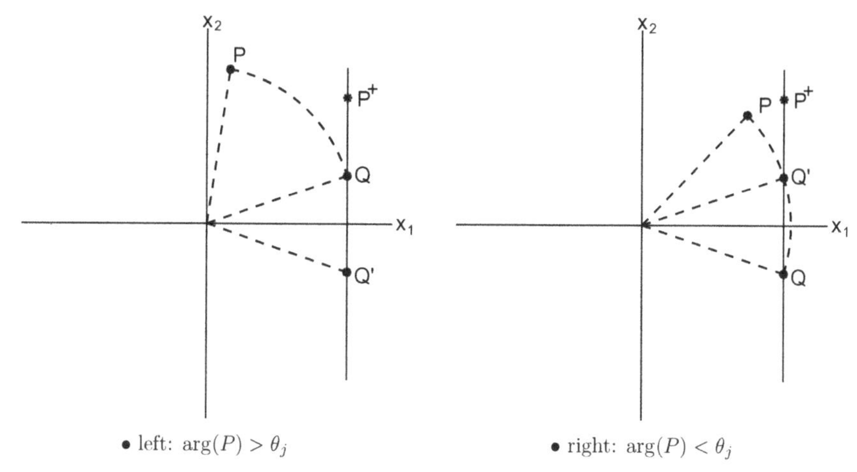

Fig. 1.2. From $P = [\xi^j; \eta^j]$ to $P^+ = [\xi^{j+1}; \eta^{j+1}]$: (1.) Q is obtained from P by rotation by the angle $\theta_j = \pi/2^{j+2}$ clockwise; (2.) Q' is the reflection of Q w.r.t. the abscissa axis; (3.) P^+ is a point on the line QQ' above both Q and Q'.

thus getting a pair symmetric to each other w.r.t. the x_1-axis points $Q = Q_j$, $Q' = Q'_j$;

— $P^+ = P_{j+1}$ is a point on the vertical line passing through Q, Q' which is above both these points (i.e., the ordinate of P^+ is $\geq$ the ordinates of Q and of Q').

(c) says that the point P_ν should belong to the triangle $\Delta = \{0 \leq x_1 \leq y_3, x_2 \leq \tan(\pi/2^{\nu+1})x_1\}$.

Now, observe that from this description it follows that points associated with feasible solution to (1.40) satisfy

$$\|[y_1; y_2]\|_2 \leq \|P_0\|_2 \leq \|P_1\|_2 \leq \dots \leq \|P_\nu\|_2 \leq \frac{1}{\cos(\pi/2^{\nu+1})} y_3,$$

which proves (ii). On the other hand, if $y_3 \geq \|[y_1, y_2]\|_2$, then, setting $P_0 = [|y_1|; |y_2|]$ and specifying for all j P_{j+1} as either Q_j, or Q'_j, depending on which one of these points has a larger ordinate, we get

$$y_3 \geq \|[y_1; y_2]\|_2 = \|P_0\|_2 = \dots = \|P_\nu\|_2. \qquad (!)$$

We claim that what we get in this way is a feasible solution to (1.40). This is clearly true when $y_1 = y_2 = 0$; assuming that the latter is not the case, observe that by construction we have satisfied the constraints (1.40(a) and (b). To prove that (1.40(c)) is satisfied as well, let us look at the arguments ϕ_j of the points P_j (the angles between the direction of the x_1-axis and the directions of vectors P_j). P_0 lives in the first quadrant, and thus $\phi_0 \leq \pi/2$. Since Q_0 is obtained by rotating P_0 clockwise by the angle $\pi/4$, Q'_0 is symmetric to Q_0 w.r.t. the x_1-axis and P_1 is the member of the pair Q_0, Q'_0 living in the first quadrant, the argument ϕ_1 of P_1 satisfies $0 \leq \phi_1 \leq \pi/4$. Similar reasoning, applied to P_1 in the role of P_0, results in $0 \leq \phi_2 \leq \pi/8$, and so on, resulting in $0 \leq \phi_\nu \leq \pi/2^{\nu+1}$. Since P_ν lives in the first quadrant, $\|P_\nu\|_2 \leq y_3$ by (!) and the argument of P_ν is $\leq \pi/2^{\nu+1}$, P_ν belongs to the triangle Δ, that is, (1.40(c)) does take place. (i) is proved.

2^0. From $n = 2$ to $n = 2^\kappa$: Now we are ready to approximate the second-order cone

$$\mathbf{L}^{n+1} = \left\{[x;\tau] \in \mathbf{R}^n \times \mathbf{R} : \tau \geq \sqrt{x_1^2 + \cdots + x_n^2}\right\}$$

where $n = 2^\kappa$. To this end, let us first add to the 2^κ variables $x_i \equiv x_i^0$ ("variables of generation 0") $2^{\kappa-1}$ variables x_i^1 ("variables of generation 1"), treating x_i^1 as the "child" of variables x_{2i-1}^0, x_{2i}^0 of generation 0, add in the same fashion $2^{\kappa-2}$ variables of generation 2, children of variables of generation 1, and so on, until arriving at two variables of generation $\kappa - 1$. By definition, the child of these two variables is $\tau \equiv x_1^\kappa$. Now let us impose on the resulting $2n - 1$ variables x_i^ℓ of all generations the constraints

$$\|[x_{2i-1}^{\ell-1}; x_{2i}^{\ell-1}]\|_2 \leq x_i^\ell,\ 1 \leq \ell \leq \kappa,\ 1 \leq i \leq 2^{\kappa-\ell}. \tag{1.43}$$

It is clear (look at the case of $n = 4$) that the projection of the solution set of this system on the plane of the original variables $x_i = x_i^0$, τ is nothing but $\mathbf{L}^{n+1}$. Now let us choose positive integers ν_1, $\nu_2,\ldots,\nu_\kappa$ and approximate the three-dimensional second-order cones given by

$$\left\|\left[x_{2i-1}^\ell; x_{2i}^\ell\right]\right\|_2 \leq x_i^{\ell+1},$$

as explained in Lemma 1.2, that is, let us replace every one of the constraints in (1.43) by the following system of linear inequalities:

$$\Pi^{(\nu_\ell)}\left[x_{2i-1}^{\ell-1}; x_{2i}^{\ell-1}; x_i^\ell; w_i^\ell\right] \leq 0.$$

As a result, (1.43) will be replaced by a system $\mathcal{S}$ of linear inequalities in $2n-1$ variables x_i^ℓ and additional variables w_i^ℓ. Denoting by $I(\nu)$ and by $V(\nu)$ the number of rows, respectively, columns in the matrix $\Pi^{(\nu)}$ (so that $I(\nu) \leq O(1)\nu$ and $V(\nu) \leq O(1)\nu$), the numbers of variables and constraints in $\mathcal{S}$ are, respectively,

$$V = 2^\kappa + 2^{\kappa-1}V(\nu_1) + 2^{\kappa-2}V(\nu_2) + \cdots + V(\nu_\kappa) \leq O(1)\sum_{\ell=1}^{\kappa} 2^{\kappa-\ell}\nu_\ell,$$

$$I = 2^{\kappa-1}I(\nu_1) + 2^{\kappa-2}V(\nu_2) + \cdots + V(\nu_\kappa) \leq O(1)\sum_{\ell=1}^{\kappa} 2^{\kappa-\ell}\nu_\ell. \tag{1.44}$$

Invoking Lemma, it is clear that the projection $\widehat{L}$ of the feasible set of $\mathcal{S}$ onto the space of x,τ-variables satisfies the inclusions in Theorem 1.3 with the factor

$$\gamma = (1+\chi(\nu_1))(1+\chi(\nu_2))...(1+\chi(\nu_\kappa))$$

in the role of the desired factor $(1+\epsilon)$. It remains to choose $\nu_1,\ldots,\nu_\ell$ in a way which ensures that $\gamma \leq 1+\epsilon$, minimizing under this condition the sum $J := \sum_{\ell=1}^{\kappa} 2^{\kappa-\ell}\nu_\ell$ which, according to (1.44), is responsible for the sizes (number of variables and number of constraints) of $\mathcal{S}$. Setting $\nu_\ell = O(1)\ln(\kappa/\epsilon)$ with appropriately chosen $O(1)$, we get $\chi(\nu_\ell) \leq \epsilon/(2\kappa)$, whence $\gamma \leq 1+\epsilon$ provided $\epsilon \leq 0.1$. With this choice of ν_ℓ, we get $J \leq O(1)n\ln(\kappa/\epsilon) \leq O(1)n\ln(\ln(n)/\epsilon)$ (recall that $\kappa = \log_2 n$), which, up to replacing $\ln(1/\epsilon)$ with $\ln(\ln(n)/\epsilon)$, is what is stated in Theorem 1.3. To get what is announced in the theorem *exactly*, one needs to carry out the optimization in ν_ℓ more carefully[19]; we leave this task to the reader. □

How it works. Setting $y_3 = 1$ in (S_ν), we get a system (T_ν) of $n_\nu \leq O(1)\nu$ linear inequalities in variables $y \in \mathbf{R}^2$ and $m_\nu \leq O(1)\nu$

[19]Our current choice of ν_ℓ distributes the overall "budget of inaccuracy" ϵ among all cascades $\ell = 1,\ldots,\kappa$ of our approximation scheme equally, and this is clearly nonoptimal. Indeed, the number $2^{\kappa-\ell}$ of three-dimensional second-order cones we should approximate in cascade ℓ rapidly decreases as ℓ grows, so that in terms of the total number of linear inequalities in the approximation, it is better to use relatively rough approximations of the 3D cones in cascades with small ℓ and gradually improve the quality of approximation as ℓ grows.

additional variables in such a way that the projection of the solution set P^ν of this system onto the 2D plane of y-variables "approximates the unit circle B_2 within accuracy $\chi(\nu)$," meaning that the projection is in between the unit circle B_2 and its $(1+\chi(\nu))$-enlargement $B_2^{1+\chi(\nu)}$. Here, we illustrate the related numbers. To get a reference point, note that the perfect n-vertex polygon circumscribed around B_2 approximates B_2 within accuracy

$$\epsilon(n) = \frac{1}{\cos(\pi/n)} - 1 \approx \frac{\pi^2}{2n^2}.$$

- P^3 lives in $\mathbf{R}^9$ and is given by 12 linear inequalities. Its projection onto 2D plane approximates B_2 within accuracy 5.e-3 (as a 16-side perfect polygon).
- P^6 lives in $\mathbf{R}^{12}$ and is given by 18 linear inequalities. Its projection onto 2D plane approximates B_2 within accuracy 3.e-4 (as a 127-side perfect polygon).
- P^{12} lives in $\mathbf{R}^{18}$ and is given by 30 linear inequalities. Its projection onto 2D plane approximates B_2 within accuracy 7.e-8 (as a 8,192-side perfect polygon).
- P^{24} lives in $\mathbf{R}^{30}$ and is given by 54 linear inequalities. Its projection onto 2D plane approximates B_2 within accuracy 4.e-15 (as a 34,209,933-side perfect polygon).

1.5 Exercises

In what follows, $*$ and † mark exercises with solutions in Online supplements and in the end of the textbook, respectively.

Exercise 1.1.* (1) Draw the feasible set of the LO program

$$\begin{array}{l} \max\limits_{x_1,x_2} x_2 \\ \text{s.t.} \\ \qquad x_1 - 2x_2 \leq 0 \\ \qquad 2x_1 - 3x_2 \leq 2 \\ \qquad x_1 - x_2 \leq 3 \\ \qquad -x_1 + 2x_2 \leq 2 \\ \qquad -2x_1 + x_2 \leq 0 \end{array}$$

and find the optimal value and an optimal solution to the problem.

(2) Now, assume that the objective is replaced with $\cos(\phi)x_1 + \sin(\phi)x_2$, where ϕ is chosen at random, according to the uniform distribution on $[0, 2\pi]$. What is the probability to get, as an optimal solution, the same point as in the original problem?

Exercise 1.2 (Rucksack problem).* There are n goods available to you; the maximal available volume of good j is $v_j \geq 0$, and the value of good j per unit of volume is $c_j \geq 0$. You have a rucksack of volume v and want to fill it with goods to get as large a total value of the rucksack as possible. Build an LO model of the resulting problem and present an elementary scheme for generating an optimal solution.

Exercise 1.3.* A 24/7 call center works as follows: every agent works 5 days in a row and has two days rest, e.g., every week works Tuesday–Saturday and rests on Sunday and Monday. The numbers of agents working every day of a week should be at least given numbers $d_1, ..., d_7$. The manager wants to meet this requirement with the minimal possible total number of agents employed, by deciding what will be the days off of the agents. Assuming that d_i are large, so that we can ignore integrality restrictions, formulate the manager's problem as an LO program.

Exercise 1.4.* The water supply system in a town includes pump station, tank and a distribution network. At every given hour, the pump station can pump the water partly into the tank, and partly – directly into the distribution network. To pump a gallon of water, it takes one unit of electrical energy, and the cost of energy consumed in hour t is c_t dollars per unit (usually, for night hours c_t is less than for day hours). The demand for water in hour t, $0 \leq t \leq 23$, is d_t gallons, and this demand can be partly satisfied from the tank, and partly — from the station, no matter what are the parts. At the beginning of hour 0, the tank is empty, same as it should be at the end of hour 23. The capacity of the tank is C gallons. We want to decide how much water should be pumped every hour t, $0 \leq t \leq 23$, into the network and into the tank in order to meet the demand at the lowest possible energy cost. Build an LO model of the situation.

Exercise 1.5.† Run experiment as follows:

(1) Pick at random in the segment $[0,1]$ two "true" parameters θ_0^* and θ_1^* of the regression model

$$y = \theta_0 + \theta_1 x;$$

(2) Generate a sample of $N = 1{,}000$ observation errors $\xi_i \sim P$, where P is a given distribution, and then generate observations y_i according to

$$y_i = \theta_0^* + \theta_1^* x_i + \xi_i,\ x_i = i/N;$$

(3) Estimate the parameters from the observations according to the following three estimation schemes:

$$\begin{array}{ll} \text{uniform fit}: & [\theta_{0,\infty};\theta_{1,\infty}] = \operatorname{argmin}_\theta \max\limits_{1\le i\le N} |y_i - [\theta_0 + \theta_1 x_i]|,\\ \text{least squares fit}: & [\theta_{0,2};\theta_{1,2}] = \operatorname{argmin}_\theta \sum\limits_{1\le i\le N} (y_i - [\theta_0 + \theta_1 x_i])^2,\\ \ell_1 \text{ fit}: & [\theta_{0,1};\theta_{1,1}] = \operatorname{argmin}_\theta \sum\limits_{1\le i\le N} |y_i - [\theta_0 + \theta_1 x_i]|, \end{array}$$

and compare the estimates with the true values of the parameters.

Run 3 series of experiments:

- P is the uniform distribution on $[-1,1]$;
- P is the standard Gaussian distribution with the density $\frac{1}{\sqrt{2\pi}}\exp\{-t^2/2\}$;
- P is the Cauchy distribution with the density $\frac{1}{\pi(1+t^2)}$.

Try to explain the results you get. When doing so, you can think about a simpler problem, where you are observing N times a scalar parameter θ^* according to $y_i = \theta^* + \xi_i$, $1 \le i \le N$, and then use the above techniques to estimate θ^*.

Exercise 1.6. In the following list, mark by **P** the polyhedral sets, and by **PR** — the polyhedral representations. For those polyhedral sets in the following list which are *not* given by polyhedral representations, point out their polyhedral representations.

- $X = \{[x_1; x_2] : x_1 + x_2 \leq 0\}$
- $X = \{[x_1; x_2] : \max[x_1, x_2] \leq 0\}$
- $X = \{[x_1, x_2] : \max[x_1, x_2] \geq 0\}$
- $X = \{[x_1; x_2] : \min[x_1, x_2] \leq 0\}$
- $X = \{[x_1; x_2] : \min[x_1, x_2] \geq 0\}$
- $X = \{[x_1; x_2] : x_1^2 + x_2^2 \leq 1\}$
- $X = \{[x_1; x_2] : x_1^2 + x_2^2 \leq -1\}$
- $X = \{[x_1; x_2] : -1 \leq x_1 \leq 1, -1 \leq x_2 \leq 1, x_1^2 + x_2^2 \leq 1\}$
- $X = \{[x_1; x_2] : -1 \leq x_1 \leq 1, -1 \leq x_2 \leq 1, x_1^2 + x_2^2 \leq 2\}$
- $X = \{[x_1; x_2] : |x_1| + |x_2| \leq 1\}$
- $X = \{[x_1; x_2] : |x_1| - |x_2| \leq 1\}$
- $X = \{[x_1; x_2] : |x_1| + x_2 \leq 1\}$
- $X = \{[x_1; x_2] : |x_1| - x_2 \leq 1\}$
- $X = \{[x_1; x_2] : x_1 - |x_2| \leq 1\}$
- $X = \{[x_1; x_2] : -x_1 - |x_2| \leq 1\}$

Exercise 1.7.* Represent the projection X of the polyhedral set

$$Y = \{x \in \mathbf{R}^3 : -1 \leq x_1 + x_2 \leq 1, -1 \leq x_2 + x_3 \leq 1,$$
$$-1 \leq x_1 + x_3 \leq 1, x_1 + x_2 + x_3 \leq 2\}$$

onto the x_1, x_2-plane by a system of linear inequalities in the variables x_1, x_2.

Exercise 1.8.* In the following list, mark by **P** the polyhedrally representable functions, and build their polyhedral representations.

- $f(x_1, x_2) \equiv 0$
- $f(x_1, x_2) = x_1 - x_2$
- $f(x_1, x_2) = \max[x_1, x_2]$
- $f(x_1, x_2) = \min[x_1, x_2]$
- $f(x_1, x_2) = 1 - \max[x_1, x_2]$
- $f(x_1, x_2) = 1 - \min[x_1, x_2]$
- $f(x_1, x_2) = \begin{cases} \max[x_1, x_2], & \max[x_1, x_2] \leq 1 \\ +\infty, & \text{otherwise} \end{cases}$
- $f(x_1, x_2) = \begin{cases} \max[x_1, x_2], & \min[x_1, x_2] \leq 1 \\ +\infty, & \text{otherwise} \end{cases}$

- $f(x_1,x_2) = \begin{cases} \min[x_1,x_2], & x_2 \geq 2, x_1 \leq 0 \\ +\infty, & \text{otherwise} \end{cases}$
- $f(x_1,x_2,x_3) = \max[x_1,x_2] + \max[x_1,x_3]$
- $f(x_1,x_2,x_3) = \max[x_1,x_2] - \max[x_1,x_3]$
- $f(x_1,x_2,x_3) = \max[x_1,x_2] + \min[x_1,x_3]$
- $f(x_1,x_2,x_3) = \max[x_1,x_2] - \min[x_1,x_3]$
- $f(x_1,x_2) = \max[x_1 + \max[x_2,x_3], x_3 + \max[x_1,x_2]]$
- $f(x_1,x_2) = \max[|x_1|,|x_2|]$
- $f(x_1,x_2) = |\max[x_1,x_2]|$

Exercise 1.9.* In the following list, some problems can be posed as LO programs. Identify these problems and reformulate them as LO programs.

(1)

$$\min_{x_1,x_2} \{\max[|2x_1 + 3x_2|, |x_1 - x_2|] : |x_1| + 2\max[x_1,x_2] \leq 1\},$$

(2)

$$\max_{x_1,x_2} \{\max[|2x_1 + 3x_2|, |x_1 - x_2|] : |x_1| + 2\max[x_1,x_2] \leq 1\},$$

(3)

$$\max_{x_1,x_2} \{2\min[x_1 + x_2, 2x_2] - |x_1 - x_2| : \max[|x_1 - 2x_2|, |x_2|] - 2x_2 \leq 1 - |x_1|\}.$$

Exercise 1.10 (Computational study).† Generate at random and solve 1,000 LO problems

$$\max_{x \in \mathbf{R}^5} \left\{ \sum_{j=1}^{5} c_j x_j : \sum_{j=1}^{5} a_{ij} x_j \leq b_j, i \leq 10 \right\}$$

to get an idea what are the chances of getting solvable/infeasible/unboundsed instances. Draw the entries in the data, independently of each other, from the uniform distribution on $\{-1, 1\}$.

Part I

Geometry of Linear Optimization

Chapter 2

Polyhedral Sets and their Geometry

An LO program

$$\text{Opt} = \max_x \left\{ c^T x : Ax \leq b \right\} \tag{2.1}$$

is the problem of maximizing a linear objective over the set

$$X = \{x \in \mathbf{R}^n : Ax \leq b\}$$

given by finitely many nonstrict linear inequalities. As we remember, sets of this form are called *polyhedral*, and their geometry plays a crucial role in the theory and algorithms of LO. In this chapter, we make the first major step toward understanding the geometry of a polyhedral set; our ultimate goal is to establish the following fundamental result.

Theorem (Structure of a polyhedral set). *A nonempty subset X in $\mathbf{R}^n$ is polyhedral if and only if it can be represented via finitely many "generators" $v_1, \ldots, v_I \in \mathbf{R}^n$, $r_1, \ldots, r_J \in \mathbf{R}^n$ according to*

$$X = \left\{ x = \sum_{i=1}^{I} \lambda_i v_i + \sum_{j=1}^{J} \mu_j r_j : \lambda_i \geq 0, \sum_i \lambda_i = 1; \mu_j \geq 0 \right\}.$$

The statement of theorem is illustrated in Fig. 2.1. Eventually, it will become clear why this result indeed is crucial, and "on the way" to it, we will learn numerous notions and techniques which form the major bulk of the theoretical core of LO.

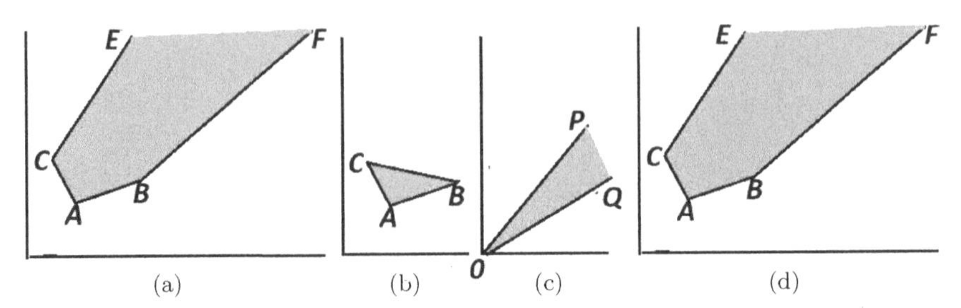

Fig. 2.1. From a polyhedral set to its generators and back: (a) polyhedral set X in $\mathbf{R}^2$; (b) the v-generators of X are the points A, B, C, and the set $\{\sum_{i=1}^{3} \lambda_i v_i : \lambda \geq 0, \sum_i \lambda_i = 1\}$ is the triangle $\triangle ABC$; (c) the r-generators of X are the directions of rays $\overline{OP}$ (parallel to $\overline{CE}$) and $\overline{OQ}$ (parallel to $\overline{BF}$), and the set $\{\sum_{j=1}^{2} \mu_j r_j : \mu \geq 0\}$ is the angle $\angle POQ$; (d) to get X back from generators, one sums up points from the triangle $\triangle ABC$ and the angle $\angle POQ$.

2.1 Preliminaries: Linear and affine subspaces, convexity

2.1.1 *Linear subspaces*

Here, we recall some basic facts of linear algebra; the reader is supposed to be well acquainted with these facts, so that we just list them.

2.1.1.1 *Linear subspaces: Definition and examples*

Recall that a *linear subspace* of $\mathbf{R}^n$ is a *nonempty* subset L of $\mathbf{R}^n$ which is closed w.r.t. taking linear operations: addition and multiplication by reals. Thus, $L \subset \mathbf{R}^n$ is a linear subspace iff it possesses the following three properties:

- $L \neq \emptyset$;
- (closedness w.r.t. additions) whenever $x, y \in L$, we have $x + y \in L$;
- (closedness w.r.t. multiplications by reals) whenever $x \in L$ and $\lambda \in \mathbf{R}$, we have $\lambda x \in L$.

From closedness w.r.t. additions and multiplications by reals, it clearly follows that a linear subspace is closed w.r.t. taking *linear combinations* of its elements: whenever $x_1, \ldots, x_k \in L$ and $\lambda_1, \ldots, \lambda_k \in \mathbf{R}$, the vector $\sum_{i=1}^{k} \lambda_i x_i$, called the *linear combination of the vectors x_i with coefficients λ_i*, also belongs to L. Linear subspaces of $\mathbf{R}^n$ are exactly the nonempty subsets of $\mathbf{R}^n$ closed w.r.t.

taking linear combinations, with whatever coefficients, of their elements (why?).

Examples of linear subspaces in $\mathbf{R}^n$:

- The entire $\mathbf{R}^n$ clearly is a linear subspace. This is the largest linear subspace in $\mathbf{R}^n$: whenever L is a linear subspace, we clearly have $L \subset \mathbf{R}^n$.
- The origin $\{0\}$ clearly is a linear subspace. This is the smallest linear subspace in $\mathbf{R}^n$: whenever L is a linear subspace, we have $\{0\} \subset L$. Indeed, since L is nonempty, there exists $x \in L$, whence $0 = 0 \cdot x \in L$ as well.
- The solution set

$$\{x : Ax = 0\} \tag{2.2}$$

of a *homogeneous* system of linear equations with an $m \times n$ matrix A. This set clearly is a linear subspace (check it!). We shall see in a while that *every* linear subspace in $\mathbf{R}^n$ can be represented in the form of (2.2) and, in particular, is a polyhedral set.

2.1.1.2 *Calculus of linear subspaces*

There are several important operations with linear subspaces:

(1) *Taking intersection*: If L_1, L_2 are linear subspaces of $\mathbf{R}^n$, so is their intersection $L_1 \cap L_2$. In fact, *the intersection $\bigcap_{\alpha \in \mathcal{A}} L_\alpha$ of a whatever family $\{L_\alpha\}_{\alpha \in \mathcal{A}}$ of linear subspaces is a linear subspace* (check it!).
Note: The *union* of two linear subspaces L_1, L_2 is *not* a linear subspace, unless one of the subspaces contains the other one (why?).
(2) *Taking arithmetic sum*: If L_1, L_2 are linear subspaces in $\mathbf{R}^n$, so is the set $L_1 + L_2 = \{x = x_1 + x_2 : x_1 \in L_1, x_2 \in L_2\}$ (check it!).
(3) *Taking orthogonal complement*: If L is a linear subspace in $\mathbf{R}^n$, so is its *orthogonal complement* — the set $L^\perp = \{y : y^T x = 0 \,\forall x \in L\}$ of vectors orthogonal to all vectors from L.
(4) *Taking linear image*: If L is a linear subspace in $\mathbf{R}^n$ and A is an $m \times n$ matrix, then the set $AL := \{Ax : x \in L\} \subset \mathbf{R}^m$ — the image of L under the linear mapping $x \mapsto Ax : \mathbf{R}^n \to \mathbf{R}^m$ – is a linear subspace in $\mathbf{R}^m$ (check it!).

(5) *Taking inverse linear image*: If L is a linear subspace in $\mathbf{R}^n$ and A is an $n \times k$ matrix, then the set $A^{-1}L = \{y \in \mathbf{R}^k : Ay \in L\}$ — the inverse image of L under the linear mapping $y \mapsto Ay : \mathbf{R}^k \to \mathbf{R}^n$ — is a linear subspace in $\mathbf{R}^k$ (check it!).
(6) *Taking direct product:* Recall that the direct product $X_1 \times \cdots \times X_k$ of sets $X_i \subset \mathbf{R}^{n_i}$, $1 \leq i \leq k$ is the set in $\mathbf{R}^{n_1+\cdots+n_k}$ composed of all block vectors $x = [x^1; \ldots; x^k]$ with blocks $x^i \in X_i$, $1 \leq i \leq k$. When all X_i are linear subspaces, so is their direct product (check it!).

The most important "calculus rules" are the following relations:

$$\begin{array}{l}
\text{(a) } (L^\perp)^\perp = L,\\
\text{(b) } L \cap L^\perp = \{0\},\ \ L + L^\perp = \mathbf{R}^n,\\
\text{(c) } (L_1 + L_2)^\perp = L_1^\perp \cap L_2^\perp,\\
\text{(d) } (AL)^\perp = (A^T)^{-1}L^\perp, \text{ that is } \{y : y^TAx = 0\ \forall x \in L\}\\
\qquad = \{y : A^Ty \in L^\perp\},\\
\text{(e) } (L_1 \times \cdots \times L_k)^\perp = L_1^\perp \times \cdots \times L_k^\perp,
\end{array} \tag{2.3}$$

where $L, L_1, L_2 \subset \mathbf{R}^n$ (in (a–d)) and $L_i \subset \mathbf{R}^{n_i}$ (in (e)) are linear subspaces and A is an arbitrary $m \times n$ matrix.

Comment on (2.3(b)). Let L be a linear subspace in $\mathbf{R}^n$. The fact that $L + L^\perp = \mathbf{R}^n$ means that every $x \in \mathbf{R}^n$ can be decomposed as $x = x_L + x_{L^\perp}$ with $x_L \in L$ and $x_{L^\perp} \in L^\perp$; in particular, x_L is orthogonal to $x_{L^\perp}$. Since $L \cap L^\perp = \{0\}$, the decomposition of x into a sum of two vectors, one from L and one from $L^\perp$, is unique (check it!), and both x_L and $x_{L^\perp}$ linearly depend on x. The components x_L and $x_{L^\perp}$ of x are called *orthogonal projections of x onto L and $L^\perp$*, respectively, and the mapping $x \mapsto x_L$ is called the *orthogonal projector onto L*. Since x_L and $x_{L^\perp}$ are orthogonal, we have the identity

$$\|x\|_2^2 := x^Tx = \|x_L\|_2^2 + \|x_{L^\perp}\|_2^2.$$

2.1.1.3 *Linear subspaces: Linear span, dimension, linear independence, bases*

Linear span: For every nonempty set $X \subset \mathbf{R}^n$, the set of all linear combinations, with finitely many terms, of vectors from X is a linear subspace (why?), and this linear subspace (called the *linear span*

Lin(X) of X) is the smallest, w.r.t. inclusion, of the linear subspaces containing X: Lin(X) is a linear subspace, and

$$X \subset \mathrm{Lin}(X) \subset L$$

whenever L is a linear subspace containing X.

Linear span of an empty set: In the above definition of Lin(X), $X \subset \mathbf{R}^n$ was assumed to be nonempty. It is convenient to assign with the linear span the empty set as well; *by definition*, $\mathrm{Lin}(\emptyset) = \{0\}$. This definition is in full accordance with the standard convention that a sum with empty set of terms is well defined and is equal to 0, same as it is consistent with the above claim that Lin(X) is the smallest linear subspace containing X.

When $L = \mathrm{Lin}(X)$, we say that L is *spanned* (or, more rigorously, *linearly spanned*) by the set $X \subset \mathbf{R}^n$; thus, a linear subspace L is spanned by a set iff L is exactly the set of all linear combinations of vectors from X (with the convention that a linear combination with empty set of terms is equal to 0).

Linear independence: Recall that a collection $x_1, \ldots, x_k$ of vectors from $\mathbf{R}^n$ is called *linearly independent* if the only linear combination of these vectors equal to 0 is the trivial one — all coefficients are equal to 0:

$$\sum_{i=1}^{k} \lambda_i x_i = 0 \Rightarrow \lambda_i = 0,\ i = 1, \ldots, k.$$

An equivalent definition (check equivalency!) is that $x_1, \ldots, x_k$ are linearly independent if and only if the coefficients in a linear combination of $x_1, \ldots, x_k$ are uniquely defined by the value of this combination:

$$\sum_{i=1}^{k} \lambda_i x_i = \sum_{i=1}^{k} \mu_i x_i \Leftrightarrow \lambda_i = \mu_i,\ i = 1, \ldots, k.$$

By definition, an empty (i.e., with $k = 0$) collection of k vectors is linearly independent. Clearly, a nonempty linearly independent collection is composed of distinct and nonzero vectors.

Dimension and bases: A given linear subspace L can be represented as a span of many sets (e.g., $L = \text{Lin}(L)$). *The* fundamental fact of linear algebra is that L always is spanned by a *finite* set of vectors. Given this fact, we immediately conclude that there exist "minimal w.r.t. inclusion finite representations of L as a linear span," that is, finite collections $\{x_1, \ldots, x_k\}$ composed of k points which linearly span L (i.e., every vector from L is a linear combination of vectors $x_1, \ldots, x_k$), such that when eliminating an element from the collection, the remaining vectors do *not* span the entire L. The following fundamental fact of linear algebra provides us with more details on this subject:

Theorem 2.1. *Let L be a linear subspace of $\mathbf{R}^n$. Then the following can be noted:*

(i) *There exist finite collections $x_1, \ldots, x_k$ which span L and are "minimal" in this respect (i.e., such that eliminating one or more elements from the collection, the remaining vectors do not span L). All minimal finite collections of vectors spanning L, when nonempty, are collections of nonzero vectors distinct from each other, have the same cardinality, called the dimension of L (notation:* dim L*), and are called linear bases of L. The only linear subspace of $\mathbf{R}^n$ spanned by empty set of vector is the trivial linear subspace $\{0\}$; its dimension, by definition, is 0.*

(ii) *Let L be a nontrivial (i.e., distinct from $\{0\}$) linear subspace in $\mathbf{R}^n$. A collection $x_1, \ldots, x_k$ of k vectors from L is a linear basis in L if and only if this collection possesses one of the following properties equivalent to each other:*

  (ii.1) *The collection spans L and the vectors of the collection are linearly independent.*

  (ii.2) *The vectors $x_1, \ldots, x_k$ form a maximal, w.r.t. inclusion, linearly independent collection of vectors from L, that is, $x_1, \ldots, x_k$ are linearly independent, but adding to the collection one more vector x_{k+1} from L yields a linearly dependent collection of vectors.*

  (ii.3) *Every vector x from L admits exactly one representation as a linear combination of vectors $x_1, \ldots, x_k$, that is, for every $x \in L$, there exists exactly one collection of coefficients $\lambda_1, \ldots, \lambda_k$ (called the coordinates of x in the basis $x_1, \ldots, x_k$) such that $x = \sum_{i=1}^{k} \lambda_i x_i$.*

(iii) *One has* $\{0\} \neq L = \mathrm{Lin}(X)$ *for certain* X *iff* L *admits a linear basis composed of vectors from* X.

(iv) *For every* $n \times n$ *nonsingular matrix* A, *vectors* $Ax_1, \ldots, Ax_k$ *form a basis in the image* $AL = \{Ax : x \in L\}$ *of* L *under the linear mapping* $x \mapsto Ax$ *iff* $x_1, \ldots, x_k$ *form a basis in* L.

The dimension of a linear subspace satisfies several basic relations:

$$\begin{array}{ll} \text{(a)} & 0 \leq \dim L \leq n, \\ \text{(b)} & L_1 \subset L_2 \Rightarrow \dim L_1 \leq \dim L_2, \text{ the inequality being strict unless } L_1 = L_2, \\ \text{(c)} & \dim(L_1 + L_2) + \dim(L_1 \cap L_2) = \dim L_1 + \dim L_2, \\ \text{(d)} & \dim L + \dim L^{\perp} = n, \\ \text{(e)} & \dim(AL) \leq \dim L, \text{ with equality when } A \text{ is square and nonsingular}, \\ \text{(f)} & \dim(L_1 \times \cdots \times L_k) = \dim L_1 + \cdots + \dim L_k, \end{array} \tag{2.4}$$

where $L, L_1, L_2 \subset \mathbf{R}^n$ (in (a)–(e)) and $L_i \subset \mathbf{R}^{n_i}$ (in (f)) are linear subspaces.

Examples: • $\dim \mathbf{R}^n = n$, and $\mathbf{R}^n$ is the *only* subspace of $\mathbf{R}^n$ of dimension n. Bases in $\mathbf{R}^n$ are exactly the collections of n linearly independent vectors (or, if your prefer, collections of columns of $n \times n$ nonsingular matrices), e.g., the *standard basis* is composed of *basic orths* $e_1, \ldots, e_n$, where e_i has a single nonzero entry, equal to 1, namely, in the position i. The coordinates of $x \in \mathbf{R}^n$ in this standard basis are just the entries of x.

- $\dim \{0\} = 0$, and $\{0\}$ is the only subspace of $\mathbf{R}^n$ of dimension 0. The only basis of $\{0\}$ is the empty set.
- The dimensions of all *proper* (distinct from $\{0\}$ and $\mathbf{R}^n$) linear subspaces of $\mathbf{R}^n$ are integers ≥ 1 and $\leq n - 1$.
- The dimension of the solution set $L = \{x : Ax = 0\}$ of a system of m linear equations with n variables x is $n - \mathrm{Rank}\, A$. One way to see it is to note that this solution set is nothing but the orthogonal complement to the linear span of the (transposes of the) rows of A; the dimension of this span is $\mathrm{Rank}(A)$ (definition of rank), and it remains to use the rule (2.4(d)).

2.1.1.4 *"Inner" and "outer" description of a linear subspace*

We have seen that a linear subspace $L \subset \mathbf{R}^n$ of dimension k can be specified (in fact, in many ways, unless $L = \{0\}$) by a finite set which spans L. This is a kind of an "inner" description of L: in order to get L, one starts with a finite number of vectors from L and then augments them by all their linear combinations.

There is another equally universal way to specify a linear subspace: to point our system of homogeneous linear equations $Ax = 0$ such that the solution set of this system is exactly L.

Proposition 2.1. (i) *$L \subset \mathbf{R}^n$ is a linear subspace if and only if L is a solution set of a (finite) system of homogeneous linear equations $Ax = 0$; in particular, a linear subspace is a polyhedral set.*

(ii) *When L is a linear subspace of $\mathbf{R}^n$, the relation $L = \{x : Ax = 0\}$ holds true if and only if the (transposes of the) rows of A span $L^\perp$.*

Proof. We have already seen that the solution set of a system of homogeneous linear equations in variables $x \in \mathbf{R}^n$ is a linear subspace of $\mathbf{R}^n$. Now, let L be a linear subspace of $\mathbf{R}^n$. It is immediately seen (check it!) that

> (!) if $a_1, \ldots, a_m$ span a linear subspace M, then the solution set of the system
> $$a_i^T x = 0,\ i = 1, \ldots, m \qquad (*)$$
> is exactly $M^\perp$.

It follows that choosing $a_1, \ldots, a_m$ to span $L^\perp$, we get, as a solution set of $(*)$, the linear subspace $(L^\perp)^\perp = L$, where the concluding equality is nothing but (2.3(a)). Thus, every linear subspace L indeed can be represented as a solution set of $(*)$, provided that m and $a_1, \ldots, a_m$ are chosen properly, namely, $a_1, \ldots, a_m$ span $L^\perp$. To see that the latter condition is not only sufficient but also *necessary* for the solution set of $(*)$ to be L, note that when $L = \{x : a_i^T x = 0,\ i = 1, \ldots, m\}$, we definitely have $a_i \in L^\perp$, and all which remains to verify is that $a_1, \ldots, a_m$ not only belong to $L^\perp$ but also span this linear subspace. Indeed, let $M := \text{Lin}\{a_1, \ldots, a_m\}$. By (!), we have $M^\perp = \{x : a_i^T x = 0,\ i = 1, \ldots, m\} = L$, whence $M = (M^\perp)^\perp = L^\perp$, as claimed. □

A representation $L = \{x : Ax = 0\}$ of a linear subspace L is a kind of "outer" description of L — one says that in order to get L, we should delete from the entire space all points violating one or more constraints in the system $Ax = 0$.

Comments: The fact that a linear subspace admits transparent "inner" and "outer" representations, however simple it looks, is crucial: in some situations, these are inner representations which help understand what is going on, while in other situations, outer representations do the work. For example, when passing from two linear subspaces L_1, L_2 to their sum $L_1 + L_2$, an inner representation of the result is readily given by inner representations of the operands (since the union of two finite sets, the first spanning L_1 and the second spanning L_2, clearly spans $L_1 + L_2$); at the same time, there is no equally simple way to get an outer representation of $L_1 + L_2$ from outer representations of the operands L_1, L_2. When passing from the sum of two subspaces to their intersection $L_1 \cap L_2$, the situation is reversed: the outer representation of $L_1 \cap L_2$ is readily given by those of L_1 and L_2 (put the systems of homogeneous equations specifying L_1 and L_2 into a single system of equations), while there is no equally simple way to build a spanning set for $L_1 \cap L_2$, given spanning sets for L_1 and L_2. In the sequel, speaking about entities more complicated than linear subspaces (affine spaces and polyhedral sets), we will systematically look for both their "inner" and "outer" descriptions.

2.1.2 *Affine subspaces*

2.1.2.1 *Affine subspace: Definition and examples*

Definition 2.1 (Affine subspace). A set $M \subset \mathbf{R}^m$ is called affine subspace if it can be represented as a shift of a linear subspace: for properly chosen linear subspace $L \subset \mathbf{R}^n$ and point $a \in \mathbf{R}^n$, we have

$$M = a + L := \{x = a + y, y \in L\} = \{x : x - a \in L\}. \tag{2.5}$$

It is immediately seen (check it!) that the linear subspace L participating in (2.5) is uniquely defined by M, specifically, $L = \{z = x - y : x, y \in M\}$; L is called the linear subspace *parallel* to the affine subspace M. In contrast to L, the shift vector a in (2.5) is not uniquely defined by M, and one can use in the role of a an arbitrary point from M (and only from M).

A linear subspace of $\mathbf{R}^n$ is a nonempty subset of $\mathbf{R}^n$ which is closed w.r.t. taking linear combinations of its elements. An affine subspace admits a similar description, with *affine* combinations in the role of just linear ones.

Definition 2.2 (Affine combination). An affine combination of vectors $x_1, \ldots, x_k$ is their linear combination $\sum_{i=1}^k \lambda_i x_i$ with unit sum of coefficients: $\sum_{i=1}^k \lambda_i = 1$.

The characteristic property of affine combinations as compared to plain linear ones is that when all vectors participating in the combination are shifted by the same vector a, the combination itself is shifted by the same vector:

$$\sum_{i=1}^k \lambda_i x_i \text{ is affine combination of } x_1, \ldots, x_k$$

$$\Downarrow$$

$$\sum_{i=1}^k \lambda_i (x_i + a) = \left(\sum_{i=1}^k \lambda_i x_i \right) + a \, \forall a.$$

Proposition 2.2. *A subset M of $\mathbf{R}^n$ is an affine subspace iff it is nonempty and is closed w.r.t. taking affine combinations of its elements.*

Proof. Assume M is an affine subspace. Then it is nonempty and is representable as $M = a + L$ with L being a subspace. If now $x_i = a + u_i$, $i = 1, \ldots, k$, are points from M (so that $u_i \in L$) and $\lambda_1, \ldots, \lambda_k$ sum up to 1, then $\sum_{i=1}^k \lambda_i x_i = \sum_{i=1}^k \lambda_i u_i + a$ and $\sum_i \lambda_i u_i \in L$ since L is a linear subspace and $u_i \in L$ for all i. Thus, M is nonempty and is closed w.r.t. taking affine combinations of its elements.

Vice versa, let M be nonempty and closed w.r.t. taking affine combinations of its elements, and let us prove that then M is an affine subspace. Let us fix a point a in M (this is possible, since M is nonempty), and let $L = M - a = \{y - a : y \in M\}$. All we need to prove is that L is a linear subspace, that is, L is nonempty (which is evident, since $0 = a - a \in L$ due to $a \in M$) and is closed w.r.t.

taking linear combinations. Indeed, when $x_1, \ldots, x_k \in L$ (that is, $y_i = x_i + a \in M$ for all i) and μ_i are reals, we have $x = \sum_i \mu_i x_i = \sum_i \mu_i (y_i - a) = (1 - \sum_i \mu_i)a + \sum_i \mu_i y_i - a$. Since $a \in M$ and all $y_i \in M$, the combination $y = (1 - \sum_i \mu_i)a + \sum_i \mu_i y_i$ (which clearly has coefficients summing up to 1) is an affine combination of vectors from M and as such belongs to M. We see that $x = y - a$ with certain $y \in M$, that is, $x \in L$. Thus, L indeed is a nonempty subset of $\mathbf{R}^n$ closed w.r.t. taking linear combinations of its elements. □

Task 2.1. Proposition 2.2 states that a nonempty subset M of $\mathbf{R}^n$ closed w.r.t. taking affine combinations of its elements is an affine subspace. Prove that the conclusion remains true when closedness w.r.t. taking all affine combinations of the elements of M is weakened to closedness w.r.t. taking only two-term affine combinations $\lambda x + (1 - \lambda)y$ of elements of M. Geometrically, a nonempty subset M of $\mathbf{R}^n$ is an affine subspace if and only if along with every two distinct points $x, y \in M$, M contains the entire line passing through these points.

Examples of affine subspaces in $\mathbf{R}^n$:

- The entire $\mathbf{R}^n$ is an affine subspace; its parallel linear subspace is also $\mathbf{R}^n$, and the shift point a can be chosen as any vector.
- A singleton $\{a\}$, with $a \in \mathbf{R}^n$, is an affine subspace; the shift vector is a, the parallel linear subspace is $\{0\}$.
- The solution set of a *solvable* system of linear equations in variable $x \in \mathbf{R}^n$ is an affine subspace in $\mathbf{R}^n$:

$$M := \{x : Ax = b\} \neq \emptyset \Rightarrow M \text{ is an affine subspace.}$$

  Indeed, as linear algebra teaches us, "the general solution to a system of linear equations is the sum of its particular solution and the general solution to the corresponding homogeneous system:" If a solves $Ax = b$ and $L = \{x : Ax = 0\}$ is the solution set of the corresponding homogeneous system, then

$$M := \{x : Ax = b\} = a + L,$$

  and L is a linear subspace by its origin. Thus, a (nonempty) solution set of a system of linear equations is indeed an affine subspace, with the parallel linear subspace being the solution of the set of the corresponding homogeneous system.

Note that the reasoning in the latter example can be inverted: If $L = \{x : Ax = 0\}$ and $a \in \mathbf{R}^n$, then $a + L = \{x : Ax = b := Aa\}$, that is, *every shift of the solution set of a homogeneous system of linear equations is the set of solutions to the corresponding inhomogeneous system with properly chosen right-hand side.* Recalling that linear subspaces in $\mathbf{R}^n$ are exactly the same as solution sets of homogeneous systems of linear equations, we conclude that *affine subspaces of* $\mathbf{R}^n$ *are nothing but the solution sets of solvable systems of linear equations in variable* $x \in \mathbf{R}^n$, and in particular, *every affine subspace is a polyhedral set.*

2.1.2.2 *Calculus of affine subspaces*

Some of the basic operations with linear subspaces can be extended to affine subspaces, specifically the following:

(1) *Taking intersection*: If M_1, M_2 are affine subspaces in $\mathbf{R}^n$ *and their intersection is nonempty*, this intersection is an affine subspace. Moreover, if $\{M_\alpha\}_{\alpha \in \mathcal{A}}$ is a *whatever* family of affine subspaces and the set $M = \bigcap_{\alpha \in \mathcal{A}} M_\alpha$ is nonempty, it is an affine subspace.

    Indeed, given a family of sets closed w.r.t. certain operation with elements of the set (e.g., taking affine combination of the elements), their intersection clearly is closed w.r.t. the same operation; thus, the intersection of an arbitrary family of affine subspaces is closed w.r.t. taking affine combinations of its elements. If, in addition, this intersection is nonempty, then it is an affine subspace (Proposition 2.2).

(2) *Taking arithmetic sum*: If M_1, M_2 are affine subspaces in $\mathbf{R}^n$, so is the set $M_1 + M_2 = \{x = x_1 + x_2 : x_1 \in M_1, x_2 \in M_2\}$ (check it!).

(3) *Taking affine image*: If M is a linear subspace in $\mathbf{R}^n$, A is an $m \times n$ matrix and $b \in \mathbf{R}^m$, then the set $AM + b := \{Ax + b : x \in M\} \subset \mathbf{R}^m$ — the image of M under the *affine* mapping $x \mapsto Ax + b : \mathbf{R}^n \to \mathbf{R}^m$ – is an affine subspace in $\mathbf{R}^m$ (check it!). In particular, a shift $M + b = \{y = x + b : x \in M\}$ ($b \in \mathbf{R}^n$) of an affine subspace $M \subset \mathbf{R}^n$ is an affine subspace.

(4) *Taking inverse affine image*: If M is a linear subspace in $\mathbf{R}^n$, A is an $n \times k$ matrix, and $b \in \mathbf{R}^n$ *and the set* $\{y \in \mathbf{R}^k : Ay + b \in M\}$ — *the inverse image of* L *under the affine mapping* $y \mapsto Ay + b :$

$\mathbf{R}^k \to \mathbf{R}^n$ — *is nonempty*, then this inverse image is an affine subspace in $\mathbf{R}^k$ (check it!).

(5) *Taking direct product*: If $M_i \subset \mathbf{R}^{n_i}$ are affine subspaces, $i = 1, \ldots, k$, then their direct product $M_1 \times \cdots \times M_k$ is an affine subspace in $\mathbf{R}^{n_1 + \cdots + n_k}$ (check it!).

2.1.2.3 *Affine subspaces: Affine span, affine dimension, affine independence, affine bases*

Affine span: Recall that the linear span of a set $X \subset \mathbf{R}^n$ is the set composed of all linear combinations of elements from X, and this is the smallest w.r.t. inclusion linear subspace containing X. Similarly, given a *nonempty* set $X \subset \mathbf{R}^n$, we can form the set $\text{Aff}(X)$ of all *affine* combinations of elements of X. Since the affine combination of affine combinations of elements of X is again an affine combination of elements of X (check it!), $\text{Aff}(X)$ is closed w.r.t. taking affine combinations of its elements. Besides this, $\text{Aff}(X) \supset X$, since every $x \in X$ is an affine combination of elements from X: $x = 1 \cdot x$. Thus, $\text{Aff}(X)$ is a nonempty set closed w.r.t. taking affine combinations of its elements and thus is an affine subspace (Proposition 2.2). As we have seen, this affine subspace contains X; it is clear (check it!) that $\text{Aff}(X)$ is the smallest, w.r.t. inclusion, affine subspace containing X:

$$\emptyset \neq X \subset M,\ M \text{ is affine subspace} \Rightarrow X \subset \text{Aff}(X) \subset M.$$

The set $\text{Aff}(X)$ is called the *affine span* (or *affine hull*) of X; we also say that X *affinely spans* an affine subspace M if $M = \text{Aff}(X)$.

As it should be clear in advance, there is a tight relation between linear and affine spans.

Proposition 2.3. *Let $\emptyset \neq X \subset \mathbf{R}^n$, and let $a \in X$. Then* $\text{Aff}(X) = a + \text{Lin}(X - a)$. *In particular, X affinely spans an affine subspace M if and only if $X \subset M$ and the set $X - a$, for some (and then — for every) $a \in X$, linearly spans the linear subspace L to which M is parallel.*

Proof. Let $a \in X$. By definition, $\text{Aff}(X)$ is composed of all vectors representable as $x = \sum_{i=1}^k \lambda_i x_i$ with some k, some $x_i \in X$ and some λ_i with $\sum_i \lambda_i = 1$; for x of the outlined form, we have $x - a = \sum_{i=1}^k \lambda_i (x_i - a)$. In other words, the linear subspace L to which Aff(X) is parallel (this linear subspace is exactly the set of all differences $x - a$ with $x \in \text{Aff}(X)$) is *exactly* the set of affine combinations of vectors from $X - a$. Since the latter set contains the origin (due to $a \in X$), the set of all *affine* combinations of vectors from $X - a$ is exactly the same as the set Lin($X - a$) of *all linear* combinations of the points from $X - a$ (take a linear combination $\sum_i \mu_i(x_i - a)$ of vectors from $X - a$ and rewrite it as the affine combination $\sum_i \mu_i(x_i - a) + (1 - \sum_i \mu_i)(a - a)$ of vectors from the same set $X - a$). Thus, $L = \text{Lin}(X - a)$, and therefore, $\text{Aff}(X) = a + \text{Lin}(X - a)$ for whatever $a \in X$. □

Affine independence: Recall that a collection of $k \geq 0$ vectors $x_1, \ldots, x_k$ is called linearly independent iff the coefficients of a linear combination of $x_1, \ldots, x_k$ are uniquely defined by the value of this combination. The notion of *affine independence* mimics this approach: specifically, a collection of $k+1$ vectors $x_0, \ldots, x_k$ is called *affine independent* if the coefficients of every *affine* combination of $x_0, \ldots, x_k$ are uniquely defined by the value of this combination:

$$\sum_{i=0}^{k} \lambda_i x_i = \sum_{i=0}^{k} \mu_i x_i, \sum_i \lambda_i = \sum_i \mu_i = 1 \Rightarrow \lambda_i = \mu_i, \; 0 \leq i \leq k.$$

Equivalent definition of affine independence (check equivalency!) is as follows: *$x_0, \ldots, x_k$ are affinely independent iff the only linear combination of $x_0, \ldots, x_k$ which is equal to 0 and has zero sum of coefficients is the trivial combination:*

$$\sum_{i=0}^{k} \lambda_i x_i = 0, \sum_{i=0}^{k} \lambda_i = 0 \Rightarrow \lambda_i = 0, \; 0 \leq i \leq k.$$

As it could be easily guessed, affine independence "reduces" to linear one.

Lemma 2.1. *Let $k \geq 0$, and let $x_0, \ldots, x_k$ be a collection of $k+1$ vectors from $\mathbf{R}^n$. This collection is affinely independent iff the collection $x_1 - x_0, x_2 - x_0, \ldots, x_k - x_0$ is linearly independent.*

Proof of Lemma. There is nothing to prove when $k = 0$, since in this case, the system of vectors $x_1 - x_0, \ldots, x_k - x_0$ is empty and thus is linearly independent, and at the same time, the only solution of the system $\sum_{i=0}^{0} \lambda_i x_i = 0, \sum_{i=0}^{0} \lambda_i = 0$ is $\lambda_0 = 0$. Now, let $k \geq 1$. Assuming that the vectors $x_1 - x_0, \ldots, x_k - x_0$ are linearly dependent, so that there exists nonzero $\mu = [\mu_1; \ldots; \mu_k]$ with $\sum_{i=1}^{k} \mu_i(x_i - x_0) = 0$, we get a nontrivial solution $\lambda_0 = -\sum_{i=1}^{k} \mu_i, \lambda_1 = \mu_1, \ldots, \lambda_k = \mu_k$ to the system

$$\sum_{i=0}^{k} \lambda_i x_i = 0, \sum_{i=0}^{k} \lambda_i = 0, \tag{$*$}$$

that is, $x_0, \ldots, x_k$ are not affinely independent. Vice versa, assuming that $x_0, \ldots, x_k$ are not affinely independent, $(*)$ has a nontrivial solution $\lambda_0, \ldots, \lambda_k$. Setting $\mu_i = \lambda_i$, $1 \leq i \leq k$, and taking into account that $\lambda_0 = -\sum_{i=1}^{k} \lambda_i$ by the last — the scalar — equality in $(*)$, the first equality in $(*)$ reads $\sum_{i=1}^{k} \mu_i(x_i - x_0) = 0$. Note that not all $\mu_i = \lambda_i$, $i = 1, \ldots, k$, are zeros (since otherwise we would also have $\lambda_0 = -\sum_{i=1}^{k} \lambda_i = 0$, that is, $\lambda_0, \ldots, \lambda_k$ are zeros, which is not the case). Thus, if $x_0, \ldots, x_k$ are not affinely independent, $x_1 - x_0, \ldots, x_k - x_0$ are not linearly independent. Lemma is proved. □

Affine dimension and affine bases: *By definition*, the *affine dimension* of an affine subspace $M = a + L$ is the linear dimension of the parallel linear subspace L. It is also convenient to define *affine dimension of an arbitrary nonempty set* $X \subset \mathbf{R}^n$ as the affine dimension of the affine subspace $\mathrm{Aff}(X)$. In our book, we will not use any other notion of dimension, and therefore, from now on, speaking about affine dimension of a set, we shall skip the adjective and call it simply "dimension."

Remark. The above definitions and convention should be checked for consistency, since with them, some sets (namely, affine subspaces M) are within the scope of *two* definitions of affine dimension: the first definition, applicable to affine subspaces, says that the affine dimension of an affine subspace M is the dimension of the parallel linear subspace, and the second definition, applicable to all nonempty sets, says that the affine dimension of M is the affine dimension of $\mathrm{Aff}(M)$, i.e., the dimension of the linear subspace parallel to $\mathrm{Aff}(M)$. Fortunately, these two definitions are consistent in the intersection of their scopes: since $M = \mathrm{Aff}(M)$ for an affine subspace M, both definitions of the affine dimension for such an M say the same.

The convention to skip "affine" when speaking about affine dimension could have its own potential dangers, since some sets — specifically, linear subspaces — again are in the scope of two definitions of dimension: the first is the dimension of a linear subspace L (the minimal cardinality of a spanning set of L) and the second is the affine dimension of a nonempty set. Fortunately, here again, the definitions coincide in their common scope: affine dimension of a linear subspace L is, by definition, the dimension of the linear subspace parallel to $\text{Aff}(L)$; but for a linear subspace L, $\text{Aff}(L) = L$ and L is the linear subspace parallel to $\text{Aff}(L)$, so that both definitions again say the same.

The bottom line is that we have associated with every *nonempty* subset X of $\mathbf{R}^n$ its *dimension* $\dim X$, which is an integer between 0 and n. For a linear subspace X, the dimension is the cardinality of a minimal finite subset of X which (linearly) spans X or, equivalently, the maximal cardinality of linearly independent subsets of X; for an affine subspace X, the dimension is the just defined dimension of the linear subspace parallel to X, and for an arbitrary nonempty subset X of $\mathbf{R}^n$, the dimension is the just defined dimension of the affine span $\text{Aff}(X)$ of X. When X is in the scope of more than one definition, all applicable definitions of the dimension are consistent with each other.

Affine bases: Let $M = a + L$ be an affine subspace. We definitely can represent M as $\text{Aff}(X)$ for a nonempty set X (which usually can be chosen in many different ways), e.g., as $M = \text{Aff}(M)$. An immediate corollary of Theorem 2.1 and Proposition 2.3 is that M *can be represented as* $\text{Aff}(X)$ *for a finite collection* $X = \{x_0, \ldots, x_k\}$, where $k \geq 0$. Indeed, by Proposition 2.3, a necessary and sufficient condition to have $M = \text{Aff}(\{x_0, x_1, \ldots, x_k\})$ is $x_0 \in M$ and $\text{Lin}(X - x_0) = L$ (check it!). In order to meet this condition, we can take as x_0 an arbitrary point of M and select k and a finite set of vectors $d_1, \ldots, d_k$ in such a way that this set linearly spans L; setting $x_i = x_0 + d_i$, $1 \leq i \leq k$, we get a finite set $X = \{x_0, \ldots, x_k\}$ which meets the above necessary and sufficient condition and thus affinely spans M. This reasoning clearly can be inverted, leading to the following result.

Proposition 2.4. *An affine subspace $M = a + L$ is affinely spanned by a finite set $X = \{x_0, \ldots, x_k\}$ if and only if $x_0 \in M$ and the k vectors $x_1 - x_0, x_2 - x_0, \ldots, x_k - x_0$ linearly span L. In particular,* *the minimal in cardinality subsets* $X = \{x_0, \ldots, x_k\}$ *which affinely span* M *are of cardinality* $k+1 = \dim M + 1$ *and are characterized by the inclusion* $x_0 \in M$ *and the fact that* $x_1 - x_0, \ldots, x_k - x_0$ *is a linear basis in* L.

Note that the "in particular" part of the latter statement is readily given by Theorem 2.1.

The minimal, w.r.t. inclusion, collections $x_0, x_1, \ldots, x_k$ of vectors from an affine subspace M which affinely span M have a name — they are called *affine bases* of the affine subspace M. It is a good and easy exercise to combine Proposition 2.4, Lemma 2.1 and Theorem 2.1 to get the following "affine" version of the latter Theorem.

Theorem 2.2. *Let* M *be an affine subspace of* $\mathbf{R}^n$. *Then the following holds true:*

(i) *There exist finite collections* $x_0, \ldots, x_k$ *which affinely span* M *and are "minimal" in this respect (i.e., such that eliminating one or more elements from the collection, the remaining vectors do not affinely span* M*). All minimal finite collections of vectors affinely spanning* M *are composed of distinct vectors from* M *and have the same cardinality, namely,* $\dim M + 1$; *these collections are called affine bases of* M.

(ii) *A collection* $x_0, \ldots, x_k$ *of* $k+1$ *vectors from* M *is an affine basis in* M *if and only if this collection possesses one of the following equivalent to each other properties:*

(ii.1) *The collection affinely spans* M, *and the vectors of the collection are affinely independent.*

(ii.2) *The vectors* $x_0, \ldots, x_k$ *form a maximal, w.r.t. inclusion, affinely independent collection of vectors from* M, *that is,* $x_0, \ldots, x_k$ *are affinely independent, but adding to the collection one more vector* x_{k+1} *from* M *yields an affinely dependent collection of vectors.*

(ii.3) *Every vector* x *from* M *admits exactly one representation as an affine combination of vectors* $x_0, \ldots, x_k$, *that is, for every* $x \in M$, *there exists exactly one collection of coefficients* $\lambda_0, \ldots, \lambda_k$ *(called the affine coordinates of* x *in the affine basis* $x_0, \ldots, x_k$*) such that* $\sum_{i=0}^k \lambda_i = 1$ *and* $x = \sum_{i=0}^k \lambda_i x_i$.

(ii.4) *The vectors* $x_1 - x_0, \ldots, x_k - x_0$ *form a linear basis in the linear subspace* L *to which* M *is parallel.*

(iii) *One has* $M = \mathrm{Aff}(X)$ *for certain* X *iff* M *admits an affine basis composed of vectors from* X.

(iv) *Let* A *be an* $n \times n$ *nonsingular matrix. Vectors* $Ax_0 + b, \ldots, Ax_k + b$ *form an affine basis of the image* $AM + b = \{y = Ax + b : x \in M\}$ *of* M *under the affine mapping* $x \mapsto Ax + b$ *iff the vectors* $x_0, \ldots, x_m$ *form an affine basis in* M.

The dimension, restricted onto affine subspaces, satisfies basic relations as follows (cf. (2.4)):

$$\begin{aligned}
&\text{(a) } 0 \le \dim M \le n,\\
&\text{(b) } M_1 \subset M_2 \Rightarrow \dim M_1 \le \dim M_2, \text{ the inequality being strict}\\
&\quad \text{unless } M_1 = M_2,\\
&\text{(c) if } M_1 \cap M_2 \neq \emptyset, \text{ then}\\
&\quad \dim(M_1 + M_2) + \dim(M_1 \cap M_2) = \dim M_1 + \dim M_2,\\
&\text{(d) } \dim(AM + b) \le \dim M, \text{ with equality in the case when } A \text{ is}\\
&\quad \text{square and nonsingular,}\\
&\text{(e) } \dim(M_1 \times \cdots \times M_k) = \textstyle\sum_{i=1}^k \dim(M_i),
\end{aligned} \tag{2.6}$$

where M, M_1, M_2, M_i are affine subspaces. Pay attention to the premise $M_1 \cap M_2 \neq \emptyset$ in (c).[1]

As regards the dimension of *arbitrary* (nonempty) subsets $X \subset \mathbf{R}^n$, seemingly the only "universal" facts here are as follows (check them!):

$$\begin{aligned}
&\text{(a) } \emptyset \neq X \subset \mathbf{R}^n \Rightarrow \dim X \in \{0, 1, \ldots, n\},\\
&\text{(b) } \emptyset \neq X \subset Y \subset \mathbf{R}^n \Rightarrow \dim X \le \dim Y.
\end{aligned} \tag{2.7}$$

[1]When $M_1 \cap M_2 = \emptyset$, the conclusion in (c) does not make sense, since the dimension of an empty set is undefined. There are good reasons for it; indeed, when trying to assign an empty set the dimension, we would like to maintain its basic properties, e.g., to ensure the validity of (c) when $M_1 \cap M_2 = \emptyset$, and this is clearly impossible: take two ℓ-dimensional distinct from each other affine subspaces parallel to the same proper linear subspace L of $\mathbf{R}^n$: $M_1 = L + a$, $M_2 = L + b$, $b \notin M_1$. Then $M_1 \cap M_2 = \emptyset$, $\dim M_1 = \dim M_2 = \dim L$, $\dim(M_1 + M_2) = \dim([a + b] + L) = \dim L$. We see that in order to meet the conclusion in (c), we should have $\dim(\emptyset) + \dim L = 2 \dim L$, i.e., $\dim(\emptyset)$ should be $\dim L$. But the latter quantity depends on what is L, and thus, there is no meaningful way to assign $\emptyset$ a dimension.

Examples:

- There exists exactly one affine subspace in $\mathbf{R}^n$ of the maximal possible dimension n — this is the entire $\mathbf{R}^n$. Affine bases in $\mathbf{R}^n$ are of the form $a, a+f_1, \ldots, a+f_n$, where $a \in \mathbf{R}^n$ and $f_1, \ldots, f_n$ are linearly independent.
- Unless $n = 0$, there exist many affine subspaces of the minimal possible dimension 0; these are singletons $M = \{a\}$, $a \in \mathbf{R}^n$. An affine basis in a singleton $\{a\}$ is composed of the only vector, namely, a.
- Two types of affine subspaces have special names — *lines* and *hyperplanes*.
    - A *line* $\ell \subset \mathbf{R}^n$ is the affine span of two distinct points a, b: $\ell = \{x = (1-\lambda)a + \lambda b, \lambda \in \mathbf{R}\} = \{x = a + \lambda(b-a), \lambda \in R\}$. The parallel linear subspace is one-dimensional and is linearly spanned by $b - a$ (same as by the difference of any other pair of two distinct points from the line). Affine bases of ℓ are exactly pairs of distinct points from ℓ.
    - A *hyperplane* $\Pi \subset \mathbf{R}^n$ is an affine subspace of dimension $n-1$ or, equivalently (why?), the solution set of a *single* nontrivial (not all coefficients at the variables are zeros) linear equation $a^T x = b$.

2.1.2.4 *"Inner" and "outer" description of an affine subspace*

We have seen that an affine subspace M can be represented (usually in many ways) as the affine span of a finite collection of vectors $x_0, \ldots, x_k$, that is, represented as the set of all affine combinations of vectors $x_0, \ldots, x_k$. This is an "inner" representation of an affine set M, and the minimal k in such a representation is $\dim M$.

An outer representation of an affine subspace M is its representation as the solution set of a solvable system of linear equations:

$$M = \{x : Ax = b\}. \tag{2.8}$$

As we have already seen, such a representation is always possible, and the (transposes of) rows of A in such a representation always linearly span the orthogonal complement $L^\perp$ of the linear subspace L to which M is parallel. The minimal number of equations in a representation (2.8) of M is $\dim L^\perp = n - \dim L = n - \dim M$.

2.1.3 *Convexity*

Linear and affine subspaces of $\mathbf{R}^n$ we have studied so far are polyhedral sets, so that the family of polyhedral sets in $\mathbf{R}^n$ is wider than the family of affine subspaces (which in turn is wider than the family of linear subspaces). The family of *convex sets* in $\mathbf{R}^n$ is, in turn, wider than the family of polyhedral sets, and convexity is, perhaps, the most important property of a polyhedral set. In this section, we intend to investigate convexity in more detail, the rationale being twofold. First, simple facts we are about to establish play an important role in understanding the geometry of polyhedral sets. Second, while polyhedral sets form only a tiny part of the family of convex sets, this "tiny part" is representable: a significant part of the results on polyhedral sets can be extended, sometimes with small modifications, on convex sets. In the sequel, we shall indicate (without proofs) the most important of these extensions, thus putting our results on polyhedral sets in a proper perspective.

We have already defined the notion of a convex set and a convex function and have claimed that all polyhedral sets are convex, leaving the verification of this claim to the reader (see Remark B in Section 1.3.2). For reader's convenience, we start with reiterating the definitions and proving the above claim.

Definition 2.3.

(i) Let X be a subset of $\mathbf{R}^n$. We say that X is convex if along with every pair x, y of its points, X contains the segment linking these points:

$$\forall (x, y \in X, \lambda \in [0,1]) : (1-\lambda)x + \lambda y \in X.$$

(ii) Let $f : \mathbf{R}^n \to \mathbf{R} \cup \{\infty\}$ be a function. f is called convex if its epigraph — the set

$$\mathrm{Epi}(f) = \{[x;\tau] \in \mathbf{R}^n \times \mathbf{R} : \tau \geq f(x)\}$$

— is convex, or, equivalently (check equivalence!), for all $x, y \in \mathbf{R}^n$ and all $\lambda \in [0,1]$, it holds

$$f((1-\lambda)x + \lambda y) \leq (1-\lambda)f(x) + \lambda f(y).^2$$

[2]To interpret the value of the right-hand side when $f(x)$, or $f(y)$, or both is/are $+\infty$, see the conventions on the arithmetics of the extended real axis in Remark A in Section 1.3.2.

Examples: For our purpose, the most important example of a convex set is a polyhedral one.

Proposition 2.5 (Convexity of a polyhedral set). *Let $X = \{x : Ax \leq b\}$ be a polyhedral set in $\mathbf{R}^n$. Then X is convex.*

Proof. Indeed, let $x, y \in X$ and $\lambda \in [0, 1]$. Then $Ax \leq b$, whence, $A(1-\lambda)x = (1-\lambda)Ax \leq (1-\lambda)b$ due to $\lambda \leq 1$, and $Ay \leq b$, whence $A(\lambda y) \leq \lambda b$ due to $\lambda \geq 0$. Summing up the vector inequalities $A(1-\lambda)x \leq (1-\lambda)b$ and $A\lambda y \leq \lambda b$, we get $A[(1-\lambda)x + \lambda y] \leq b$. Thus, $(1-\lambda)x + \lambda y \in X$, and this is so whenever $x, y \in X$ and $\lambda \in [0, 1]$, meaning that X is convex. □

Corollary 2.1. *A polyhedrally representable function* (see Section 1.3.2) *is convex.*

Indeed, the epigraph of a polyhedrally representable function, by definition of polyhedral representability and by Theorem 1.2, is a polyhedral set. □

Recall that both Proposition 2.5 and Corollary 2.1 were announced already in Chapter 1.

Of course, there exist nonpolyhedral convex sets, e.g., the *Euclidean ball* $\{x : \|x\|_2^2 \leq 1\}$, same as there exist convex functions which are not polyhedrally representable, e.g., the Euclidean norm $f(x) = \|x\|_2$.[3]

2.1.3.1 *Calculus of convex sets and functions*

There are several important convexity-preserving operations with convex sets and convex functions quite similar to the calculus of polyhedral representability and, as far as sets are concerned, to the calculus of linear/affine subspaces.

---

[3]Of course, both the convexity of Euclidean ball in $\mathbf{R}^n$ and the fact that this set is nonpolyhedral, unless $n = 1$, need to be proved. We, however, can ignore this task — in our book, we are interested in what is polyhedral and not in what is *not* polyhedral.

Calculus of convex sets: The basic calculus rules are as follows (check their validity!):

(1) *Taking intersection*: If X_1, X_2 are convex sets in $\mathbf{R}^n$, so is their intersection $X_1 \cap X_2$. In fact, *the intersection* $\bigcap_{\alpha\in\mathcal{A}} X_\alpha$ *of a whatever family* $\{X_\alpha\}_{\alpha\in\mathcal{A}}$ *of convex sets in* $\mathbf{R}^n$ *is a convex set.*
Note: The *union* of two convex sets most often than not is nonconvex.
(2) *Taking arithmetic sum*: If X_1, X_2 are convex sets $\mathbf{R}^n$, so is the set $X_1 + X_2 = \{x = x_1 + x_2 : x_1 \in X_1, x_2 \in X_2\}$.
(3) *Taking affine image*: If X is a convex set in $\mathbf{R}^n$, A is an $m \times n$ matrix and $b \in \mathbf{R}^m$, then the set $AX + b := \{Ax + b : x \in X\} \subset \mathbf{R}^m$ — the image of X under the affine mapping $x \mapsto Ax + b : \mathbf{R}^n \to \mathbf{R}^m$ — is a convex set in $\mathbf{R}^m$.
(4) *Taking inverse affine image*: If X is a convex set in $\mathbf{R}^n$, A is an $n \times k$ matrix and $b \in \mathbf{R}^n$, then the set $\{y \in \mathbf{R}^k : Ay + b \in X\}$ — the inverse image of X under the affine mapping $y \mapsto Ay + b : \mathbf{R}^k \to \mathbf{R}^n$ — is a convex set in $\mathbf{R}^k$.
(5) *Taking direct product*: If the sets $X_i \subset \mathbf{R}^{n_i}$, $1 \le i \le k$, are convex, so is their direct product $X_1 \times \cdots \times X_k \subset \mathbf{R}^{n_1+\cdots+n_k}$.

Calculus of convex functions: The basic rules (check their validity!) are as follows:

(1) *Taking linear combinations with positive coefficients*: If functions $f_i : \mathbf{R}^n \to \mathbf{R} \cup \{+\infty\}$ are convex and $\lambda_i > 0$, $1 \le i \le k$, then the function $f(x) = \sum_{i=1}^k \lambda_i f_i(x)$ is convex.
(2) *Direct summation*: If functions $f_i : \mathbf{R}^{n_i} \to \mathbf{R} \cup \{+\infty\}$, $1 \le i \le k$, are convex, so is their direct sum

$$f([x^1; \ldots; x^k]) = \sum_{i=1}^k f_i(x^i) : \mathbf{R}^{n_1+\cdots+n_k} \to \mathbf{R} \cup \{+\infty\}.$$

(3) *Taking supremum*: The supremum $f(x) = \sup_{\alpha\in\mathcal{A}} f_\alpha(x)$ of a whatever (nonempty) family $\{f_\alpha\}_{\alpha\in\mathcal{A}}$ of convex functions is convex.
(4) *Affine substitution of argument*: If a function $f(x) : \mathbf{R}^n \to \mathbf{R} \cup \{+\infty\}$ is convex and $x = Ay + b : \mathbf{R}^m \to \mathbf{R}^n$ is an affine

mapping, then the function $g(y) = f(Ay+b) : \mathbf{R}^m \to \mathbf{R} \cup \{+\infty\}$ is convex.

(5) *Theorem on superposition: Let $f_i(x) : \mathbf{R}^n \to \mathbf{R} \cup \{+\infty\}$ be convex functions, and let $F(y) : \mathbf{R}^m \to \mathbf{R} \cup \{+\infty\}$ be a convex function which is nondecreasing w.r.t. every one of the variables $y_1, \ldots, y_m$. Then the superposition*

$$g(x) = \begin{cases} F(f_1(x), \ldots, f_m(x)), & f_i(x) < +\infty, 1 \leq i \leq m, \\ +\infty & \textit{otherwise} \end{cases}$$

of F and $f_1, \ldots, f_m$ is convex.

Note that if some of f_i, say, $f_1, \ldots, f_k$, are affine, then the superposition theorem remains valid when we require the monotonicity of F w.r.t. the variables $y_{k+1}, \ldots, y_m$ only.

2.1.3.2 *Convex combinations and convex hull, dimension*

A linear subspace is a nonempty set closed w.r.t. taking linear combinations of its elements; an affine subspace is a nonempty set closed with respect to taking affine combinations of its elements. Convex sets in $\mathbf{R}^n$ admit a similar characterization: these are subsets of $\mathbf{R}^n$ (not necessarily nonempty) closed w.r.t. taking *convex combinations* of its elements, where a convex combination of vectors $x_1, \ldots, x_k$ is defined as their *linear* combination with *nonnegative* coefficients *summing up to one*:

$$\sum_{i=1}^{k} \lambda_i x_i \text{ is a convex combination of } x_i \;\Leftrightarrow \lambda_i \geq 0, 1 \leq i \leq k,$$
$$\text{and } \sum_{i=1}^{k} \lambda_i = 1.$$

Note that convexity of X, by definition, means closedness of X with respect to taking *2-term* convex combinations of its elements. It is an easy exercise to check (do it!) that the following two statements hold true.

Proposition 2.6. *A set $X \subset \mathbf{R}^n$ is convex iff this set is closed w.r.t. taking all convex combinations of its elements.*

Corollary 2.2 (Jensen's inequality). *If $f : \mathbf{R}^n \to \mathbf{R} \cup \{\infty\}$ is a convex function, then the value of f at a convex combination of points is less than or equal to the corresponding convex combination of the values of f at the points: whenever $x_1, \ldots, x_k \in \mathbf{R}^n$ and $\lambda_i \geq 0$, $1 \leq i \leq k$, are such that $\sum_{i=1}^k \lambda_i = 1$, one has*

$$f\left(\sum_{i=1}^{k} \lambda_i x_i\right) \leq \sum_{i=1}^{k} \lambda_i f(x_i).$$

Jensen's inequality is one of the most useful tools in Mathematics.[4]

Given a set $X \subset \mathbf{R}^n$, we can form a set composed of all (finite) convex combinations of vectors from X; this set is called the *convex hull* of X (notation: $\mathrm{Conv}(X)$). Since a convex combination of convex combinations of certain vectors is again a convex combination of these vectors (why?), $\mathrm{Conv}(X)$ is a convex set; invoking Proposition 2.6, this is the smallest, w.r.t. inclusion, convex set containing X:

$$X \subset Y, Y \text{ is convex } \Rightarrow X \subset \mathrm{Conv}(X) \subset Y.$$

2.1.3.3 *Relative interior*

We have already assigned every nonempty subset X of $\mathbf{R}^n$ with dimension, defined as affine dimension of $\mathrm{Aff}(X)$ or, equivalently, as linear dimension of the linear subspace parallel to $\mathrm{Aff}(X)$; in particular, all nonempty convex sets in $\mathbf{R}^n$ are assigned with their dimensions, which are integers from the range $0, 1, \ldots, n$. Now, a set X of a large dimension can be very "tiny," e.g., taking the union of $n + 1$ points affinely spanning $\mathbf{R}^n$, we get a *finite* set X of dimension n; clearly, in this case, the dimension does not say much about how "massive" our set is (and in fact has nothing in common with topological properties of the set). The situation changes dramatically when the set X is convex; here, the dimension, as defined above,

[4] A reader could ask how such an easy-to-prove fact as Jensen's inequality could be that important. Well, it is indeed easy to apply this tool; a nontrivial part of the job, if any, is to prove that the function the inequality is applied to is convex.

quantifies properly the "massiveness" of the set due to the following result which we shall use on different occasions in the sequel.

Theorem 2.3. *Let X be a nonempty convex set in $\mathbf{R}^n$. Then X has a nonempty interior in its affine span: there exists $\bar{x} \in X$ and $r > 0$ such that*

$$y \in \mathrm{Aff}(X), \|y - \bar{x}\|_2 \leq r \Rightarrow y \in X.$$

In particular, if X is full-dimensional (i.e., dim $X = n$ *or, which is the same,* $\mathrm{Aff}(X) = \mathbf{R}^n$*), then X contains a Euclidean ball of positive radius.*

Proof. Let $M = \mathrm{Aff}(X)$. When $k := \dim M = 0$, we have $X = M = \{a\}$, and we can take $\bar{x} = a$ and, say, $r = 1$. Now, let $k > 0$. By Theorem 2.2(iii), we can find a collection $x_0, x_1, \ldots, x_k$ of points from X which is an affine basis of M, meaning that, by item (ii) of the same theorem, every vector from M can be represented, in a unique way, as an affine combination of vectors $x_0, \ldots, x_k$. In other words, the system of linear equations in variables $\lambda_0, \ldots, \lambda_k$

$$\begin{aligned} \sum_{i=0}^{k} \lambda_i x_i &= x, \\ \sum_{i=0}^{k} \lambda_i &= 1 \end{aligned} \tag{2.9}$$

has a solution if and only if $x \in M$, and this solution is unique. Now, linear algebra says that in such a situation, the solution $\lambda = \lambda(x)$ is a continuous function of $x \in M$. Setting $\bar{x} = \frac{1}{k+1}\sum_{i=0}^{k} x_i$, the solution is $\lambda(\bar{x}) = [1/(k+1); \ldots 1/(k+1)]$, that is, it is strictly positive; by continuity, there is a neighborhood of positive radius r of $\bar{x}$ in M, where the solution is nonnegative:

$$x \in M, \|x - \bar{x}\|_2 \leq r \quad \Rightarrow \quad \lambda(x) \geq 0.$$

Looking at our system, we see that all vectors $x \in M$ for which it has a nonnegative solution are convex combinations of $x_0, \ldots, x_k$ and thus belong to X (since X is convex and $x_0, \ldots, x_k \in X$). Thus, X indeed contains a neighborhood of $\bar{x}$ in M.

For those who do not remember why "Linear Algebra says...," here is the reminder. System (2.9) is solvable for $x \in M \neq \emptyset$ and the solution is unique, meaning that the matrix $A = [[x_0; 1], \ldots, [x_k; 1]]$ of the system has rank $k+1$, and thus, we can extract from A $k+1$ rows which form a nonsingular $(k+1) \times (k+1)$ submatrix $\widehat{A}$. In other words, the solution, when it exists, is $\widehat{A}^{-1}P(x)$, where $P(x)$ is the part of the right-hand side vector $[x; 1]$ in (2.9). Thus, the solution $\lambda(x)$ is merely an affine function of $x \in M$ and thus is continuous in $x \in M$. □

Relative interior: Let $X \subset \mathbf{R}^n$ be nonempty. A point $x \in X$ is called *relative interior point* of X if all close enough to x points *from the affine span of* X belong to X, that is, if

$$\exists r > 0 : \left(x' \in \mathrm{Aff}(X) \;\&\; \|x' - x\|_2 \leq r\right) \Rightarrow x' \in X.$$

The set of all relative interior points of X is called the *relative interior* (notation: $\mathrm{rint}\, X$) of X. Theorem 2.3 underlies the following useful statement.

Proposition 2.7. *Let $X \subset \mathbf{R}^n$ be a nonempty convex set. Then the relative interior* $\mathrm{rint}\, X$ *of X is nonempty, convex, and whenever $x \in \mathrm{rint}\, X$, $y \in X$ and $\lambda \in [0, 1)$, we have $x + \lambda(y - x) \in \mathrm{rint}\, X$.*

Proof. Let X be convex and nonempty. Nonemptiness of $\mathrm{rint}\, X$ is given by Theorem 2.3. Next, let L be the linear space parallel to $\mathrm{Aff}(X)$. If $x, y \in \mathrm{rint}\, X$, then for some $r > 0$ and all $h \in L$ with $\|h\|_2 \leq r$, we have $x + h \in X$, $y + h \in X$, whence for $\lambda \in [0, 1]$ and h as above, it holds $[(1-\lambda)x + \lambda y] + h = (1-\lambda)[x+h] + \lambda[y+h] \in X$ (X is convex!), implying that $(1-\lambda)x + \lambda y \in \mathrm{rint}\, X$. Thus, $\mathrm{rint}\, X$ is convex. Finally, when $x \in \mathrm{rint}\, X$, so that for some $r > 0$ and all $h \in L$ with $\|h\|_2 \leq r$, it holds $x + h \in X$, and $y \in X$, for $\lambda \in [0, 1)$ and h as above, we have $[(1-\lambda)x + \lambda y] + (1-\lambda)h = (1-\lambda)[x+h] + \lambda y \in X$, that is, $[(1-\lambda)x + \lambda y] + h' \in X$ for all $h' \in L$, such that $\|h'\|_2 \leq (1-\lambda)r$. Since $r' = (1-\lambda)r > 0$, we conclude that $(1-\lambda)x + \lambda y \in \mathrm{rint}\, X$. □

2.1.3.4 *"Inner" and "outer" representations of convex sets*

An "inner" representation of a convex set X is its representation as a convex hull: to get X, you should choose appropriately a set $Y \subset \mathbf{R}^n$ (e.g., $Y = X$) and augment it by all convex combinations of elements of Y. While you indeed can get in this fashion a whatever convex set in $\mathbf{R}^n$, this result is incomparably less useful than its linear/affine subspace counterparts known to us: in the latter case, the set of "generators" Y could be chosen to be finite and even not too large (at most n elements in the case of a linear subspace and at most $n+1$ element in the case of an affine subspace in $\mathbf{R}^n$). In the convex case, a finite set of "generators" is not always enough.

"Good" — closed — convex sets in $\mathbf{R}^n$ also admit "outer" description, specifically, a closed convex set in $\mathbf{R}^n$ always is a solution set of a system $a_i^T x \leq b_i$, $i = 1, 2, \ldots$ of *countably many* nonstrict linear inequalities. Here again, the result is instructive, but incomparably less useful than its linear/affine subspace counterparts, since our abilities to handle infinite (even countable) systems of linear inequalities are severely restricted.

2.1.4 *Cones*

Cones: Definition. A particular type of convex sets important for us is formed by *cones*. By definition, *a cone is a nonempty convex set X composed of rays emanating from the origin, that is, such that $tx \in X$ whenever $x \in X$ and $t \geq 0$.* An immediate observation (check it!) is that X *is a cone if and only if X is nonempty and is closed w.r.t. taking conic combinations* (linear combinations with nonnegative coefficients) *of its elements.*

Examples: Important for us will be *polyhedral cones* which, by definition, are the solution sets of finite systems of *homogeneous* linear inequalities:

$$X \text{ is a polyhedral cone } \Leftrightarrow \exists A : X = \{x \in \mathbf{R}^n : Ax \leq 0\}.$$

In particular, the *nonnegative orthant* $\mathbf{R}^n_+ = \{x \in \mathbf{R}^n : x \geq 0\}$ is a polyhedral cone.

Calculus of cones in its most important (and most elementary) part is as follows (check the validity of the claims!):

(1) *Taking intersection*: If X_1, X_2 are cones in $\mathbf{R}^n$, so is their intersection $X_1 \cap X_2$. In fact, *the intersection $\bigcap_{\alpha \in \mathcal{A}} X_\alpha$ of a whatever family $\{X_\alpha\}_{\alpha \in \mathcal{A}}$ of cones in $\mathbf{R}^n$ is a cone.*

(2) *Taking arithmetic sum*: If X_1, X_2 are cones in $\mathbf{R}^n$, so is the set $X_1 + X_2 = \{x = x_1 + x_2 : x_1 \in X_1, x_2 \in X_2\}$.
(3) *Taking linear image*: If X is a cone in $\mathbf{R}^n$ and A is an $m \times n$ matrix, then the set $AX := \{Ax : x \in X\} \subset \mathbf{R}^m$ — the image of X under the linear mapping $x \mapsto Ax : \mathbf{R}^n \to \mathbf{R}^m$ — is a cone in $\mathbf{R}^m$.
(4) *Taking inverse linear image*: If X is a cone in $\mathbf{R}^n$ and A is an $n \times k$ matrix, then the set $\{y \in \mathbf{R}^k : Ay \in X\}$ — the inverse image of X under the linear mapping $y \mapsto Ay : \mathbf{R}^k \to \mathbf{R}^n$ – is a cone in $\mathbf{R}^k$.
(5) *Taking direct products*: If $X_i \subset \mathbf{R}^{n_i}$ are cones, $1 \leq i \leq k$, so is the direct product $X_1 \times \cdots \times X_k \subset \mathbf{R}^{n_1 + \cdots + n_k}$.
(6) *Passing to the dual cone*: If X is a cone in $\mathbf{R}^n$, so is its *dual cone* defined as

$$X_* = \{y \in \mathbf{R}^n : y^T x \geq 0 \ \forall x \in X\}.$$

Conic hulls: Similar to what we observed above, given a nonempty set $X \subset \mathbf{R}^n$ and taking all conic combinations of vectors from X, we get a cone (called the *conic hull* of X, notation: $\text{Cone}(X)$). By definition (and in full accordance with the standard convention that an empty sum of vectors from $\mathbf{R}^n$ has value, namely, the origin), the conic hull of an empty set in $\mathbf{R}^n$ is the trivial cone $\{0\}$. Note that a conic hull of $X \subset \mathbf{R}^n$ is the smallest, w.r.t. inclusion, among the cones containing X. For example, the nonnegative orthant $\mathbf{R}^n_+$ is the conic hull of the set composed of all n basic orths in $\mathbf{R}^n$.

2.2 Preparing tools

In this section, we develop technical tools to be used later.

2.2.1 *Caratheodory theorem*

We start with the following statement, highly important by its own right.

Theorem 2.4 (Caratheodory theorem). *Let $x, x_1, \ldots, x_k$ be vectors from $\mathbf{R}^n$. If x is a convex combination of vectors $x_1, \ldots, x_k$, then*

x is a convex combination of at most $n+1$ properly chosen vectors from the collection $x_1, \ldots, x_k$.

Moreover, if the dimension of the set $X = \{x_1, \ldots, x_k\}$ is m, then $n+1$ in the above conclusion can be replaced with $m+1$.

Proof. Let us look at all possible representations $x = \sum_{i=1}^{k} \lambda_i x_i$ of x as a convex combination of vectors $x_1, \ldots, x_k$. Given that the corresponding family is nonempty, it definitely contains a minimal in the number of actual terms (i.e., those with $\lambda_i \neq 0$) representation. All we should prove is that the number of actual terms in this minimal representation is $\leq n+1$. We can assume without loss of generality that the minimal representation in question is $x = \sum_{i=1}^{s} \lambda_i x_i$; from minimality, of course, $\lambda_i > 0$, $1 \leq i \leq s$. Assume, on the contrary to what should be proved, that $s > n+1$, and let us lead this assumption to a contradiction. Indeed, when $s > n+1$, the homogeneous system of linear equations

$$\sum_{i=1}^{s} \delta_i x_i = 0 \quad [n \text{ scalar equations}],$$

$$\sum_{i=1}^{s} \delta_i = 0 \quad [1 \text{ scalar equation}]$$

in $s > n+1$ variables δ has more variables than equations and thus has a nontrivial (with not all entries equal to zero) solution $\bar{\delta}$. Observe that we have

$$\sum_{i=1}^{s} \underbrace{[\lambda_i + t\bar{\delta}_i]}_{\lambda_i(t)} x_i = x, \sum_{i=1}^{s} \lambda_i(t) = 1 \qquad (!)$$

for all $t \geq 0$. Now, since $\sum_i \bar{\delta}_i = 0$ and not all $\bar{\delta}_i$ are zeros, among the reals $\bar{\delta}_i$, there are strictly negative, meaning that for large enough values of t, not all of the coefficients $\lambda_i(t)$ are nonnegative. At the same time, when $t = 0$, these coefficients are positive. It follows that there exists the largest $t = \bar{t} \geq 0$ such that all $\lambda_i(t)$ are still nonnegative, and since for a larger value of t, not all $\lambda_i(t)$ are nonnegative, the nonnegative reals $\lambda_i(\bar{t})$, $i = 1, \ldots, s$, include one or more zeros.[5]

[5]In fact we could, of course, skip this explanation and point out $\bar{t}$ explicitly: $\bar{t} = \min_{i:\bar{\delta}_i<0}[-\lambda_i/\bar{\delta}_i]$.

When $t = \bar{t}$, (!) says that x is a convex combination of $x_1, \ldots, x_s$, the coefficients being $\lambda_i(\bar{t})$; but some of these coefficients are zeros, thus we managed to represent x as a convex combination of *less than* s of the vectors $x_1, \ldots, x_k$, which contradicts the origin of s. We have arrived at the required contradiction.

The second part of the statement is in fact nothing but the first part in a slight disguise. Indeed, let $M = a + L$ be the affine span of $x_1, \ldots, x_k$; then clearly $x \in M$. Shifting all vectors $x_1, \ldots, x_k, x$ by $-a$ (which affects neither the premise nor the conclusion of the statement we want to prove), we can assume that the affine span of $x_1, \ldots, x_k$ is the linear subspace L of dimension $m = \dim\{x_1, \ldots, x_k\}$ to which M is parallel. Note that $x \in L$ along with $x_1, \ldots, x_k$, and we lose nothing when thinking about all vectors $x, x_1, \ldots, x_k$ as of *vectors in* $\mathbf{R}^m$ (since as far as linear operations are concerned — and these are the *only* operations underlying our assumptions and targets, L is nothing but $\mathbf{R}^m$). Invoking the already-proved part of the statement, we arrive at its remaining part.[6] □

Remark 2.1. Note that the result stated in Caratheodory theorem is sharp: Without additional assumptions on $x_1, \ldots, x_k$, you cannot replace $n+1$ and $m+1$ with smaller numbers, e.g., given $m \leq n$, consider collection of $m+1$ vectors in $\mathbf{R}^n$ as follows: $x_0 = 0$, $x_1 = e_1, \ldots, x_m = e_m$, where $e_1, \ldots, e_n$ are the basic orths (see p. 87). These vectors clearly are affinely independent (why?), so that $\dim\{x_0, x_1, \ldots, x_m\} = m$ and the vector $x = \frac{1}{m+1}[x_0 + \cdots + x_m]$ – which is a convex combination of $x_0, \ldots, x_m$, admits exactly one representation as an affine (and thus as a convex) combination of $x_0, \ldots, x_m$ (Theorem 2.2). Thus, in any representation of x as a convex combination of $x_0, x_1, \ldots, x_m$, all these $m+1$ vectors should be present with positive coefficients.

Illustration: Let us look at the following story:

> In the nature, there are 26 "pure" types of tea, denoted A, B,..., Z; all other types are mixtures of these "pure" types. What they sell in the

[6] We believe the concluding reasoning should be carried out exactly once; in the sequel in similar situations, we will just write as follows: "We can assume without loss of generality that the affine span of $x_1, \ldots, x_k$ is the entire $\mathbf{R}^n$."

market, are certain blends of tea, not the pure types; there are totally 111 blends which are sold.

John prefers a specific blend of tea which is not sold in the market; from long experience, he found that in order to get the blend he prefers, he can buy 93 of the 111 market blends and mix them in certain proportion.

An OR student (known to be good in this subject) pointed out that in fact John could produce his favorite blend of tea by appropriate mixing of just 27 of the properly selected market blends. Another OR student pointed out that only 26 of market blends are enough, and the third student said that 24 also is enough. John did not believe in neither of these recommendations, since no one of the students asked him what his favorite blend is. Is John right?

The answer is that the first two students are definitely right, while the third can be wrong. Indeed, we can identify a unit (in weight) amount of a blend of tea with $n = 26$-dimensional vector $x = [x_A; x_B; \ldots; x_Z]$, where entries in x are the weights of the corresponding pure components of the blend; the resulting vector is nonnegative and the sum of its entries equals to 1 (which is the total weight of our unit of blend). With this identification, let $\bar{x}$ be the blend preferred by John, and let $x^1, \ldots, x^N$ be the marketed blends ($N = 111$). What we know is that John can get a unit amount of his favorite blend by buying marking blends in certain amounts λ_i, $i = 1, \ldots, N$, and putting them together, that is, we know that

$$\bar{x} = \sum_{i=1}^{N} \lambda_i x^i. \qquad (*)$$

Of course, λ_i are nonnegative due to their origin; comparing "weights" (sums of entries) of the right-hand side and the left-hand side in the vector equality $(*)$, we see that $\sum_i \lambda_i = 1$. Thus, the story tells us that the 26-dimensional vector $\bar{x}$ is a convex combination of N vectors $x^1, \ldots, x^N$. By the first part of the Caratheodory theorem, x can be represented as a convex combination of $26 + 1 = 27$ properly chosen vectors x^i, which justifies the conclusion of the first student. Noting that in fact $x^1, \ldots, x^N, \bar{x}$ belong to the hyperplane $\{x : x_A + x_B + \cdots + x_Z = 1\}$ in $\mathbf{R}^{26}$, we conclude that the dimension of $\{x^1, \ldots, x^N\}$ is at most the dimension of this hyperplane, which is $m = 25$. Thus, the second part of Caratheodory theorem says that just $m + 1 = 26$ properly chosen market blends will do, so that the second student is right as well. As regards the third student, whether he is right or not, it depends on what the vectors $\bar{x}, x^1, \ldots, x^N$ are e.g., it may happen than $\bar{x} = x^1$ and John merely missed this fact — then the third student is right. On the other hand, it may happen that $x^1, \ldots, x^N$ are just the 26 "pure" teas (basic orths), with some of them repeated (who told us that the same blend cannot be sold under several different names?). If every basic orth is present in the collection $x^1, \ldots, x^N$ and the favorite blend of John is $\bar{x} = [1/26; \ldots; 1/26]$, it can indeed be obtained as mixture of market blends, but such a mixture should contain at least 26 of these blends. Note that the third student would definitely be right if the dimension of $\{x^1, \ldots, x^N\}$ were < 24 (why?).

2.2.1.1 *Caratheodory theorem in conic form and Shapley–Folkman Theorem*

Caratheodory theorem speaks about the representation of a vector as convex combination — linear combination with nonnegative coefficients summing up to 1 — of another vectors. When passing from convex combinations to *conic* ones — linear combinations with nonnegative coefficients — Theorem can be improved.

Theorem 2.5 (Caratheodory theorem in conic form). *Let $x, x_1, \ldots, x_k$ be vectors from $\mathbf{R}^n$. If x is a conic combination (linear combination with nonnegative coefficients) of vectors $x_1, \ldots, x_k$, then x is a conic combination of at most n properly chosen vectors from the collection $x_1, \ldots, x_k$.*

Proof is given by straightforward simplification of the proof of the plain Caratheodory theorem and is left to the reader.

Caratheodory theorem in conic form admits nice and useful consequence, *Shapley–Folkman theorem*, as follows.

Theorem 2.6 (Shapley–Folkman theorem). *Let V_i be nonempty sets in $\mathbf{R}^d$ with convex hulls $\overline{V}_i = \mathrm{Conv}(V_i)$, $i = 1, \ldots, k$, and let $x \in \mathbf{R}^n$ belong to the arithmetic sum of $\overline{V}_i$:*

$$x = x_1 + \cdots + x_k, \quad x_i \in \overline{V}_i, i \le k. \tag{2.10}$$

Then there exists representation

$$x = y_1 + \cdots + y_k, \quad y_i \in \overline{V}_i, i \le k$$

in which at least $k - d$ of vectors y_i belong to V_i.

Proof. Given representation (2.10) and applying Caratheodory theorem, we can find vectors $x_{ij} \in V_i$, $j = 1, \ldots, d+1$, such that $x_i = \sum_j \overline{\lambda}_{ij} x_{ij}$ with $\overline{\lambda}_{ij} \geq 0$ such that $\sum_j \overline{\lambda}_{ij} = 1$, $i = 1, \ldots, k$. Consider the system of linear equations

$$\begin{aligned} \sum_{i,j} \lambda_{ij} x_{ij} &= x \qquad (a), \\ \sum_j \lambda_{ij} &= 1 \qquad (b_i), \\ i &= 1, \ldots, k \end{aligned} \tag{S}$$

in variables λ_{ij}, $i \leq k$, $j \leq d+1$, and let $A_{ij} \in \mathbf{R}^{d+k}$ be the vector of coefficients at λ_{ij} in this system and b be its right-hand side, so that the system reads $\sum_{i,j} \lambda_{ij} A_{ij} = b$. The collection $\{\overline{\lambda}_{ij}, i \leq k, j \leq d+1\}$ is a nonnegative solution to this system, implying that b is a conic combination of A_{ij}'s. Applying Caratheodory theorem in conic form, we conclude that there exists a nonnegative solution $\{\lambda_{ij}\}$ to system (S) in which at most $d+k$ entries are positive, and the remaining are zeros. Now, let n_i be the number of nonzeros among λ_{ij}, $j = 1, \ldots, d+1$, and m be the number of i's for which $n_i > 1$. Note that equations (b_i) say that $n_i \geq 1$ for every i, so that the total number n of nonzeros among the collection $\{\lambda_{ij}, i \leq k, j \leq d+1\}$ is at least $k+m$. On the other hand, this number is at most $d+k$, implying that $m \leq d$. At the same time, λ_{ij} solve (S), that is,

$$x = \sum_{i=1}^{k} \underbrace{\sum_j \lambda_{ij} x_{ij}}_{=:y_i}.$$

By (S) and due to $\lambda_{ij} \geq 0$, we have $y_i \in \overline{V}_i$ for every i and clearly $y_i \in V_i$ whenever $n_i = 1$ (since for such an i, exactly one of λ_{ij}'s, $1 \leq j \leq d+1$, is nonzero and therefore is equal to 1 by (b_i), and $x_{ij} \in V_i$ for all j). Thus, in the resulting representation of x as sum of vectors $y_i \in \overline{V}_i$ at least $k - m \geq k - d$ vectors in fact belong to the corresponding V_i's. □

Qualitatively speaking, Shapley–Folkman theorem states that arithmetic summation of sets possesses certain convexification properties. For example, when $V_i \subset \mathbf{R}^d$ are of Euclidean diameters

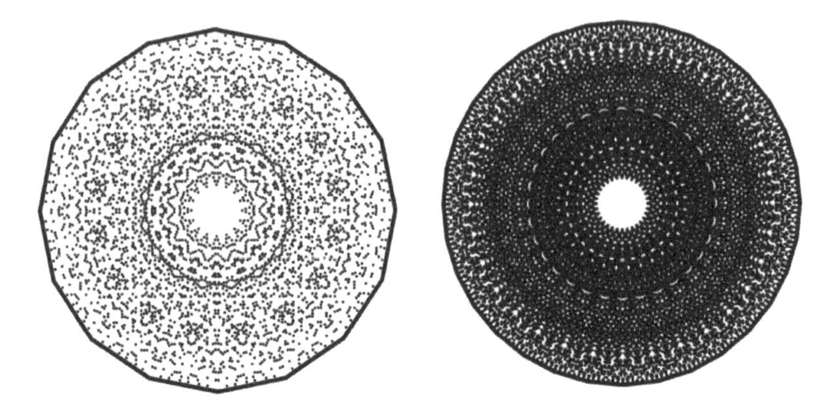

Fig. 2.2. Black dots: the finite set $V = V_1 + V_2 + V_3$, where V_i, $i = 1, 2, 3$, are the vertices of concentric perfect m-side polygons with ratio of linear sizes $4 : 2 : 1$. Bold broken line: boundary of the perfect m-side polygon $\text{Conv}(V_1) + \text{Conv}(V_2) + \text{Conv}(V_3)$. Left: $m = 16$; right: $m = 32$. Think where the white inner "cogwheels" come from and what, approximately, the ratios of their diameters to the diameters of the respective V's are.

(i.e., maximal pairwise $\|\cdot\|_2$-distances) ρ_i, $\rho_1 \geq \rho_2 \geq \cdots \geq \rho_k$, Shapley–Folkman theorem says that the set $V = V_1 + \cdots + V_k$ is "$\rho^d := \rho_1 + \cdots + \rho_d$ - dense" in the convex set $\overline{V} = \overline{V}_1 + \cdots + \overline{V}_k$, meaning that $V \subset \overline{V}$ and every point of $\overline{V}$ is at the $\|\cdot\|_2$-distance $\leq \rho^d$ from some point of V. Since the linear sizes of $\overline{V}$ can be of order of $\rho^k = \rho_1 + \cdots + \rho_k$, in the case of $\rho^k \gg \rho^d$ (which happens, e.g., when ρ_i are of the same order of magnitude and $k/d \gg 1$), the convex set $\overline{V}$ is a reasonably good outer approximation of (by itself, perhaps, highly nonconvex) set V, see Fig. 2.2. For instructive application of the Shapley–Folkman theorem, see illustration in Section 2.4.4.3.

Remark 2.2. Results presented so far in this section were formulated as descriptive statements — as existence theorems, *if something exists* (e.g., representation of $x \in \mathbf{R}^n$ as convex combination of k vectors x_i), *then something else exists* (e.g., representation of x as the convex combination of at most $n+1$ vectors properly selected among x_i's). In fact, taken along with their proofs, these results become operational — they provide us with simple algorithms allowing one to efficiently convert the above "something one" into "something else." For example, the proof of Caratheodory theorem describes a simple linear algebra construction which allows us to extract from the representation of a given vector x as a convex combination of vectors from a given collection $\mathcal{C}$ of cardinality $k > n+1$ similar representation of x via sub-collection of $\mathcal{C}$ with cardinality $< k$. Iterating this construction, we can efficiently find representation of x as a convex combination of at most $n+1$ members of the initial collection. The situations with conic form of Caratheodory theorem and with Shapley–Folkman theorem are completely similar.

2.2.2 *Radon Theorem*

Theorem 2.7 (Radon theorem). *Let $x_1, \ldots, x_k$ be a collection of $k > n+1$ vectors from $\mathbf{R}^n$. Then one can split the index set $I = \{1, \ldots, k\}$ into two nonempty and nonoverlapping subsets I_1 and I_2 such that the convex hulls of the sets $\{x_i : i \in I_1\}$ and $\{x_i : i \in I_2\}$ intersect.*

Proof follows the same idea as for Caratheodory theorem. Specifically, consider the homogeneous system of linear equations in variables $\delta_1, \ldots, \delta_k$:

$$\sum_{i=1}^{k} \delta_i x_i = 0 \text{ [} n \text{ scalar equations]},$$

$$\sum_{i=1}^{k} \delta_i = 0 \text{ [1 scalar equation]}.$$

Since $k \geq n+2$, the number of equations in this system is less than the number of variables; since the system is homogeneous, it therefore has a nontrivial solution $\bar{\delta}$. Now, let $I_1 = \{i : \delta_i > 0\}$ and $I_2 = \{i : \delta_i \leq 0\}$. The sets I_1 and I_2 clearly form a partition of I, and both of them are nonempty; indeed, since $\sum_i \bar{\delta}_i = 0$ and not all $\bar{\delta}_i$ are zeros, there definitely are strictly positive and strictly negative among $\bar{\delta}_i$.

Now, let us set $S = \sum_{i \in I_1} \bar{\delta}_i$. The equations from our system read

$$\sum_{i \in I_1} \bar{\delta}_i x_i = \sum_{i \in I_2} [-\bar{\delta}_i] x_i,$$

$$\sum_{i \in I_1} \bar{\delta}_i = S = \sum_{i \in I_2} [-\bar{\delta}_i].$$

Setting $\lambda_i = \bar{\delta}_i / S$, $i \in I_1$, $\mu_i = -\bar{\delta}_i / S$, $i \in I_2$, we get $\lambda_i \geq 0$, $\mu_i \geq 0$, and the above reads

$$\sum_{i \in I_1} \lambda_i x_i = \sum_{i \in I_2} \mu_i x_i,$$

$$\sum_{i \in I_1} \lambda_i = 1 = \sum_{i \in I_2} \mu_i,$$

meaning that $\text{Conv}\{x_i : i \in I_1\} \cap \text{Conv}\{x_i : i \in I_2\} \neq \emptyset$. □

2.2.3 *Helly Theorem*

Theorem 2.8 (Helly theorem). *Let $A_1, \ldots, A_N$ be nonempty convex sets in $\mathbf{R}^n$.*

(i) *If every $n+1$ of the sets A_i have a point in common, then all N sets A_i have a point in common.*
(ii) *If the dimension of $A_1 \cup \cdots \cup A_N$ is m and every $m+1$ of the sets A_i have a point in common, then all N sets A_i have a point in common.*

Proof. By exactly the same reasons as in the proof of Caratheodory theorem, (ii) is a straightforward consequence of (i) (replace the "universe" $\mathbf{R}^n$ with $\text{Aff}\{\bigcup_i A_i\}$), so that all we need is to prove (i). This will be done by induction in the number N of the sets. There is nothing to prove when $N \leq n+1$; thus, all we need is to verify that the statement is true for a family of $N > n+1$ convex sets A_i, *given that the statement is true for every family of $< N$ convex sets* (this is our inductive hypothesis). This is easy: Given N nonempty convex sets A_i such that every $n+1$ of them have a point in common, let us form the N sets

$$B_1 = A_2 \cap A_3 \cap \cdots \cap A_N, B_2 = A_1 \cap A_3 \cap \cdots \cap A_N, \ldots, B_N$$
$$= A_1 \cap A_2 \cap \cdots \cap A_{N-1},$$

that is, B_i is the intersection of $N-1$ sets $A_1, \ldots, A_{i-1}, A_{i+1}, \ldots, A_N$. By inductive hypothesis, these sets are nonempty (and are convex along with $A_1, \ldots, A_N$). Let us choose in the (nonempty!) set B_i a point x_i, $i = 1, \ldots, N$, thus ending up with a collection $x_1, \ldots, x_N$ of $N > n+1$ points from $\mathbf{R}^n$. Applying Radon Theorem, we can split this collection into two parts with intersecting convex hulls; to save notation — and of course without loss of generality — we can assume that this partition is $I_1 = \{1, \ldots, m\}$, $I_2 = \{m+1, \ldots, N\}$, so that

$$\exists b \in \mathrm{Conv}\{x_1, \ldots, x_m\} \cap \mathrm{Conv}\{x_{m+1}, \ldots, x_N\}.$$

We claim that $b \in A_i$ for all i, so that the intersection of all A_i is nonempty; this would complete the inductive step and thus the proof of Helly Theorem. To justify our claim, note that when $i \leq m$, the point x_i belongs to B_i and thus to every one of the sets $A_{m+1}, \ldots, A_N$ (by construction of B_i). Thus, every one of the points $x_1, \ldots, x_m$ belongs to every set A_i with $i > m$; but then the latter set, being convex, contains $b \in \mathrm{Conv}\{x_1, \ldots, x_m\}$. The bottom line is that $b \in A_i$ whenever $i > m$. By "symmetric" reasoning, $b \in A_i$ whenever $i \leq m$. Indeed, whenever $j > m$, the point x_j belongs to B_j and thus to every one of the sets $A_1, \ldots, A_m$ (by construction of B_j). Thus, every one of the points $x_{m+1}, \ldots, x_N$ belongs to every set A_i with $i \leq m$; being convex, the latter set also contains the point $b \in \mathrm{Conv}\{x_{m+1}, \ldots, x_N\}$, as claimed. □

Remark 2.3. Note that the Helly Theorem is "as sharp as it could be:" we can easily point out $n+1$ nonempty convex sets $A_1, \ldots, A_{n+1}$ in $\mathbf{R}^n$ such that every n of them have a point in common, while the intersection of all $n+1$ sets is empty. An example is given by $A_i = \{x \in \mathbf{R}^n : x_i \leq 0\}$, $i = 1, \ldots, n$, $A_{n+1} = \{x \in \mathbf{R}^n : \sum_i x_i \geq 1\}$.

Illustration 1: Let S be a set composed of 1,000,000 distinct points on the axis and $f(s)$ be a real-valued function on this set. Assume that for every 7-point subset S' of S, there exists an algebraic polynomial $p(s)$ of degree ≤ 5 such that $|p(s) - f(s)| \leq 0.01$ whenever

$s \in S'$. We now want to find a spline of degree ≤ 5 (a piecewise polynomial function on the axis with pieces — algebraic polynomials of degree ≤ 5) which approximates f at every point of S within accuracy 0.01. How many pieces should we take?

The answer is: just one. Indeed, we can identify a polynomial $p(s) = p_0+p_1s+\cdots+p_5s^5$ of degree ≤ 5 with its vector of coefficients $[p_0;\ldots;p_5] \in \mathbf{R}^6$. For a fixed $s \in S$, the set of (vectors of coefficients) of polynomials which approximate $f(s)$ within accuracy 0.01 is $P_s = \{[p_0;\ldots;p_5] : |f(s)-p_0-p_1s-\cdots-p_5s^5| \leq 0.01\}$; we see that the set is polyhedral and is thus convex. What is given to us is that every 7 sets from the 1,000,000-set family $\{P_s : s \in S\}$ have a point in common. By Helly Theorem, *all* 1,000,000 sets have a point p_* in common, and the corresponding polynomial of degree ≤ 5 approximates f at every point from S within accuracy of 0.01.

Illustration 2: The daily functioning of a plant is described by the system of linear constraints

$$\begin{array}{ll} \text{(a)} & Ax \leq f \in \mathbf{R}^{10}, \\ \text{(b)} & Bx \geq d \in \mathbf{R}^{2009}, \\ \text{(c)} & Cx \leq c \in \mathbf{R}^{2000}, \end{array} \tag{!}$$

where

- x is the decision vector — production plan for the day;
- f is *nonnegative* vector of resources (money, manpower, electric power, etc.) available for the day;
- d is the vector of daily demands on different kinds on plant's production.

The plant works as follows. In the evening of day $t-1$, the manager should order the resources f for the next day; when doing so, he does not know exactly what will be the vector of demands for the next day, but has a collection $D = \{d^1,\ldots,d^{1,000,000}\}$ of 1,000,000 demand scenarios and knows that the actual demand of day t will be a point from this set. The goal of the manager is to order the resources f in the evening of day $t-1$ in such a way that when in the next morning the actual vector of demands of day t will become known, it will be possible to find a production plan x which satisfies the constraints (!).

Finally, the manager knows that every scenario demand $d^i \in D$ can be "served," in the aforementioned sense, by properly chosen vector $f^i = [f^i_1; \ldots; f^i_{10}] \geq 0$ of resources at the cost $\sum_{j=1}^{10} c_j f^i_j$ not exceeding \$1 ($c_j \geq 0$ are given prices of the resources). How much money should a smart manager spend on the next-day resources in order to guarantee that the tomorrow demand will be satisfied?

The answer is: \$11 is enough. Indeed, let F_i be the set of all resource vectors $f \geq 0$ which allow us to satisfy demand $d^i \in D$. This set is convex (check it!), and we know that it contains a vector f^i which costs at most \$1. Now, let A_i be the set of all nonnegative resource vectors f which cost at most \$11 and allow us to satisfy the demand d^i. This set is also convex (as the intersection of the convex set F_i and the polyhedral — and thus convex — set $\{f : f \geq 0, c^T f \leq 11\}$). We claim that every 11 of the convex sets $A_i \subset \mathbf{R}^{10}$ have a point in common. Indeed, given these 11 sets $A_{i_1}, \ldots, A_{i_{11}}$, consider the vector of resources $f = f^{i_1} + \cdots + f^{i_{11}}$. Since f^{i_p} allows us to satisfy the demand d^{i_p} and $f \geq f^{i_p}$, f allows as to satisfy the same demand as well (look at (!)); since every f^{i_p} costs at most \$1, f costs at most \$11. Thus, $f \geq 0$ allows us to satisfy demand d^{i_p} and costs at most \$11 and this belongs to A_{i_p}, $1 \leq p \leq 11$. By Helly Theorem, all A_i have a point in common, let it be denoted f_*. By its origin, this is a nonnegative vector with cost ≤ 11, and since it belongs to every A_i, it allows us to satisfy every demand $d^i \in D$.

2.2.4 *Homogeneous Farkas Lemma*

Our next statement is *the* key which will unlock basically all the locks to be opened in the sequel.

Consider the situation as follows: we are given a finite system of homogeneous linear inequalities

$$a_i^T x \geq 0, \quad i = 1, \ldots, m \tag{2.11}$$

in variables $x \in \mathbf{R}^n$, along with another homogeneous linear inequality

$$a^T x \geq 0, \tag{2.12}$$

which we will call the *target* one. The question we are interested is: *When* (2.12) *is a consequence of the system* (2.11), *meaning that* *whenever x satisfies the system, it satisfies the target inequality as well?*

There is a trivial *sufficient* condition for the target inequality to be a consequence of the system, specifically the representability of the target inequality as a weighted sum, with nonnegative coefficients, of the inequalities from the system. Specifically, we have the following observation:

Observation 2.1. *If* a is a conic combination of $a_1, \ldots, a_m$,

$$\exists \lambda_i \geq 0 : a = \sum_{i=1}^{m} \lambda_i a_i, \tag{2.13}$$

then the target inequality (2.12) is a consequence of the system (2.11).

Indeed, let λ_i be as in (2.13), and let x be a solution of (2.11). Multiplying the inequalities $a_i^T x \geq 0$ by the nonnegative weights λ_i and summing up, we get $a^T x \equiv [\sum_i \lambda_i a_i]^T x = \sum_i \lambda_i a_i^T x \geq 0$. □

The homogeneous Farkas lemma is an incomparably deeper result with states that the above trivial sufficient condition is in fact *necessary* for the target inequality to be a consequence of the system.

Theorem 2.9 (Homogeneous Farkas lemma). *The target inequality* (2.12) *is a consequence of the system* (2.11) *iff a is a conic combination of $a_1, \ldots, a_m$.*

Equivalent, and sometimes more instructive, form of this statement reads as follows:

It is easy to certify both that a is a conic combination of $a_1, \ldots, a_m$, and that a is not a conic combination of $a_1, \ldots, a_m$:

- to certify the first fact, it suffices to point out nonnegative λ_i such that $a = \sum_i \lambda_i a_i$;
- to certify the second fact, it suffices to point out x such that $a_i^T x \geq 0$, $i = 1, \ldots, m$, and $a^T x < 0$.

a is a conic combination of $a_1, \ldots, a_m$ iff the certificate of the first kind exists, and a *is not* a conic combination of $a_1, \ldots, a_m$ iff the certificate of the second kind exists.

Proof of HFL. We have already seen that *if* a is a conic combination of $a_1, \ldots, a_m$, *then* the target inequality is a consequence of the system. All we need is to prove the inverse statement:

(#) *if* the target inequality is a consequence of the system, *then* a is a conic combination of $a_1, \ldots, a_m$.

Intelligent proof of (#). Observe that the set $X = \{x = \sum_{i=1}^m \lambda_i a_i : \lambda_i \geq 0\}$ is polyhedrally representable: $X = \{x : \exists \lambda : x = \sum_i \lambda_i a_i, \lambda \geq 0\}$ and as such is polyhedral (Fourier–Motzkin elimination, Theorem 1.2):

$$X = \{x \in \mathbf{R}^n : d_\ell^T x \geq \delta_\ell,\ \ell = 1, \ldots, L\} \tag{$*$}$$

for some d_ℓ, δ_ℓ; observe that $\delta_\ell \leq 0$ due to the evident inclusion $0 \in X$. Now, let the target inequality $a^T x \geq 0$ be a consequence of the system $a_i^T x \geq 0$, $1 \leq i \leq m$, and let us lead to a contradiction the assumption that a is not a conic combination of a_i or, which is the same, the assumption that $a \notin X$. Assuming $a \notin X$ and looking at $(*)$, there exists ℓ_* such that $d_{\ell_*}^T a < \delta_{\ell_*}$ and thus $d_{\ell_*}^T a < 0$ due to $\delta_{\ell_*} \leq 0$. On the other hand, we have $\lambda a_i \in X$ for all $\lambda \geq 0$ and all i, meaning that $\lambda d_{\ell_*}^T a_i \geq \delta_{\ell^*}$ for all $\lambda \geq 0$ and all i, whence $d_{\ell_*}^T a_i \geq 0$ for all i (look what happens when $\lambda \to +\infty$). Thus, $\bar{x} := d_{\ell_*}$ satisfies the system $a_i^T x \geq 0$, $1 \leq i \leq m$, and violates the target inequality $a^T x \geq 0$, which under the premise of (#) is impossible; we have arrived at a desired contradiction. $\square$

***Alternative proof of (#) based on Helly theorem.** There is nothing to prove when $a = 0$, since then a is indeed a conic combination of $a_1, \ldots, a_m$. Thus, from now on, we assume that $a \neq 0$ and that (2.12) is a consequence of (2.11); our goal is to derive from these assumptions that a is a conic combination of $a_1, \ldots, a_m$.

1^0. Let us set $A_i = \{x \in \mathbf{R}^n : a_i^T x \geq 0,\ a^T x = -1\}$. Note that every A_i is a polyhedral set and as such is convex (perhaps empty) and that the intersection of all m sets $A_1, \ldots, A_m$ is definitely empty (indeed, a vector x from this intersection, if exists, solves the system and does not solve the target inequality, which is impossible). Let us extract from the family of sets $A_1, \ldots, A_m$ with empty intersection a minimal, w.r.t. inclusion, subfamily with the same property. Without loss of generality, we can assume that this subfamily is $A_1, \ldots, A_k$. Thus, $k \geq 1$, the intersection of $A_1, \ldots, A_k$ is empty, and either $k = 1$

(meaning that A_1 is empty) or the intersection of every $k-1$ sets from $A_1, \ldots, A_k$ is nonempty.

2^0. We claim that $a \in L := \text{Lin}\{a_1, \ldots, a_k\}$. Indeed, otherwise a has a nonzero projection h onto $L^\perp$, so that $h^T a_i = 0$, $i = 1, \ldots, k$, and $h^T a = h^T h > 0$. Setting $x = -\frac{1}{h^T h} h$, we get $a_i^T x = 0$, $1 \le i \le k$, and $a^T x = -1$; thus, h belongs to $A_1, \ldots, A_k$, and these sets have a point in common, which in fact is not the case.

3^0. We claim — and this is the central component of the proof — that $a_1, \ldots, a_k$ are linearly independent or, which is the same, that $\dim L = k$. Since L is linearly spanned by $a_1, \ldots, a_k$, the only alternative to $\dim L = k$ is $\dim L < k$; we assume that the latter is the case, and let us lead this assumption to a contradiction. Observe, first, that $k > 1$, since otherwise $L = \{0\}$ due to $\dim L < k$, implying that $a = 0$ (we already know that $a \in L$), which is not the case. Further, consider the hyperplane $\Pi = \{x \in L : a^T x = -1\}$ in L. Since $0 \neq a \in L$, Π is indeed a hyperplane, and thus, $\dim \Pi = \dim L - 1 < k - 1$. Now, the orthogonal projections B_i of the sets $A_i = \{x : a_i^T x \ge 0, a^T x = -1\}$ onto L, $1 \le i \le k$, clearly belong to Π and to A_i (since when projecting orthogonally $x \in A_i$ onto L, the inner products with a and a_i remain intact due to $a \in L$, $a_i \in L$). B_i, $i = 1, \ldots, k \ge 1$, are convex subsets of Π, and the intersection of every $k - 1 > 0$ of the sets B_i is nonempty (indeed, it contains the projection onto L of every point from the intersection of the corresponding A_i, and this intersection is nonempty). Since $\dim \Pi + 1 < k$, we conclude that the intersection of every $\dim \Pi + 1$ of the sets $B_1, \ldots, B_k$ is nonempty, whence the intersection of all k of the sets $B_1, \ldots, B_k$ is nonempty (Helly theorem). Recalling that $B_i \subset A_i$, we conclude that the intersection of $A_1, \ldots, A_k$ is nonempty, which is a desired contradiction (recall how k and $A_1, \ldots, A_k$ were chosen).

4^0. Now, it is easy to complete the proof. By 2^0, $a \in L = \text{Lin}\{a_1, \ldots, a_k\}$, that is, $a = \sum_{i=1}^k \lambda_i a_i$ with certain λ_i; we are about to prove that all λ_i here are nonnegative, which will bring us to our goal — to prove that a is a conic combination of $a_1, \ldots, a_m$. In order to verify that $\lambda_i \ge 0$, $1 \le i \le k$, assume, on the contrary, that not all λ_i are nonnegative, say, that $\lambda_1 < 0$. Since $a_1, \ldots, a_k$ are linearly independent by 3^0, by linear algebra there exists a vector $\bar{x} \in \mathbf{R}^n$ such that $a_1^T \bar{x} = 1/|\lambda_1|$, $a_2^T \bar{x} = 0, \ldots, a_k^T \bar{x} = 0$. We have $a^T \bar{x} = \sum_i \lambda_i a_i^T \bar{x} = \lambda_1/|\lambda_1| = -1$, while, by construction, $a_i^T \bar{x} \ge 0$,

$i = 1, \dots, k$. We see that $\bar{x} \in A_1 \cap \cdots \cap A_k$, which is impossible, since the right-hand side intersection is empty. Thus, assuming that not all of $\lambda_1, \dots, \lambda_k$ are nonnegative, we arrive at a contradiction. $\square$

Instructive application: The HFL is the key instrument in further developments. As for now, here is one of fundamental results readily given by HFL (in fact, this result is an equivalent reformulation of HFL).

Theorem 2.10. *Let $K = \{x \in \mathbf{R}^n : Ax \le 0\}$ be a polyhedral cone and K^* be its dual cone:*

$$K^* = \{y \in \mathbf{R}^n : y^T x \ge 0 \, \forall x \in K\}.$$

Then K^* is the conic hull of the (transposes of the) rows of $-A$:

$$K^* = \{y = A^T\lambda : \lambda \le 0\}. \tag{2.14}$$

In particular, K is the cone dual to K^*:

$$K = \{x \in \mathbf{R}^n : x^T y \ge 0 \, \forall y \in K^*\}.$$

Proof. Let the rows of A be $a_1^T, \dots, a_m^T$. By definition, vectors $a \in K^*$ are vectors which have nonnegative inner products with all vectors from K, that is, with all vectors which have nonnegative inner products with the vectors $-a_1, \dots, -a_m$. Equivalently, $a \in K^*$ if and only if the linear inequality $a^T x \ge 0$ is a consequence of the system of linear inequalities $-a_i^T x \ge 0$, $i = 1, \dots, m$. By HFL, this is the case if and only if a is a conic combination of $-a_1, \dots, -a_k$, and we arrive at (2.14). To get the "in particular" part of the statement, note that in view of (2.14), we have

$$\begin{aligned}(K^*)^* &= \{x : x^T A^T \lambda \ge 0 \, \forall \lambda \le 0\} = \{x : \lambda^T [Ax] \ge 0 \, \forall \lambda \le 0\} \\ &\underbrace{=}_{(*)} \{x : Ax \le 0\} = K,\end{aligned}$$

where $(*)$ is given by the following evident observation: *a vector has nonnegative inner products with all vectors which are ≤ 0 if and only if this vector itself is ≤ 0.* $\square$

Note that the relation $(K^*)^* = K$ is in fact true for all *closed* cones, not necessarily polyhedral ones; this is a far-reaching extension of the rule $(L^\perp)^\perp = L$ for linear subspaces L (note that a linear subspace is a very special case of a cone, and the cone dual to a linear subspace L is $L^\perp$ (why?)).

2.3 Faces, vertices, recessive directions, extreme rays

In this section, we prepare tools which will allow us to prove the theorem on the structure of a polyhedral set announced at the beginning of this chapter. We focus on a polyhedral set

$$X = \{x \in \mathbf{R}^n : Ax \leq b\}, A = \begin{bmatrix} a_1^T \\ \cdots \\ a_m^T \end{bmatrix} \in \mathbf{R}^{m \times n}. \tag{2.15}$$

Unless otherwise explicitly stated, we assume in the sequel that X is nonempty, denote by M the affine span of X,

$$M = \mathrm{Aff}(X),$$

and denote by $\mathcal{I}$ the set of indices of the constraints, $\mathcal{I} = \{1, 2, \ldots, m\}$.

2.3.1 *Faces*

A *face* of the nonempty polyhedral set X given by (2.15), by definition, is a *nonempty subset of X composed of all points where the inequalities $a_i^T x \leq b_i$ with indices from certain subset I of $\mathcal{I}$ are satisfied as equalities* ("are active" in the optimization slang). Thus, a face of X is a *nonempty* subset of X which can be represented as

$$X_I = \left\{ x : \begin{array}{l} a_i^T x = b_i, \ i \in I \\ a_i^T x \leq b_i, \ i \in \overline{I} \end{array} \right\}, \tag{2.16}$$

where I is a subset of the set $\mathcal{I} = \{1, \ldots, m\}$ of indices of the linear inequalities defining X, and $\overline{I}$ is the complement of I in $\mathcal{I}$. Note that by (2.16), we have the following:

- *a face X_I of a nonempty polyhedral set is itself a nonempty polyhedral set;*
- *the intersection of two faces X_{I_1}, X_{I_2} of X is the set $X_{I_1 \cup I_2}$ and thus is a face of X, provided that it is nonempty.*

Besides this, we have the following:

- *A face of a face X_I of a polyhedral set X can be represented as a face of X itself.*

  Indeed, X_I can be represented as the polyhedral set

$$X_I = \left\{ x \in \mathbf{R}^n : \begin{array}{c} a_i^T x \le b_i, i \in I \\ -a_i^T x \le -b_i, \, i \in I \\ a_i^T x \le b_i, i \in \overline{I} \end{array} \right\}.$$

  By definition, a face of the latter polyhedral set is obtained from the above description by replacing some of the inequalities with their equality versions in such a way that the resulting system of inequalities and equalities is feasible. When turning one of the inequalities $a_i^T x \le b_i$, $i \in I$, or $-a_i^T x_i \le -b_i$, $i \in I$, into an equality, we do not change X_I, so that we lose nothing by assuming that the inequalities we are turning into equalities to get a face of X_I are from the group $a_i^T x \le b_i$, $i \in \overline{I}$; in the latter case, the result is as if we were replacing with equalities some of the inequalities defining X, so that the result, being nonempty, is indeed a face of X.

Warning: For the time being, our definition of a face of a polyhedral set is *not* geometric: a face is defined in terms of a particular description of X by a system of linear inequalities rather than in terms of X itself. This is where "can be represented as a face" instead of "is a face" in the latter statement comes from. In the mean time, we shall see that faces can be defined solely in terms of X.

Example: Consider the *standard simplex*:

$$\begin{array}{c} \Delta_n = \{x \in \mathbf{R}^n : x_i \ge 0, 1 \le i \le n, \sum_i x_i \le 1\} \\ \left[\begin{array}{c} \text{in the format of (2.15): } m = n+1, a_i = -e_i, \\ b_i = 0 \text{ for } i = 1, \ldots, n \,, a_{n+1} = [1; \ldots; 1], b_{n+1} = 1 \end{array} \right]. \end{array} \quad (2.17)$$

In this case, every subset I of $\mathcal{I} = \{1, \ldots, n+1\}$, *except for $\mathcal{I}$ itself*, defines a face of Δ_n; such a face is composed of all points from Δ_n which have prescribed coordinates equal to zero (indices of these coordinates form the intersection of I with $\{1, \ldots, n\}$) and perhaps the unit sum of entries (the latter is the case when $n+1 \in I$).

According to our definition, the entire X is a face of itself (this face corresponds to $I = \emptyset$). All faces which are distinct from X (and thus form proper — distinct from the entire X and from $\emptyset$ — subsets of X) are called *proper*, e.g., all faces of the standard simplex Δ_n ($n > 0$) corresponding to proper ($I \neq \emptyset$, $I \neq \mathcal{I}$) subset I of $\mathcal{I}$ are proper.

Proposition 2.8. *Let X be a nonempty polyhedral set given by (2.15). Then every proper face X_I of X has dimension strictly less than the one of X.*

Proof. Let X_I be a proper face of X; then definitely $I \neq \emptyset$. Let us look at the linear equations $a_i^T x = b_i$, $i \in I$. If every one of these equations is satisfied by all points from $M = \text{Aff}(X)$, then it is satisfied everywhere on X, whence, comparing (2.16) and (2.15), $X_I = X$, which is impossible, since X_I is a proper face of X. Thus, there exists $i = i_*$ such that the linear equation $a_{i_*}^T x = b_{i_*}$, which is satisfied everywhere on X_I, is violated somewhere on M. Setting $M^+ = \{x \in M : a_{i_*}^T x = b_{i_*}\}$, we get a nonempty set (it contains X_I which is nonempty) and as such is an affine subspace (since it is a nonempty intersection of two affine subspaces M and $\{x : a_{i_*}^T x = b_{i_*}\}$). Since $M^+ \subset M$ by construction and $M^+ \neq M$, we have $\dim M^+ < \dim M$ (see (2.6(b))). It remains to be noted that $X_I \subset M^+$ due to $X_I \subset X \subset M$ and $a_{i_*}^T x = b_{i_*}$ for all $x \in X_I$, whence $\dim X_I \leq \dim M^+ < \dim M$. □

2.3.2 *Vertices, a.k.a. extreme points*

2.3.2.1 *Definition and characterization*

The definition of a vertex is as follows.

Definition 2.4. A vertex (another name — an extreme point) of a nonempty polyhedral set X given by (2.15) is a point $v \in \mathbf{R}^n$ such that the singleton $\{v\}$ is a face of X.

This definition refers to the particular *representation* of the set X (since for the time being, the notion of a face of X is defined in terms of the representation (2.15) of X rather than in terms of X itself). It is easy, however, to express the notion of a vertex in terms of X. The following proposition presents both algebraic and geometric characterizations of vertices.

Proposition 2.9. *Let $X \subset \mathbf{R}^n$ be a nonempty polyhedral set given by* (2.15).

(i) [Algebraic characterization of vertices]. *A singleton $\{v\}$ is a face of X if and only if v satisfies all inequalities $a_i^T x \le b_i$ defining X and among those inequalities which are active (i.e., satisfied as equalities) at v, there are n with linearly independent a_i.*

(ii) [Geometric characterization of vertices]. *A point v is a vertex of X iff $v \in X$, and from $v \pm h \in X$, it follows that $h = 0$ (geometrically, v belongs to X and is not the midpoint of a nontrivial — not reducing to a point — segment belonging to X).*

Proof. (i) Let v be a vertex of X. By definition, this means that for a properly chosen $I \subset \mathcal{I}$, v is the unique solution of the system of equality and inequality constraints in variables x:

$$a_i^T x = b_i,\ i \in I,\ \ a_i^T x \le b_i,\ i \in \overline{I}. \tag{$*$}$$

Without loss of generality, we may assume that all inequalities $a_i^T x \le b_i$, $i \in \overline{I}$, are satisfied at v strictly (indeed, otherwise we could move all indices $i \in \overline{I}$ with $a_i^T v = b_i$ into I; this transformation of the partition $\mathcal{I} = I \cup \overline{I}$ keeps v a solution of the transformed system ($*$) and can only decrease the number of solutions to the system; thus, v is the unique solution to the transformed system). Given that all the inequality constraints in ($*$) are satisfied at v strictly, all we should verify is that among the vectors a_i, $i \in I$, there are n linearly independent. Assuming that it is not the case, there exists a nonzero vector h which is orthogonal to all vectors a_i, $i \in I$. Setting $v_t = v + th$, observe that for all small enough in modulus values of t v_t solves ($*$). Indeed, the constraints in ($*$) are satisfied when $t = 0$, whence the *equality* constraints $a_i^T v_t = b_i$, $i \in I$, are satisfied for all t due to $a_i^T h = 0$, $i \in I$. The inequality constraints $a_i^T v_t \le b_i$, $i \in \overline{I}$, are satisfied *strictly* when $t = 0$, and therefore, every one of them remains valid in an appropriate neighborhood of $t = 0$; since the number of constraints is finite, there exists a single neighborhood which "serves" in this sense all the constraints with $i \in \overline{I}$. The bottom line is that indeed v_t solves ($*$) for all t's close enough to 0; since $h \neq 0$, we see that the solution to $(*)$ is not unique, thus arriving at the desired contradiction.

We have proved that *if* v is a vertex of X, *then* the characterization from (i) holds true. To prove the inverse, assume that $v \in X$ and that among the inequalities $a_i^T x \leq b_i$ active at v, there are n with linearly independent a_i. Denoting by I the set of indices of the inequalities active at v, we have $v \in X_I$, and therefore, X_I is nonempty and is thus a face of X. It remains to be noted that v is the only point in X_I, since every point x in this set must satisfy the system of equations

$$a_i^T x = b_i, \; i \in I,$$

and the matrix of this system with n variables is of rank n, so that its solution (which does exist — the system is solved by v!) is unique. (i) is proved.

(ii) Proof of (ii) more or less repeats the above reasoning. In one direction, let v be a vertex of X. By (i), among the inequalities $a_i^T v \leq b_i$, $i \in \mathcal{I}$, n inequalities with linearly independent a_i's are equalities, let their indices form a set I. If now h is such that $v \pm h \in X$, then one has $a_i^T [v \pm h] \leq b_i$, $i \in I$, which combines with $a_i^T v = b_i$, $i \in I$, to imply that $a_i^T h = 0$ for all $i \in I$. Since among the vectors a_i, $i \in I$, there are n linearly independent (and thus spanning the entire $\mathbf{R}^n$), the only vector h orthogonal to all a_i, $i \in I$, is $h = 0$. Thus, *if* v is a vertex, *then* $v \in X$ and $v \pm h \in X$ implies that $h = 0$. To prove the inverse, let $v \in X$ be such that $v \pm h \in X$ implies that $h = 0$, and let us prove that v is a vertex. Indeed, let I be the set of indices of all inequalities $a_i^T x \leq b_i$, which are active at v. Assuming that there are no n linearly independent among the vectors a_i, $i \in I$, we can find a nonzero f which is orthogonal to all a_i, $i \in I$, so that all points $v_t = v + tf$ satisfy the inequalities $a_i^T v_t \leq b_i$, $i \in I$. Since the inequalities $a_i^T x \leq b_i$, $i \notin I$, are satisfied at v strictly (due to how we defined I), the points v_t, same as in the proof of (i), satisfy these inequalities for all t small enough in absolute value. The bottom line is that under our assumption that the set of a_i's with $i \in I$ does not contain n linearly independent vectors, there exist small positive t and nonzero vector f such that $v \pm tf$ satisfies all inequalities in (2.15) and thus belongs to X; thus, with $h = tf \neq 0$, we have $v \pm h \in X$, which is impossible. We conclude that among a_i's with $i \in I$, there are n linearly independent; invoking (i), v is a vertex. □

Quiz: What are the vertices of the polyhedral set shown in Fig. 2.1(a)?

Answer: The vertices of the triangle shown in Fig. 2.1(b); the triangle itself is the convex hull of these vertices.

Corollary 2.3. *The set of extreme points of a nonempty polyhedral set is finite.*

Indeed, by Proposition 2.9, an extreme point should solve a subsystem of the $m \times n$ system of linear equations $Ax = b$ composed of n linearly independent equations; the solution to such a subsystem is unique, and the number of subsystems is finite. □

Task 2.2. Prove the following claim:

> Let v be a vertex of a polyhedral set X and $v = \sum_{i=1}^{k} \lambda_i x_i$ be a representation of v as a convex combination of points $x_i \in X$, where all the coefficients are strictly positive. Then $x_i = v$ for all i. As a corollary, whenever $\mathrm{Conv}(Y) \subset X$ and $v \in \mathrm{Conv}(Y)$, we have $v \in Y$.

Task 2.3. Prove that whenever $X_1 \subset X_2$ are polyhedral sets, every extreme point of X_2 which belongs to X_1 is an extreme point of X_1.

We add to the above results the following simple and important proposition:

Proposition 2.10. *A vertex of a face of a nonempty polyhedral set X is a vertex of X itself.*

Proof. We have seen that a face of a face X_I of a polyhedral set X can be represented as a face of X itself. It follows that a vertex of a face X_I of X (which, by definition, is a singleton face of X_I) can be represented as a singleton face of X and is thus a vertex of X. □

Discussion. The crucial role played by vertices of polyhedral sets stems from the following facts (which we shall prove in the mean time):

A. *Every bounded nonempty polyhedral set X is the convex full of the (finite, by Corollary 2.3) set of its extreme points; moreover, whenever X is represented as $X = \mathrm{Conv}(Y)$, Y contains the set of extreme points of X.*

We shall also see that extreme points form a basic "building block" in the description of an arbitrary nonempty polyhedral set not containing lines, not necessarily a bounded one.

Note that the notion of an extreme point can be extended to the case of convex sets: an extreme point of a convex set Q is a point v of Q which cannot be represented as a midpoint of a nontrivial segment belonging to Q: $v \in Q$ and $v \pm h \in Q$ implies that $h = 0$. The "convex analogy" of **A** is the (finite-dimensional version of the) *Krein–Milman theorem: Every nonempty, bounded and closed convex set Q in $\mathbf{R}^n$ is the convex hull of the set of its extreme points* (which now can be infinite)*; moreover, whenever $Q = \mathrm{Conv}(Y)$, Y contains all extreme points of Q.*

B. *If the feasible set of a solvable LO program* (which is always a polyhedral set) *does not contain lines, then among the optimal solutions there are those which are vertices of the feasible set.*

C. *Vertices of polyhedral set, their algebraic characterization as given by Proposition* 2.9*, and Corollary* 2.3 *are instrumental for simplex-type algorithms of LO.*

From the outlined (and many other similar) facts it should be clear that understanding the structure of extreme points of a polyhedral set contributes a lot to understanding of this set and to our abilities to work with it. As an instructive exercise, let us describe the extreme points of two important families of polyhedral sets.

2.3.2.2 *Example: Extreme points of the intersection of $\|\cdot\|_\infty$- and $\|\cdot\|_1$ balls*

Consider the polyhedral set

$$X = \left\{ x \in \mathbf{R}^n : -1 \leq x_i \leq 1,\, 1 \leq i \leq n, \sum_i |x_i| \leq k \right\}, \tag{2.18}$$

where k is a nonnegative integer $\leq n$.

Note: (2.18) is *not* a representation of X by a finite set of linear inequalities, since the inequality $\sum_i |x_i| \leq k$ is not linear. To get a polyhedral description of X, we should replace this nonlinear inequality with the system of 2^n linear inequalities $\sum_i \epsilon_i x_i \leq k$ corresponding to all collections of ± 1 coefficients ϵ_i. We, however, have seen that a vertex of a polyhedral set is a geometric notion — a notion which can be expressed solely in terms of the set X without referring to its polyhedral description.

Note also that geometrically, X is the intersection of the unit box $B_\infty(1) = \{x \in \mathbf{R}^n : -1 \le x_i \le 1\,\forall i\}$ and the *ℓ_1-ball of radius k* $B_1(k) = \{x \in \mathbf{R}^n : \sum_{i=1}^n |x_i| \le k\}$.

We claim that *the extreme points of X are nothing but the vectors from $\mathbf{R}^n$ with exactly k nonzero coordinates, equal each either to* 1 *or to* -1. Indeed, we have the following:

(a) Let v be a vector of the outlined type, and let us prove that it is an extreme point of X. By Proposition 2.9(ii), we should prove that if h is such that $v \pm h \in X$, then $h = 0$. Indeed, let $v \pm h \in X$ and let I be the set of indices of the k nonzero coordinates of v. When $i \in I$, then either $v_i = 1$ – and then $v \pm h \in X$ implies that $1 \ge v_i \pm h_i = 1 \pm h_i$, that is, $h_i = 0$, or $v_i = -1$, and then $v \pm h \in X$ implies that $-1 \le v_i \pm h_i = -1 \pm h_i$, and we again arrive at $h_i = 0$. With this in mind, $v + h \in X$ implies that $k + \sum_{i \notin I} |h_i| = \sum_{i \in I} |v_i| + \sum_{i \notin I} |h_i| \le k$, whence $h_i = 0$ when $i \notin I$, meaning that $h = 0$, as required.

(b) Let v be an extreme point of X, and let us prove that v has exactly k nonzero coordinates, equal to ± 1. Indeed, let J be the set of indices i, for which v_i is neither 1 nor -1. Observe that J cannot have less than $n - k$ indices, since otherwise v has $> k$ coordinates of magnitude 1 and thus does not belong to X (due to the constraint $\sum_i |x_i| \le k$ participating in the definition of X). When J contains exactly $n-k$ indices, the same constraint says that $v_i = 0$ when $i \in J$, and then v has exactly k nonzero entries, equal to ± 1, which is what we claimed. It remains to be verified that J cannot contain more than $n - k$ indices. Assuming that this is the case, Card $J > n - k$, note that then v has at least one entry which is neither -1 nor 1 (since $n - k \ge 0$); without loss of generality, we can assume that this entry is v_1. Since $v \in X$, this entry should belong to $(-1, 1)$. Now, it may happen that $\sum_i |v_i| < k$. In this case, the vectors $[v_1 - \delta; v_2; \ldots; v_n]$ and $[v_1 + \delta; v_2; \ldots; v_n]$ with small enough positive δ belong to X, and thus, v is the midpoint of a nontrivial segment belonging to X, which is impossible. Thus, $\sum_i |v_i| = k$. Since the number of entries of magnitude 1 in v is $\ell = n - \mathrm{Card}(J) < k$, we have $\sum_{i \in J} |v_i| = k - \sum_{i \notin J} |v_i| = k - \ell \ge 1$, and since $|v_i| < 1$ for $i \in J$, the relation $\sum_{i \in J} |v_i| \ge 1$ implies that *at least two of the entries v_i, $i \in J$, are nonzero* (and are of magnitude < 1, as all entries v_i of v with indices from J). Without loss of generality, we can assume that the two entries in question are v_1 and v_2: $v_1 \neq 0$, $v_2 \neq 0$, $|v_1| < 1$, $|v_2| < 1$. Then the vectors $[v_1 - \text{sign}(v_1)\delta; v_2 + \text{sign}(v_2)\delta; v_3; \ldots; v_n]$ and $[v_1 + \text{sign}(v_1)\delta; v_2 - \text{sign}(v_2)\delta; v_3; \ldots; v_n]$ for all small positive δ belong to X, and v again is a midpoint of a nontrivial segment belonging to X, which is impossible. □

Particular cases: When $k = n$, X is nothing but the unit box $\{x \in \mathbf{R}^n : -1 \leq x_i \leq 1, 1 \leq i \leq n\}$, and the above result reads as follows: *the vertices of the unit box are exactly the* ± 1 *vectors.* Similar statement for an arbitrary box $\{x \in \mathbf{R}^n : p_i \leq x_i \leq q_i \forall i \leq n\}$ which is nonempty ($p \leq q$) reads as follows: *the vertices of the box are all vectors* v *with "extreme" coordinates (i.e.,* $v_i = p_i$ *or* $v_i = q_i$ *for every* i*)* (check it!).

When $k = 1$, X becomes the unit ℓ_1-ball $\{x \in \mathbf{R}^n : \sum_i |x_i| \leq 1\}$, and our result reads as follows: *the vertices of the unit* ℓ_1*-ball in* $\mathbf{R}^n$ *are exactly the* $2n$ *vectors* $\pm e_i$, $i = 1, \ldots, n$ (as always, e_i are the basic orths).

Modifications: Slightly modifying the above reasoning, one arrives at the useful facts stated in the following.

Task 2.4. Verify that whenever k is an integer such that $1 \leq k \leq n$, then the following apply:

(i) The extreme points of the polyhedral set $\{x \in \mathbf{R}^n : 0 \leq x_i \leq 1, 1 \leq i \leq n, \sum_i x_i \leq k\}$ are exactly the Boolean vectors (i.e., vectors with coordinates 0 and 1) with at most k coordinates equal to 1.

(ii) The extreme points of the polyhedral set $\{x \in \mathbf{R}^n : 0 \leq x_i \leq 1, 1 \leq i \leq n, \sum_i x_i = k\}$ are exactly the Boolean vectors (i.e., vectors with coordinates 0 and 1) with exactly k coordinates equal to 1.

Note that the second set is a face of the first one.

2.3.2.3 *Example: Extreme points of the set of double stochastic matrices*

An $n \times n$ matrix is called *doubly stochastic* if it has nonnegative entries which sum up to 1 in every row and every column. The set of these matrices is

$$\Pi_n = \left\{ x = [x_{ij}] \in \mathbf{R}^{n\times n} = \mathbf{R}^{n^2} : x_{ij} \geq 0 \quad \forall i, j, \sum_i x_{ij} = 1, \right.$$
$$\left. \sum_j x_{ij} = 1, \ \forall i, j \right\};$$

we see that this is a polyhedral set in $\mathbf{R}^{n^2} = \mathbf{R}^{n\times n}$ given by n^2 inequality constraints and $2n$ equality constraints. In fact, we can reduce the number of equality constraints by 1 by dropping one of them, say, $\sum_i x_{i1} = 1$; indeed, if the sum of x_{ij} in every row is 1, then the total sum of entries is n; and if, in addition, the column sums in all columns except for the first one are equal to 1, then the sum of entries in the first column automatically equals to $n - (n-1) = 1$.

The following fact has an extremely wide variety of applications.

Theorem 2.11 (Birkhoff). *The extreme points of the set of double stochastic $n \times n$ matrices are exactly the $n \times n$ permutation matrices (exactly one nonzero entry, equal to 1, in every row and every column).*

Proof. The fact that permutation matrices are extreme points of the set Π_n of doubly stochastic matrices is easy: By the above result, these matrices are extreme points of the box $\{[x_{ij}] \in \mathbf{R}^{n\times n} : 0 \leq x_{ij} \leq 1 \, \forall i, j\}$, which contains Π_n, and the statement of Task 2.3 remains to be used.

To prove that every vertex v of Π_n is a permutation matrix, let us use induction in n. The base $n = 1$ is evident: Π_1 is a singleton $\{1\}$, and clearly, 1 is the only extreme point of this set (see geometric characterization of extreme points, Proposition 2.9(ii)), and this indeed is an 1×1 permutation matrix. Inductive step $n-1 \Rightarrow n$ is as follows. Let v be an extreme point of Π_n; as it was already explained, we can think that the latter set is given by n^2 inequalities $x_{ij} \geq 0$ and $2n - 1$ linear equality constraints. Since Π_2 "lives" in $\mathbf{R}^{n^2}$, algebraic characterization of vertices (Proposition 2.9(i)) says that n^2 linearly independent constraints from the description of Π_n should become active at v. $2n-1$ linear equality constraints participating in the description of Π_n contribute $2n - 1$ linearly independent active constraints, and the remaining $n^2 - 2n + 1$ active at v constraints should come from the n^2 constraints $x_{ij} \geq 0$. In other words, *at least* $n^2 - 2n + 1 = (n-1)^2$ *entries in* v *are zeros.* It follows that *there is a column in* v *with at most one nonzero entry*, since otherwise the number of zero entries in every column was $\leq n-2$, and the total number of zero entries would be $\leq n(n-2) < (n-1)^2$. Now, let j_1 be the index of the column where v has at most one nonzero entry. Since the corresponding column sum is 1, this "at most one nonzero entry" means "exactly one nonzero entry equal to 1." Since the row sums in v are equal to 1 and the entries are nonnegative, this nonzero entry, equal to 1, is the only nonzero entry in its row, let this row be i_1. Eliminating from the doubly stochastic $n \times n$ matrix v the column j_1 and the row i_1, we get a $(n-1) \times (n-1)$ matrix $\widehat{v}$ which is clearly doubly stochastic; since v is a vertex in Π_n, it is completely straightforward to verify (do it!) that $\widehat{v}$ is a vertex in Π_{n-1}. By the inductive hypothesis, $\widehat{v}$ is an $(n-1) \times (n-1)$ permutation matrix; recalling the relation between v and $\widehat{v}$, v is a permutation matrix itself. □

2.3.3 *Recessive directions*

2.3.3.1 *Definition and characterization*

Definition 2.5. Let X be a nonempty polyhedral set. A vector $e \in \mathbf{R}^n$ is called a recessive direction of X if there exists $\bar{x} \in X$ such that the entire ray $\{\bar{x} + te : t \geq 0\}$ emanating from $\bar{x}$ and directed by e belongs to X. The set of all recessive directions of X is called the recessive cone of X, denoted $\mathrm{Rec}(X)$.[7]

Note that this definition is geometric: It does not refer to a representation of X in the form of (2.15).

Examples:

- $e = 0$ is a recessive direction of every nonempty polyhedral set.
- Recessive directions of an affine subspace M are exactly the vectors from the parallel linear subspace.

[7]We shall see in a while that the terminology is consistent — the set of all recessive directions of a polyhedral set X is indeed a cone.

Algebraic characterization of recessive directions: Given a polyhedral description (2.15) of a nonempty polyhedral set X, it is easy to characterize its recessive directions.

Proposition 2.11 (Algebraic characterization of recessive directions). *Let X be a nonempty polyhedral set given by $X = \{x : Ax \leq b\}$, see (2.15). Then we have the following:*

(i) *The recessive directions of X are exactly the vectors from the polyhedral cone*

$$\mathrm{Rec}(X) := \{e : Ae \leq 0\}. \tag{2.19}$$

(ii) *One has $X = X + \mathrm{Rec}(X)$, that is, whenever $x \in X$ and $e \in \mathrm{Rec}(X)$, the ray $\{x + te : t \geq 0\}$ is contained in X.*[8]

Proof. If e is a recessive direction, then there exists $\bar{x} \in X$ such that $a_i^T\bar{x} + ta_i^Te = a_i^T[\bar{x} + te] \leq b_i$ for all $t > 0$ and all $i \in \mathcal{I}$, which clearly implies that $a_i^Te \leq 0$ for all i, that is, $Ae \leq 0$. Vice versa, if e is such that $Ae \leq 0$, then for every $x \in X$ and every $t \geq 0$, we have $A(x + te) = Ax + tAe \leq Ax \leq b$, that is, the ray $\{x + te : t \geq 0\}$ belongs to X. □

Remark. The definition of a recessive direction can be straightforwardly extended from polyhedral to convex sets. The "convex analogy" of Proposition 2.11 states that the set of recessive directions of a nonempty *closed* convex set Q is a closed cone, and a ray, starting at a point of Q and directed by a recessive direction, belongs to Q.

Quiz: What is the recessive cone of the polyhedral set X shown in Fig. 2.1(a)? What is the recessive cone of the triangle shown in Fig. 2.1(b)?

Answer: The recessive cone of X is the angle between the two rays shown in Fig. 2.1(c). The recessive cone of the triangle is trivial — it is the origin.

---

[8] Note the difference between the information provided in (ii) and the definition of a recessive direction e: By the latter, e is a recessive direction if *some* ray directed by e belongs to X; (ii) says that whenever e possesses this property, *every* ray directed by e and emanating from a point of X is contained in X.

2.3.3.2 *Recessive subspace and decomposition*

Recall that a *line* ℓ in $\mathbf{R}^n$ is an affine subspace of $\mathbf{R}^n$ of dimension 1; this is exactly the same as to say that $\ell = \{x + te : t \in \mathbf{R}\}$ with some x and a *nonzero* vector e. It is easy to understand when a polyhedral set X contains lines and what these lines are.

Proposition 2.12. *Let X be a nonempty polyhedral set given by $X = \{x : Ax \leq b\}$, see (2.15). Then X contains a set of the form $\{x + te : t \in \mathbf{R}\}$ if and only if $x \in X$ and $Ae = 0$. In particular, we have the following:*

- *X contains a line, directed by vector $e \neq 0$, iff e belongs to the kernel $\operatorname{Ker} A := \{e : Ae = 0\}$ of A and in this case every line directed by e and passing through a point of X belongs to X.*
- *X contains lines iff $\operatorname{Ker} A \neq \{0\}$ or, equivalently, iff $\operatorname{Rec}(X) \cap [-\operatorname{Rec}(X)] \neq \{0\}$.*

Proof. The set $\{x + te : t \in \mathbf{R}\}$ belongs to X iff both the rays $\{x + te : t \geq 0\}$ and $\{x - te : t \geq 0\}$ emanating from x and directed by e and $-e$ belong to X; by Proposition 2.11, this is the case iff $x \in X$ and both e and $-e$ belong to $\operatorname{Rec}(X)$, that is, $Ae \leq 0$ and $-Ae \leq 0$ or, which is the same, $Ae = 0$. □

In many aspects, polyhedral sets containing lines are less suited for analysis than those not containing lines, e.g., a polyhedral set containing lines definitely does not have extreme points (why?), whereas, as we shall see in a while, a polyhedral set not containing lines does have extreme points. Fortunately, it is easy to "get rid" of the lines.

Proposition 2.13 (Decomposition). *A nonempty polyhedral set $X = \{x : Ax \leq b\}$, see (2.15), can be represented as the arithmetic sum of a linear subspace $L = \operatorname{Ker} A$ and a nonempty polyhedral set $\bar{X}$ which does not contain lines:*

$$X = \bar{X} + \operatorname{Ker} A. \tag{2.20}$$

One can take as $\bar{X}$ the intersection $X \cap L^{\perp}$.

Proof. Let us set $\bar{X} = X \cap L^{\perp}$; we claim that this is a nonempty polyhedral set satisfying (2.20). Indeed, X is nonempty by assumption; with $x \in X$ and every $e \in \operatorname{Ker} A$, the vector $x - e$ belongs to X (Proposition 2.12). Specifying e as the orthogonal projection of x

onto L (so that $x - e$ is the orthogonal projection $\bar{x}$ of x onto $L^\perp$), we conclude that $x - e$ belongs to $\bar{X}$, and thus, $\bar{X}$ is nonempty and, moreover, is such that $x = \bar{x} + e \in \bar{X} + \mathrm{Ker}\, A$, whence $\bar{X} + \mathrm{Ker}\, A \supset X$. Since $\bar{X}$ is nonempty, it is a polyhedral set; indeed, to get a polyhedral description of $\bar{X}$, one should add to the inequalities $Ax \leq b$ specifying X a system of homogeneous linear inequalities specifying $L^\perp$, say, the system $a_{m+i}^T x = 0$, $1 \leq i \leq \dim \mathrm{Ker}\, A$, where vectors a_{m+i}, $1 \leq i \leq \dim \mathrm{Ker}\, A$, span the linear subspace $\mathrm{Ker}\, A$. By construction, $\bar{X}$ is a part of X, whence $\bar{X} + \mathrm{Ker}\, A \subset X$ by Proposition 2.12. We have already seen that $\bar{X} + \mathrm{Ker}\, A \supset X$ as well, whence $\bar{X} + \mathrm{Ker}\, A = X$. Finally, $\bar{X}$ does not contain lines, since the direction e of a line in $\bar{X}$, by Proposition 2.12, should be a nonzero vector satisfying the equations $a_i^T e = 0$, $i \in \mathcal{I}$, and $a_{m+i}^T e = 0$, $1 \leq i \leq \dim \mathrm{Ker}\, A$; the first group of these equations says that $e \in \mathrm{Ker}\, A$, and the second that $e \in (\mathrm{Ker}\, A)^\perp$, meaning that $e \in \mathrm{Ker}\, A \cap (\mathrm{Ker}\, A)^\perp = \{0\}$, which is impossible. $\square$

Proposition 2.13 is the first step toward our goal — describing the structure of a polyhedral set; it says that investigating this structure can be reduced to the case when the set does not contain lines. Before passing to the remaining steps, we need one more element of "equipment" — the notions of a *base* and *extreme ray* of a polyhedral cone.

2.3.4 *Bases and extreme rays of a polyhedral cone*

Consider a polyhedral cone:

$$R = \{x \in \mathbf{R}^n : Bx \leq 0\} \quad B = \begin{bmatrix} b_1^T \\ \cdots \\ b_k^T \end{bmatrix} \in \mathbf{R}^{k \times n}. \tag{2.21}$$

Pointed cones: Cone (2.21) is called *pointed* if and only if it does not contain lines; invoking Proposition 2.12, this is the case if and only if $\mathrm{Ker}\, B = \{0\}$, or, which is the same, there is no nonzero vector e such that $Be \leq 0$ and $B[-e] \leq 0$. Geometrically, the latter means that the only vector e such that $e \in K$ and $-e \in K$ is the zero vector, e.g., the nonnegative orthant $\mathbf{R}^n_+ := \{x \in \mathbf{R}^n : x \geq 0\} = \{x : [-I]x \leq 0\}$ is pointed.

Base of a cone: Assume that the cone K given by (2.21) is nontrivial — does not reduce to the singleton $\{0\}$.

Consider a hyperplane Π in $\mathbf{R}^n$ which does not pass through the origin. Without loss of generality, we can represent such a hyperplane by linear equation with unit right-hand side:

$$\Pi = \{x \in \mathbf{R}^n : f^T x = 1\} \quad [f \neq 0]. \tag{2.22}$$

It may happen that this hyperplane intersects all nontrivial rays spanned by vectors from K, that is,

$$x \in K, x \neq 0 \Rightarrow \exists s \geq 0 : sx \in \Pi.$$

This is exactly the same as to say that f has positive inner products with all nonzero vectors from K (why?). When f possesses this property, the polyhedral set $Y = K \cap \Pi = \{x : Bx \leq 0, f^T x = 1\}$ is nonempty and "remembers" K:

$$K = \{x = ty : t \geq 0, y \in Y\}.$$

In this situation, we shall call Y a *base* of K; thus, a base of $K = \{x : Bx \leq 0\}$ is a nonempty set of the form

$$Y = \{x : Bx \leq 0, f^T x = 1\},$$

where f has positive inner products with all nonzero vectors from K. For example, the bases of nonnegative orthant are "produced" by strictly positive vectors f and only by those vectors (why?)

Now, not every polyhedral cone admits a base; say, the trivial cone $\{0\}$ does not admit it. Another "bad" in this respect cone is a cone which is *not* pointed. Indeed, if K is not pointed, then, as we have already seen, there is a nonzero e such that both e and $-e$ belong to K; but then there cannot exist a vector f forming positive inner products with all nonzero vectors from K. A useful (and even crucial, as we shall see in the sequel) fact is that the outlined two obstacles — triviality and nonpointedness — are the only obstacles to existing a base.

Proposition 2.14. *Let a polyhedral cone $K = \{x : Bx \leq 0\}$ in $\mathbf{R}^n$ be nontrivial ($K \neq \{0\}$) and pointed. Then there exists $f \in \mathbf{R}^n$ such that $f^T x > 0$ for all $x \in K \backslash \{0\}$, that is, K admits a base.*

Proof. Let $K^* = \{y \in \mathbf{R}^n : y^T x \geq 0 \, \forall x \in K\}$ be the cone dual to K. We claim that K^* has a nonempty interior: There exists a vector f such that an Euclidean ball of certain positive radius r, centered at f, is contained in K^*. Note that this claim implies the result we seek to prove: Indeed, if f is as above and $x \in K\backslash\{0\}$, then $(f-h)^T x \geq 0$ whenever $\|h\|_2 \leq r$, whence $f^T x \geq \sup_{h:\|h\|_2 \leq r} h^T x > 0$.

It remains to support our claim. Assume, on the contrary to what should be proved, that the interior of K^* is empty. Since K^* is a convex set, we conclude that $\text{Aff}(K^*)$ is less than the entire $\mathbf{R}^n$ (Theorem 2.3). Since $0 \in K^*$, $\text{Aff}(K^*) \ni 0$, and thus, $L = \text{Aff}(K^*)$ is a linear subspace in $\mathbf{R}^n$ of the dimension $< n$. Thus, there exists $e \neq 0$ such that $e \in L^\perp$, whence $e^T y = 0$ for all $y \in K^*$. But then both e and $-e$ belong to $(K^*)^*$, and *the latter cone is nothing but* K (Theorem 2.10). Thus, K is not pointed, which is the desired contradiction. □

The importance of the notion of a base of a cone K is that on the one hand, it "remembers" K and thus bears full information on the cone, and on the other hand, it is in certain respects simpler than K and is thus easier to investigate, e.g., we shall see in the mean time that a base of a cone is a *bounded* nonempty polyhedral set, and those sets admit simple description — they are convex hulls of the (nonempty and finite) sets of their vertices. For the time being, we shall prove a simpler statement.

Proposition 2.15. *The recessive cone of a base Y of a polyhedral cone K is trivial:* $\text{Rec}(Y) = \{0\}$.

Proof. Let $Y = \{x : Bx \leq 0, f^T x = 1\}$ be a base of a polyhedral cone $K = \{x : Bx \leq 0\}$, so that $K \neq \{0\}$ and f has positive inner products with all nonzero vectors from K. By Proposition 2.11, $\text{Rec}(Y) = \{e : B^T e = 0, f^T e = 0\}$; assuming that the latter set contains a nonzero vector e, we see that e is a nonzero vector from K, and this vector has zero inner product with f, which is impossible. □

Remark. The notion of a base makes sense, and the results stated in Propositions 2.14 and 2.15 hold true for arbitrary *closed* cones and not only polyhedral ones.

Extreme rays of a polyhedral cone. Extreme rays of polyhedral cones are "cone analogies" of extreme points of polyhedral sets.

Definition 2.6. Let K be a cone. A ray in K is the set of all vectors $\{tx : t \geq 0\}$, where $x \in K$; x is called a generator, or direction, of the ray. A ray is called nontrivial, if it contains nonzero vectors.

From the definition, it follows that generators of a nontrivial ray R are all nonzero points on this ray, so that all generators of a nontrivial ray R are positive multiples of each other.

Definition 2.7 (Extreme ray). An extreme ray of a cone K is a nontrivial ray $R = \{tx : t \geq 0\}$ in K with the following property: Whenever a generator x of the ray is represented as $x = u + v$ with $u, v \in K$, both vectors u and v belong to R.

Task 2.5. Let K be a pointed cone and $v_1, \ldots, v_k$ be points of K. Prove the following:

(1) If $v_1 + \cdots + v_k = 0$, then $v_1 = \cdots = v_k = 0$.
(2) If $v_1 + \cdots + v_k = e$ is a generator of an extreme ray of K, then v_i, $1 \leq i \leq k$, are nonnegative multiples of e.

Note that the notion of extreme ray is geometric — it does not refer to a particular description of the cone in question. Note also that if the property characteristic for a generator of an extreme ray holds true for one of the generators of a given nontrivial ray, it automatically holds true for all other generators of the ray.

Example: A ray in the nonnegative orthant $\mathbf{R}^n_+$ is the set composed of all nonnegative multiples of a nonnegative vector. Nontrivial rays are composed of all nonnegative multiples of *nonzero* nonnegative vectors, e.g., the set $\{t[1;1;0] : t \geq 0\}$ is a nontrivial ray in $\mathbf{R}^3$ (draw it!). On the closest inspection, *the extreme rays of* $\mathbf{R}^n_+$ *are the nonnegative parts of the n coordinate axes, that is, the rays generated by basic orths* $e_1, \ldots, e_n$. To see that, say, the ray $R_1 = \{te_1 : t \geq 0\}$ is extreme, we should take a whatever generator of the ray, e.g., e_1, and check that when $e_1 = u + v$ with nonnegative vectors u, v, then both u and v are nonnegative multiples of e_1, which is evident (since from $u_i + v_i = (e_1)_i$ and $u_i, v_i \geq 0$, it follows that $u_i = v_i = 0$ when $i = 2, \ldots, n$). To see that the nonnegative parts of the coordinate axes are the only extreme rays of $\mathbf{R}^n_+$, assume that x is a generator of an extreme ray R. Then x is a nonzero nonnegative vector. If x has at least two positive entries, say, $x_1 > 0$ and $x_2 > 0$, then we have $x = [x_1; 0; \ldots; 0] + [0; x_2; x_3; \ldots; x_n]$, thus getting a decomposition of x into the sum of two nonnegative vectors which clearly are not nonnegative multiples of x. Thus, x has exactly one positive entry and is thus a positive multiple of certain basic orth, meaning that R is the nonnegative part of the corresponding coordinate axis.

The role of extreme rays in the geometry of polyhedral cones is similar to the role of vertices in the geometry of polyhedral sets, e.g., we have already mentioned the fact we intend to prove in the sequel: *a nonempty and bounded polyhedral set is the convex hull of the (finite) set of its vertices.* The "cone analogy" of this statement reads as follows: *A pointed and nontrivial ($K \neq \{0\}$) polyhedral cone K possesses extreme rays, their number is finite, and K is the conic hull of the (generators of the) extreme rays.* Here, we present an important result on algebraic characterization of extreme rays and their relation to extreme points.

Proposition 2.16. *Let $K \subset \mathbf{R}^n$ be a pointed nontrivial ($K \neq \{0\}$) polyhedral cone given by* (2.21). *Then we have the following:*

(i) *A nonzero vector $e \in K$ generates an extreme ray of K if and only if among the inequalities $b_i^T x \leq 0$, $i = 1, \ldots, k$, defining K there are at least $n-1$ linearly independent inequalities which are active (i.e., are equalities) at e.*

(ii) *Let $Y = \{x : Bx \leq 0, f^T x = 1\}$ be a base of K* (note that K admits a base by Proposition 2.14) *so that every nontrivial ray in K intersects Y. A nontrivial ray R in K is extreme if and only if its intersection with Y is an extreme point of Y.*
Item (ii) *is illustrated in* Fig. 2.3.

Proof. (i) Let e generate a nontrivial ray R in K, and let I be the set of indices of vectors b_i which are orthogonal to e. We should prove that R is extreme iff among the vectors b_i, $i \in I$, there are $n-1$ linearly independent.

In one direction, assume that the set $\{b_i : i \in I\}$ contains $n-1$ linearly independent vectors, say, $b_1, \ldots, b_{n-1}$, and let us prove that R is an extreme ray. Indeed, let $e = u+v$ with $u, v \in K$; we should prove that then $u, v \in R$. Since $e = u+v$, we have $b_i^T(u+v) = b_i^T e = 0$, $1 \leq i \leq n-1$, on the one hand, and $b_i^T u \leq 0$, $b_i^T v \leq 0$, on the

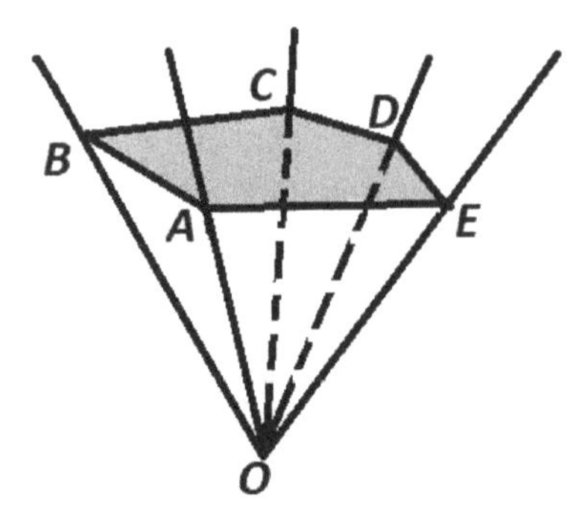

Fig. 2.3. Polyhedral cone, its base (pentagon $ABCDE$) and extreme rays of the cone.

other hand, due to $u, v \in K$. We conclude that $b_i^T u = b_i^T v = 0$, $1 \leq i \leq n-1$, so that every one of the vectors u, v, e is a solution to the system of homogeneous linear equations

$$b_i^T x = 0,\ 1 \leq i \leq n-1$$

in variables x. Since $b_1, \ldots, b_{n-1}$ are linearly independent, the solution set L of this system is a one-dimensional linear subspace, and since e is a nonzero vector from this linear subspace, all other vectors from L are multiples of e. In particular, $u = \lambda e$ and $v = \mu e$ with some real λ, μ. Now, recall that K is pointed and $u, v, e \in K$, meaning that $\lambda \geq 0$, $\mu \geq 0$ (e.g., assuming $\lambda < 0$, we get $0 \neq u \in K$ and $-u = |\lambda| e \in K$, which is impossible). Thus, u and v are nonnegative multiples of e and thus belong to R, as claimed.

In the other direction, assume that R is an extreme ray, and let us prove that the set $\{b_i : i \in I\}$ contains $n-1$ linearly independent vectors. Assume that the latter is not the case. Then the dimension of the linear span of the set is $\leq n-2$, meaning that the dimension of the orthogonal complement to this linear span is ≥ 2. Thus, the linear space of all vectors orthogonal to all b_i, $i \in I$, is of dimension ≥ 2. Therefore, this space (which contains e due to the origin of I) also contains a vector h which is not proportional to e. Now, note that $b_i^T(e + th) = 0$ for all $i \in I$ and $b_i^T(e + th) < 0$ when $i \notin I$ and t is small enough in absolute value (since for i in question, $b_i^T e < 0$). It follows that there exists a small positive t such that $b_i^T(e \pm th) \leq 0$ for all $i = 1, \ldots, k$, meaning that the vectors $u = \frac{1}{2}[e + th]$ and $v = \frac{1}{2}[e - th]$ belong to K. We clearly have $u + v = e$; since R is an extreme ray, both u and v should be nonnegative multiples of e, which, due to $t > 0$, would imply that h is proportional to e, which is not the case. We have arrived at a desired contradiction. (i) is proved.

(ii) Let $R = \{te : t \geq 0\}$ be a nontrivial ray in K; then e is a nonzero vector from K, and from the definition of a base, it follows that the ray R intersects Y at certain nonzero point e_R, which is also a generator of R. We should prove that R is an extreme ray iff e_R is an extreme point of Y.

In one direction, let R be an extreme ray. Let us lead to a contradiction the assumption that e_R is not an extreme point of Y. Indeed, under this assumption, there exists a nonzero vector h such that both $e_R + h$ and $e_R - h$ belong to Y and thus belong to K. Setting $u = \frac{1}{2}[e_R + h]$, $v = \frac{1}{2}[e_R - h]$, we get $u, v \in K$ (K is a cone!) and $u + v = e_R$. Since R is an extreme ray, it follows that u and v are proportional to e_R, whence also h is proportional to e_R. The latter is impossible: since e_R and $e_R + h$ belong to Y, we have $f^T e_R = 1 = f^T(e_R + h)$, whence $f^T h = 0$; since h is nonzero and is proportional to e_R, we conclude that $f^T e_R = 0$, while in fact $f^T e_R = 1$ due to $e_R \in Y$. We have arrived at a desired contradiction.

In the other direction, let e_R be an extreme point of Y, and let us prove that then R is an extreme ray of K. Indeed, let $e_R = u+v$ with $u, v \in K$. We should prove that in this case, both v and u belong to R. Assume that this is not the case, say, u does not belong to R. Then $u \neq 0$, and thus, $u \notin -R$ (since K is pointed and $u \in K$). Thus, u is not a real multiple of e_R, which combines with $e_R = u + v$ to imply that v is also not a real multiple of e_R. In particular, both u and v are nonzero, and since $u, v \in K$, we have $\bar{u} = \lambda u \in Y$ and $\bar{v} = \mu v \in Y$ for properly chosen $\lambda, \mu > 0$. From equations $f^T e_R = f^T \bar{u} = f^T \bar{v} = 1$ (given by $e_R, \bar{u}, \bar{v} \in Y$) and $f^T e_R = f^T u + f^T v$ (due to $e_R = u + v$), it follows that $\lambda f^T u = 1$, $\mu f^T v = 1$ and $f^T u + f^T v = 1$, whence $\frac{1}{\lambda} + \frac{1}{\mu} = 1$. Since λ and μ are positive, we conclude that $e_R = u+v = \frac{1}{\lambda}\bar{u} + \frac{1}{\mu}\bar{v}$ is a convex combination of the vectors $\bar{u}, \bar{v}$ from Y, the coefficients in the combination being positive. Since e_R is an extreme point of Y, we should have $e_R = \bar{u} = \bar{v}$ (according to the statement in Task 2.2), whence u and v are proportional to e_R, which is not the case. We have arrived at a desired contradiction. $\square$

Corollary 2.4. *Let K be a nontrivial ($K \neq \{0\}$) and pointed polyhedral cone. Then the number of extreme rays in K is finite.*

Proof. Indeed, by Proposition 2.14, K admits a base Y; by Proposition 2.16(ii), the number of extreme rays in K is equal to the number of vertices in Y, and the latter number, by Corollary 2.3, is finite. □

In fact, in the context of Corollary 2.4, the assumptions that K is nontrivial and pointed are redundant, by a very simple reason: neither trivial nor nonpointed cones have extreme rays. Indeed, the trivial cone cannot have extreme rays since the latter, by definition, are nontrivial. The fact that a nonpointed cone K has no extreme rays can be verified as follows. Let $R = \{te : t \geq 0\}$ be a nontrivial ray in K, and let f be a nonzero vector such that $\pm f \in K$. If f is not proportional to e, then the vectors $u = e + f$ and $v = e - f$ belong to K due to $\pm f \in K$ and $e \in K$, and do not belong to R (since they are not proportional to e). If f is proportional to e, say, $f = \tau e$, then $\tau \neq 0$ (since $f \neq 0$). Setting $u = \frac{1}{2}[e+\frac{2}{\tau}f]$, $v = \frac{1}{2}[e-\frac{2}{\tau}f]$, we, as above, get $u, v \in K$ and $u + v = e$; at the same time, $v = \frac{1}{2}[e - \frac{2}{\tau}\tau e] = -\frac{1}{2}e$, that is, $v \notin R$. In both cases, the ray R is not extreme.

The following proposition establishes a property of extreme rays similar to the one of extreme points (cf. Proposition 2.10).

Proposition 2.17. *Let K be a polyhedral cone given by (2.21), and let $K_I = \{x : Bx \leq 0, b_i^T x = 0, i \in I\}$ be a face of K, which is clearly a polyhedral cone along with K. Every extreme ray of K_I is an extreme ray of K.*

Proof is left to the reader. □

2.4 Structure of polyhedral sets

Now we are well prepared to attack directly our goal — to prove the theorem on structure of polyhedral sets stated in the preface to this chapter.

2.4.1 *First step*

We start with the following fundamental statement.

Theorem 2.12. *Let X be a nonempty polyhedral set, given by* (2.15), *which does not contain lines. Then we have the following*:

(i) *The set* $\mathrm{Ext}(X)$ *of extreme points of X is nonempty and finite, and X is the arithmetic sum of the convex hull of this set and the recessive cone of X*:

$$X = \mathrm{Conv}(\mathrm{Ext}(X)) + \mathrm{Rec}(X). \tag{2.23}$$

(ii) *If the recessive cone of X is nontrivial* ($\mathrm{Rec}(X) \neq \{0\}$), *then this cone possesses extreme rays, the number of these rays is finite, and the cone is the conic hull of its extreme rays: if $R_j = \{te_j : t \geq 0\}$, $1 \leq j \leq J$, are the extreme rays of* $\mathrm{Rec}(X)$, *then*

$$\mathrm{Rec}(X) = \mathrm{Cone}\{e_1, \ldots, e_J\} := \left\{ x = \sum_{j=1}^{J} \mu_j e_j : \mu_j \geq 0 \,\forall j \right\}. \tag{2.24}$$

(iii) *As a straightforward corollary of* (i) *and* (ii), *we get that*

$$X = \left\{ x = \sum_{i=1}^{I} \lambda_i v_i + \sum_{j=1}^{J} \mu_j e_j : \lambda_i \geq 0, \sum_i \lambda_i = 1,\, \mu_j \geq 0 \right\}. \tag{2.25}$$

Proof. (i) The proof is by induction in the dimension k of X (that is, the affine dimension of $M = \mathrm{Aff}(X)$ or, which is the same, linear dimension of the linear space L to which M is parallel).

Base $k = 0$ is trivial, since in this case X is a singleton: $X = \{v\}$, and we clearly have $\mathrm{Ext}\{X\} = \{v\}$, $\mathrm{Rec}(X) = \{0\}$, so that $\mathrm{Ext}\{X\}$ is indeed nonempty and finite, and (2.23) indeed holds true.

Inductive step $k \Rightarrow k+1$. Assume that (i) holds true for all nonempty polyhedral sets, not containing lines, of dimension $\leq k$, and let X be a nonempty polyhedral set, not containing lines, of dimension $k+1$. Let us prove that (i) holds true for X.

The linear subspace L parallel to $\mathrm{Aff}(X)$ is of dimension $k+1 \geq 1$ and thus contains a nonzero vector e. Since X does not contain lines, its recessive cone is pointed (Proposition 2.12); thus, either e or $-e$ is not a recessive direction of X. Swapping, if necessary, e and $-e$, we can assume without loss of generality that $e \notin \mathrm{Rec}(X)$.

Now, two options are possible:

A: $-e$ is a recessive direction of X;
B: $-e$ is *not* a recessive direction of X.

Let $\bar{x} \in X$. In order to prove that $\mathrm{Ext}(X) \neq \emptyset$ and that $\bar{x} \in \mathrm{Conv}(\mathrm{Ext}(X)) + \mathrm{Rec}(X)$, we act as follows:

In the case of A, consider the ray $R = \{x_t = \bar{x} + te : t \geq 0\}$. Since e is not a recessive direction, this ray is not contained in X, meaning that some of the inequalities $a_i^T x \leq b_i$, $i \in \mathcal{I}$, are not satisfied somewhere on R. Now, let us move along the ray R, starting at $\bar{x}$. At the beginning, we are at $\bar{x}$, i.e., are in X; eventually, we shall leave X, meaning that there is $\bar{t} \geq 0$ such that "at time $\bar{t}$, the point x_t is about to leave X," meaning that $\bar{t}$ is the largest $t \geq 0$ such that $x_t \in X$ for $0 \leq t \leq \bar{t}$. It is clear that $x_{\bar{t}}$ belongs to a proper face $\bar{X}$ of X.

Here is the "algebraic translation" of the above reasoning (which by itself appeals to geometric intuition). Since $x_0 = \bar{x} \in X$, all the inequalities $a_i^T x_t \leq b_i$, $i \in \mathcal{I}$, are satisfied at $t = 0$, and since $a_i^T x_t$ is an affine function of t, an inequality $a_i^T x_t \leq b_i$ which is violated at some $t > 0$ (and we know that such an inequality exists) is such that the affine function of t $a_i^T x_t = a_i^T x + t a_i^T e$ is increasing with t, that is, $a_i^T e > 0$. The bottom line is that there exists i such that $a_i^T e > 0$, so that the set $I = \{i : a_i^T e > 0\}$ is nonempty. Let us set

$$\bar{t} = \min_{i \in I} \frac{b_i - a_i^T \bar{x}}{a_i^T e}, \tag{$*$}$$

and let $i_* \in I$ be the value of i corresponding to the above minimum. Taking into account that $a_i^T \bar{x} \leq b_i$ for all i and that $a_i^T x_t \leq b_i$ for all $t \geq 0$ whenever $i \notin I$ (why?), we conclude that $\bar{t} \geq 0$ and that $a_i^T x_{\bar{t}} \leq b_i$ for all i, and on the top of it, $a_{i_*}^T x_{\bar{t}} = b_{i_*}$. Now, consider the set $\bar{X} = X_{\{i_*\}} = \{x : a_{i_*}^T x = b_i, a_i^T x \leq b_i \, \forall i \neq i_*\}$. This set is nonempty (it contains $x_{\bar{t}}$) and is thus a face of X. This face is proper, since otherwise $a_{i_*}^T x$ would be equal to b_{i_*} on the entire X, meaning that X is contained in the affine subspace $N = \{x : a_{i_*}^T x = b_i\}$. But then $\mathrm{Aff}(X) \subset N$, meaning that $a_{i_*}^T x \equiv b_i$ for all $x \in M = \mathrm{Aff}(X)$; since e belongs to the parallel to M linear subspace, the latter implies that $a_{i_*}^T e = 0$, while by construction $a_{i_*}^T e > 0$. Thus, $x_{\bar{t}}$ belongs to a *proper* face $\bar{X}$ of X, as claimed.

Now, by Proposition 2.8, the dimension of $\bar{X}$ is $< \dim X$, that is, it is at most k. Besides this, $\bar{X}$ is a nonempty polyhedral set, and this set does not contain lines (since $\bar{X}$ is a part of X, and X does not contain lines). Applying inductive hypothesis, we conclude that $\text{Ext}(\bar{X})$ is a finite nonempty set such that $x_{\bar{t}} \in \text{Conv}(\text{Ext}(\bar{X})) + r$ with $r \in \text{Rec}(\bar{X})$. Now, by Proposition 2.10, $\text{Ext}(\bar{X}) \subset \text{Ext}(X)$, and by the definition of a recessive direction, r, being recessive direction of a part of X, is a recessive direction of X. We have arrived at the following conclusion:

> (!) $\text{Ext}(X)$ is nonempty (and is finite by Corollary 2.3) and $x_{\bar{t}} \in \text{Conv}(\text{Ext}(X)) + r$ for some $r \in \text{Rec}(X)$.

Note that our reasoning did not yet use the assumption that we are in the case A, that is, $-e$ is a recessive direction of X; all we use till now is that $e \in L$ is not a recessive direction of X.

Now, recall that we are in the case of A, so that $-e$ is a recessive direction of X, whence

$$\bar{x} = x_{\bar{t}} - \bar{t}e \in \text{Conv}\{\text{Ext}(X)\} + r', r' = r + \bar{t}(-e).$$

Since r and $-e$ are recessive directions of X and $\text{Rec}(X)$ is a cone, we have $r' \in \text{Rec}(X)$ and thus $\bar{x} \in \text{Conv}(\text{Ext}(X)) + \text{Rec}(X)$.

In the case of B, the intermediate conclusion (!) still holds true — as we have mentioned, its validity stems from the fact that $e \in L$ is not a recessive direction of X. But now similar conclusion can be extracted from considering the ray $R_- = \{\bar{x} + t[-e] : t \geq 0\}$, which, by the exactly same reasoning as above, "hits" certain proper face of X. Thus, now, in addition to (!), there exists $\widetilde{t} \geq 0$ such that $\widetilde{x}_{\widetilde{t}} = \bar{x} - \widetilde{t}e$ satisfies

$$\widetilde{x}_{\widetilde{t}} \in \text{Conv}(\text{Ext}(X)) + \text{Rec}(X)$$

or, equivalently, there exist nonnegative weights $\widetilde{\lambda}_v$, $v \in \text{Ext}(X)$, summing up to 1, and a vector $\widetilde{r} \in \text{Rec}(X)$ such that

$$\bar{x} - \widetilde{t}e = \sum_{v \in \text{Ext}(X)} \widetilde{\lambda}_v v + \widetilde{r}.$$

Moreover, by (!), there exist nonnegative weights $\bar{\lambda}_v$, $v \in \text{Ext}(X)$, summing up to 1, and a vector $\bar{r} \in \text{Rec}(X)$ such that

$$\bar{x} + \bar{t}e = \sum_{v \in \text{Ext}(X)} \bar{\lambda}_v v + \bar{r}.$$

Since $\bar{t}$ and $\widetilde{t}$ are nonnegative, we have

$$\bar{x} = \mu[\bar{x} + \bar{t}e] + (1-\mu)[\bar{x} - \widetilde{t}e],$$

with properly chosen $\mu \in [0,1]$, so that

$$\bar{x} = \underbrace{\sum_{v \in \mathrm{Ext}(X)} [\mu\bar{\lambda}_v + (1-\mu)\widetilde{\lambda}_v]v}_{\in \mathrm{Conv}(\mathrm{Ext}(X))} + \underbrace{[\mu\bar{r} + (1-\mu)\widetilde{r}]}_{\in \mathrm{Rec}(X)}.$$

The bottom line is that in both cases A, B $\mathrm{Ext}(X)$ is finite and nonempty and $X \subset \mathrm{Conv}(\mathrm{Ext}(X)) + \mathrm{Rec}(X)$. The inverse inclusion is trivial: The polyhedral set X is convex and contains $\mathrm{Ext}(X)$ and therefore contains $\mathrm{Conv}(\mathrm{Ext}(X))$ and $X + \mathrm{Rec}(X) = X$ by Proposition 2.11, whence $\mathrm{Conv}(\mathrm{Ext}(X)) + \mathrm{Rec}(X) \subset X$. Thus, we have proved that $X = \mathrm{Conv}(\mathrm{Ext}(X)) + \mathrm{Rec}(X)$ and that $\mathrm{Ext}(X)$ is nonempty and finite. Inductive step and thus the verification of Theorem 2.12(i) are complete.

(ii) Assume that $\mathrm{Rec}(X) = \{x : Ax \le 0\}$ is a nontrivial cone. Since X does not contain lines, $\mathrm{Rec}(X)$ is also pointed (Proposition 2.12). Invoking Proposition 2.14, $\mathrm{Rec}(X)$ admits a base Y, which is clearly a nonempty polyhedral set. By Proposition 2.15, the recessive cone of Y is trivial, whence, in particular, Y does not contain lines. Applying (I) to Y, we conclude that $\mathrm{Ext}(Y)$ is nonempty and finite, and $Y = \mathrm{Conv}(\mathrm{Ext}(Y)) + \mathrm{Rec}(Y) = \mathrm{Conv}(\mathrm{Ext}(Y)) + \{0\} = \mathrm{Conv}(\mathrm{Ext}(Y))$. Further, by Proposition 2.16(ii), the extreme points $v_1, \ldots, v_p$ of Y are generators of extreme rays of $\mathrm{Rec}(X)$, and every extreme ray of $\mathrm{Rec}(X)$ is generated by one of these extreme points; thus, $\mathrm{Rec}(X)$ has extreme rays, and their number is finite. It remains to be proved that $\mathrm{Rec}(X)$ is the conic hull of the set of generators of its extreme rays. The validity of this statement is clearly independent of how we choose the generators; choosing them as $v_1, \ldots, v_p$, we arrive at the necessity to prove that $\mathrm{Rec}(X) = \mathrm{Cone}(\{v_1, \ldots, v_p\})$, that is, every $x \in \mathrm{Rec}(X)$ can be represented as a linear combination, with nonnegative coefficients, of $v_1, \ldots, v_p$. There is nothing to prove when $x = 0$; if $x \neq 0$, then, by the construction of a base, $x = \lambda\bar{x}$ with $\lambda > 0$ and $\bar{x} \in Y$. As we have already seen, $Y = \mathrm{Conv}(\{v_1, \ldots, v_p\})$, meaning that $\bar{x}$ is a convex combination of $v_1, \ldots, v_p$; but then $x = \lambda\bar{x}$ is a linear combination of $v_1, \ldots, v_p$ with nonnegative coefficients (recall that $\lambda > 0$), as claimed. (ii) is proved.

(iii) is readily given by (i) and (ii). $\square$

Remark. For an arbitrary nonempty *closed* convex X, the following analogy of Theorem 2.12 holds true: if X does not contain lines, then the set of extreme points of X is nonempty (but not necessary finite), and X is the arithmetic sum of $\mathrm{Conv}(\mathrm{Ext}(X))$ and the recessive cone of X.

2.4.1.1 *Immediate corollaries*

Theorem 2.12 readily implies a lot of important information on polyhedral sets.

Corollary 2.5. *The set* $\mathrm{Ext}(X)$ *of vertices of a bounded nonempty polyhedral set* X *is nonempty and finite, and* $X = \mathrm{Conv}(\mathrm{Ext}(X))$.

Indeed, the recessive cone of a *bounded* nonempty polyhedral set is clearly trivial and, in particular, the set does not contain lines. Theorem 2.12(i), remains to be applied. □

Corollary 2.6. *A pointed nontrivial polyhedral cone* $K = \{x : Bx \leq 0\}$ *has extreme rays, their number is finite, and the cone is the conic hull of (generators of) the extreme rays.*

Indeed, for a cone K, we clearly have $K = \mathrm{Rec}(K)$; when K is pointed, K is a nonempty polyhedral set not containing lines. Theorem 2.12(ii), with K in the role of X, remains to be applied. □

Corollary 2.7. *Every nonempty polyhedral set* $X \subset \mathbf{R}^n$ *admits a description as follows: It is possible to point out a finite and nonempty set* $\mathcal{V} = \{v_1, \ldots, v_I\} \subset \mathbf{R}^n$ *and a finite (possibly, empty) set* $\mathcal{R} = \{r_1, \ldots, r_J\} \subset \mathbf{R}^n$ *in such a way that*

$$\begin{aligned} X &= \mathrm{Conv}(\mathcal{V}) + \mathrm{Cone}\,(\mathcal{R}) \\ &= \left\{ x = \sum_{i=1}^{p} \lambda_i v_i + \sum_{j=1}^{J} \mu_j r_j : \lambda_i \geq 0, \sum_i \lambda_i = 1, \mu_j \geq 0 \right\}. \end{aligned} \tag{2.26}$$

For every representation of X *of this type, it holds that*

$$\mathrm{Rec}(X) = \mathrm{Cone}\,(\mathcal{R}) := \left\{ \sum_j \mu_j r_j : \mu_j \geq 0 \right\}. \tag{2.27}$$

In connection with possible emptiness of $\mathcal{R}$, recall that according to our convention, "a sum of vectors with empty set of terms equals to 0," $\text{Cone}\{\emptyset\} = \{0\}$ *is the trivial cone, so that* (2.26) *makes sense when* $\mathcal{R} = \emptyset$ *and reads in this case as* $X = \text{Conv}(\mathcal{V})$.

Proof. By Proposition 2.13, a nonempty polyhedral set X can be represented as the sum of a nonempty polyhedral set $\bar{X}$ not containing lines and a linear subspace L. We can find a finite set $f_1, \ldots, f_p$ which linearly spans L, so that $L = \text{Cone}(\{g_1, \ldots, g_{2p}\})$, where $g_1, \ldots, g_p$ are the same as $f_1, \ldots, f_p$ and $g_{p+1}, \ldots, g_{2p}$ are the same as $-f_1, \ldots, -f_p$. By Theorem 2.12(iii), we can find a nonempty finite set $\mathcal{V}$ and a finite set $\mathcal{M}$ such that $\bar{X} = \text{Conv}(\mathcal{V}) + \text{Cone}(\mathcal{M})$, so that

$$\begin{aligned} X = \bar{X} + L &= [\text{Conv}(\mathcal{V}) + \text{Cone}(\mathcal{M})] + \text{Cone}(\{g_1, \ldots, g_{2p}\}) \\ &= \text{Conv}(\mathcal{V}) + [\text{Cone}(\mathcal{M}) + \text{Cone}(\{g_1, \ldots, g_{2p}\})] \\ &= \text{Conv}(\mathcal{V}) + \text{Cone}(\mathcal{M} \cup \{g_1, \ldots, g_{2p}\}). \end{aligned}$$

It remains to be proved that in every representation,

$$X = \text{Conv}(\mathcal{V}) + \text{Cone}(\mathcal{R}), \tag{$*$}$$

with finite nonempty $\mathcal{V} = \{v_1, \ldots, v_I\}$ and finite $\mathcal{R} = \{r_1, \ldots, r_J\}$, it holds $\text{Cone}(\mathcal{R}) = \text{Rec}(X)$. The inclusion $\text{Cone}(\mathcal{R}) \subset \text{Rec}(X)$ is evident (since every r_j is clearly a recessive direction of X). To prove the inverse inclusion, assume that it does not take place, that is, there exists $r \in \text{Rec}(X) \backslash \text{Cone}(\mathcal{R})$, and let us lead this assumption to a contradiction. Since r is not a conic combination of $r_1, \ldots, r_J$, the HFL says that there exists f such that $f^T r_j \geq 0$ for all j and $f^T r < 0$. The first of these two facts implies that the linear form $f^T x$ of x is below bounded on X; indeed, by ($*$), every point $x \in X$ is of the form $\sum_i \lambda_i v_i + \sum_j \mu_j r_j$ with nonnegative λ_i, μ_j such that $\sum_i \lambda_i = 1$, whence

$$f^T x = \sum_i \lambda_i f^T v_i + \underbrace{\sum\nolimits_j \mu_j f^T r_j}_{\geq 0} \geq \sum_i \lambda_i \left[\min_\ell f^T v_\ell\right] = \min_\ell f^T v_\ell > -\infty.$$

The second fact implies that same form $f^T x$ is *not* below bounded on X. Indeed, since $r \in \text{Rec}(X)$, taking $\bar{x} \in X$, the ray $R = \{\bar{x} + tr :$

$t \geq 0\}$ is contained in X, and since $f^T r < 0$, $f^T x$ is not bounded below on R and thus on X. We got a desired contradiction. $\square$

Corollary 2.8. *A nonempty polyhedral set X possesses extreme point iff X does not contain lines. In addition, the set of extreme points of X is finite.*

Indeed, if X does not contain lines, X has extreme points and their number is finite by Theorem 2.12. Now, assume that X contains lines, and let $e \neq 0$ be a direction of such a line, then for every $x \in X$, the vectors $x \pm e$ belong to X (Proposition 2.11; note that both e and $-e$ are recessive directions of X), and thus, x is not an extreme point of X. Thus, $\mathrm{Ext}(X) = \emptyset$. $\square$

Remark. The first claim in the corollary is valid for every nonempty closed convex set.

Corollary 2.9. *Let X be a nonempty polyhedral set. Then X is bounded iff the recessive cone of X is trivial:* $\mathrm{Rec}(X) = \{0\}$, *and in this case, X is the convex hull of a nonempty finite set (e.g., the set* $\mathrm{Ext}(X)$).

Indeed, if $\mathrm{Rec}(X) \neq \{0\}$, then X contains a ray $\{\bar{x} + te : t \geq 0\}$ with $e \neq 0$; this ray, and therefore X, is an unbounded set. Vice versa, if $\mathrm{Rec}(X) = \{0\}$, then X clearly does not contain lines, and by Theorem 2.12(i), $X = \mathrm{Conv}(\mathrm{Ext}(X))$ and $\mathrm{Ext}(X)$ is finite and is thus bounded; the convcx hull of a bounded set clearly is bounded as well, so that $\mathrm{Rec}(X) = \{0\}$ implies that X is bounded. $\square$

Remark. The statement in Corollary, modulo "finite," is valid for every nonempty *closed* convex set: *A nonempty closed convex set X is bounded iff its recessive cone is trivial, and in this case* $X = \mathrm{Conv}(\mathrm{Ext}(X))$.

Corollary 2.10. *A nonempty polyhedral set is bounded iff it can be represented as the convex full of a finite nonempty set $\{v_1, \ldots, v_p\}$. When this is the case, every vertex of X is among the points $v_1, \ldots, v_p$.*

Indeed, if X is bounded, then $\mathrm{Ext}(X)$ is nonempty, finite and $X = \mathrm{Conv}(\mathrm{Ext}(X))$ by Corollary 2.9. Vice versa, if a nonempty polyhedral set X is represented as $\mathrm{Conv}(V)$, $V = \{v_1, \ldots, v_N\}$, then V is nonempty and X is bounded (since the convex hull of a bounded set is clearly bounded (why?)). Finally, if $X = \mathrm{Conv}(\{v_1, \ldots, v_N\})$

and v is a vertex of X, then $v \in \{v_1, \ldots, v_N\}$ by the result stated in Task 2.2. $\square$

The next immediate corollary of Theorem 2.12 is the first major fact of the LO theory.

Corollary 2.11. *Consider an LO program* $\max_x \{c^T x : Ax \leq b\}$, *and let the feasible set* $X = \{x : Ax \leq b\}$ *of the problem be nonempty. Then we have the following:*

(i) *The program is bounded from above iff c has nonpositive inner products with all recessive directions of X (or, which is the same, $-c$ belongs to the cone dual to the recessive cone* $\text{Rec}(X) = \{x : Ax \leq 0\}$ *of X). Whenever this is the case, the program is solvable (i.e., admits an optimal solution).*
(ii) *If X does not contain lines* (or, which is the same due to the nonemptiness of X, if $\text{Ker}\, A = \{0\}$, see Proposition 2.12) *and the program is bounded from above, then among its optimal solutions, there are extreme points of X.*

Indeed, if c has a positive inner product with some $r \in \text{Rec}(X)$, then, taking $\bar{x} \in X$ and observing that the ray $R = \{\bar{x} + tr : t \geq 0\}$ is contained in X and that $c^T x$ is not bounded above on R (why?), the objective is not bounded above on the feasible set. Now, let c have nonpositive inner products with all vectors from $\text{Rec}(X)$; let us prove that the problem is solvable. Indeed, by Corollary 2.7, there is a representation

$$X = \text{Conv}(\{v_1, \ldots, v_I\}) + \text{Cone}\,(\mathcal{R}), \qquad (*)$$

with certain nonempty $\mathcal{V} = \{v_1, \ldots, v_I\}$ and finite $\mathcal{R} = \{r_1, \ldots, r_J\}$. Then, of course, $v_i \in X$ and $r_j \in \text{Rec}(X)$ for all i, j. Since every $x \in X$ can be represented as

$$x = \sum_i \lambda_i v_i + \sum_j \mu_j r_j \qquad [\lambda_i \geq 0, \mu_j \geq 0, \sum_i \lambda_i = 1],$$

we have

$$\begin{aligned} c^T x &= \sum_i \lambda_i c^T v_i + \underbrace{\sum_j \mu_j c^T r_j}_{\leq 0} \leq \sum_i \lambda_i c^T v_i \\ &\leq \sum_i \lambda_i \left[\max\left[c^T v_1, \ldots, c^T v_I\right]\right] = \max\left[c^T v_1, \ldots, c^T v_I\right]. \end{aligned}$$

We see that the objective is bounded above everywhere on X by the maximum of its values at the points $v_1, \ldots, v_I$. It follows that the best (with the largest value of the objective) of the (belonging to X!) points $v_1, \ldots, v_I$ is an optimal solution to the program. (i) is proved. To prove (ii), it remains to be noted that when X does not contain lines, one can take, as the above $\mathcal{V}$, the set of extreme points of X. $\square$

Remark. The fact that a feasible and bounded LO program admits an optimal solution, stated in Corollary 2.11(i), is already known to us; we obtained it (via a quite different tool, the Fourier–Motzkin elimination scheme) already in Chapter 1.

2.4.1.2 *Minimality of the representation stated in Theorem* 2.12

Corollary 2.7 states that every nonempty polyhedral set X can be represented in the form of (2.26):

$$X = \text{Conv}(\mathcal{V}) + \text{Cone}\,(\mathcal{R}),$$

with certain nonempty finite $\mathcal{V}$ and finite, possibly, empty, set $\mathcal{R}$. When X does not contain lines, Theorem 2.12 states that in such a representation, one can take as $\mathcal{V}$ the set of extreme points of X and as $\mathcal{R}$ — the set of generators of the extreme rays of $\text{Rec}(X)$. There are, of course, other options: We can add to the just defined $\mathcal{V}$ any extra point of X and add to $\mathcal{R}$ any vector from $\text{Rec}(X)$. It turns out, however, that the representation stated by Theorem 2.12 is the only "minimal" one.

Proposition 2.18. *Let X be a nonempty polyhedral set not containing lines, let $\mathcal{V}_* = \text{Ext}(X) = \{v_1, \ldots, v_p\}$, and let $\mathcal{R}_* = \{r_1, \ldots, r_q\}$ be the set of generators of the extreme rays of* $\text{Rec}(X)$. *Then for every representation*

$$X = \text{Conv}(\mathcal{V}) + \text{Cone}\,(\mathcal{R}), \qquad (*)$$

where $\mathcal{V} = \{u_1, \ldots, u_P\}$ *is a nonempty finite set and* $\mathcal{R} = \{e_1, \ldots, e_Q\}$ *is a finite set, the following is true:*

(i) *every point* $v_i \in \mathcal{V}_*$ *of* X *is one of the points* $u_1, \ldots, u_P$;
(ii) $\mathrm{Cone}\,(\mathcal{R}) = \mathrm{Rec}(X)$, *and every vector* $r_i \in \mathcal{R}_*$ *is positive multiple of one of the vectors* $e_1, \ldots, e_Q$ *(that is, every extreme ray of* $\mathrm{Rec}(X)$, *if any exists, is among the rays generated by* $e_1, \ldots, e_Q$*).*

Proof. (i) Let v be a vertex of X, and let us prove that $v \in \mathcal{V}$. Indeed, by $(*)$, we have

$$v = \underbrace{\sum_{i=1}^{P} \lambda_i u_i}_{u} + \underbrace{\sum_{j=1}^{Q} \mu_j e_j}_{e} \qquad \left[\lambda_j \geq 0, \mu_j \geq 0, \sum_i \lambda_i = 1\right].$$

We clearly have $u_i \in X$ and $e_j \in \mathrm{Rec}(X)$ for all i, j, whence $v \in X$ and $e \in \mathrm{Rec}(X)$. We claim that $e = 0$; indeed, otherwise, we would have $v - e = u \in X$ and $v + e \in X$ due to $v \in X$ and $e \in \mathrm{Rec}(X)$. Thus, $v \pm e \in X$ and $e \neq 0$, which is impossible for the extreme point v (see geometric characterization of extreme points, Proposition 2.9). Thus, $v \in \mathrm{Conv}(\{u_1, \ldots, u_P\}) \subset X$, whence v, being a vertex of X, is one of the points $u_1, \ldots, u_P$ by the result stated in Task 2.2(i). Item (i) is proved.

(ii) The fact that $\mathrm{Cone}\,(\mathcal{R}) = \mathrm{Rec}(X)$ was established in Corollary 2.7. It follows that if e is a generator of an extreme ray R in $\mathrm{Rec}(X)$, then $e = \sum_{j=1}^{Q} \lambda_j e_j$ with some $\lambda_j \geq 0$. Since $e_j \in \mathrm{Rec}(X)$, applying the result stated in Task 2.5(2), those of the vectors $\lambda_j e_j$ which are nonzero (and there are such vectors, since $\sum_j \lambda_j e_j = e \neq 0$) are positive multiples of e. Thus, R admits a generator which is one of the vectors $e_1, \ldots, e_Q$. □

Note that the "minimality" result stated by Proposition 2.18 heavily exploits the fact that X does not contain lines. While a nonempty polyhedral set containing lines is still the sum of the convex hull of a nonempty finite set $\mathcal{V}$ and the conic hull of a finite set $\mathcal{R}$, there definitely exist pairs $\mathcal{V}, \mathcal{R}$ and $\mathcal{V}'$, $\mathcal{R}'$ which "produce" the same set X and at the same time, say, $\mathcal{V}' \cap \mathcal{V} = \emptyset$ (and even $\mathrm{Conv}(\mathcal{V}) \cap \mathrm{Conv}(\mathcal{V}') = \emptyset$). What is uniquely defined by a *whatever* nonempty polyhedral set X is the conic hull of $\mathcal{R}$: this is nothing but the recessive cone of X. To arrive at this conclusion, you should repeat word by word the reasoning which we used to demonstrate that $\text{Conv}(\mathcal{R}) = \text{Rec}(X)$ when proving Proposition 2.18(ii).

2.4.2 *Second step*

We have proved that every nonempty polyhedral set X, along with its "outer" description $X = \{x : Ax \leq b\}$, admits a simple "inner" representation: it is generated by two properly chosen finite sets $\mathcal{V}$ (this set is nonempty) and $\mathcal{R}$ (this set can be empty) according to (2.26). What is missing yet is the inverse statement — that *every* set representable in the just outlined form is polyhedral. Or local goal is to establish this missing element, thus ending up with a nice outer (solution set of a solvable system of linear inequalities) and inner (given by (2.26)) representation of nonempty polyhedral sets. Here is this missing element.

Theorem 2.13. *Let $\mathcal{V} = \{v_1, \ldots, v_I\}$ be a finite nonempty subset of $\mathbf{R}^n$, and $\mathcal{R} = \{r_1, \ldots, r_J\}$ be a finite subset of $\mathbf{R}^n$, and let*

$$X = \text{Conv}(\mathcal{V}) + \text{Cone}\,(\mathcal{R}). \tag{2.28}$$

Then X is a nonempty polyhedral set.

Proof. We are about to present two alternative proofs: one immediate and another one more involving and more instructive.

Immediate proof: X is clearly nonempty and polyhedrally representable (as the image of a clearly polyhedral set $\{[\lambda; \mu] \in \mathbf{R}^I \times \mathbf{R}^J : \lambda \geq 0, \sum_i \lambda_i = 1, \mu \geq 0\}$ under the linear mapping $[\lambda; \mu] \mapsto \sum_i \lambda_i v_i + \sum_j \mu_j r_j$). By Fourier–Motzkin elimination (Theorem 1.2), X is a polyhedral set. □

Alternative proof: The fact that X is nonempty is evident: $\mathcal{V} \subset X$. We can assume without loss of generality that $0 \in X$. Indeed, shifting all vectors from $\mathcal{V}$ by $-v_1$, we shift by the same vector the set X given by (2.28), thus ensuring that the shifted X contains the origin. At the same time, a shift of a set is clearly polyhedral iff the set is so.

Now, the *polar* $\text{Polar}\,(Y)$ of a set $Y \subset \mathbf{R}^n$ which contains the origin is, by definition, the set of all vectors $f \in \mathbf{R}^n$ such that $f^T y \leq 1$ for all $y \in Y$.

For example (check what follows!), we have the following:

- The polar of $\{0\}$ is $\mathbf{R}^n$, and the polar of $\mathbf{R}^n$ is $\{0\}$.
- The polar of a linear subspace L is its orthogonal complement $L^\perp$.
- The polar of a cone is minus its dual cone.

The polar clearly is nonempty — it contains the origin.

Our plan of attack is as follows:

A. Assuming that the set X given by (2.28) contains the origin, we shall prove that its polar $X^* = \text{Polar}\,(X)$ is a nonempty polyhedral set. As such, it admits the representation of the form similar to (2.28):

$$X^* = \text{Conv}(\mathcal{V}^*) + \text{Cone}\,(\mathcal{R}^*) \qquad (!)$$

($\mathcal{V}^*$ is finite and nonempty, and $\mathcal{R}^*$ is finite).

B. X^* clearly contains the origin and by A admits a representation (!), whence, by the same A, the set $(X^*)^*$ is polyhedral. On the other hand, we shall prove that $(X^*)^* = X$, thus arriving at the desired conclusion that X is polyhedral.

Let us execute our plan.

A is immediate. Observe that f satisfies $f^T x \le 1$ for all $x \in X$ iff $f^T r_j \le 0$ for all $j \le J$ and $f^T v_i \le 1$ for all $i \le I$.

Indeed, if $f^T r_j > 0$ for some j, then $f^T[v_1 + tr_j] \to +\infty$ as $t \to \infty$ and since $v_1 + tr_j \in X$ for every $t \ge 0$ due to $(*)$, $f \notin X^*$; and of course $f \notin X^*$ when $f^T v_i > 1$ for some i, since all $v_i \in X$. Thus, *if* $f \in X^*$, then $f^T r_j \le 0$ for all j and $f^T v_i \le 1$ for all i. Vice versa, assume that $f^T r_j \le 0$ and $f^T v_i \le 1$ for all i; then $f^T x \le 1$ for all $x \in X$ by exactly the same argument as used in the proof of item (i) in Corollary 2.11.

Now, the above observation proves that X^* is the solution set of the system $\{r_j^T f \le 0, 1 \le j \le J, v_i^T f \le 1, 1 \le i \le I\}$ of linear inequalities in variables f, and since the system is solvable (e.g., 0 is its feasible solution), X^* is a nonempty polyhedral set, as stated in A.

B. All we need is to prove that the set X given by (2.28) is the polar of its polar: $X = (X^*)^*$. By definition of the polar, every set is *part* of the polar of its polar, so that $X \subset (X^*)^*$. To prove equality, let us prove that $(X^*)^* \subset X$. To this end, we need the following fact.

Lemma 2.2. *Let X be given by (2.28) and $y \in \mathbf{R}^n$ be such that $y \notin X$. Then there exists $f \in \mathbf{R}^n$ and $\epsilon > 0$ such that*

$$f^T y > f^T x + \epsilon \; \forall x \in X.$$

Lemma ⇒ The required conclusion: Taking Lemma for granted, assume, on the contrary to what we want to prove, that there exists $\bar{x} \in (X^*)^* \backslash X$, and let us lead this assumption to a contradiction. Indeed, applying Lemma to a vector $\bar{x} \in (X^*)^* \backslash X$, we conclude that there exists f such that $f^T \bar{x} > f^T x + \epsilon \; \forall x \in X$. Since $0 \in X$, it follows that $0 \leq \sup_{x \in X} f^T x < f^T \bar{x} - \epsilon$. Since $\epsilon > 0$, multiplying f by an appropriate positive real, we can ensure that $0 \leq \sup_{x \in X} f^T x \leq 1 < f^T \bar{x}$, meaning that $f \in X^*$ and $\bar{x}^T f > 1$; the latter is the desired contradiction, since $\bar{x} \in (X^*)^*$, and thus, we should have $\bar{x}^T g \leq 1$ for all $g \in X^*$, in particular, for $g = f$.

Proof of Lemma. Consider vectors

$$v_i^+ = [v_i; 1], 1 \leq i \leq I, \; v_{i+j}^+ = [r_j; 0], \; 1 \leq j \leq J, \; y^+ = [y; 1]$$

in $\mathbf{R}^{n+1}$. We claim that y^+ is *not* a conic combination of v_i^+, $1 \leq i \leq I + J$.

Indeed, assuming that

$$y^+ = \sum_{\ell=1}^{I+J} \lambda_\ell v_\ell^+ \text{ with } \lambda_\ell \geq 0,$$

and looking at the last coordinates of all vectors involved, we get $1 = \sum_{i=1}^{I} \lambda_i$; looking at the first $n - 1$ coordinates, we conclude that $y = \sum_{i=1}^{I} \lambda_\ell v_\ell + \sum_{j=1}^{J} \lambda_{I+j} \lambda_j r_j$. The bottom line is that y is a convex combination of $v_1, \ldots, v_I$ plus a conic combination of $r_1, \ldots, r_J$, that is, $y \in X$ by (2.28), which is not the case.

Now, since y^+ is not a conic combination of $v_1^+, \ldots, v_{I+J}^+$, the HFL says to us that there exists $f^+ = [f; \alpha]$ such that $f^T y + \alpha = [f^+]^T y^+ =: \epsilon > 0$, while $[f^+]^T v_\ell^+ \leq 0$, $1 \leq \ell \leq I + J$. In other words, $f^T v_i \le -\alpha$ and $f^T r_j \le 0$ for all i, j. Now, if $x \in X$, then by (2.28), we have

$$x = \sum_i \lambda_i v_i + \sum_j \mu_j r_j$$

for certain $\lambda_j \ge 0$ summing up to 1 and $\mu_j \ge 0$. It follows that

$$f^T x = \sum_i \lambda_i f^T v_i + \sum_j \mu_j f^T r_j \le \sum_i \lambda_i[-\alpha] + \sum_j \mu_j \cdot 0 = -\alpha.$$

The bottom line is that $f^T x \le -\alpha$ everywhere on X, while $f^T y = -\alpha + \epsilon$ with $\epsilon > 0$, that is, $f^T x + \epsilon \le f^T y$ for all $x \in X$, as claimed in the lemma. □

Note that in the case of a polyhedral set $X = \{x : Ax \le b\}$, the result stated by Lemma 2.2 is evident: if $y \notin X$, then there exists a row a_i^T of A such that $a_i^T y > b_i$, while $a_i^T x \le b_i$ everywhere on X. In contrast to this, Lemma 2.2 is a nonevident and powerful statement (which, on the closest inspection, inherited its power from the HFL) — It provides us with a separation-type result at times when we do not know whether the set X in question is or is not polyhedral.

2.4.2.1 *Separation theorem for convex sets*

The new "driving force" in the previous proof — Lemma 2.2 — admits a far-reaching extension onto general convex sets, which is the "in particular" part in Theorem 2.14 we are about to present. We start with the following.

Definition 2.8. Let X, Y be two nonempty sets in $\mathbf{R}^n$. We say that a linear function $e^T x$ separates these sets if the function is nonconstant on $X \cup Y$ and everywhere on X is the same as or larger than everywhere on Y, that is, $e^T x \ge e^T y$ whenever $x \in X$ and $y \in Y$.

Equivalent definition of separation is (check equivalency!):

e separates X and Y iff

$$\inf_{x \in X} e^T x \ge \sup_{y \in Y} e^T y \;\&\; \sup_{x \in X} e^T x > \inf_{y \in Y} e^T y. \tag{2.29}$$

Recall that the *relative interior* $\mathrm{rint}\,X$ of a nonempty set $X \subset \mathbf{R}^n$ is composed of all relative interior points of the set — points $x \in X$ such that the intersection of the ball of some positive radius r centered at x with the affine span of X is contained in X:

$$\exists r > 0 : y \in \mathrm{Aff}(X), \|y - x\|_2 \leq r \Rightarrow y \in X.$$

Proposition 2.7 states that relative interior of a nonempty convex set is convex, nonempty, and is *dense* in X, meaning that every point $x \in X$ is the limit of an appropriately chosen sequence of points from $\mathrm{rint}\,X$.

The separation theorem for convex sets, which is an extremely powerful (if not *the* most powerful) tool of convex analysis, reads as follows.

Theorem 2.14. *Let X, Y be nonempty convex sets in $\mathbf{R}^n$. These sets can be separated iff their relative interiors do not intersect.*

In particular, if Y is a nonempty closed convex set in $\mathbf{R}^n$ and $x \in \mathbf{R}^n$ does not belong to X, then there exists a linear form which strictly separates x and Y:

$$e^T x > \sup_{y \in Y} e^T y.$$

Proof. While we usually do not prove "convex extensions" of our polyhedral results, we cannot skip the proof of the separation theorem, since we will use this result when speaking about conic duality. The proof is easy — the major part of the task "sits" already in Lemma 2.2.

1^0. *Separating a point and a convex set not containing this point*: Let us prove that if $X = \{x\}$ is a singleton and Y is a nonempty convex set which does not contain x, then X and Y can be separated. By shifting x and Y by the same vector, which of course does not affect neither the premise ($x \notin Y$) nor the conclusion ("x and Y can be separated") of the statement we are proving, we can assume without loss of generality that $0 \in Y$. The claim is easy to verify when $x \notin \mathrm{Aff}(Y)$. Indeed, since $0 \in Y$, $\mathrm{Aff}(Y)$ is a linear subspace; when $x \notin \mathrm{Aff}(Y)$, taking, as e, the orthogonal projection of x onto the orthogonal complement to this linear space, we get $e^T x = e^T e > 0$ and $e^T y = 0$ for all $y \in \mathrm{Aff}(Y) \supset Y$, as required in (2.29). Now, let $x \in \mathrm{Aff}(Y)$. Replacing, if necessary, $\mathbf{R}^n$ with $\mathrm{Lin}(Y) = \mathrm{Aff}(Y)$, we can assume without loss of generality that $\mathrm{Aff}(Y) = \mathrm{Lin}(Y) = \mathbf{R}^n$.

Now let $y_1, y_2, \ldots$ be a sequence of vectors from Y which is dense in Y, meaning that every point in Y is the limit of certain converging subsequence of $\{y_i\}_{i=1}^{\infty}$; the existence of such a dense sequence $\{y_i \in Y\}$ for a nonempty set $Y \in \mathbf{R}^n$ is the standard and simple fact of real analysis (it is called *separability of* $\mathbf{R}^n$). Now, let $Y_k = \text{Conv}\{y_1, \ldots, y_k\}$, $k = 1, 2, \ldots$. Since Y is convex and $x \notin Y$, $x \notin Y_k$, whence, by Lemma 2.2, for every k, there exists a linear form e_k which strictly separates x and Y_k:

$$e_k^T x > \sup_{y \in Y} e_k^T y. \qquad (!)$$

We clearly have $e_k \neq 0$, and since (!) remains intact when e_k is multiplied by a positive real, we can assume without loss of generality that $\|e_k\|_2 = 1$. Now, a sequence of unit vectors in $\mathbf{R}^n$ always contains a converging subsequence, and the limit of this subsequence is a unit vector; thus, we can extract from $\{e_k\}$ a subsequence $\{e_{k_i}\}_{i=1}^{\infty}$ converging to certain unit vector e. For every i, the vector y_i belongs to all but finitely many sets Y_k, whence the inequality

$$e_{k_j}^T x \geq e_{k_j}^T y_i$$

holds true for all but finitely many values of j. Passing to limit as $j \to \infty$, we get

$$e^T x \geq e^T y_i$$

for all i, and since $\{y_i\}_{i=1}^{\infty}$ is dense in Y, we conclude that $e^T x \geq e^T y$ *for all* $y \in Y$. All that remains to check in order to conclude that the linear form given by e separates $\{x\}$ and Y is that $e^T z$ is not constant on $\{x\} \cup Y$, which is immediate: e by construction is a unit vector and Y is full-dimensional, so that $e^T z$ is nonconstant already on Y.

2^0. *Separating two nonintersecting nonempty convex sets:* Let now X and Y be nonempty nonintersecting convex sets. In order to separate them, we note that the set $Y - X = \{y - x : y \in Y, x \in X\}$ is nonempty and convex, and does not contain the origin (the latter — since X and Y do not intersect). By the previous item, we can separate 0 and $Y - X$, that is, there exists e such that

$$0 = e^T 0 \geq e^T(y - x) \ \forall (x \in X, y \in Y) \ \&$$

$$0 > \inf\{e^T(y - x) : y \in Y, x \in X\},$$

which is nothing but (2.29).

3^0. *Separating two nonempty convex sets with nonintersecting relative interiors*: Let X and Y be nonempty convex sets with nonintersecting relative interiors. As mentioned, $\mathrm{rint}\, X$ and $\mathrm{rint}\, Y$ are convex nonempty sets, so that they can be separated by the previous item: there exists e such that

$$\inf_{x\in \mathrm{rint}\, X} e^T x \geq \sup_{y\in \mathrm{rint}\, Y} e^T y \;\&\; \sup_{x\in \mathrm{rint}\, X} e^T x > \inf_{y\in \mathrm{rint}\, Y} e^T y.$$

Since the relative interior of a convex set is dense in this set, the sup and inf in the above relation remain intact when the relative interiors of X and Y are replaced with X, Y themselves, so that separating the relative interiors of X, Y, we automatically separate X and Y.

4^0. We have proved that *if* the relative interiors of nonempty convex sets do not intersect, *then* the sets can be separated. The inverse statement is nearly evident. Indeed, assume that $\mathrm{rint}\, X \cap \mathrm{rint}\, Y \neq \emptyset$, and let us lead to a contradiction the assumption that X and Y can be separated. Let e separate X and Y, and let $a \in \mathrm{rint}\, X \cap \mathrm{rint}\, Y$. By the first inequality in (2.29), the linear function $e^T x$ everywhere on X should be $\geq e^T a$ (since $a \in Y$), and since $a \in X$, a should be a minimizer of $e^T x$ on X. But a linear function $f^T w$ can attain its minimum on a set Z at a point z of the relative interior of this set only when the function is constant on the set:

Indeed, by definition of a relative interior point, Z contains the intersection D of $\mathrm{Aff}(Z)$ with a ball of positive radius, so that restricted on D, the linear function in question should attain its minimum on D at the center z of D. The latter is possible only when the function is constant on D, since D is symmetric w.r.t. z. When $f^T w$ is constant on D, then $f^T h$ is constant on $D - z$, and the latter set is centered at the origin ball of positive radius in the linear space L to which $\mathrm{Aff}(Z)$ is parallel. Thus, f should be orthogonal to all small enough vectors from L and should consequently be orthogonal to the entire L, whence $f^T w$ is constant on $\mathrm{Aff}(Z)$.

We see that the function $e^T y$ is constant, and equal to $e^T a$, on the entire $\mathrm{Aff}(X)$ and thus on the entire X. By "symmetric" reasoning, $e^T y$ attains its maximum on Y at the point $a \in \mathrm{rint}\, Y$, whence $e^T y$ is identically equal to $e^T a$ on Y. We see that $e^T w$ is constant on $X \cup Y$, which contradicts the origin of e. Thus, the convex sets with intersecting relative interiors cannot be separated.

5^0. The "in particular" part of the separation theorem remains to be proved. When $x \notin Y$ and Y is nonempty, closed and convex, then there exists $r > 0$ such that the ball B of radius r centered at x does not intersect Y (why?). Applying the already-proved part of the separation theorem to $X = B$ and Y, we get a nonzero (why?) vector e such that

$$\inf_{x' \in B} e^T x' \geq \max_{y \in Y} e^T y.$$

Taking into account that the left-hand side inf is $e^T x - r\|e\|_2$ (why?), we see that e strongly separates $\{x\}$ and Y. □

Remark. A mathematically oriented reader could note that when proving separation theorem for convex sets, we entered a completely new world. Indeed, aside of this proof and the last section in Chapter 1, *all our constructions were purely rationally algebraic* — we never used square roots, exponents, convergence, facts like "every bounded sequence of reals/vectors admits extracting a subsequence which has a limit," etc. More specifically, all our constructions, proofs and results would remain intact if we were

- replacing our "universes" $\mathbf{R}$ and $\mathbf{R}^n$ with the field of rational numbers $\mathbf{Q}$ and the vector space $\mathbf{Q}^n$ of n-dimensional vectors with rational coordinates, the linear operations being vector addition and multiplication of vectors by *rational* scalars;
- replacing polyhedral sets in $\mathbf{R}^n$ — sets of real vectors solving finite systems of nonstrict linear inequalities with real coefficients — with their "rational" counterparts — sets of rational vectors solving finite systems of linear inequalities with rational coefficients;
- allowing for *rational* rather than real scalars when speaking about linear/convex/conic combinations and in all other places where we use operations involving scalars.

In fact, aside of the last section in Chapter 1 and the story about separation theorem for Convex Sets, we could use in the role of our basic field of scalars $\mathbf{R}$ not only the field of rational numbers $\mathbf{Q}$ but also every *sub-field* of $\mathbf{R}$ — a nonempty subset of $\mathbf{R}$ which does not reduce to $\{0\}$ and is closed w.r.t. the four arithmetic operations. Different from $\mathbf{Q}$ and $\mathbf{R}$ examples of sub-fields are, say, real numbers which can be represented as $p + r\sqrt{2}$ with rational p, q, or *algebraic numbers* — reals which are roots of algebraic polynomials with rational coefficients.

Note that we will follow the "fully rationally algebraic" approach till the concluding chapters on ellipsoid method (Chapter 6) and conic programming/interior point algorithms (Chapter 7), where working with reals becomes a must. In particular, till then, we will not use neither Theorem 2.14 nor the approximation results from Section 1.4.

2.4.3 *Immediate corollaries*

Theorem 2.13 allows us to "complete" some of the results we already know. For example, we have the following:

(1) Corollary 2.5 tells us that *a nonempty bounded polyhedral set X is the convex hull of a nonempty finite set* (e.g., the set $\mathrm{Ext}(X)$). Theorem 2.13 adds to this that the inverse is also true: *The convex hull of a finite nonempty set is a nonempty (and clearly bounded) polyhedral set.*
(2) Corollary 2.6 tells us that every pointed and nontrivial polyhedral cone is the conic hull of a nonempty finite set. Clearly, the trivial cone is also the conic hull of a nonempty finite set (specifically, the singleton $\{0\}$), same as it is the conic hull of the empty set. Taking into account that every polyhedral cone is the sum of a linear subspace and a pointed polyhedral cone (this is an immediate corollary of Proposition 2.13), we conclude that *every polyhedral cone is the conic hull of a finite, perhaps empty, set.* Theorem 2.13 adds to this that the inverse is also true: *The conic hull of a finite set $\{r_1, \dots, r_j\}$ is a polyhedral cone* (look what is given by (2.28) when $\mathcal{V} = \{0\}$).
(3) Corollary 2.7 says that *every nonempty polyhedral set admits representation* (2.28). Theorem 2.13 says that the inverse is also true.

In addition, Theorems 2.12 and 2.13 allow us to make the following important conclusion (which we will enrich in Section 3.3.4):

> *If X is a polyhedral set containing the origin, so is its polar $\{y : y^T x \leq 1\, \forall x \in X\}$, and X is the polar of its polar.*
>
> We have proved this statement when proving Theorem 2.13 (check it!).

It is very instructive to look how the inner and outer representations of a polyhedral set, or, if you prefer, Theorems 2.12 and 2.13, complement each other when justifying different facts about polyhedral sets, e.g., using inner representations of the operand(s), it is immediate to justify the claims that the arithmetic sum of polyhedral sets or an affine image of such a set are polyhedral; this task looks completely intractable when using the outer descriptions of the operands (recall that when carrying out the latter task in Section 1.3.2, we "used a cannon" — the Fourier–Motzkin elimination scheme). Similarly, with the outer descriptions, it is absolutely clear that the intersection of two polyhedral sets, or the inverse affine image of such a set, are polyhedral, while inner descriptions give absolutely no hint why these operations preserve polyhedrality.

2.4.4 ***Extending calculus of polyhedral representability**

With our current knowledge, we can add to the calculus of polyhedral representability from Section 1.3.3 two useful rules.

2.4.4.1 **Polyhedral representation of recessive cone*

Proposition 2.11 states that when $X = \{x : Ax \leq b\}$ is a nonempty polyhedral set, then the recessive cone of X is $\mathrm{Rec}(X) = \{h : Ah \leq 0\}$ and is therefore a polyhedral set. A natural question is: Assume that $X \neq \emptyset$ is given by polyhedral representation

$$X = \{x : \exists w : Px + Qw \leq r\}.$$

Is there an easy way to convert this representation to p.r. of $\mathrm{Rec}(X)$?

A natural guess is that

S.6.

$$\emptyset \neq X = \{x : \exists w : Px + Qw \leq r\} \Rightarrow \mathrm{Rec}(X) = \{h : \exists u : Ph + Qu \leq 0\}. \tag{2.30}$$

This guess is correct although this is not so evident. What is evident is that the concluding set in (2.30) is contained in $\mathrm{Rec}(X)$. Indeed, if $x \in X$, so that $Px + Qw \leq r$ for some w, and h is such that

$Ph + Qu \le 0$, then

$$P[x + th] + Q[w + tu] \le r \quad \forall t \ge 0,$$

that is, $x + th \in X$ for all $t \ge 0$, and therefore, $h \in \mathrm{Rec}(X)$. The opposite direction is more involved. Thus, assume that $h \in \mathrm{Rec}(X)$, and let us prove that $Ph + Qu \le 0$ for some u. Without loss of generality, we can assume that Q is nontrivial (otherwise, the required result is given by Proposition 2.11) and has trivial kernel (otherwise, we can restrict Q onto the orthogonal complement to this kernel without affecting what we are given and what we want to prove). Consider the set

$$W(x) = \{u : Qu \le r - Px\}.$$

Since $\mathrm{Ker}\, Q = \{0\}$, this polyhedral set does not contain lines, and $x \in X$ if and only if $W(x)$ is nonempty. Whenever this is the case, $W(x)$ has an extreme point $w(x)$, and this point, by algebraic characterization of extreme points, is such that $Q_I w(x) = [r - Px]_I$, where I is the set of indexes of cardinality $m := \dim u$ such that the submatrix Q_I composed of rows of Q with indexes from I is nonsingular, and $[r - Px]_I$ is the subvector of $r - Px$ composed of entries with indexes from I. Now, $x_j := x + jh \in X$, $j = 1, 2, \ldots$, implying that $w_j = w(x_j)$ are well defined. Let us look at the corresponding index sets I_j. These are m-element subsets of $(\dim r)$-element set, and therefore some of these sets are met in the sequence $I_1, I_2, \ldots$ infinitely many times, that is, for properly selected $j_1 < j_2 < \cdots$, we have $I_{j_s} = I$, $s = 1, 2, \ldots$. Consequently,

$$w(x_{j_s}) = Q_I^{-1}[r - Px_{j_s}]_I = Q_I^{-1}[r - Px]_I - j_s \underbrace{Q_I^{-1}[Ph]_I}_{-u}.$$

We have $Px_{j_s} + Qw_{j_s} \le r$, so that

$$Px + QQ_I^{-1}[r - Px]_I + j_s[Ph + Qu] \le r, \; s = 1, 2, \ldots$$

implying that $Ph + Qu \le 0$, as required. □

2.4.4.2 **Polyhedral representation of perspective transform*

Let $X \subset \mathbf{R}^n$ be a nonempty convex set. The (closed) *perspective transform* of X is the set

$$\overline{X} = \mathrm{cl}\{[x;t] \in \mathbf{R}^{n+1} : t > 0, t^{-1}x \in X\}.$$

Geometrically, to get $\overline{X}$, we

- add to $\mathbf{R}^n$ t-axis and lift $X \subset \mathbf{R}^n$ by 1 along this axis, thus getting the set $X_+ = \{[x;1] : x \in X\}$;
- take the union of rays in $\mathbf{R}^{n+1}$ emanating from the origin and crossing X_+ and pass from this set to its closure.

It is easily seen that $\overline{X}$ is a closed convex cone, for example,

- when $p \in \{1, 2, \infty\}$ and $X = \{x : \|x\|_p \leq 1\}$, we have $\overline{X} = \{[x;t] : \|x\|_p \leq t\}$;
- when $X = \mathbf{R}^n$, we have $\overline{X} = \{[x,t] : t \geq 0\}$;
- when $X = \mathbf{R}^n_+$, we have $\overline{X} = \mathbf{R}^{n+1}_+$.

The question we want to address is: Given a polyhedral representation

$$X = \{x : \exists w : Px + Qw \leq r\}$$

of a nonempty polyhedral set X, what is the polyhedral representation of $\overline{X}$? The answer is immediate:

S.7.

$$\begin{aligned} \emptyset \neq X = \{x : \exists w : Px + Qw \leq r\} &\Rightarrow \overline{X} \\ &= \{[x;t] : \exists w : Px - tr + Qw \leq 0 \,\&t \geq 0\}. \end{aligned} \tag{2.31}$$

Justification is as follows: What we should prove is that $[x;t]$ can be represented as $\lim_{j\to\infty}[x_j;t_j]$ with $t_j > 0$ and $t_j^{-1}x_j \in X$ if and only if $[x;t]$ belongs to the right-hand side set in (2.31). In other words, we should prove the following:

(a) if $t_j > 0$, $t_j^{-1}Px_j + Qw_j - r \leq 0$ for some w_j and $[x_j;t_j] \to [x;t]$, $j \to \infty$, then $t \geq 0$ and $Px + Qw - tr \leq 0$ for some w;
(b) vice versa, if $t \geq 0$ and $Px + Qw - tr \leq 0$, then there exist $t_j > 0$, x_j and w_j such that $t_j^{-1}Px_j + Qw_j - r \leq 0$ and $[x_j;t_j] \to [x;t]$ as $j \to \infty$.

(b) is evident. Indeed, under the premise of (b),

- in the case of $t > 0$, we can take $t_j \equiv t$, $x_j = t^{-1}x$, $w_j = t^{-1}w$;
- in the case of $t = 0$, since $X \neq \emptyset$, there exist $\bar{x}$ and $\bar{w}$ such that $P\bar{x} + Q\bar{w} - r \leq 0$, so that $P[\lambda x + \bar{x}] + Q[\lambda w + \bar{w}] - r \leq 0$ for all $\lambda > 0$; selecting somehow $t_j \to +0$, $j \to \infty$, and setting $x_j = x + t_j\bar{x}$, $w_j = t_j^{-1}w + \bar{w}$, we get $t_j > 0$, $t_j^{-1}Px_j + Qw_j - r \leq 0$ and $[x_j; t_j] \to [x; 0]$ as $j \to \infty$, as required.

To prove (a), note that this claim is trivially true when $Q = 0$. Assuming Q nonzero, the same argument as in the previous section shows that we lose nothing by assuming that the kernel of Q is trivial and that $w_j = Q_{I_j}^{-1}[r - t_j^{-1}Px_j]_{I_j}$ (for notation, see the previous section). In this case, the premise of (a) states that $t_j^{-1}Px_j + Qw_j - r \leq 0$, whence

$$Px_j + QQ_{I_j}^{-1}[t_jr - Px_j]_{I_j} - t_jr \leq 0. \qquad (*)$$

Similar to when justifying S.6, the sets I_{j_s} for properly selected $j_1 < j_2 < \cdots$ coincide with some set I, so that passing in $(*)$ to limit, as $s \to \infty$, along $j = j_s$, we get $Px + QQ_I^{-1}[tr - Px]_I - tr \leq 0$, justifying the conclusion in (a). □

Example: Perspective transform of a proper p.r.f. Let $f(x) : \mathbf{R}^n \to \mathbf{R} \cup \{+\infty\}$ be a polyhedrally representable proper function:

$$\mathrm{Epi}\{f\} := \{[x;\tau] : \tau \geq f(x)\} = \{[x;\tau] : \exists w : Px + \tau p + Qw \leq r\}. \qquad (2.32)$$

The perspective transform of Epi$\{f\}$ by definition is the set

$$\overline{X} = \mathrm{cl}\left\{[[x;t];\tau] : t > 0, \underbrace{t^{-1}\tau \geq f(t^{-1}x)}_{\Leftrightarrow \tau \geq tf(x/t)}\right\},$$

and by S.7, we have

$$\begin{aligned}\overline{X} &= \mathrm{cl}\{[[x;t];\tau] : t > 0, \underbrace{t^{-1}\tau \geq f(t^{-1}x)}_{\Leftrightarrow \tau \geq tf(x/t)}\} \\ &= \{[[x;t];\tau] : \exists w : Px + \tau p + Qw - tr \leq 0, t \geq 0\}.\end{aligned} \qquad (2.33)$$

We claim that $\overline{X}$ is the epigraph of some polyhedrally representable proper function of $[x;t]$, called *perspective transform* $\overline{f}(x,t)$ of f, and built as follows:

- for $[x;t]$ with $t > 0$, $\overline{f}(x,t) = tf(x/t)$;
- for $t < 0$, $\overline{f}(x,t) = +\infty$;
- at a point $[x;0]$, the function is $\liminf_{t'>0,[x';t']\to[x;0]} t'f(x'/t')$.

Examples:

- $f(x) = a^T x + b$, $\mathrm{Epi}\{f\} = \{[x;\tau] : \tau \geq a^T x + b\}$,

$$\overline{X} = \{[x;t];\tau] : a^T x + tb - \tau \leq 0,\ t \geq 0\} = \mathrm{Epi}\{\overline{f}(x,t)\},$$
$$\overline{f}(x,t) = \begin{cases} a^T x + bt, & t \geq 0, \\ +\infty, & t < 0. \end{cases}$$

- $f(x) = \|x\|_1 : \mathbf{R}^n \to \mathbf{R}$, $\mathrm{Epi}\{f\} = \{[x;\tau] : \exists w : -w_i \leq x_i \leq w_i,\ i \leq n, \tau \geq \sum_i w_i\}$,

$$\overline{X} = \left\{[[x;t],\tau] : \exists w : -w_i \leq x_i \leq w_i,\ i \leq n, \tau \geq \sum_i w_i, t \geq 0\right\}$$
$$= \mathrm{Epi}\{\overline{f}(x,t)\},$$
$$\overline{f}(x,t) = \begin{cases} \|x\|_1, & t \geq 0, \\ +\infty, & t < 0. \end{cases}$$

- $f(x) = \|x\|_\infty - 1 : \mathbf{R}^n \to \mathbf{R}$, $\mathrm{Epi}\{f\} = \{[x;\tau] : \exists w : -w \leq x_i \leq w,\ i \leq n, \tau \geq w - 1\}$,

$$\overline{X} = \{[[x;t],\tau] : \exists w : -w \leq x_i \leq w,\ i \leq n, \tau \geq w - t, t \geq 0\}$$
$$= \mathrm{Epi}\{\overline{f}(x,t)\},$$
$$\overline{f}(x,t) = \begin{cases} \|x\|_\infty - t, & t \geq 0, \\ +\infty, & t < 0. \end{cases}$$

The above claim has two components:

A. *The set* $\overline{X}$ *in* (2.33) *is indeed the epigraph of a proper function of* $[x;t]$ *taking values in* $\mathbf{R}\cup\{+\infty\}$ (since this set is nonempty and given by polyhedral representation, this function automatically is proper p.r.f.).

B. *The function in* **A** *is exactly the perspective transform of* f *as defined above.*

Justification of **A** is given by the following useful by its own right proposition.

Proposition 2.19. *A system of linear inequalities*

$$\overline{P}z + \tau\overline{p} + \overline{Q}w \leq \overline{r} \tag{2.34}$$

in variables $z \in \mathbf{R}^N$, $\tau \in \mathbf{R}$, $w \in \mathbf{R}^k$ *represents a proper polyhedrally representable function* $f(z)$, *that is,* $\mathrm{Epi}\{f\} = \{[z;\tau] : \exists w : \overline{P}x + \tau\overline{p} + \overline{Q}w \leq \overline{r}\}$ *if and only if the system is solvable, all solutions to the homogeneous system of linear inequalities* $\tau\overline{p} + \overline{Q}w \leq 0$ *in variables* τ, w *have* $\tau \geq 0$, *and there is a solution to the latter system with* $\tau > 0$.

With the proposition (to be proved at the end of this section) at our disposal, A becomes evident: by (2.33), we have

$$\begin{array}{rr} \overline{X} = \{[[x;t];\tau] : \exists w : \overline{P}[x;t] + \tau\overline{p} + \overline{Q}w \leq \overline{r}\}, & \text{(a)} \\ \overline{P}[x;t] \equiv [Px - tr; -t], \, \overline{p} = [p;0], \, \overline{Q} = [Q;0,\ldots,0], \, \overline{r} = 0. & \text{(b)} \end{array} \tag{2.35}$$

Applying the "only if" part of the proposition to (2.32) (that is, setting $z \equiv x$ and $\overline{P} = P$, $\overline{p} = p$, $\overline{Q} = Q$, $\overline{r} = r$), we conclude that on the set $\{[\tau;w] : \tau\overline{p} + \overline{Q}w \leq 0\} = \{[\tau;w] : \tau p + Qw \leq 0\}$, we have $\tau \geq 0$ and that this set contains a point with $\tau > 0$. Besides this, the set (2.33) is nonempty, since every pair $[x;\tau] \in \mathrm{Epi}\{f\}$ generates a triple $[[x;1];\tau] \in \overline{X}$ and $\mathrm{Epi}\{f\}$ is nonempty. We conclude that the data in (2.35(b)) satisfy the premise of the "if" part of Proposition 2.19; applying proposition to these data, we arrive at A.

B. By A, $\overline{X}$ is the epigraph of certain p.r.f., and by (2.33), on the intersection of the domain of this p.r.f. with half-space $t > 0$ (this intersection is nonempty!), the function is $tf(x/t)$. Invoking Proposition 1.1, we conclude that the p.r.f. in question is indeed the function $\overline{f}$ as defined above. □

Note that the perspective transform (same as its convexity-preserving property) extends from polyhedrally representable functions to convex ones.

Proof of Proposition 2.19. In one direction, assume that (2.34) represents a proper function f. Then, of course, (2.34) is feasible. Next, if $z \in \mathrm{Dom}\, f$, the set $\mathcal{T}(z) := \{\tau : \exists w : \tau\overline{p} + \overline{Q}w \leq r - \overline{P}z\}$ is the ray $\{\tau \geq f(z)\} \subset \mathbf{R}$, and the recessive cone of this set is the nonnegative ray; on the other hand, by S.6, this recessive cone is the set $\mathcal{R} := \{\tau \in \mathbf{R} : \exists w : \tau\overline{p} + \overline{Q}w \leq 0\}$, that is, the cone $\mathcal{K} := \{[\tau, w] : \tau\overline{p} + \overline{Q}w \leq 0\}$ is contained in the half-space $\{\tau \geq 0\}$ and contains a point with positive τ-component, as claimed.

Now assume that the set $\mathcal{E} = \{[z; \tau] : \exists w : \overline{P}z + \tau\overline{p} + \overline{Q}w \leq r\}$ is nonempty and that the cone $\mathcal{K}$ belongs to the half-space $\tau \geq 0$ and contains a point with positive τ-component, and let us prove that $\mathcal{E}$ is the epigraph of a function, specifically the function f with the domain which is the projection E of $\mathcal{E}$ onto the z-plane, and the value at a point x from this domain equal to $\min_{\tau \in \mathcal{T}(z)} \tau$. All we need to verify is that when $z \in E$, the set $\mathcal{T}(z)$ is a ray of the form $\{\tau : \tau \geq a\}$ with a real a. Indeed, when $z \in E$, $\mathcal{T}(z)$ is nonempty and is a polyhedral set on the real axis; by S.6, the recessive cone of this set is the projection of $\mathcal{K}$ onto the τ-axis, and we are in the situation when this projection is $\mathbf{R}_+$. Thus, the recessive cone of the polyhedral subset $\mathcal{T}(z)$ of $\mathbf{R}$ is $\mathbf{R}_+$, so that $\mathcal{T}(z)$ indeed is a ray. □

2.4.4.3 ** When is the convex hull of finite union of polyhedral sets polyhedral?*

Let $X_1, \ldots, X_K$ be nonempty polyhedral sets in $\mathbf{R}^n$. As is immediately seen, their union $X = \bigcup_{k=1}^K X_k$ is not necessarily convex and is thus not necessarily polyhedral. What is convex is the convex hull $\overline{X} = \text{Conv}(\bigcup_{k=1}^K X_k)$ of the union. Is it polyhedral? Let us look at two examples where $K = 2$, $n = 2$:

- $X_1 = \{[x_1; x_2] : 0 \leq x_1 \leq 1, x_2 = 0\}$, $X_2 = \{[0; 1]\}$. Here, $\overline{X} = \{[x_1; x_2] : x \geq 0, x_1 + x_2 \leq 1\}$ is a nice polyhedral set;
- $X_1 = \{[x_1; x_2] : 0 \leq x_1, x_2 = 0\}$, $X_2 = \{[0; 1]\}$. Here, $\overline{X} = \{[x_1; x_2] : x \geq 0, x_2 \left\{\begin{array}{l} \leq 1\ , x_1 = 0 \\ < 1\ , x_1 > 0 \end{array}\right\}$ is *not* polyhedral, but the closure of $\overline{X}$ is.

The following result explains what is going on here.

Proposition 2.20. *Let X_k, $k \leq K$, be nonempty polyhedral sets in $\mathbf{R}^n$ given by polyhedral representations*

$$X_k = \{x : \exists u^k : P_k x + Q_k u^k \leq r_k\}.$$

Consider the set $\widehat{X}$ given by polyhedral representation

$$\widehat{X} = \left\{ x : \exists x^k, u^k, \lambda_k, k \le K : \begin{cases} P_k x^k + Q_k u^k - \lambda_k r_k \le 0, k \le K & \text{(a)} \\ \lambda_k \ge 0, \sum_k \lambda_k = 1 & \text{(b)} \\ \sum_k x^k = x & \text{(c)} \end{cases} \right\}.$$

The polyhedral set $\widehat{X}$ is the closure of the convex hull $\overline{X}$ of the union of $X_1, \dots, X_K$. If the latter sets are bounded, then $\widehat{X}$ is $\overline{X}$ itself, so that in the case of bounded sets X_k, the convex hull of their union is polyhedral.

Proof. Observe, first, that the convex hull $\overline{X}$ of the union $X = \bigcup_{k \le K} X_k$ is exactly the set of vectors representable as $x = \sum_k \lambda_k x^k$ with $x^k \in X_k$, $\lambda_k \ge 0$, and $\sum_k \lambda_k = 1$.

Indeed, all we need to verify is that if $x = \sum_{s=1}^{S} \mu_s y^s$ with $\mu_s > 0$, $y^s \in X$, $s \le S$, and $\sum_s \mu_s = 1$, then $x = \sum_k \lambda_k x^k$ with properly selected $\lambda_k \ge 0$, $x^k \in X_k$, $k \le K$, and $\sum_k \lambda_k = 1$. To see that it is indeed the case, note that we can clearly split the index set $\{1, 2, \dots, S\}$ into K nonoverlapping subsets S_k, $k \le K$ (some of these subsets can be empty) in such a way that $s \in S_k$ implies $\mu_s > 0$ and $y^s \in X_k$. For k with nonempty S_k, let us set $\lambda_k = \sum_{s \in S_k} \mu_s$ and $x^k = \sum_{s \in S_k} \frac{\mu_s}{\lambda_k} y^s$, so that $x^k \in X_k$ due to the convexity of the latter set. For k with empty S_k, let us set $\lambda_k = 0$ and select somehow x^k in the (nonempty!) set X_k. As a result, we get $x = \sum_s \mu_s y^s = \sum_k \lambda_k x^k$ with $x^k \in X_k$, $\lambda_k \ge 0$, and $\sum_k \lambda_k = 1$; verification is complete. Note that so far we did not use polyhedrality of X_k's, just nonemptiness and convexity of these sets.

Let $\widetilde{X}$ be the set of all vectors representable as convex combinations, *with positive coefficients*, of vectors from $X_1, \dots, X_K$; note that $\widetilde{X} \subset \overline{X}$.

Observe that

$$\widetilde{X} = \left\{ x : \exists x^k, u^k, \lambda_k, k \le K : \begin{cases} P_k x^k + Q_k u^k - \lambda_k r_k \le 0, & (a') \\ \quad k \le K & \\ \lambda > 0, \sum_k \lambda_k = 1 & (b') \\ x = \sum_k x^k & (c') \end{cases} \right\}$$

(!)

Indeed, when x belongs to the right-hand side set in (!), we have $y^k := \lambda_k^{-1}x^k \in X_k$ due to $P_k y^k + Q_k[\lambda_k^{-1}u^k] \leq r_k$ and $x = \sum_k \lambda_k y^k$. Vice versa, when $x \in \widetilde{X}$, we have $x = \sum_k \lambda_k y^k$ with positive λ_k summing up to 1 and $y^k \in X_k$. The latter means that there exist v^k such that $P_k y^k + Q_k v^k \leq r_k$. Setting $x^k = \lambda_k y^k$, $u^k = \lambda_k v^k$, we ensure the validity of (a′)–(c′), so that x belongs to the right-hand side set in (!).

Observe that $\widetilde{X}$ is dense in $\overline{X}$, meaning that every point $x \in \overline{X}$ is the limit of a sequence of points from $\widetilde{X}$. Indeed, when $x = \sum_k \lambda_k x^k$ with nonnegative λ_k summing up to 1 and $x^k \in X_k$, we have $x = \lim_{i\to\infty} \sum_k \frac{\lambda_k + 1/i}{1+K/i} x^k$, and the points from the right-hand side sequence belong to $\widetilde{X}$. Finally, observe that $\widehat{X}$ is closed (it is polyhedrally representable and thus polyhedral) and $\widetilde{X}$ is dense in $\widehat{X}$. Indeed, by (!), we have $\widetilde{X} \subset \widehat{X}$. On the other hand, let us fix somehow $\overline{x}^k \in X_k$ and $\overline{\lambda}_k > 0$ such that $\sum_k \overline{\lambda}_k = 1$, and let $\overline{u}^k$ be such that $P_k\overline{x}^k + Q_k\overline{u}^k \leq r_k$. Given $x \in \widehat{X}$, there exist x^k, u^k and λ_k satisfying (a)–(c). Setting

$$x^{ki} = (1-1/i)x^k + (1/i)[\overline{\lambda}_k\overline{x}^k],\ u^{ki} = (1-1/i)u^k + (1/i)[\overline{\lambda}_k\overline{u}^k],$$
$$\lambda_{ki} = (1-1/i)\lambda_k + (1/i)\overline{\lambda}_k,\ i = 1, 2, \ldots$$

and taking into account that

$$P_k x^k + Q_k u^k \leq \lambda_k r_k, \quad P_k[\overline{\lambda}_k\overline{x}^k] + Q_k[\overline{\lambda}_k\overline{u}^k] \leq \overline{\lambda}_k r_k, k \leq K,$$

we ensure that $P_k x^{ki} + Q_k u^{ki} \leq \lambda_{ki} r_k$, $\lambda_{ki} > 0$, $\sum_k \lambda_{ki} = 1$, implying that $x^{(i)} := \sum_k x^{ki} \in \widetilde{X}$. As $i \to \infty$, we clearly have $x^{(i)} \to x$, so that $\widetilde{X}$ is indeed dense in $\widehat{X}$. The latter combines with closedness of $\widehat{X}$ to imply that $\widehat{X}$ is the closure of $\widetilde{X}$, and the latter set, due to the fact that $\widetilde{X}$ is dense in $\overline{X}$, is the same as the closure of $\overline{X}$. Thus, $\widehat{X} = \mathrm{cl}\,\overline{X}$.

It remains to be proved that when X_k are bounded, then $\overline{X}$ is closed. This is immediate: Assuming that $x = \lim_{i\to\infty} \sum_k \lambda_{ki} x^{ki}$ with nonnegative λ_{ki}, $\sum_k \lambda_{ki} = 1$, and $x^{ki} \in X_k$, boundedness of X_k, $k \leq K$, allows us to find a subsequence $i_1 < i_2 < \cdots$ of indexes such that for some λ_k and x^k, $k \leq K$, it holds $\lambda_{ki_s} \to \lambda_k$ and $x^{ki_s} \to x^k$ for every k as $s \to \infty$. Since X_k are polyhedral and thus closed, we have $x^k \in X_k$, and of course $\lambda_k \geq 0$, $\sum_k \lambda_k = 1$, that is, $x = \lim_{s\to\infty} \sum_k \lambda_{ki_s} x^{ki_s} = \sum_k \lambda_k x^k \in \overline{X}$. □

Illustration: Consider the situation as follows. We are given n nonempty and bounded polyhedral sets $X_j \subset \mathbf{R}^r$, $j = 1, \ldots, n$; think about them as "resource sets" of production units: entries in $x \in X_j$ are amounts of various resources and X_j — the set of vectors of resources available, in principle, for jth unit. The vector of products of jth unit depends on the vector x_j of resources consumed by the unit and production plan utilized in the unit; there exists $K_j < \infty$ of these plans, and the vector of products $y_j \in \mathbf{R}^p$ stemming from resources x_j under kth plan can be picked by us, at our will, from the set

$$Y_j^k[x_j] = \{y_j \in \mathbf{R}^p : z_j := [x_j; -y_j] \in V_j^k\},$$

where V_j^k, $k \leq K_j$, are given to us bounded polyhedral "technological sets" of the units with projections onto the x_j-plane equal to X_j, so that for every $k \leq K_j$, it holds

$$x_j \in X_j \Leftrightarrow \exists y_j : [x_j; -y_j] \in V_j^k. \tag{2.36}$$

We assume that all the sets V_j^k are given by polyhedral representations and set

$$V_j = \bigcup_{k \leq K_j} V_j^k.$$

Given total vector $R \in \mathbf{R}^r$ of resources available to all n units and the vector $P \in \mathbf{R}^p$ of total demands on the products, for $j \leq n$, we want to select $x_j \in X_j$, $k_j \leq K_j$, and $y_j \in Y_j^{k_j}[x_j]$ in such a way that

$$\sum_j x_j \leq R \;\&\; \sum_j y_j \geq P$$

or, which is the same, select $z_j = [x_j; v_j] \in V_j$, $j \leq n$, in such a way that $\sum_j z_j \leq [R; -P]$. The presence of "combinatorial part" in our decision — selection of production plans in finite sets — makes the problem difficult (NP-hard, see Chapter 6). Let us try to apply

Shapley–Folkman theorem to overcome, to some extent, this difficulty. Observe that our problem reads as follows:

$$\text{Find } z_j \in V_j \text{ such that } \sum_{j=1}^{n} z_j \leq s := [R; -P]. \tag{P}$$

Note that given polyhedral representations of V_j^k, we, in view of Proposition 2.20, can build efficiently explicit polyhedral representations of the convex hulls

$$\overline{V}_j = \mathrm{Conv}(V_j)$$

of the sets V_j. Let us relax the problem of interest (P) to the following problem:

$$\text{Find } z_j \in \overline{V}_j \text{ such that } \sum_{j=1}^{n} z_j \leq s := [R; -P]. \tag{$\overline{P}$}$$

By calculus of polyhedral representations, $(\overline{P})$ is problem of the following form:

> Given polyhedral representation of nonempty polyhedral set $Z \subset \mathbf{R}^{r+p}$ and vector $s \in \mathbf{R}^{r+p}$, find $z \in Z$ such that $z \leq s$,

which is an explicit LO feasibility problem. Thus, we can apply LO to check whether $(\overline{P})$ is solvable, and if it is the case, find a solution $\{z_j, j \leq n\}$ to $(\overline{P})$. Applying Shapley–Folkman theorem, we can convert, in a computationally efficient fashion, this solution into another feasible solution, $\{[x_j; v_j], j \leq n\}$, for which for all but at most

$$d := \min[r + p, n]$$

components $[x_j; v_j]$ belong to V_j, that is, "are implementable" — for the corresponding j, $x_j \in X_j$ and $y_j = -v_j \in Y_j^{k_j}[x_j]$ with properly selected $k_j \leq K_j$. For "bad" indexes j — those for which $[x_j; v_j] \in \overline{V}_j \backslash V_j$, let the set of these indexes j be denoted by J — we still have $x_j \in X_j$. We can correct the corresponding y_j, passing from $[x_j; v_j]$ to $[x_j; \overline{v}_j]$ with $\overline{v}_j \in -Y_j^1[x_j]$ or, better, $\overline{v}_j$ defined as the optimal solution to the "best" — with the smallest optimal value — of the convex optimization problem

$$\min_{u_k}\{\|v_j - u_k\|_q : [x_j; u_k] \in V_j^k\},\ k \leq K_j.$$

As a result, we get "fully implementable" solution $\{[x_j; \overline{v}_j], j \leq n\}$, where $\overline{v}_j = v_j$ for $j \notin J$, to problem (P). This solution, in general, is infeasible. However, selecting $q \in [1, \infty]$, setting

$$D_j = \max_{[x;v] \in V_j, [x;v'] \in V_j} \|v - v'\|_q,\ D = \max_j D_j$$

and taking into account that $\mathrm{Card}(J) \leq d = \min[r + p, n]$, we have $\sum_j [x_j; \overline{v}_j] \leq s + \delta$, $\|\delta\|_q \leq dD$, and $\sum_j x_j \leq R$. In the case of "mass production," when $\|P\|_q$ is large, the violation of the constraint $\sum_j \overline{v}_j \leq -P$ as quantified by $\|\delta\|_q$ is a small fraction of the magnitude of P, and our implementable solution has chances to be a good, from "practical perspective," surrogate of a feasible solution to (P).

***Refinement by randomization:** The above corrections $v_j \mapsto \overline{v}_j$, $j \in J$ were aimed at getting implementable solution to (P) which satisfies the resource constraint $\sum_j x_j \leq R$, albeit perhaps violating the production constraint $\sum_j v_j \leq -P$. If we are ready to violate both of the constraints, we can use alternative correction possessing certain theoretical advantages. Specifically, let us set $q = 2$,[9] redefine D_j and D as

$$D_j = \max_{z, z' \in V_j} \|z - z'\|_2,\ D = \max_j D_j$$

and relax the assumption (2.36) to

$$\forall j \leq n : k \leq K_j, [u_j; v_j] \in V_j^k \Rightarrow u_j \in X_j.$$

For $j \in J$, the vectors $z_j = [x_j; v_j]$ we have built when solving $(\overline{P})$ are convex combinations of certain vectors $z_j^k = [x_j^k; v_j^k] \in V_j^k$:

$$[x_j; v_j] = \sum_{k=1}^{K_j} \lambda_k^j z_j^k \qquad \left[\lambda^j \geq 0, \sum_{k=1}^{K_j} \lambda_k^j = 1\right].$$

[9]With some minor modifications, we could allow for q to lie in $[2, \infty]$.

Treating λ^j as probability distribution on the finite set $\{1,\ldots,K_j\}$, let us generate random collections of vectors $[\xi_j;\eta_j]$, $j\in J$, by generating at random, independent of each other, indexes k_j, $j\in J$, with k_j drawn from the distribution λ^j on $\{1,\ldots,K_j\}$. Observe that all realizations $[\xi_j,\eta_j]$ belong to V_j and are thus implementable by jth production unit. Besides this,

$$\mathbf{E}\{[\xi_j;\eta_j]\}=z_j:=[x_j;v_j]$$

($\mathbf{E}$ stands for expectation) and

$$\mathbf{E}\{\|[\xi_j;\eta_j]-z_j\|_2^2\}\leq D^2,$$

implying, in view of mutual independence of $[\xi_j;\eta_j]$ over j and since the random vectors $[\xi_j;\eta_j]-z_j$ are zero mean, that

$$\mathbf{E}\Bigg\{\Bigg\|\underbrace{\sum_{j\in J}([\xi_j;\eta_j]-z_j)}_{\delta}\Bigg\|_2^2\Bigg\}=\sum_{j\in J}\mathbf{E}\{\|[\xi_j;\eta_j]-z_j\|_2^2\}\leq dD^2.$$

It follows that the typical $\|\cdot\|_2$-norm of the random vector δ is of the order of $D\sqrt{d}$. At the same time, random solution ζ to (P) for which blocks ζ_j with $j\notin J$ are $[x_j;v_j]$ and blocks ζ_j with $j\in J$ are $[\xi_j;\eta_j]$ is implementable — $\zeta_j\in V_j$ for all j — and satisfies $\sum_j\zeta_j\leq[R;-P]+\delta$. Thus, the typical $\|\cdot\|_2$-norm of violation of the target constraint $\sum_j z_j\leq[R;-P]$ by our implementable random solution is of the order of $D\sqrt{d}$ — much better than for our original deterministic corrections, where the bound on violation of the constraint was proportional to d. To get a reasonable implementable solution to (P), we can generate at random, say, 100 realizations of ζ and select the one with the smallest $\|\cdot\|_2$-norm of constraint violation.

2.5 Exercises

Exercise 2.1.* In the following list, mark the sets which are linear subspaces and point out their dimensions:

- $\mathbf{R}^n$,
- $\{0\}$,

- $\emptyset$,
- $\{x \in \mathbf{R}^n : \sum_{i=1}^n ix_i = 0\}$,
- $\{x \in \mathbf{R}^n : \sum_{i=1}^n ix_i^2 = 0\}$,
- $\{x \in \mathbf{R}^n : \sum_{i=1}^n ix_i = 1\}$,
- $\{x \in \mathbf{R}^n : \sum_{i=1}^n ix_i^2 = 1\}$.

Exercise 2.2.* Point out a linear and an affine basis in the linear subspace

$$\left\{x \in \mathbf{R}^n : \sum_{i=1}^n x_i = 0\right\}$$

and the orthogonal complement to this subspace.

Exercise 2.3.* Point out the linear subspace parallel to the affine subspace,

$$M = \left\{x \in \mathbf{R}^n : \sum_{i=1}^n x_i = 1\right\} \subset \mathbf{R}^n,$$

and an affine basis in M.

Exercise 2.4.* In the following list, point out the dimensions of the sets and mark those sets which are convex:

(1) $\mathbf{R}^n$,
(2) $\{0\}$,
(3) $\{x \in \mathbf{R}^n : \sum_{i=1}^n ix_i = 0\}$,
(4) $\{x \in \mathbf{R}^n : \sum_{i=1}^n ix_i \leq 0\}$,
(5) $\{x \in \mathbf{R}^n : \sum_{i=1}^n ix_i \geq 0\}$,
(6) $\{x \in \mathbf{R}^n : \sum_{i=1}^n ix_i^2 = 1\}$,
(7) $\{x \in \mathbf{R}^n : \sum_{i=1}^n ix_i^2 \leq 1\}$,
(8) $\{x \in \mathbf{R}^n : \sum_{i=1}^n ix_i^2 \geq 1\}$,
(9) $\{x \in \mathbf{R}^2 : |x_1| + |x_2| \leq 1\}$,
(10) $\{x \in \mathbf{R}^2 : |x_1| - |x_2| \leq 1\}$,
(11) $\{x \in \mathbf{R}^2 : -|x_1| - |x_2| \leq 1\}$.

Exercise 2.5.* For the sets to follow, point out their linear and affine spans and their convex hulls:

- $X = \{[0;1],[1;1],[2;1]\}$,
- $X = \{[0;0];[1;0];[1;1];[0;1]\}$,
- $X = \{x \in \mathbf{R}^2 : x_2 = 0, |x_1| \leq 1\}$,
- $X = \{x \in \mathbf{R}^2 : x_2 = 1, |x_1| \leq 1\}$,
- $X = \{x \in \mathbf{R}^2 : |x_1| - |x_2| = 1\}$.

Exercise 2.6.* Mark by **T** those of the following claims which are always true:

- The linear image $Y = \{y = Ax : x \in X\}$ of a linear subspace X is a linear subspace.
- The linear image $Y = \{y = Ax : x \in X\}$ of an affine subspace X is an affine subspace.
- The linear image $Y = \{y = Ax : x \in X\}$ of a convex set X is convex.
- The affine image $Y = \{y = Ax + b : x \in X\}$ of a linear subspace X is a linear subspace.
- The affine image $Y = \{y = Ax + b : x \in X\}$ of an affine subspace X is an affine subspace.
- The affine image $Y = \{y = Ax + b : x \in X\}$ of a convex set X is convex.
- The intersection of two linear subspaces in $\mathbf{R}^n$ is a linear subspace.
- The intersection of two affine subspaces in $\mathbf{R}^n$ is an affine subspace.
- The intersection of two affine subspaces in $\mathbf{R}^n$, when nonempty, is an affine subspace.
- The intersection of two convex sets in $\mathbf{R}^n$ is a convex set.
- The intersection of two convex sets in $\mathbf{R}^n$, when nonempty, is a convex set.

Exercise 2.7.* Given are n distinct from each other sets

$$E_1 \subset E_2 \subset \cdots \subset E_n$$

in $\mathbf{R}^{100}$. How large can be n, if

- every one of E_i is a linear subspace,
- every one of E_i is an affine subspace,
- every one of E_i is a convex set.

Exercise 2.8.† Let X be a nonempty convex set in $\mathbf{R}^n$ and $x \in X$. Prove that x is an extreme point of X

- if and only if the set $X \backslash \{x\}$ is convex,
- if and only if for every representation $x = \sum_{i=1}^{k} \lambda_i x_i$ of x as a convex combination of points from X with positive coefficients λ_i, it holds $x_i = x$, $i = 1, \ldots, k$.

Exercise 2.9.[†] Let $X = \{x \in \mathbf{R}^n : a_i^T x \leq b_i, 1 \leq i \leq m\}$ be a polyhedral set. Prove the following:

(1) If X' is a face of X, then there exists a linear function $e^T x$ such that

$$X' = \operatorname*{Argmax}_{x \in X} e^T x := \left\{ x \in X : e^T x = \sup_{x' \in X} e^T x' \right\}.$$

(2) If v is a vertex of X, then there exists a linear function $e^T x$ such that v is the unique maximizer of this function on X.

Exercise 2.10.[*] Describe all extreme points of the following convex sets:

(1) $X = \text{Conv}\{1, 2, 3, 4, 5\}$,
(2) $X = \text{Conv}\{[0; 0], [1; 1], [1; 0], [0.5; 0.5]\}$,
(3) $X = \{x \in \mathbf{R}^n : 0 \leq x_1 \leq 1, 1 \leq i \leq n, \sum_{i=1}^{n} x_i \leq 3/2\}$,
(4) $X = \{x \in \mathbf{R}^n : \|x\|_2 \leq 1\}$,
(5) $X = \{x \in \mathbf{R}^n : x \geq 0, \sum_{i=1}^{n} x_i = 1, \sum_{i=1}^{n} a_i x_i = 1\}$, where $a_1 < a_2 < \cdots < a_n$ and $n \geq 2$.

Exercise 2.11.[†] By Birkhoff Theorem, the extreme points of the polytope $\Pi_n = \{[x_{ij}] \in \mathbf{R}^{n \times n} : x_{ij} \geq 0, \sum_i x_{ij} = 1 \,\forall j, \sum_j x_{ij} = 1 \,\forall i\}$ are exactly the Boolean (with entries 0 and 1) matrices from this set. Prove that the same holds true for the "polytope of sub-doubly-stochastic" matrices $\Pi_{m,n} = \{[x_{ij}] \in \mathbf{R}^{m \times n} : x_{ij} \geq 0, \sum_i x_{ij} \leq 1 \,\forall j, \sum_j x_{ij} \leq 1 \,\forall i\}$.

Exercise 2.12 (follow-up to Exercise 2.11).[†] Let x be an $n \times n$ entrywise nonnegative matrix with all row and all column sums ≤ 1. Is it true that for some doubly-stochastic matrix $\overline{x}$, the matrix $\overline{x} - x$ is entrywise nonnegative?

Exercise 2.13.[*] Point out the recessive cones and the extreme points of the polyhedral sets:

(1) $X = \{x \in \mathbf{R}^3 : x_1 \geq 0, x_2 \geq 0, x_3 \geq 0\}$,
(2) $X = \{x \in \mathbf{R}^2 : x_1 \geq 0, x_2 \geq 0, x_1 + x_2 \leq 2\}$,
(3) $X = \{x \in \mathbf{R}^3 : x_1 \geq 0, x_2 \geq 0, x_1 + x_2 \leq 2\}$,
think what is the general form of the result on $\mathrm{Ext}(X)$ you got,
(4) $X = \{x \in \mathbf{R}^3 : x_1 \geq 0, x_2 \geq 0, x_1 + x_2 \leq 2, x_3 \geq 0\}$.

Exercise 2.14.* Prove that if a nonempty polyhedral set X is represented as

$$X = \mathrm{Conv}(\{v_1, \ldots, v_N\}) + \mathrm{Cone}(\{r_1, \ldots, r_M\}),$$

then $\mathrm{Rec}(X) = \mathrm{Cone}(\{r_1, \ldots, r_M\})$.

Hint: Assuming that a recessive direction e of X does not belong to $\mathrm{Cone}(\{r_1, \ldots, r_M\})$, use the homogeneous Farkas lemma to verify that there exists d such that $d^T e < 0$ and $d^T r_j \geq 0$ for all j and think of whether the linear function $d^T x$ of x is bounded below on X.

Exercise 2.15.† Prove that if K is a polyhedral cone, then the dual cone K_* is so and $(K_*)_* = K$.

Exercise 2.16.† Prove that if $K = \{x \in \mathbf{R}^n : a_i^T x \geq 0, i = 1, \ldots, m\}$, then $K_* = \mathrm{Cone}(\{a_1, \ldots, a_m\})$.

Exercise 2.17.† Prove that if K is a polyhedral cone and d is a generator of an extreme ray of K, then in every representation,

$$d = d_1 + \cdots + d_M, \; d_i \in K \, \forall i,$$

the vectors d_i are nonnegative multiples of d.

Exercise 2.18.† Let

$$K = \mathrm{Cone}(\{r_1, \ldots, r_M\}).$$

Prove that if R is an extreme ray of K, then one of r_j can be chosen as a generator of R. What is the "extreme point" analogy of this statement?

Exercise 2.19.† Let $K_1, \ldots, K_m$ be polyhedral cones in $\mathbf{R}^n$. Prove that

(1) $K_1 + \cdots + K_m$ is a polyhedral cone in $\mathbf{R}^n$.
(2) $(K_1 \cap K_2 \cap \cdots \cap K_m)_* = (K_1)_* + \cdots + (K_m)_*$.

Exercise 2.20.* For the polyhedral cones to follow, point out a base (if it exists), extreme rays (if they exist) and a polyhedral representation of the dual cone:

(1) $K = \{0\} \subset \mathbf{R}$,
(2) $K = \mathbf{R}$,
(3) $K = \{x \in \mathbf{R}^2 : x_1 \geq 0, x_2 \geq 0, x_1 \leq x_2\}$,
(4) $K = \{x \in \mathbf{R}^3 : x_1 + x_2 \geq 0, x_2 + x_3 \geq 0, x_1 + x_3 \geq 0\}$,
(5) $K = \{x \in \mathbf{R}^3 : x_1 + x_2 \geq x_3, x_2 + x_3 \geq x_1, x_1 + x_3 \geq x_2\}$.

Exercise 2.21.* For the polyhedral sets X to follow, find their representations in the form of

$$X = \text{Conv}(\{v_1, \ldots, v_N\}) + \text{Cone}\,(\{r_1, \ldots, r_M\})$$

(you may skip the derivation and present the results only):

(1) $X = \{x \in \mathbf{R}^n : x \geq 0, \sum_i x_i \leq 1\}$,
(2) $X = \{x \in \mathbf{R}^n : x \geq 0, \sum_i x_i \geq 1\}$,
(3) $X = \{x \in \mathbf{R}^n : x \geq 0, 1 \leq \sum_i x_i \leq 2\}$,
(4) $X = \{x \in \mathbf{R}^3 : x \geq 0, x_1 + x_2 - x_3 \geq 0\}$,
(5) $X = \{x \in \mathbf{R}^3 : x \geq 0, x_1 + x_2 - x_3 \geq 1\}$.

Exercise 2.22 (Computational study).† Implement in software the approaches presented in the illustration in Section 2.4.4.3 and run simulations. Recommended setup (for notation, see Section 2.4.4.3):

- $n = 20$, $p = r = 10$,
- $X_j = \{x \in \mathbf{R}^p : 0 \leq x \leq [1; \ldots; 1]\}$, $j = 1, \ldots, n$,
- $Y_j[x_j] = \{y \in \mathbf{R}^p : 0 \leq y \leq P^{i,j} x_j\}$, where $P^{i,j}$, $1 \leq i \leq p$, $1 \leq j \leq n$, are generated at random permutation matrices,
- R, P: generated at random with entries P_i, R_i uniformly distributed on $[0, n]$.

Quantify the inaccuracy of implementable solution $\{[x_j; y_j], j \leq n\}$ to (P) by the quantity

$$\epsilon = \frac{[\sum_j x_j - R]_+ + [P - \sum_j y_j]_+}{\sum_i P_i + \sum_i R_i},$$

where $[z]_+$ is the sum, over i, of positive parts $\max[z_i, 0]$ of entries z_i in vector z.

Apply both deterministic and randomized techniques for building implementable solutions to problem (P) described in illustration, select the result with smaller inaccuracy and register the resulting inaccuracies in a series of simulations.

Exercise 2.23 (Semi-computational study).* Let n and m be positive integers with $n \geq 2$. Let us call an (m, n)*-bundle* a collection $\mathcal{E}$ of m $(n-1)$-dimensional linear subspaces $E_1, \ldots, E_m$ in $\mathbf{R}^n$. Such a collection can be represented by m nonzero vectors $e_i \in \mathbf{R}^n$, $i \leq n$, according to $E_i = \{x \in \mathbf{R}^n : e_i^T x = 0\}$. Let us call an (m, n)-bundle $\mathcal{E} = \{E_i, i \leq m\}$ *regular* if for every $k \leq \min[m, n]$, the intersection of every k of the linear subspaces from the collection $\mathcal{E}$ is of dimension $n - k$. Equivalently, (m, n)-bundle $\mathcal{E} = \{E_i = \{x \in \mathbf{R}^n : e_i^T x = 0\}, i \leq m\}$ is regular if and only if for every $k \leq \min[m, n]$, any k among the vectors $e_1, \ldots, e_m$ are linearly independent.

Given (m, n)-bundle, we can partition $\mathbf{R}^n$ as follows: At every point x outside of $\cup_{i \leq m} E_i$, the m reals $e_1^T x, e_2^T x, \ldots, e_m^T x$ are nonzero; let us call the sequence of their signs the *signature* of x. The set of points $x \notin \cup_{i \leq m} E_i$ with a given signature is clearly convex and open, and these sets do not overlap; let us call a nonempty set of this type the *cell* of the partition given by $\mathcal{E}$ *with signature in question.* Thus, given $\mathcal{E}$, we can partition $\mathbf{R}^n$ into the union of hyperplanes E_i and a finite number of cells — nonempty open convex sets; their closures are cones bounded by hyperplanes E_i.

As a simple example, let $1 \leq m \leq n$ with $n > 1$ and E_i, $1 \leq i \leq m$, be coordinate hyperplanes: $E_i = \{x \in \mathbf{R}^n : x_i = 0\}$. The cells of the partition associated with the just described regular (m, n)-bundle are the 2^m sets $\{x \in \mathbf{R}^n : \epsilon_i x_i > 0, i \leq m\}$ associated with 2^m sequences $\vec{\epsilon} = (\epsilon_1, \ldots, \epsilon_m)$ with $\epsilon_i = \pm 1$.

The goal of this exercise is to understand how many cells are there in a partition associated with (m, n)-bundle and extract some consequences.

(1) Prove the following fact:

> (!) The number of cells in the partition associated with regular (m, n)-bundle, $m \geq 1$, $n \geq 2$, depends solely on m, n, but not on the specific (m, n)-bundle in question. This number, let it be denoted

$N(m,n)$, satisfies the relation

$$\begin{aligned} n \geq 2, m = 1 &\Rightarrow N(m,n) = 2, \\ n \geq 2, m \geq 2 &\Rightarrow N(m,n) = N(m-1,n) + N(m-1,n-1), \end{aligned} \tag{2.37}$$

where by definition $N(\mu,1) = 2$ when $\mu \geq 1$.

(2) Given (m,n)-bundle with $n \geq 2$, we can subject the vectors $e_1,\ldots,e_m$ to whatever small perturbations to make the perturbed bundle regular; on the other hand, small enough perturbations of the data of an (m,n)-bundle cannot reduce the number of cells in the associated partition. Derive from these observations the following:

Specifying the function $N(m,n)$ of $m \geq 1, n \geq 2$ by the recurrence,

$$N(m+1,n+1) = N(m,n+1) + N(m,n), \quad m \geq 1, n \geq 2$$

and "initial conditions" $N(m,1) \equiv 2$, $m \geq 1$ and $N(1,n) = 2$, $n \geq 1$, we get an upper bound on the number of cells in a partition associated with (m,n)-bundle; this bound coincides with the actual number of cells whenever the bundle is regular.

Use the recurrence to compute $N(5,10)$ and $N(10,20)$.

(3) Imagine you draw at random, in a symmetric with respect to orthogonal rotation fashion, 11-dimensional linear subspace of $\mathbf{R}^{22}$.[10] What are the chances to get a linear subspace containing strictly positive vector? Answer the same question when you draw your 11-dimensional subspace at random from $\mathbf{R}^{21}$.

(4) Consider m hyperplanes $F_i = \{x \in \mathbf{R}^n : a_i^T x = b_i\}$, $a_i \neq 0$, $b_i \neq 0$; let us call such a collection an *inhomogeneous* (m,n)*-bundle*. For a point $x \in \mathbf{R}^n \backslash (\cup_i F_i)$, the reals $a_i^T x - b_i$, $i = 1,\ldots,m$, are

[10] To get a random k-dimensional subspace E in $\mathbf{R}^m$ in a fashion invariant w.r.t. rotations of the space, it suffices to generate an $(n-k)\times n$ *Gaussian matrix* G — random matrix with entries drawn at random, independent of each other, from the standard Gaussian distribution, and to take $E = \operatorname{Ker} G$. Indeed, when U is an $n \times n$ orthogonal matrix, GU has exactly the same distribution as G, so that with our construction, the distribution of $U^{-1}E$ is the same as the distribution of E. Alternatively, you can generate $n \times k$ Gaussian matrix A and take its image space as E.

nonzero; let us call the sequence of their signs *the signature* of x. Similar to that in the case when the hyperplanes pass through the origin, the entire $\mathbf{R}^n$ is partitioned into the union $\cup_i F_i$ of F_i and a number of cells — convex sets with nonempty interior composed of points with common signature. Prove that with $N(m,n)$ given by (!), the function

$$\frac{1}{2}N(m+1,n+1)$$

is a tight upper bound on the number of cells in the partition of $\mathbf{R}^n$ associated with an inhomogeneous (m,n)-bundle.

(5) Given positive integers m, n, consider random polyhedral set

$$P = \{x \in \mathbf{R}^n : Ax \geq b\},$$

with $m \times n$ matrix A and entries in A, b drawn, independent of each other, from the standard Gaussian distribution. What is the probability $p(m,n)$ for this set to have a nonempty interior? Fill the following table:

$p(10,10)$	$p(20,10)$	$p(30,10)$	$p(40,10)$	$p(50,10)$	$p(60,10)$	$p(70,10)$

(6) Consider standard form LO program with n variables and $m \leq n$ equality constraints:

$$\max\left\{\sum_{j=1}^{n} c_j x_i : \sum_{j=1}^{n} A_{ij}x_j = b_i, i \leq m, \ x \geq 0\right\}.$$

Assuming A_{ij} and b_i drawn at random, independent of each other, from the standard Gaussian distribution, what is the probability $q(m,n)$ for the problem to have strictly positive feasible solution? Fill the following table:

$q(10,10)$	$q(10,20)$	$q(10,30)$	$q(10,40)$	$q(10,50)$
$q(100,100)$	$q(100,150)$	$q(100,200)$	$q(100,250)$	$q(100,300)$

Exercise 2.24 (Computational study).[†] Find the number of extreme points in the polyhedral set

$$Q = \left\{ x \in \mathbf{R}^5 : \sum_{j=1}^{5} \cos(i \cdot j) x_j \le 10,\ i = 1, \ldots, 10 \right\}$$

and compare it with the upper bound $\binom{10}{5} = 252$ given by algebraic characterization of extreme points.

Exercise 2.25 (Computational study).[†] Generate 1000 polyhedral sets

$$\left\{ x \in \mathbf{R}^5 : \sum_{j=1}^{5} a_{ij} x_j \le b_i,\ i \le 10 \right\}$$

with the data entries drawn at random, independent of each other, from the uniform distribution on $\{-1, 0, 1\}$ and find percentage of instances which are

- empty,
- with lines,
- bounded,
- unbounded,
- nonempty and bounded.

Chapter 3

Theory of Systems of Linear Inequalities and Duality

With all due respect to the results on polyhedral sets we became acquainted with, there still are pretty simple questions about these sets (or, which is basically the same, about finite systems of linear inequalities) which we do not know how to answer. Examples of these questions are as follows:

- *How to recognize whether or not a polyhedral set $X = \{x : Ax \leq b\}$ is empty?*
- *How to recognize that a polyhedral set is bounded/unbounded?*
- *How to recognize whether or not two polyhedral sets $X = \{x : Ax \leq b\}$ and $X' = \{x : A'x \leq b'\}$ coincide with each other?* More generally, *with X, X' as above, how to recognize that $X \subset X'$?*
- *How to recognize whether or not a given LO program is feasible/bounded/solvable?*

  This list can be easily extended...

Now, there are two ways to pose and to answer questions like those mentioned above:

A. (Descriptive approach) One way is to ask what are "easy to certify," necessary and sufficient conditions for a candidate answer to be valid? For example, it is easy to certify that the solution set $X = \{x \in \mathbf{R}^n : Ax \leq b\}$ of a system of linear inequalities $Ax \leq b$ is nonempty — a certificate is given by any feasible solution to the system. Given a candidate certificate of this type — a vector $x \in \mathbf{R}^n$ — it is easy to check whether it is a valid certificate (plug x into the system and check whether it becomes satisfied); if it is so, the system definitely is solvable. And of course, vice versa — if the system is solvable, this property can be certified by a just outlined certificate. In contrast to this, it is unclear how to certify that the solution set of a system $Ax \leq b$ is empty, which makes the question "whether $X = \{x : Ax \leq b\}$ is or is not empty" difficult.

B. (Operational approach) After we know *what* are "simple" certificates for candidate answers to the question under consideration, it is natural to ask *how to generate* appropriate certificates. For example, we know what is a simple certificate for nonemptiness of the solution set of a system $Ax \leq b$ — this is (any) feasible solution to this system. This being said, it is unclear how to build such a certificate given the system, even when it is known in advance that such a certificate exists.

In this chapter, we focus on A; specifically, we will find out what are simple certificates for various properties of polyhedral sets. The questions of how to build the certificates (which, essentially, is an algorithmic question) will be answered in the part of the book devoted to LO algorithms.

3.1 General theorem on alternative

3.1.1 *GTA: Formulation, proof, different versions*

Consider a general finite system of m linear inequalities, strict and nonstrict, in variables $x \in \mathbf{R}^n$. Such a system can always be written down in the form

$$a_i^T x \begin{cases} < b_i, & i \in I \\ \leq b_i, & i \in \overline{I}, \end{cases} \tag{$\mathcal{S}$}$$

where $a_i \in \mathbf{R}^n$, $b_i \in \mathbf{R}$, $1 \leq i \leq m$, I is a certain subset of $\mathcal{I} := \{1, \ldots, m\}$, and $\overline{I}$ is the complement of I in $\mathcal{I}$.

The fact that $(\mathcal{S})$ is a "universal" form of a finite system of strict and nonstrict linear inequalities and linear equations is evident: a linear equality $a^T x = b$ can be equivalently represented by the system of two opposite inequalities $a^T x \leq b$ and $a^T x \geq b$, thus, we can think that our system comprises inequalities only; every one of these inequalities can be written in the form $a^T x \Omega b$, where Ω is a relation from the list $<, \leq, \geq, >$.

Finally, we can make the signs Ω of all inequalities either $\leq$, or $<$, by replacing an inequality of the form $a^T x > b$ ($a^T x \geq b$) by its equivalent $[-a]^T x < [-b]$ ($[-a]^T x \leq [-b]$).

In what follows we assume that the system is nonempty, otherwise all kinds of questions about the system become trivial: an empty system of equations is solvable, and its solution set is the entire $\mathbf{R}^n$.

The most basic descriptive question about $(\mathcal{S})$ is whether or not the system is solvable. It is easy to certify that $(\mathcal{S})$ is solvable: as we have already explained, any feasible solution is a certificate. For example, to certify solvability of the system

$$\begin{array}{rrrl} -4u & -9v & +5w & > 1.9 \\ -2u & +6v & & \geq -2 \\ 7u & & -5w & \geq \ \ 1 \end{array}$$

it suffices to point out a solution, e.g., $u = 1/7, v = -2/7, w = 0$; every one who knows arithmetics can plug this solution into the system and see that the solution is feasible, thus certifying that the system is solvable.

A much more difficult question is how to certify that the system is *un*solvable.[1] In mathematics, the typical way to prove impossibility of something is to assume that this something does take place and then to lead this assumption to a contradiction. It turns out that finite systems of linear inequalities are simple enough to allow for unified — and extremely simple — scheme for leading the assumption of solvability of the system in question to a contradiction. Specifically, assume that $(\mathcal{S})$ has a solution $\bar{x}$, and let us try to lead this assumption to a contradiction by "linear aggregation of the inequalities of the system."

[1]In real life, it was realized long ago that certifying a *negative* statement is, in general, impossible. Say, a French-speaking man can easily certify his knowledge of language: he can start speaking French, thus proving his knowledge to everybody who speaks French. But how could a man certify that he does *not* know French? The consequences of understanding that it is difficult or impossible to certify negative statements are reflected in rules like "a person is not guilty until proved otherwise," and this is why in the court of law the accused is not required to certify the negative statement "I did not commit this crime;" somebody else is required to prove the positive statement "the accused did commit the crime."

Let us start with an example: To certify that the system

$$\begin{array}{rrrl} -4u & -9v & +5w & > \ \ 2 \\ -2u & +6v & & \geq -2 \\ 7u & & -5w & \geq \ \ 1 \end{array}$$

has no solutions, it suffices to point out that aggregating the inequalities of the system with weights 2, 3, 2, we get a contradictory inequality as follows:

$$\begin{array}{rr|rrrl} & 2\times & -4u & -9v & +5w & > \ \ 2 \\ + & & & & & \\ & 3\times & -2u & +6v & & \geq -2 \\ + & & & & & \\ & 2\times & 7u & & -5w & \geq \ \ 1 \\ \hline & & 0\cdot u & +0\cdot v & +0\cdot w & > \ \ 0 \end{array}$$

By how we aggregate, every solution to the system *must* solve the aggregated inequality. The latter clearly has no solutions, implying that so is the system.

This example suggests the way to certify insolvability of $(\mathcal{S})$ as follows. Let λ_i, $1 \leq i \leq m$, be *nonnegative* "aggregation weights." Let us multiply the inequalities of system $(\mathcal{S})$ by scalars λ_i and sum up the results. We shall arrive at the "aggregated inequality"

$$\left[\sum_i \lambda_i a_i\right]^T x \ \Omega \ \sum_i \lambda_i b_i, \tag{3.1}$$

where the sign Ω of the inequality is either $<$ (this is the case when at least one strict inequality in $(\mathcal{S})$ gets a positive weight, i.e., $\lambda_i > 0$ for some $i \in I$), or $\leq$ (this is the case when $\lambda_i = 0$ for all $i \in I$). Due to its origin and due to the elementary properties of the relations "$\leq$" and "$<$" between reals, the aggregated inequality (3.1) is a *consequence* of the system — it is satisfied at every solution to the system. It follows that *if the aggregated inequality has no solutions at all, then so does the system* $(\mathcal{S})$. Thus, *every collection* λ *of aggregation weights* $\lambda_i \geq 0$ *which results in unsolvable aggregated inequality* (3.1) *can be considered a certificate of insolvability of* $(\mathcal{S})$.

Now, it is easy to say when the aggregated inequality (3.1) has no solutions at all. First, the vector of coefficients $\sum_i \lambda_i a_i$ of the variables should be zero, otherwise the left-hand side with properly chosen x can be made as negative as you want, meaning that the inequality is solvable. Now, whether the inequality $0^T x \Omega a$ is solvable or not depends on what is Ω and what is a. When $\Omega =$ " $<$," this inequality is unsolvable iff $a \leq 0$, and when $\Omega =$ " $\leq$," it is unsolvable iff $a < 0$. We have arrived at the following simple

Proposition 3.1. *Assume that one of the systems of linear inequalities*

$$\text{(I)} \quad \left\{ \begin{array}{l} \lambda_i \geq 0, i \in \mathcal{I} \\ \sum\limits_{i \in I} \lambda_i > 0 \\ \sum\limits_{i \in \mathcal{I}} \lambda_i a_i = 0 \\ \sum\limits_{i \in \mathcal{I}} \lambda_i b_i \leq 0 \end{array} \right., \quad \text{(II)} \quad \left\{ \begin{array}{l} \lambda_i \geq 0, i \in \mathcal{I} \\ \sum\limits_{i \in I} \lambda_i = 0 \\ \sum\limits_{i \in \mathcal{I}} \lambda_i a_i = 0 \\ \sum\limits_{i \in \mathcal{I}} \lambda_i b_i < 0 \end{array} \right. \tag{3.2}$$

in variables $\lambda_1, \ldots, \lambda_m$ is solvable. Then system $(\mathcal{S})$ is insolvable.

Indeed, if λ solves (I), then, aggregating the inequalities in $(\mathcal{S})$ with the weights λ_i, one gets a contradictory inequality of the form $0^T x <$ "something nonpositive." When λ solves (II), the same aggregation results in a contradictory inequality of the form $0^T x \leq$ "something negative." In both cases, $(\mathcal{S})$ admits a consequence which is a contradictory inequality (i.e., inequality with no solutions), whence $(\mathcal{S})$ itself has no solutions. □

Remark 3.1. Both systems in (3.2) are homogeneous, and therefore their solvability/insolvability is equivalent to solvability/insolvability of their "normalized" versions

$$\text{(I}'\text{)} \quad \left\{ \begin{array}{l} \lambda_i \geq 0, i \in \mathcal{I} \\ \sum\limits_{i \in I} \lambda_i \geq 1 \\ \sum\limits_{i=1}^{m} \lambda_i a_i = 0 \\ \sum\limits_{i=1}^{i} \lambda_i b_i \leq 0 \end{array} \right., \quad \text{(II}'\text{)} \left\{ \begin{array}{l} \lambda_i \geq 0, i \in \mathcal{I} \\ \sum\limits_{i \in I} \lambda_i = 0 \\ \sum\limits_{i=1}^{m} \lambda_i a_i = 0 \\ \sum\limits_{i=1}^{i} \lambda_i b_i \leq -1 \end{array} \right.$$

which contain only equalities and nonstrict inequalities. Thus, Proposition 3.1 says that if either (I′) or (II′) is solvable, then $(\mathcal{S})$ is unsolvable.

One of the major results of the LO theory is that the simple *sufficient* condition for insolvability of $(\mathcal{S})$ stated by Proposition 3.1 is in fact *necessary and sufficient*:

Theorem 3.1 (General theorem on alternative (GTA)). *A finite system $(\mathcal{S})$ of linear inequalities is insolvable if and only if it can be led to a contradiction by admissible $(\lambda \geq 0)$ aggregation. In other words, system $(\mathcal{S})$ of linear inequalities in variables x has no solutions iff one of the systems* (I), (II) *of linear inequalities in variables λ has a solution.*

Postponing the proof of GTA, we see that both solvability and insolvability of $(\mathcal{S})$ admit simple certificates: to certify solvability, it suffices to point out a feasible solution x to the system; to certify insolvability, it suffices to point out a feasible solution λ to either (I) or (II). In both cases, $(\mathcal{S})$ possesses the certified property iff an indicated certificate exists, and in both cases it is easy to check whether a candidate certificate is indeed a certificate.

Proof of GTA. Proposition 3.1 justifies the GTA in one direction — it states that *if* either (I) or (II) or both the systems are solvable, *then* $(\mathcal{S})$ is insolvable. It remains to verify that the inverse is also true. Thus, assume that $(\mathcal{S})$ has no solutions, and let us prove then that at least one of the systems (I), (II) has a solution. Observe that we already know that this result takes place in the special case when all inequalities of the system are homogeneous (i.e., $b = 0$) and the system contains exactly one strict inequality, let it be the first one. Indeed, if the system of the form

$$p_1^T x < 0, p_2^T x \leq 0, \ldots, p_k^T x \leq 0 \tag{S}$$

with $k \geq 1$ inequalities has no solutions, then the homogeneous linear inequality $p_1^T x \geq 0$ is a consequence of the system of homogeneous linear inequalities $-p_2^T x \geq 0$, $-p_3^T x \geq 0, \ldots, -p_k^T x \geq 0$. By Homogeneous Farkas Lemma (HFL), in this case, there exist nonnegative $\lambda_2, \ldots, \lambda_k$ such that $p_1 = \sum_{i=2}^k \lambda_i[-p_i]$, or, setting $\lambda_1 = 1$,

$$\lambda_i \geq 0,\ 1 \leq i \leq k, \sum_{i=1}^k \lambda_i p_i = 0.$$

We see that λ solves system (I) associated with (S).

Now, let us derive GTA from the above particular case of it. To this end, given system $(\mathcal{S})$, consider the following system of homogeneous linear inequalities in variables x, y, z (y, z are scalar variables):

$$\begin{array}{rl} -z < 0, & \text{(a)} \\ a_i^T x - b_i y + z \leq 0,\ i \in I, & \text{(b)} \\ a_i^T x - b_i y \leq 0,\ i \in \overline{I}, & \text{(c)} \\ z - y \leq 0. & \text{(d)} \end{array} \tag{3.3}$$

We claim that this system is insolvable. Indeed, assuming that this system has a solution $[x; y; z]$, we have $z > 0$ by (a), whence $y > 0$ by (d). Setting $x' = y^{-1}x$ and dividing (b) and (c) by y, we get

$$a_i^T x' - b_i \leq -(z/y) < 0,\ i \in I, \quad a_i^T x' - b_i \leq 0, i \in \overline{I},$$

that is, x' solves $(\mathcal{S})$, which is impossible, since we are under the assumption that $(\mathcal{S})$ is insolvable.

Now, system (3.3) is in the form of (S), whence, as we have seen, it follows that there exist nonnegative $\lambda_0, \ldots, \lambda_{m+1}$ with $\lambda_0 = 1$ such that the combination of the right-hand sides in (3.3) with coefficients $\lambda_0, \ldots, \lambda_{m+1}$ is identically zero. This amounts to

$$\begin{array}{lll} \sum_{i=1}^m \lambda_i a_i = 0 & \text{[look at the coefficients at } x\text{]}, & \text{(a)} \\ \sum_{i=1}^m \lambda_i b_i = -\lambda_{m+1} & \text{[look at the coefficients at } y\text{]}, & \text{(b)} \\ \sum_{i \in I} \lambda_i + \lambda_{m+1} = 1 & \text{[look at the coefficients at } z\text{]}. & \text{(c)} \end{array} \tag{3.4}$$

Recalling that $\lambda_1, \ldots, \lambda_{m+1}$ are nonnegative, we see that

- In the case of $\lambda_{m+1} > 0$, (3.4(a),(b)) say that $\lambda_1, \ldots, \lambda_m$ solve (II);
- In the case of $\lambda_{m+1} = 0$, (3.4(a),(b),(c)) say that $\lambda_1, \ldots, \lambda_m$ solve (I).

Thus, in all cases either (I) or (II) is solvable. □

Several remarks are in order.

A. There is no necessity to memorize the specific forms of systems (I), (II); what you should memorize is the *principle* underlying GTA, and this principle is pretty simple:

> *A finite system of (strict and nonstrict) linear inequalities has no solutions if and only if one can lead it to contradiction by admissible aggregation of the inequalities of the system, that is, by assigning the inequalities weights making it legitimate to take the weighted sum of the inequalities and making this weighted sum a contradictory inequality.*

In this form, this principle is applicable to every system, not necessarily to one in the standard form $(\mathcal{S})$. What does it actually mean that the aggregated inequality is contradictory? That depends on the structure of the original system, but in all cases it is straightforward to understand what "contradictory" means and thus it is straightforward to understand how to certify insolvability.

For example, the recipe to realize that the system

$$\begin{aligned} x_1 + x_2 &> 0 \\ x_1 - x_2 &= 3 \\ 2x_1 &\leq 2 \end{aligned}$$

is insolvable is as follows: let us multiply the first inequality by a nonnegative weight λ_1, the second equality — by a *whatever* weight λ_2, and the third inequality — by a nonpositive weight λ_3; then it is legitimate to sum the resulting relations up, arriving at

$$\lambda_1[x_1 + x_2] + \lambda_2[x_1 - x_2] + \lambda_3[2x_1] \;\Omega\; \lambda_1 \cdot 0 + \lambda_2 \cdot 3 + \lambda_3 \cdot 2 \qquad (*)$$

where Ω is ">" when $\lambda_1 > 0$ and Ω is "$\geq$" when $\lambda_1 = 0$. Now, let us impose on λ's the requirement that the left-hand side in the aggregated inequality as a function x_1, x_2 is identically zero. Adding to this the above restrictions on the signs of λ_i's, we arrive at the system of restrictions on λ, specifically,

$$\begin{aligned} &\lambda_1 + \lambda_2 + 2\lambda_3 = 0, \\ &\lambda_1 - \lambda_2 = 0, \\ &\lambda_1 \geq 0, \lambda_3 \leq 0, \end{aligned} \qquad (!)$$

which expresses equivalently the fact that the aggregation is "legitimate" and results in an identically zero left-hand side in the aggregated inequality $(*)$. The next step is to consider separately the cases where $\Omega =$ ">" and $\Omega =$ "$\geq$." The first case takes place when $\lambda_1 > 0$, and here, under assumption (!), $(*)$ reads $0 > 3\lambda_2 + 2\lambda_3$; thus, here the fact that the aggregated inequality is contradictory boils down to

$$\lambda_1 > 0, \quad 3\lambda_2 + 2\lambda_3 \geq 0. \qquad \text{(a)}$$

We get a system of constraints on λ such that its feasibility implies insolvability of the original system in x-variables. The second case to be considered is that $\Omega =$ "$\geq$," which corresponds to $\lambda_1 = 0$; here the aggregated inequality, under assumption (!), reads $0 \geq 3\lambda_2 + 2\lambda_3$, thus, in the case in question the fact that the aggregated inequality is contradictory boils down to

$$\lambda_1 = 0, \quad 3\lambda_2 + 2\lambda_3 > 0. \tag{b}$$

Now, GTA says the original system in x-variables is insolvable iff augmenting (!) by either (a), or (b), we get a solvable system of constraints on λ. Since the system (!) implies that $\lambda_1 = \lambda_2 = -\lambda_3$, and thus $3\lambda_2 + 2\lambda_3 = -\lambda_3$, these augmentations are equivalent to

$$\lambda_3 < 0,$$

and

$$\lambda_3 = 0, \quad -\lambda_3 > 0.$$

The first system clearly is solvable, meaning that the original system in variables x is insolvable; as an insolvability certificate, one can use $\lambda_3 = -1, \lambda_1 = \lambda_2 = -\lambda_3 = 1$.

B. It would be easier to use the GTA if we knew in advance that one of the systems (I), (II), e.g., (II), is insolvable, thus implying that $(\mathcal{S})$ is insolvable iff (I) is solvable. Generic cases of this type are as follows:

- *there are no strict inequalities in* $(\mathcal{S})$. In this case, (I) definitely is insolvable (since the strict inequality in (I) clearly is impossible), and thus $(\mathcal{S})$ is insolvable iff (II) is solvable;
- *the subsystem of* S *comprising all nonstrict inequalities in* $(\mathcal{S})$ *is solvable*. In this case, (II) definitely is insolvable[2]; thus, $(\mathcal{S})$ is insolvable iff (I) is solvable.

The reasoning in the footnote can be extended as follows: *if we know in advance that a certain subsystem* (S) *of* $(\mathcal{S})$ *is solvable, we can be sure that the admissible aggregation weights* λ_i*, which lead* $(\mathcal{S})$ *to a contradiction, definitely include positive weights associated with some of the*

[2]Look at (II). This system in fact is not affected by the presence of strict inequalities in $(\mathcal{S})$ and would remain intact if we were dropping from $(\mathcal{S})$ all strict inequalities. Thus, if (II) were solvable, already the subsystem of nonstrict inequalities from $(\mathcal{S})$ would be insolvable, which is not the case.

inequalities outside of (S). Indeed, otherwise the contradictory aggregation, given by λ, of the inequalities from $(\mathcal{S})$ would be a contradictory aggregation of the inequalities from (S), which is impossible, since (S) is solvable.

3.1.1.1 *Corollaries of GTA*

Specifying somehow the structure of $(\mathcal{S})$ and applying GTA, we get instructive "special cases" of GTA. Here are several most-renowned special cases (on a closer inspection, every one of them is equivalent to GTA).

Homogeneous Farkas lemma is obtained from GTA when restricting $(\mathcal{S})$ to be a system of homogeneous inequalities ($b_i = 0$ for all i) and allowing for exactly one strict inequality in the system (check it!). Of course, with our way to derive GTA, this observation adds nothing to our body of knowledge — we used HFL to obtain GTA.

Inhomogeneous Farkas lemma: The next statement does add much to our body of knowledge:

Theorem 3.2 (Inhomogeneous Farkas lemma). *A nonstrict linear inequality*

$$a_0^T x \le b_0 \tag{3.5}$$

is a consequence of a solvable *system of nonstrict inequalities*

$$a_i^T x \le b_i,\ 1 \le i \le m \tag{3.6}$$

iff the target inequality (3.5) *is a weighted sum, with nonnegative coefficients, of the inequalities from the system and the identically true inequality* $0^T x \le 1$, *that is, iff there exist nonnegative coefficients* $\lambda_1, \ldots, \lambda_m$ *such that*

$$\sum_{i=1}^{m} \lambda_i a_i = a_0,\ \sum_{i=1}^{m} \lambda_i b_i \le b_0. \tag{3.7}$$

To see the role of the identically true inequality, note that to say that there exist nonnegative $\lambda_1, \ldots, \lambda_m$ *satisfying* (3.7) *is exactly the same as to say that there exist nonnegative* $\lambda_1, \ldots, \lambda_{m+1}$ *such that*

$$[a_0; b_0] = \sum_{i=1}^{m} \lambda_i [a_i; b_i] + \lambda_{m+1}[0; ...; 0; 1].$$

Proof. The fact that the existence of $\lambda_i \geq 0$ satisfying (3.7) implies that the target inequality (3.5) is a consequence of the system (3.6) is evident — look at the weighted sum $\sum_i \lambda_i a_i^T x \leq \sum_i \lambda_i b_i$ of the inequalities of the system and compare it with the target inequality; note that here the solvability of the system is irrelevant. To prove the inverse, assume that the target inequality is a consequence of the system and the system is solvable, and let us prove the existence of $\lambda_i \geq 0$ satisfying (3.7). Indeed, since the target inequality is a consequence of the system (3.6), the system of linear inequalities

$$-a^T x < -b,\ a_1^T x \leq b_1, \ldots, a_m^T x \leq b_m \tag{$*$}$$

is insolvable. The GTA says that then an appropriate weighted sum, with nonnegative weights $\mu_0, \ldots, \mu_m$, of the inequalities from the latter system is a contradictory inequality. It follows that μ_0 is nonzero (since otherwise the weights $\mu_1, \ldots, \mu_m$ would certify that the system (3.6) is insolvable, which is not the case). When $\mu_0 > 0$, the fact that the weighted sum

$$\left[\mu_0[-a_0] + \sum_{i=1}^m \mu_i a_i\right]^T x < \mu_0[-b_0] + \sum_{i=1}^m \mu_i b_i$$

of the inequalities from $(*)$ is contradictory reads

$$\mu_0 a_0 = \sum_{i=1}^m \mu_i a_i,\ \mu_0 b_0 \geq \sum_{i=1}^m \mu_i b_i,$$

meaning that $\lambda_i = \mu_i/\mu_0$ are nonnegative and satisfy (3.7). □

For two other renowned equivalent reformulations of the GTA, see Exercise 3.1.

Discussion. It is time now to explain why GTA is indeed a deep fact. Consider the following solvable system of linear inequalities:

$$-1 \leq u \leq 1,\ -1 \leq v \leq 1. \tag{!}$$

From this system it clearly follows that $u^2 \leq 1$, $v^2 \leq 1$, whence $u^2 + v^2 \leq 2$. Applying the Cauchy inequality, we have

$$u_1 + u_2 = 1 \cdot u + 1 \cdot v \leq \sqrt{1^2 + 1^2}\sqrt{u^2 + v^2} = \sqrt{2}\sqrt{u^2 + v^2},$$

which combines with the already proved $u^2 + v^2 \leq 2$ to imply that $u + v \leq \sqrt{2}\sqrt{2} = 2$. Thus, the linear inequality $u + v \leq 2$ is a consequence of the solvable system of linear inequalities (!). GTA says that we could get the same target inequality by a very simple process, free of taking squares and Cauchy inequality — merely by taking an admissible weighted sum of the inequalities from the original system. In our toy example, this is evident: we should just sum up the inequalities $u \leq 1$ and $v \leq 1$ from the system. However, a derivation of the outlined type could involve 1,000 highly nontrivial (and "highly nonlinear") steps; a statement, like GTA, capable of predicting in advance that a chain, now sophisticated, of derivations of this type, which starts with a solvable system of linear inequalities and ends with a linear inequality, can be replaced with just taking the weighted sum, with nonnegative coefficients, of the inequalities from the original system, should indeed be deep...

It should be added that in both Homogeneous and Inhomogeneous Farkas Lemma, it is crucial that we speak about linear inequalities. Consider, for example, a target homogeneous quadratic inequality

$$x^T A_0 x \leq 0 \qquad (*)$$

along with a system

$$x^T A_i x \leq 0,\ 1 \leq i \leq m, \qquad (!)$$

of similar inequalities, and let us ask ourselves when the target inequality is a consequence of the system. A straightforward attempt to extend HFL to the quadratic case would imply the conjecture "(*) is a consequence of (!) if and only if the symmetric matrix A_0 is a conic combination of the matrices A_i, $i = 1, \ldots, m$." On a closer inspection, we realize that to expect the validity of this conjecture would be too much, since there exist nontrivial (i.e., with nonzero symmetric matrix A_0) *identically true* homogeneous quadratic inequalities $x^T A_0 x \leq 0$, e.g., $-x^T x \leq 0$; these are inequalities produced by the so-called *negative semidefinite* symmetric matrices. Clearly, such an inequality is a consequence of an empty system (!) (or a system where all the matrices $A_1, \ldots, A_k$ are zero), while the matrix A_0 in question definitely is not a conic combination of an empty collection of matrices, or a collection comprising zero matrices. Well, if there exist identically true homogeneous quadratic inequalities, why not think about them as additional inequalities participating in (!)? Such a viewpoint leads us to an improved conjecture "a homogeneous quadratic inequality (*) is a consequence of a finite system (!) of similar inequalities if and only if A_0 can be represented as a conic combination of $A_1, \ldots, A_m$ *plus a negative semidefinite matrix*?" (Note that in the case of homogeneous linear inequalities similar correction of HFL "is empty," since the only homogeneous *identically true* linear inequality is trivial — all the coefficients are zero). Unfortunately, the corrected conjecture fails to be true in general; its "if" part "*if* A_0 is a conic combination of matrices $A_1, \ldots, A_m$ plus a negative semidefinite matrix" is trivially true, but the "only if" part fails to be true. Were it not so, there would be no difficult optimization problems at all (e.g., P would be equal to NP), but we are not that lucky... This being said, it should be noted that

- already the trivially true "if" part of the (improved) conjecture is extremely useful — it underlies what is called *semidefinite relaxations* of difficult combinatorial problems;
- there is a special case when the improved conjecture indeed is true — this is the case when $m = 1$ and the "system" (in fact, just a single inequality) (!) is strictly feasible — there exists $\bar{x}$ such that $\bar{x}^T A_1 \bar{x} < 0$. The fact that in the special case in question the improved conjecture is true is called $\mathcal{S}$-Lemma, which reads

  Let A, B be symmetric matrices of the same size. A homogeneous quadratic inequality $x^T Bx \leq 0$ is a consequence of a strictly feasible homogeneous quadratic inequality $x^T Ax \leq 0$ iff there exists $\lambda \geq 0$ such that the matrix $B - \lambda A$ is negative semidefinite.

However poor this fact looks compared to its "linear analogy" HFL, $\mathcal{S}$-lemma is one of the most useful facts of Optimization.

3.1.2 *Answering questions*

Now, we are in a good position to answer the questions we posed in the preface of the chapter. All the questions were on how to certify such and such a property; to save words when presenting answers, let us start with setting up some terminology. Imagine that we are interested in a certain property of an object and know how to certify this property, that is, we have at our disposal a family of candidate certificates and an easy-to-verify condition $\mathcal{C}$ on a pair ("object," "candidate certificate") such that whenever a candidate certificate and the object under consideration fit the condition $\mathcal{C}$, the object does possess the property of interest. In this situation, we shall say that we have at our disposal a *certification scheme* for the property.

For example, let the objects to be considered be polyhedral sets in $\mathbf{R}^n$, given by their descriptions of the form $\{Ax \leq b\}$, the property of interest is the nonemptiness of a set, candidate certificates are vectors from $\mathbf{R}^n$, and the condition $\mathcal{C}$ is "a candidate certificate $\bar{x}$ that satisfies the system of constraints $Ax \leq b$ specifying the object under consideration." What we have described clearly is a certification scheme.

By definition of a certification scheme, whenever it allows to certify that a given object X possesses the property of interest (that is, whenever there exists a candidate certificate S which makes $\mathcal{C}(X, S)$ valid), X indeed possesses the property. We say that a certification scheme is *complete* if the inverse is also true: whenever an object possesses the property of interest, this fact can be certified by the certification scheme. For example, the outlined certification scheme for nonemptiness of a polyhedral set clearly is complete. When strengthening the condition $\mathcal{C}$ underlying this scheme to "a candidate certificate $\bar{x}$ has zero first coordinate and satisfies the system of constrains $Ax \leq b$ specifying the object under consideration," we still have a certification scheme, but this scheme clearly is incomplete.

Finally, we note that certifying the presence of a property and certifying the *absence* of this property are, in general, two completely different tasks, this is why in the sequel we should (and will) consider both how to certify a property and how to certify its absence.[3]

When a polyhedral set is empty/nonempty? Here are the answers:

Corollary 3.1. *A polyhedral set* $X = \{x \in \mathbf{R}^n : Ax \leq b\}$ *is nonempty iff there exists* x *such that* $Ax \leq b$*. The set is empty iff there exists* $\lambda \geq 0$ *such that* $A^T\lambda = 0$ *and* $b^T\lambda < 0$.

Indeed, the first part is a plain tautology. The second part is verified as follows: by GTA, X is empty iff there exists a weighted sum $\lambda^T Ax \leq \lambda^T b$ of the inequalities, the weights being nonnegative, which is a contradictory inequality; the latter clearly is the case iff $A^T\lambda = 0$ and $b^T\lambda < 0$. □

[3]Of course, if a certain property admits a complete certification scheme $\mathcal{S}$, the absence of this property in an object X is fully characterized by saying that X does *not* admit a certificate required by $\mathcal{S}$; this, however, is a negative statement, and not a certification scheme!

When a polyhedral set contains another polyhedral set? The answer is given by

Corollary 3.2. *A polyhedral set* $Y = \{x : p_i^T x \le q_i, 1 \le i \le k\}$ *contains a* nonempty *polyhedral set* $X = \{x : A^T x \le b\}$ *iff for every* $i \le k$ *there exist* $\lambda^i \ge 0$ *such that* $p_i = A^T \lambda^i$ *and* $q_i \ge b^T \lambda^i$.

Indeed, Y contains X iff every inequality $p_i^T x \le q_i$ defining Y is satisfied everywhere on X, that is, this inequality is a consequence of the system of inequalities $Ax \le b$; it remains to use the Inhomogeneous Farkas Lemma. □

Now, $Y = \{x : p_i^T x \le q_i, 1 \le i \le k\}$ contains $X = \{x : Ax \le b\}$ iff either $X = \emptyset$, or $X \ne \emptyset$ and $X \subset Y$. It follows that in order to certify the inclusion $X \subset Y$, it suffices to point out either a vector λ satisfying $\lambda \ge 0, A^T\lambda = 0, b^T\lambda < 0$, thus certifying that $X = \emptyset$ (Corollary 3.1), or to certify that $X \ne \emptyset$ *and* point out a collection of vectors λ^i, satisfying $\lambda^i \ge 0, p_i = A^T\lambda^i, q_i \ge b^T\lambda^i$, $1 \le i \le k$, thus certifying that every one of the inequalities defining Y is valid everywhere on X. In fact, in the second case one should not bother to certify that $X \ne \emptyset$, since the existence of λ^i as in Corollary 3.2 is *sufficient* for the validity of the inclusion $X \subset Y$ independently of whether X is or is not empty (why?). It should be added that

- the outlined certification scheme for the inclusion $X \subset Y$ is complete;
- it is trivial to certify that X is *not* contained in Y; to this end it suffices to point out an $x \in X$ which violates one or more of the linear inequalities defining Y; this certification scheme is also complete.

As a direct consequence of our results, we get a complete certification scheme for checking whether two polyhedral sets $X = \{x \in \mathbf{R}^n : Ax \le b\}$ and $Y = \{x \in \mathbf{R}^n : Cx \le d\}$ are/are not identical. Indeed, $X = Y$ iff either both sets are empty or both are nonempty and both the inclusions $X \subset Y$ and $Y \subset X$ hold true, and we already know how to certify the presence/absence of all properties we have just mentioned.[4]

[4]At first glance, the very question that we have answered seems to be fully scholastic; in comparison, even the question "how many angels can sit at the tip of

When is a polyhedral set bounded/unbounded? *Certifying boundedness.* A polyhedral set $X = \{x \in \mathbf{R}^n : Ax \leq b\}$ is bounded iff there exists R such that X is contained in the box $B_R = \{x \in \mathbf{R}^n : e_i^T x \leq R, -e_i^T x \leq R, 1 \leq i \leq n\}$, where $e_1, \ldots, e_n$ are the standard basic orths. Thus, certificate of boundedness is given by a real R augmented by a certificate that $X \subset B_R$. Invoking the previous item, we arrive at the following "branching" scheme: we either certify that X is empty, a certificate being a vector λ satisfying $\lambda \geq 0, A^T\lambda = 0, b^T\lambda < 0$, or point out a collection $(R, \lambda^1, \mu^1, \ldots, \lambda^n, \mu^n)$ satisfying $\lambda^i \geq 0, A^T\lambda^i = e_i, b^T\lambda^i \leq R$, $\mu^i \geq 0, A^T\mu^i = -e_i, b^T\mu^i \leq R$, $1 \leq i \leq m$, thus certifying that $X \subset B_R$. The resulting certification scheme is complete (why?).

Note that when certifying boundedness of a nonempty set, one should not bother about selecting R; if for every i there exist $\lambda^i \geq 0$, $\mu^i \geq 0$ such that $A^T\lambda^i = e_i$, $A^T\mu^i = -e_i$, X is bounded, since augmenting $\{\lambda^i, \mu^i\}$ by large enough R ensures that $b^T\lambda^i \leq R$, $b^T\mu^i \leq R$.

Certifying unboundedness. An unbounded set clearly should be nonempty. By Corollary 2.9, a polyhedral set is unbounded iff its recessive cone is nontrivial. Applying Proposition 2.11, we conclude that in order to certify unboundedness of a polyhedral set $X = \{x \in \mathbf{R}^n : Ax \leq b\}$, it suffices to point out a vector $\bar{x}$ such that $A\bar{x} \leq b$ and a *nonzero* vector y such that $Ay \leq 0$, and this certification scheme is complete.

A useful corollary of the results in this item is that the properties of a *nonempty* polyhedral set $X = \{x : Ax \leq b\}$ to be bounded/unbounded are independent of a particular value of b, *provided that with this value of b, the set X is nonempty.*

---

a needle" discussed intensively by the Medieval scholars seems to be practical. In fact, the question is very deep, and the possibility to answer it affirmatively (that is, by indicating a complete certification scheme where the validity of candidate certificates can be efficiently verified) is "an extremely rare commodity." Indeed, recall that we wand to certify that two sets are/are not identical looking at the *descriptions* of these sets, and not at these sets as abstract mathematical beasts in the spirit of Plato. To illustrate the point, the Fermat Theorem just asks whether the set of positive integer quadruples x, y, z, p satisfying $x^p + y^p - z^p = 0$ and $p \geq 3$ is or is not equal to the set of positive integer quadruples (x, y, z, p) satisfying $x = 0$; this form of the theorem makes it neither a trivial, nor a scholastic statement.

How to certify that the dimension of a polyhedral set $X = \{x \in \mathbf{R}^n : Ax \leq b\}$ is $\geq / \leq$ a given d? First of all, only nonempty sets possess well-defined dimensions, so that the "zero step" is to certify that the set in question is nonempty. We know how to certify both this property and its absence; thus, we can work under the assumption that the nonemptiness of X is already certified.

Let us start with certifying the fact that $\dim X \geq d$, where d is a given integer. By Theorem 2.2(iii) and Lemma 2.1, $\dim X = \dim \mathrm{Aff}(X) \geq d$ iff one can point out $d+1$ vectors $x_0, \ldots, x_d \in X$ such that the d vectors $x_1 - x_0, x_2 - x_0, \ldots, x_d - x_0$ are linearly independent. Both the inclusions $x_i \in X$, $i = 0, 1, \ldots, d$, and the linear independence of $x_1 - x_0, \ldots, x_d - x_0$ are easy to verify (how?), so that we can think of a collection $x_0, \ldots, x_d$ with the outlined properties as a certificate for the relation $\dim X \geq d$. The resulting certification scheme clearly is complete.

Now, let us consider how to certify the relation $\dim X \leq d$, d being a given integer. There is nothing to certify if $d \geq n$, so that we can assume that $d < n$. For a nonempty $X \subset \mathbf{R}^n$, the relation $\dim X \leq d$ holds iff there exists an affine subspace M in $\mathbf{R}^n$ such that $X \subset M$ and $\dim M \leq d$; the latter means *exactly* that there exists a system of $n - d$ linear equations $a_i^T x = b_i$ with linearly independent $a_1, \ldots, a_{n-d}$ such that X is contained in the solution set of the system (see Section 2.1.2). The intermediate summary is that $\dim X \leq d$ iff there exist $n-d$ pairs (a_i, b_i) with linearly independent $a_1, \ldots, a_{n-d}$ such that for every i $a_i^T x \equiv b_i$ on X, or, which is the same, both $a_i^T x \leq b_i$ and $-a_i^T x \leq -b_i$ everywhere on X. Recalling how to certify that a linear inequality $\alpha^T x \leq \beta$ is a consequence of the solvable system $Ax \leq b$ of linear inequalities defining X, we arrive at the following conclusion: In order to certify that $\dim X \leq d$, where $X = \{x \in \mathbf{R}^n : A^T x \leq b\}$, A being an $m \times n$ matrix, and X being nonempty, it suffices to point out $n - d$ vectors $a_i \in \mathbf{R}^n$, $n - d$ reals b_i and $2(n - d)$ vectors $\lambda^i, \mu^i \in \mathbf{R}^m$ such that

- $a_1, \ldots, a_{n-d}$ are linearly independent,
- for every $i \leq n - d$, one has $\lambda^i \geq 0$, $A^T \lambda_i = a_i$ and $b^T \lambda^i \leq b_i$,
- for every $i \leq n - d$, one has $\mu^i \geq 0$, $A^T \mu^i = -a_i$, $b^T \mu^i \leq -b_i$.

Every one of the above conditions is easy to verify, so that we have defined a certification scheme for the relation $\dim X \leq d$, and this scheme clearly is complete.

3.1.3 *Certificates in linear optimization*

Consider an LO program of the form

$$\text{Opt} = \max_x \left\{ c^T x : \left\{ \begin{array}{l} Px \le p \ (\ell) \\ Qx \ge q \ (g) \\ Rx = r \ (e) \end{array} \right. \right\} \tag{3.8}$$

("ℓ" from "less or equal," "g" from "greater or equal," "e" from "equal"). Of course, we could stick to a "uniform" format where all the constraints are, say, the "$\le$"-inequalities; we prefer, however, to work with a more flexible format, reflecting how the LO's look "in reality." Our goal is to show how to certify the basic properties of an LO program.

3.1.3.1 *Certifying feasibility/infeasibility*

Certificate for feasibility of (3.8) is, of course, a feasible solution to the problem. Certificate for infeasibility is, according to our theory, a collection of aggregation weights $\lambda = [\lambda_\ell; \lambda_g; \lambda_e]$ associated with the constraints of the program (so that $\dim \lambda_\ell = \dim p$, $\dim \lambda_g = \dim q$, $\dim \lambda_e = \dim r$) such that first, it is legitimate to take the weighted sum of the constraints, and, second, the result of aggregation is a contradictory inequality. The restriction to be legitimate amounts to

$$\lambda_\ell \ge 0, \lambda_g \le 0, \text{ no restrictions on } \lambda_e;$$

when aggregation weights λ satisfy these restrictions, the weighted by λ sum of the constraints is the inequality

$$[P^T\lambda_\ell + Q^T\lambda_g + R^T\lambda_e]^T x \le p^T\lambda_\ell + q^T\lambda_g + r^T\lambda_e,$$

and this inequality is contradictory if and only if the vector of coefficients of x in the left-hand side vanishes, and the right-hand side is negative. Thus, λ certifies infeasibility iff

$$\lambda_\ell \ge 0, \lambda_g \le 0, P^T\lambda_\ell + Q^T\lambda_g + R^T\lambda_e = 0, p^T\lambda_\ell + q^T\lambda_g + r^T\lambda_e < 0. \tag{3.9}$$

According to our theory, the outlined certification schemes for feasibility and for infeasibility are complete.

3.1.3.2 *Certifying boundedness/unboundness*

Certifying boundedness of (3.8): An LO program (3.8) is bounded iff it is either infeasible, or it is feasible and the objective is bounded above on the feasible set, that is, there exists a real μ such that the inequality $c^T x \leq \mu$ is a consequence of the constraints. We already know how to certify infeasibility; to certify boundedness for a feasible problem, we should certify feasibility (which we again know how to do) and certify that the inequality $c^T x \leq \mu$ is a consequence of a *feasible* system of constraints, which, by the principle expressed by the Inhomogeneous Farkas Lemma, amounts to pointing out a collection of weights $\lambda = [\lambda_\ell; \lambda_g; \lambda_e]$ which makes it legitimate to take the weighted sum of the constraints and is such that this weighted sum is of the form $c^T x \leq$ constant. Thus, λ in question should satisfy

$$\lambda_\ell \geq 0, \lambda_g \leq 0, P^T \lambda_\ell + Q^T \lambda_g + R^T \lambda_e = c. \tag{3.10}$$

The resulting certification scheme for boundedness — "either certify infeasibility according to (3.9), or point out a certificate for feasibility and a λ satisfying (3.10)" — is complete. On a closer inspection, there is no need to bother about certifying feasibility in the second "branch" of the scheme, since pointing out a λ satisfying (3.10) certifies boundedness of the program independently of whether the program is or is not feasible (why?).

As an important consequence, we get the following:

Corollary 3.3. *For an LO program* $\max_x\{c^T x : P^T x \leq p, Q^T x \geq q, R^T x = r\}$ *the property to be or not to be bounded is independent of the value of the right-hand side vector* $b = [p; q; r]$, *provided that with this* b *the problem is feasible.*

Indeed, by completeness of our certification scheme for boundedness, a *feasible* LO program is bounded if and only if there exists λ satisfying (3.10), and the latter fact is or is not valid independently of what is the value of b. □

Note that Corollary 3.3 "mirrors" the evident fact that for the property of an LO program to be or not to be feasible is independent of what is the objective.

Certifying unboundedness of (3.8): Program (3.8) is unbounded iff it is feasible and the objective is not bounded above on the feasible set. By Corollary 2.8, the objective of a feasible LO is unbounded from above on the feasible set iff this set has a recessive direction y along which the objective grows: $c^T y > 0$. It follows that a certificate for unboundedness can be specified as a pair x, y such that

$$\begin{array}{ll} Px \le p, Qx \ge q, Rx = r, & \text{(a)} \\ Py \le 0, Qy \ge 0, Ry = 0, & \text{(b)} \\ c^T y > 0. & \text{(c)} \end{array} \tag{3.11}$$

Here (a) certifies the fact that the program is feasible, and (b) expresses equivalently the fact that y is a recessive direction of the feasible set (cf. Proposition 2.11). The resulting certification scheme is complete (why?).

3.1.3.3 *Certifying solvability/insolvability*

An LO program (3.8) is solvable iff it is feasible and above bounded (Corollary 2.11), and we already have at our disposal complete certification schemes for both these properties.

Similarly, (3.8) is insolvable iff it is either infeasible or is feasible and is (above) unbounded, and we already have at our disposal complete certification schemes for both these properties.

3.1.3.4 *Certifying optimality/nonoptimality*

A candidate solution $\bar{x}$ to (3.8) is optimal if and only if it is feasible and the linear inequality $c^T x \le c^T \bar{x}$ is a consequence of the (feasible!) system of constraints in (3.8). Invoking Inhomogeneous Farkas Lemma, we conclude that a certificate for optimality of $\bar{x}$ can be obtained by augmenting $\bar{x}$ by a $\lambda = [\lambda_\ell; \lambda_g; \lambda_e]$ satisfying the relations

$$\begin{array}{ll} \lambda_\ell \ge 0, \lambda_g \le 0, & \text{(a)} \\ P^T \lambda_\ell + Q^T \lambda_g + R^T \lambda_e = c, & \text{(b)} \\ p^T \lambda_\ell + q^T \lambda_g + r^T \lambda_e \le c^T \bar{x}, & \text{(c)} \end{array} \tag{3.12}$$

and that the resulting certification scheme for optimality of $\bar{x}$ is complete *provided that* $\bar{x}$ *is feasible.*

Observe that for whatever λ satisfying (3.12(a)), we have

$$\lambda_\ell^T P\bar{x} \le p^T \lambda_\ell, \quad \lambda_g^T Q\bar{x} \le q^T \lambda_g, \quad \lambda_e^T R\bar{x} = r^T \lambda_e, \tag{$*$}$$

and the first two inequalities can be equalities iff the entries in λ_ℓ and λ_g corresponding to the *nonactive* at $\bar{x}$ inequality constraints — those which are satisfied at $\bar{x}$ as strict inequalities — are zeros. Summing up the inequalities in $(*)$, we end up with

$$\lambda_\ell^T P\bar{x} + \lambda_g^T Q\bar{x} + \lambda_e^T R\bar{x} \le p^T \lambda_\ell + q^T \lambda_g + r^T \lambda_e. \tag{!}$$

On the other hand, if λ satisfies (3.12(c)), then the inequality opposite to (!) takes place, which is possible iff the inequalities in $(*)$ are equalities, which, as we have seen, are equivalent to the fact that the entries in λ associated with nonactive at $\bar{x}$ inequality constraints are zero. Vice versa, when the entries in λ satisfying the nonactive at $\bar{x}$ inequality constraints are zero, then (3.12(c)) is satisfied as an equality as well. The bottom line is that λ *satisfies* (3.12) *iff* λ *satisfies the first two relations in* (3.12) *and, in addition, the entries in* λ *associated with nonactive at* $\bar{x}$ *inequality constraints of* (3.8) *are zeros.* We have arrived at the following:

Proposition 3.2 (Karush–Kuhn–Tucker Optimality Conditions in LO). *A feasible solution* $\bar{x}$ *to an LO program* (3.8) *is optimal for the program iff* $\bar{x}$ *can be augmented by Lagrange multipliers* $\lambda = [\lambda_\ell; \lambda_g; \lambda_e]$ *in such a way that the following two facts take place:*

- *multipliers corresponding to the "$\le$"-inequality constraints* (3.8.ℓ) *are nonnegative, multipliers corresponding to the "$\ge$"-inequality constraints* (3.8(g)) *are nonpositive, and, in addition, multipliers corresponding to the nonactive at* $\bar{x}$ *inequality constraints are zero* (*all this is called "complementary slackness"*);
- *one has*

$$c = P^T \lambda_\ell + Q^T \lambda_g + R^T \lambda_e.$$

We have seen how to certify that a feasible candidate solution to (3.8) is optimal for the program. As about certifying nonoptimality, this is immediate: $\bar{x}$ is not optimal iff it either is not feasible for the program, or there exists a feasible solution y with $c^T y > c^T x$, and such a y certifies the nonoptimality of $\bar{x}$. The resulting "two-branch" certification scheme clearly is complete.

3.1.3.5 *A corollary: Faces of a polyhedral set revisited*

Recall that a face of a polyhedral set $X = \{x \in \mathbf{R}^n : a_i^T x \leq b, 1 \leq i \leq m\}$ is a *nonempty* subset of X which is cut off X by converting some of the inequalities $a_i^T x \leq b_i$ into equalities $a_i^T x = b_i$. Thus, every face of X is of the form $X_I = \{x : a_i^T x \leq b_i, i \notin I, a_i^T x = b_i, i \in I\}$, where I is a certain subset of the index set $\{1, \ldots, m\}$ (which should be such that $X_I \neq \emptyset$). As we have already mentioned, a shortcoming of this definition is that it is not geometric — it is expressed in terms of a particular representation of X rather than in terms of X as a set. Now, we are in a position to eliminate this drawback.

Proposition 3.3. *Let $X = \{x \in \mathbf{R}^n : a_i^T x \leq b_i, 1 \leq i \leq m\}$, be a nonempty polyhedral set.*

(i) *If $c^T x$ is a linear form which is bounded from above on X (and thus attains its maximum on X), then the set $\mathrm{Argmax}_X\, c^T x$ of maximizers of the form is a face of X.*
(ii) *Vice versa, every face X_I of X can be represented in the form $X_I = \mathrm{Argmax}_X\, c^T x$ for an appropriately chosen linear form $c^T x$.*

In particular, every vertex of X is the unique maximizer, taken over $x \in X$, of an appropriately chosen linear form.

Since the sets of the form $\mathrm{Argmax}_{x \in X}\, c^T x$ are defined in terms of X as a set, with no reference to a particular representation of X, the proposition indeed provides us with a purely geometric characterization of the faces.

Proof of Proposition 3.3 is easy. To verify (i), let $\bar{x}$ be a maximizer of $c^T x$ over X, and let J be the set of indices of all constraints $a_i^T x \leq b_i$ which are active at $\bar{x}$: $J = \{i : a_i^T \bar{x} = b_i\}$. Invoking the "only if" part of Proposition 3.2 and noting that we are in the situation of $P^T = [a_1; ...; a_m]$ with Q and R being empty, we conclude that there exist *nonnegative* λ_i, $i \in J$, such that $c = \sum_{i \in J} \lambda_i a_i$. Let us set $I = \{i \in J : \lambda_i > 0\}$, so that in fact $c = \sum_{i \in I} \lambda_i a_i$, and let us verify that $\mathrm{Argmax}_X\, c^T x = X_I$ (so that $\mathrm{Argmax}_X\, c^T x$ is a face in X, as claimed). If I is empty, then $c = \sum_{i \in I} \lambda_i a_i = 0$, so that

$\operatorname{Argmax}_X c^T x = X_\emptyset = X_I$, as required. Now, let $I \neq \emptyset$. We have

$$\forall x \in X : c^T x = \left[\sum_{i\in I} \lambda_i a_i\right]^T x = \sum_{i\in I} \lambda_i [a_i^T x] \underbrace{\leq}_{(*)} \sum_{i\in I} \lambda_i b_i$$
$$\max_{x\in X} c^T x = c^T \bar{x} = \sum_{i\in I} [\lambda_i a_i^T \bar{x}] = \sum_{i\in I} \lambda_i b_i.$$

(we have taken into account that $\lambda_i > 0$). From these relations it is clear that $x \in X$ is a maximizer of the linear form given by c iff the inequality $(*)$ is an equality; the latter, for $x \in X$, takes place iff $a_i^T x = b_i$ for all $i \in I$ (since for such an x, $a_i^T x \leq b_i$ and λ_i are strictly positive, $i \in I$), that is, iff $x \in X_I$, as required, (i) is proved.

To prove (ii), consider a face X_I, and let us prove that it is nothing but $\operatorname{Argmax}_X c^T x$ for a properly chosen c. There is nothing to prove when $I = \emptyset$, so that $X_I = X$; in this case, we can take $c = 0$. Now, let I be nonempty, let us choose whatever strictly positive λ_i, $i \in I$, and set $c = \sum_{i\in I} \lambda_i a_i$. By the "if" part of Proposition 3.2, every point of the (nonempty!) face X_I of X is a maximizer of $c^T x$ over $x \in X$, and since all $\lambda_i, i \in I$, are strictly positive, the reasoning used to prove (i) shows that vice versa, every maximizer of $c^T x$ over X belongs to X_I. Thus, $X_I = \operatorname{Argmax}_X c^T x$ for the c we have built. □

3.2 LO duality

We are about to develop the crucial concept of the *dual* to an LO program. The related constructions and results are mostly already known to us, so that this section is a kind of "intermediate summary."

3.2.1 *The dual of an LO program*

Consider an LO program in the form of (3.8) (we reproduce the formulation for reader's convenience):

$$\operatorname{Opt}(P) = \max_x \left\{ c^T x : \begin{cases} Px \leq p \ (\ell) \\ Qx \geq q \ (g) \\ Rx = r \ (e) \end{cases} \right\}. \qquad (P)$$

From now on we refer to this problem as the *primal* one. The origin of the problem *dual* to (P) stems from the desire to find a systematic way to bound from above the optimal value of (P). The approach we intend to use is already well known to us — this is the aggregation of the constraints. Specifically, let us associate with the constraints of (P) the vector of *dual variables* (called also *Lagrange multipliers*) $\lambda = [\lambda_\ell; \lambda_g; \lambda_e]$ restricted to satisfy sign constraints

$$\lambda_\ell \geq 0; \lambda_g \leq 0. \tag{3.13}$$

Using such a λ as the vector of aggregation weights for the constraints of (P), we get the scalar inequality

$$[P^T\lambda_\ell + Q^T\lambda_g + R^T\lambda_e]^T x \leq p^T\lambda_\ell + q^T\lambda_g + r^T\lambda_e, \tag{$*$}$$

which, by its origin, is a consequence of the system of constraints of (P). Now, if we are lucky to get in the left-hand side our objective, that is, if

$$P^T\lambda_\ell + Q^T\lambda_g + R^T\lambda_e = c,$$

then $(*)$ says to us that $p^T\lambda_\ell + q^T\lambda_g + r^T\lambda_e$ is an upper bound on $\mathrm{Opt}(P)$. The dual problem asks us to build the best possible — the smallest — bound of this type. Thus, the dual problem reads

$$\mathrm{Opt}(D) = \min_{\lambda=[\lambda_\ell;\lambda_g;\lambda_e]} \left\{ d^T\lambda := p^T\lambda_\ell + q^T\lambda_g + r^T\lambda_e : \left\{ \begin{array}{r} \lambda_\ell \geq 0 \\ \lambda_g \leq 0 \\ P^T\lambda_\ell + Q^T\lambda_g + R^T\lambda_e = c \end{array} \right\} \right. . \tag{D}$$

This is again an LO program.

3.2.2 *Linear programming duality theorem*

The most important relations between the primal and the dual LO problems are presented in the following.

Theorem 3.3 (Linear programming duality theorem). *Consider primal LO program (P) along with its dual (D). Then*

(i) [Symmetry] *Duality is symmetric: (D) is an LO program, and its dual is (equivalent to) (P).*

(ii) [Weak duality] $\mathrm{Opt}(D) \geq \mathrm{Opt}(P)$, *or, equivalently, for every pair (x, λ) of feasible solutions to (P) and (D) one has*

$$\mathrm{DualityGap}(x, \lambda) := d^T\lambda - c^T x = [p^T\lambda_\ell + q^T\lambda_g + r^T\lambda_e] - c^T x \geq 0.$$

(iii) [Strong duality] *The following properties are equivalent to each other:*

(iii.1) *(P) is feasible and bounded from above,*
(iii.2) *(P) is solvable,*
(iii.3) *(D) is feasible and bounded from below,*
(iii.4) *(D) is solvable,*
(iii.5) *Both (P) and (D) are feasible,*

and whenever one of these equivalent-to-each-other properties exists, we have $\mathrm{Opt}(P) = \mathrm{Opt}(D)$.

In addition, whenever one of the problems is feasible, the optimal values of (P) and (D) are equal to each other.

Remark 3.2. LP Duality Theorem states that the duality is symmetric, and that "solvability status" of a *feasible* problem from a primal–dual pair fully specifies this status for the other problem from the pair: if the feasible problem is bounded (or, which is the same for a feasible LO, is solvable), the other problem from the pair is solvable, and if the feasible problem is unbounded, the other problem of the pair is infeasible; in both these cases, the optimal values in the problems are the same. In contrast, an *infeasible* component of a primal–dual pair does not know what is the exact solvability status of the other problem of the pair: the latter can be

(a) unbounded, as is the case for the pair

$$\max_{x=[x_1;x_2]} \{x_1 : x_1 - x_2 \leq 0\} \qquad (P)$$

$$\min_{\lambda} \left\{ 0 \cdot \lambda : \lambda \geq 0, \left\{ \begin{array}{c} \lambda = 1 \\ -\lambda = 0 \end{array} \right\} \right. \qquad (D)$$

((D) is infeasible, (P) is feasible and unbounded, $\mathrm{Opt}(P) = \mathrm{Opt}(D) = +\infty$), and

(b) infeasible, as is the case for the pair

$$\max_{x=[x_1;x_2;x_3]} \left\{ x_1 : \left\{ \begin{array}{r} x_1 - x_2 \leq 0 \\ x_3 \leq 1 \\ -x_3 \leq -2 \end{array} \right\} \right. \qquad (P)$$

$$\min_{\lambda=[\lambda_1;\lambda_2;\lambda_3]} \left\{ \lambda_2 - 2\lambda_3 : \lambda \geq 0, \left\{ \begin{array}{r} \lambda_1 = 1 \\ -\lambda_1 = 0 \\ \lambda_2 - \lambda_3 = 0 \end{array} \right\} \right. \quad (D)$$

(both (P) and (D) are infeasible, $\mathrm{Opt}(P) = -\infty$, $\mathrm{Opt}(D) = +\infty$).

Proof of LP Duality Theorem. (i) In order to apply the recipe for building the dual to (D), we should first write it down as a maximization problem in the format of (P), that is, as

$$-\mathrm{Opt}(D) = \max_{\lambda=[\lambda_\ell;\lambda_g;\lambda_e]} \left\{ -[p^T\lambda_\ell + q^T\lambda_g + r^T\lambda_e] : \left\{ \begin{array}{r} \lambda_g \leq 0 \ (\ell) \\ \lambda_\ell \geq 0 \ (g) \\ P^T\lambda_\ell + Q^T\lambda_g + R^T\lambda_e = c \ (e) \end{array} \right\} \right. .$$

This is the problem in the form of (P), with the matrices $[0, I, 0]$, $[I, 0, 0]$, $[P^T, Q^T, R^T]$ in the role of P, Q, R, respectively; here I and 0 stand for the unit and the zero matrices of appropriate sizes (not necessarily the same in different places). Applying to the latter problem the recipe for building the dual, we arrive at the LO program

$$\min_{y=[y_\ell;y_g;y_e]} \left\{ 0^Ty_\ell + 0^Ty_g + c^Ty_e : \left\{ \begin{array}{r} y_\ell \geq 0 \\ y_g \leq 0 \\ [y_g + Py_r; y_\ell + Qy_r; Ry_e] = [-p; -q; -r] \end{array} \right\} \right. .$$

We can immediately eliminate the variables y_ℓ and y_g. Indeed, y_ℓ does not affect the objective, and what the constraints want of y_ℓ is $y_\ell \geq 0$, $y_\ell + Qy_e = -q$, which amounts to $-Qy_e \geq q$. The situation with y_g is similar. Eliminating y_ℓ and y_g, we arrive at the following equivalent reformulation of the problem dual to (D):

$$\min_{y_r}\left\{c^T y_e : \left\{\begin{array}{l} Py_e \geq -p \\ Qy_e \leq -q \\ Ry_e = -r \end{array}\right\}\right\},$$

which is nothing but (P) (set $x = -y_e$). (i) is proved.

(ii) Weak duality is readily given by the construction of (D).

(iii) Let us first verify that all five properties (iii.1)–(iii.5) are equivalent to each other.

(iii.1)⇔(iii.2) a solvable LO program clearly is feasible and bounded, the inverse is true due to Corollary 2.11(i).

(iii.3)⇔(iii.4) follows from the already proved equivalence (iii.1)⇔(iii.2) due to the fact that the duality is symmetric.

(iii.2)⇒(iii.5) If (P) is solvable, (P) is feasible. In order to verify that (D) is feasible as well, note that the inequality $c^T x \leq \mathrm{Opt}(P)$ is a consequence of the system of the constraints in (P), and this system is feasible; applying the Inhomogeneous Farkas Lemma, the inequality $c^T x \leq \mathrm{Opt}(P)$ can be obtained by taking admissible weighted sum of the constraints of (P) and the identically true inequality $0^T x \leq 1$, that is, there exists $\lambda_0 \geq 0$ and $\lambda = [\lambda_\ell; \lambda_g; \lambda_e]$ satisfying the sign constraints (3.13) such that the aggregated inequality

$$\lambda_0[0^T x] + \lambda_\ell^T Px + \lambda_g^T Qx + \lambda_e^T Rx \leq \lambda_0 \cdot 1 + \lambda_\ell^T p + \lambda_g^T q + \lambda_e^T r$$

is exactly the inequality $c^T x \leq \mathrm{Opt}(P)$, meaning that

$$P^T \lambda_\ell + q^T \lambda_g + R^T \lambda_e = c$$

and

$$\lambda_\ell^T p + \lambda_g^T q + \lambda_e^T r = \mathrm{Opt}(P) - \lambda_0. \qquad (*)$$

Since λ satisfies the sign constraints from (D), we conclude that λ is feasible for (D), so that (D) is feasible, as claimed. As a byproduct, we see from $(*)$ that the dual objective at λ is $\leq \mathrm{Opt}(P)$ (since $\lambda_0 \geq 0$), so that $\mathrm{Opt}(D) \leq \mathrm{Opt}(P)$. The strict inequality is impossible by Weak duality, and thus $\mathrm{Opt}(P) = \mathrm{Opt}(D)$.

(iii.4)⇒(iii.5) this is the same as the previous implication due to the fact that the duality is symmetric.

(iii.5)⇒(iii.1) in the case of (iii.5), (P) is feasible; since (D) is feasible as well, (P) is bounded by Weak duality, and thus (iii.1) takes place.

(iii.5)⇒(iii.3) this is the same as the previous implication due to the fact that the duality is symmetric.

We have proved that

$$\text{(iii.1)} \Leftrightarrow \text{(iii.2)} \Rightarrow \text{(iii.5)} \Rightarrow \text{(iii.1)},$$

and that (iii.3)⇔(iii.4)⇒(iii.5)⇒(iii.3). Thus, all five properties are equivalent to each other. We have also seen that (iii.2) implies that $\mathrm{Opt}(P) = \mathrm{Opt}(D)$.

The "in addition" conclusion is immediate: by primal–dual symmetry we can assume that the feasible problem is (P). If this problem is also bounded, both (P) and (D) are solvable with equal optimal values by Strong Duality. and if (P) is unbounded, we have $\mathrm{Opt}(P) = +\infty$, and the dual problem is infeasible by Weak duality, that is, $\mathrm{Opt}(D) = +\infty$ as well. □

3.3 Immediate applications of duality

In this section, we outline several important applications of the LO duality theorem.

3.3.1 *Optimality conditions in LO*

The following statement (which is just a reformulation of Proposition 3.2) is the standard formulation of optimality conditions in LO:

Proposition 3.4 (Optimality conditions in LO). *Consider the primal–dual pair of problems*

$$\mathrm{Opt}(P) = \max_x \left\{ c^T x : \begin{cases} Px \leq p \ (\ell) \\ Qx \geq q \ (g) \\ Rx = r \ (e) \end{cases} \right\}; \tag{P}$$

$$\mathrm{Opt}(D) = \min_{\lambda=[\lambda_\ell;\lambda_g;\lambda_e]} \left\{ d^T\lambda := p^T\lambda_\ell + q^T\lambda_g + r^T\lambda_e : \right.$$

$$\left. \times \left\{ \begin{array}{r} \lambda_\ell \geq 0 \\ \lambda_g \leq 0 \\ P^T\lambda_\ell + Q^T\lambda_g + R^T\lambda_e = c \end{array} \right\} \right. \tag{D}$$

and assume that both of them are feasible. A pair (x, λ) of feasible solutions to (P) and to (D) comprises of optimal solutions to the respective problems

- [Zero Duality Gap] *iff the duality gap, as evaluated at this pair, is* 0:

$$\text{DualityGap}(x, \lambda) := [p^T\lambda_\ell + q^T\lambda_g + r^T\lambda_e] - c^T x = 0$$

as well as

- [Complementary Slackness] *iff all the products of Lagrange multipliers λ_i associated with the inequality constraints of (P) and the residuals in these constraints, as evaluated at x, are zeros:*

$$[\lambda_\ell]_i[Px - p]_i = 0\,\forall i, \quad [\lambda_g]_j[Qx - q]_j = 0\ \forall j.$$

Proof. (i) Under the premise of the proposition, we have $\text{Opt}(P) = \text{Opt}(D) \in \mathbf{R}$ by the LP Duality Theorem, meaning that

$$\begin{aligned}\text{DualityGap}(x, \lambda) = & \left[[p^T\lambda_\ell + q^T\lambda_g + r^T\lambda_e] - \text{Opt}(D)\right] \\ & + \left[\text{Opt}(P) - c^T x\right].\end{aligned}$$

Since x, λ are feasible for the respective problems, the quantities in brackets are nonnegative, so that the duality gap can be zero iff both these quantities are zeros; the latter is the same as to say that x is primal–, and λ is dual optimal.

(ii) This is the same as (i) due to the following useful observation: *Whenever x is primal–, and λ is dual feasible, we have*

$$\text{DualityGap}(x, \lambda) = \lambda_\ell^T(p - Px) + \lambda_g^T(q - Qx). \tag{3.14}$$

Indeed,

$$\begin{aligned}\text{DualityGap}(x, \lambda) &= \lambda_\ell^T p + \lambda_g^T q + \lambda_e^T r - c^T x, \\ &= \lambda_\ell^T p + \lambda_g^T q + \lambda_e^T r - [P^T\lambda_\ell + Q^T\lambda_g + R^T\lambda_e]^T x \\ &\quad [\text{since } \lambda \text{ is dual feasible}], \\ &= \lambda_\ell^T(p - Px) + \lambda_g^T(q - Qx) + \lambda_e^T[r - Rx], \\ &= \lambda_\ell^T(p - Px) + \lambda_g^T(q - Qx) \\ &\quad [\text{since } x \text{ is primal feasible}].\end{aligned}$$

It remains to note that the right-hand side in (3.14) is

$$\sum_i [\lambda_\ell]_i[p - Px]_i + \sum_j [\lambda_g]^T[q - Qx]_j, \tag{!}$$

and all terms in these two sums are nonnegative due to sign restrictions on λ coming from (D) and the fact that x is primal feasible. Thus, DualityGap(x, λ) is zero iff all terms in (!) are zeros, as claimed. □

3.3.2 *Geometry of a primal–dual pair of LO programs*

Consider an LO program

$$\mathrm{Opt}(P) = \max_x \left\{c^T x : Px \le p, Rx = r\right\} \qquad (P)$$

along with its dual program

$$\mathrm{Opt}(D) = \min_{\lambda=[\lambda_\ell;\lambda_e]} \left\{p^T \lambda_\ell + r^T \lambda_e : \lambda_\ell \ge 0, P^T \lambda_\ell + R^T \lambda_e = c\right\}. \qquad (D)$$

Note that to save the notation (and of course w.l.o.g.) we have assumed that all the inequalities in the primal problem are "$\le$."

Our goal here is to rewrite both the problems in a "purely geometric" form, which, as we shall see, reveals a beautiful, simple and instructive geometry of the pair.

First, assume that the systems of linear equations participating in (P) and in (D) are solvable, and let $\bar{x}$, $-\bar{\lambda} = -[\bar{\lambda}_\ell; \bar{\lambda}_e]$ be solutions to the respective systems:

$$\begin{array}{ll} R\bar{x} = r; & \text{(a)} \\ P^T \bar{\lambda}_\ell + R^T \bar{\lambda}_e = -c. & \text{(b)} \end{array} \qquad (3.15)$$

Note that the assumption of solvability we have made is much weaker then feasibility of (P) and (D), since at this point we do not bother about inequality constraints.

Observe that for every x such that $Rx = r$, we have

$$\begin{aligned} c^T x &= -[P^T \bar{\lambda}_\ell + R^T \bar{\lambda}_e]^T x = -\bar{\lambda}_\ell^T [Px] - \bar{\lambda}_e^T [Rx] \\ &= \bar{\lambda}_\ell^T [p - Px] + \underbrace{\left[-\bar{\lambda}_\ell^T p - \bar{\lambda}_e^T r\right]}_{\mathrm{const}_P} \end{aligned}$$

so that (P) is nothing but the problem

$$\mathrm{Opt}(P) = \max_x \left\{\bar{\lambda}_\ell^T [p - Px] + \mathrm{const}_P : p - Px \ge 0, Rx = r\right\}.$$

Let us pass in this problem from the variable x to the variable

$$\xi = p - Px$$

("primal slack" — vector of residuals in the primal inequality constraints). Observe that the dimension m of this vector, same as the dimension of λ_ℓ, is equal to the number of inequality constraints in (P). We have already expressed the primal objective in terms of this new variable, and it remains to understand what are the restrictions on this vector imposed by the constraints of (P). The inequality constraints in (P) want ξ to be nonnegative; the equality constraints along with the definition of ξ read equivalently that ξ should belong to the image $\mathcal{M}_P$ of the affine subspace $M_P = \{x : Rx = r\} \subset \mathbf{R}^{\dim x}$ under the affine mapping $x \mapsto p - Px$. Let us compute $\mathcal{M}_P$. The linear subspace $\mathcal{L}_P$ to which M_P is parallel is the image of the linear subspace $L = \{x : Rx = 0\}$ (this is the subspace M_P is parallel to) under the linear mapping $x \mapsto -Px$, or, which is the same, under the mapping $x \mapsto Px$. As a shift vector for $\mathcal{M}_P$, we can take the image $p - P\bar{x}$ of the vector $\bar{x} \in M_P$ under affine mapping which maps M_P onto $\mathcal{M}_P$. We have arrived at the following intermediate result:

(!) *Problem (P) can be reduced to the problem*

$$\begin{aligned} \mathrm{Opt}(\mathcal{P}) &= \max_{\xi \in \mathbf{R}^m} \left\{ \bar{\lambda}_\ell^T \xi : \xi \geq 0,\ \xi \in \mathcal{L}_P + \bar{\xi} \right\}, \\ \mathcal{L}_P &= \{Py : Ry = 0\}, \bar{\xi} = p - P\bar{x}. \end{aligned} \qquad (\mathcal{P})$$

The optimal values of (P) and $(\mathcal{P})$ are linked by the relation

$$\mathrm{Opt}(P) = \mathrm{Opt}(\mathcal{P}) + \mathrm{const}_P = \mathrm{Opt}(\mathcal{P}) - \bar{\lambda}_\ell^T p - \bar{\lambda}_e^T r.$$

Now, let us look at the dual problem (D), and, similar to what we did with (P), let us represent it solely in terms of the "dual slack" λ_ℓ. To this end, observe that if $\lambda = [\lambda_\ell; \lambda_e]$ satisfies the equality constraints in (D), then

$$\begin{aligned} p^T\lambda_\ell + r^T\lambda_e &= p^T\lambda_\ell + [R\bar{x}]^T\lambda_e = p^T\lambda_\ell + \bar{x}^T[R^T\lambda_e] \\ &= p^T\lambda_\ell + \bar{x}^T[c - P^T\lambda_\ell] = [p - P\bar{x}]^T\lambda_\ell + \underbrace{\bar{x}^T c}_{\mathrm{const}_D} \\ &= \bar{\xi}^T\lambda_\ell + \mathrm{const}_D; \end{aligned}$$

we managed to express the dual objective on the dual feasible set solely in terms of λ_ℓ. Now, the restrictions imposed by the constraints of (D) on λ_ℓ are the nonnegativity: $\lambda_\ell \geq 0$ and the possibility to be extended, by properly chosen λ_e, to a solution to the system of linear equations $P^T\lambda_\ell + R^T\lambda_e = c$. Geometrically, the latter restriction says that λ_ℓ should belong to the affine subspace $\mathcal{M}_D$ which is the image of the affine subspace $M_D = \{[\lambda_\ell; \lambda_e] : P^T\lambda_\ell + R^T\lambda_e = c\}$ under the projection $[\lambda_\ell; \lambda_e] \mapsto \lambda_\ell$. Let us compute $\mathcal{M}_D$. The linear subspace $\mathcal{L}_D$ to which $\mathcal{M}_D$ is parallel clearly is given as

$$\mathcal{L}_D = \{\lambda_\ell : \exists \lambda_e : P^T\lambda_\ell + R^T\lambda_e = 0\};$$

as a shift vector for $\mathcal{M}_D$, we can take an arbitrary point in this affine space, e.g., the point $-\bar{\lambda}_\ell$. We have arrived at the following result:

(!!) *Problem (D) can be reduced to the problem*

$$\begin{aligned} \mathrm{Opt}(\mathcal{D}) &= \min_{\lambda_\ell \in \mathbf{R}^m} \left\{ \bar{\xi}^T\lambda_\ell : \lambda_\ell \geq 0, \lambda_\ell \in \mathcal{L}_D - \bar{\lambda}_\ell \right\}, \\ \mathcal{L}_D &= \{\lambda_\ell : \exists \lambda_e : P^T\lambda_\ell + Q^T\lambda_e = 0\}. \end{aligned} \tag{$\mathcal{D}$}$$

The optimal values of (D) and $(\mathcal{D})$ are linked by the relation

$$\mathrm{Opt}(D) = \mathrm{Opt}(\mathcal{D}) + \mathrm{const}_D = \mathrm{Opt}(\mathcal{D}) + \bar{x}^T c.$$

Now, note that the linear subspaces $\mathcal{L}_P = \{Py : Ry = 0\}$ and $\mathcal{L}_D = \{\lambda_\ell : \exists \lambda_e : P^T\lambda_\ell + R^T\lambda_e = 0\}$ of $\mathbf{R}^n$ are closely related — *they are orthogonal complements of each other.* Indeed,

$$(\mathcal{L}_P)^\perp = \{\lambda_\ell : \lambda_\ell^T Py = 0 \ \forall y : Ry = 0\},$$

that is, $\lambda_\ell \in \mathcal{L}_P^\perp$ if and only if the homogeneous linear equation $(P^T\lambda_\ell)^T y$ in variable y is a consequence of the system of homogeneous linear equations $Ry = 0$; by Linear Algebra, this is the case iff $P^T\lambda_\ell = R^T\mu$ for certain μ, or, setting $\lambda_e = -\mu$, iff $P^T\lambda_\ell + R^T\lambda_e = 0$ for properly chosen λ_e; but the latter is nothing but the description of $\mathcal{L}_D$.

We arrive at a wonderful geometric picture:

Problems (P) and (D) are reducible, respectively, to

$$\mathrm{Opt}(\mathcal{P}) = \max_{\xi \in \mathbf{R}^m} \left\{ \bar{\lambda}_\ell^T \xi : \xi \geq 0, \xi \in \mathcal{M}_P = \mathcal{L}_P + \bar{\xi} \right\}, \tag{$\mathcal{P}$}$$

and

$$\mathrm{Opt}(\mathcal{D}) = \min_{\lambda_\ell \in \mathbf{R}^m} \left\{ \bar{\xi}^T\lambda_\ell : \lambda_\ell \geq 0, \lambda_\ell \in \mathcal{M}_D = \mathcal{L}_P^\perp - \bar{\lambda}_\ell \right\}.$$

Both $(\mathcal{P})$ and $(\mathcal{D})$ are the problems of optimizing a linear objective over the intersection of an affine subspace (called primal, resp., dual feasible *plane) in* $\mathbf{R}^m$ *and the nonnegative orthant* $\mathbf{R}^m_+$*, and the "geometric data" (the objective and the affine subspace) of the problems are closely related to each other: they are given by a pair of vectors* $\bar{\xi}$, $\bar{\lambda}_\ell$ *in* $\mathbf{R}^m$ *and a pair of linear subspaces* $\mathcal{L}_P$, $\mathcal{L}_D$ *which are orthogonal complements to each other. Specifically,*

- *the feasible planes of* $(\mathcal{P})$ *and* $(\mathcal{D})$ *are shifts of the linear subspaces* $\mathcal{L}_P = \mathcal{L}_D^\perp$ *and* $\mathcal{L}_D = \mathcal{L}_P^\perp$*, respectively;*
- *the objective (to be maximized) in* $(\mathcal{P})$ *is* $\bar{\lambda}_\ell^T \xi$*, and minus the vector* $\bar{\lambda}_\ell$ *is also a shift vector for the dual feasible plane* $\mathcal{M}_D$*. Similarly, the objective (to be minimized) in* $(\mathcal{D})$ *is* $\bar{\xi}^T \lambda_\ell$*, and the vector* $\bar{\xi}$ *is also a shift vector for the primal feasible plane* $\mathcal{M}_P$*.*

The picture is quite symmetric: geometrically, the dual problem is of the same structure as the primal one, and we see that the problem dual to dual is the primal. Slight asymmetry (the vector $\bar{\lambda}_\ell$ responsible for the primal objective is *minus* the shift vector for the dual problem, while the vector $\bar{\xi}$ responsible for the dual objective is *the* shift for the primal) matches the fact that the primal problem is a maximization, and the dual problem is a minimization one; if we were writing the primal and the dual problem both as maximization or both as minimization programs, the symmetry would be "ideal."

Now, let us look at what are, geometrically, the optimal solutions we are looking for. To this end, let us express the duality gap via our slack variables ξ and λ_ℓ. Given feasible for $(\mathcal{P})$ and $(\mathcal{D})$ values of ξ and λ_ℓ, we can associate with them x and λ_e according to

$$\xi = p - Px, \;\; Rx = r, \;\; P^T\lambda_\ell + R^T\lambda_e = c;$$

clearly, x will be a feasible solution to (P), and $\lambda = [\lambda_\ell; \lambda_e]$ will be a feasible solution to (D). Let us compute the corresponding duality gap:

$$\begin{aligned}
\mathrm{DualityGap}(x, \lambda) &= [p^T\lambda_\ell + r^T\lambda_e] - c^T x = p^T\lambda_\ell + r^T\lambda_e \\
&\quad -[P^T\lambda_\ell + R^T\lambda_e]x \\
&= (p - Px)^T\lambda_\ell + [r - Rx]^T\lambda_e \\
&= \xi^T\lambda_\ell.
\end{aligned}$$

Thus, the duality gap is just the inner product of the primal and the dual slacks. Optimality conditions say that feasible solutions to (P) and (D) are optimal for the respective problems iff the duality gap, as evaluated at these solutions, is zero; in terms of $(\mathcal{P})$ and $(\mathcal{Q})$, this means that *a pair of feasible solutions to* $(\mathcal{P})$ *and* $(\mathcal{D})$ *comprises optimal solutions to the respective problems iff these solutions are orthogonal to each other.* Thus,

> *To solve* $(\mathcal{P})$ *and* $(\mathcal{D})$ *means to find a pair of nonnegative vectors that are orthogonal to each other, with the first member of the pair belonging to the affine subspace* $\mathcal{M}_P$*, and the second member belonging to the affine subspace* $\mathcal{M}_D$*. Here,* $\mathcal{M}_P$ *and* $\mathcal{M}_D$ *are given affine subspaces in* $\mathbf{R}^m$ *parallel to the linear subspaces* $\mathcal{L}_P$*,* $\mathcal{L}_D$ *which are orthogonal complements of each other.*
>
> Duality theorem says that this task is feasible iff both $\mathcal{M}_P$ and $\mathcal{M}_D$ do contain nonnegative vectors.

It is a kind of miracle that the purely geometric problem at which we have arrived, with formulation free of any numbers, a problem which, modulo its multi-dimensional nature, looks like an ordinary exercise from a high-school textbook on Geometry, has been the focus of research of at least three generations of first-rate scholars, is the subject of dedicated university courses worldwide and, last but not least, possesses a huge spectrum of applications ranging from poultry production to signal processing.

A careful reader at this point should raise the alarm: We indeed have shown that a primal–dual pair of LO programs with m inequality constraints in the primal problem and feasible systems of equality constraints can be reduced to the above geometric problem; but how do we know that every instance of the latter problem can indeed be obtained from a primal–dual pair of LO's of the outlined type? This is a legitimate question, and it is easy to answer it affirmatively. Indeed, assume we are given two affine subspaces $\mathcal{M}_P$ and $\mathcal{M}_D$ in $\mathbf{R}^m$ such that the linear subspaces, $\mathcal{L}_P$ and $\mathcal{L}_D$, to which $\mathcal{M}_P$ and $\mathcal{M}_D$ are parallel, are orthogonal complements of each other: $\mathcal{L}_D = \mathcal{L}_P^\perp$. Then

$$\mathcal{M}_P = \mathcal{L}_P + \bar{\xi} \ \& \ \mathcal{M}_D = \mathcal{L}_P^\perp - \bar{\lambda}_\ell; \tag{$*$}$$

for properly chosen $\bar{\xi}$ and $\bar{\lambda}_\ell$. We can represent $\mathcal{L}_P$ as the kernel of an appropriately chosen matrix R. Consider the LO in the following standard form:

$$\max_\xi \left\{ \bar{\lambda}_\ell^T \xi : -\xi \leq 0, R\xi = R\bar{\xi} \right\},$$

and let us take it as our primal problem (P). The dual problem (D) then will be

$$\min_{\lambda=[\lambda_\ell;\lambda_e]} \left\{ [R\bar{\xi}]^T \lambda_e : \lambda_\ell \geq 0, -\lambda_\ell + R^T \lambda_e = \bar{\lambda}_\ell \right\}.$$

Clearly, the systems of linear equality constraints in these problems are solvable, one of the solutions to the system of linear constraints in (P)

being $\bar{\xi}$ (which allows to take $\bar{x} = \bar{\xi}$), and one of the solutions to the system of linear constraints in (D) being $[-\bar{\lambda}_\ell; 0; ...; 0]$. It allows to apply to (P) and (D) our machinery to reduce them to problems $(\mathcal{P})$, $(\mathcal{D})$, respectively. From observations on what can be chosen as feasible solutions to the systems of equality constraints in (P) and in (D), it is immediate to derive (do it!) that the problem $(\mathcal{P})$ associated with (P) is nothing but (P) itself, so that its primal feasible affine plane is $\{\xi : R\xi = R\bar{\xi}\}$, which is nothing but $\mathcal{L}_P + \bar{\xi} = \mathcal{M}_P$. Thus, $(P) \equiv (\mathcal{P})$ is nothing but the problem

$$\max_{\xi}\{\bar{\lambda}_\ell^T \xi : \xi \geq 0, \xi \in \mathcal{M}_P = \mathcal{L}_P + \bar{\xi}\}.$$

According to the links we have established between $(\mathcal{P})$ and $(\mathcal{D})$, $(\mathcal{D})$ is nothing but the problem

$$\min_{\lambda_\ell}\left\{\bar{\xi}^T \lambda_\ell : \lambda_\ell \geq 0,\ \lambda_\ell \in \mathcal{L}_P^\perp - \bar{\lambda}_\ell\right\}.$$

The affine feasible plane in $(\mathcal{D})$ is therefore $\mathcal{M}_D$ (see $(*)$), and our previous analysis shows that the geometric problem associated with (P), (D) is exactly the problem "from the nonnegative parts of $\mathcal{M}_P$ and $\mathcal{M}_D$ pick vectors which are orthogonal to each other."

The last question to be addressed is: the pair of problems $(\mathcal{P})$, $(\mathcal{D})$ is specified by a pair $\mathcal{M}_P$, $\mathcal{M}_D$ of affine subspaces in $\mathbf{R}^m$ (this pair cannot be arbitrary: the linear subspaces parallel to $\mathcal{M}_P$, $\mathcal{M}_D$ should be orthogonal complements to each other) *and* a pair of vectors $\bar{\xi} \in \mathcal{M}_P$ and $\bar{\lambda}_\ell \in [-\mathcal{M}_D]$. What happens when we shift these vectors along the corresponding affine subspaces? The answer is: essentially, nothing happens. When $\bar{\lambda}_\ell$ is shifted along $-\mathcal{M}_D$, that is, is replaced with $\bar{\lambda}_\ell + \Delta$ with $\Delta \in \mathcal{L}_D$, the dual problem $(\mathcal{D})$ remains intact, and the objective of the primal problem $(\mathcal{P})$ *restricted on the feasible plane* $\mathcal{M}_P$ *of this problem* is shifted by a constant depending on Δ (check both these claims!); this affects the optimal value of $(\mathcal{P})$, but does not affect the optimal set of $(\mathcal{P})$. Similarly, when shifting $\bar{\xi}$ along $\mathcal{M}_P$, the primal problem $(\mathcal{P})$ remains intact, and the objective of the dual problem $(\mathcal{D})$ on the feasible plane of this problem is shifted by a constant depending on the shift in $\bar{\xi}$. We could predict this "irrelevance of particular choices of $\bar{\xi} \in \mathcal{M}_P$ and $\bar{\lambda}_\ell \in -\mathcal{M}_D$" in advance, since the geometric form of the primal–dual pair $(\mathcal{P})$, $\mathcal{D})$ of problems: "find a pair of nonnegative vectors that are orthogonal to each other, one from $\mathcal{M}_P$ and one from $\mathcal{M}_D$" is posed solely in terms of $\mathcal{M}_P$ and $\mathcal{M}_D$.

Our geometric findings are illustrated in Fig. 3.1.

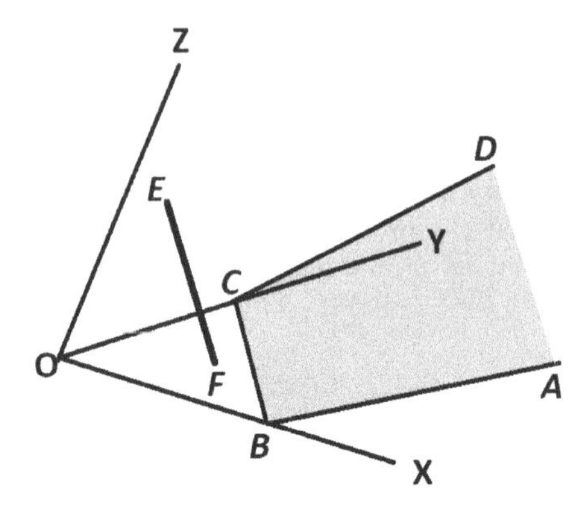

Fig. 3.1. Geometry of primal–dual pair of LO programs.
Notes:

- OXYZ: Nonnegative orthant $\mathbf{R}^3_+$.
- $ABCD$: (Visible part of the) feasible set of $(\mathcal{P})$ — intersection of the 2D primal feasible plane $\mathcal{M}_P$ with the nonnegative orthant $\mathbf{R}^3_+$.
- EF: Feasible set of $(\mathcal{D})$ — intersection of the 1D dual feasible plane $\mathcal{M}_D$ with the nonnegative orthant $\mathbf{R}^3_+$. Pay attention to orthogonality of the plane $ABCD$ and the line EF.
- B: Optimal solution to $(\mathcal{P})$, E: optimal solution to $(\mathcal{D})$. Pay attention to orthogonality of B (which lays on the X-axis) and E, (which belongs to the YZ coordinate plane).

3.3.3 *Antagonistic bilinear games*

3.3.3.1 *Games: Preliminaries*

Consider an *antagonistic game* as follows: we are given two nonempty sets $X \subset \mathbf{R}^m$, $Y \subset \mathbf{R}^n$ and a (bounded, for the sake of simplicity) *cost function* $F(x, y) : X \times Y \to \mathbf{R}$. These data give rise to the game of two players, you and me. You select a point $y \in Y$, I select a point $x \in X$; as a result, I pay to you the sum $F(x, y)$. I am interested in minimizing my loss $F(x, y)$, and you — in maximizing your gain $F(x, y)$. How should we make our choices x, y? Consider two situations:

- I make my choice $x \in X$ first, and it becomes known to you when you make your choice $y \in Y$.

  In this case, I should be ready to pay for a choice $x \in X$ the sum as large as $\overline{F}(x) := \sup_{y \in Y} F(x, y)$: selecting x, I never will pay more than $\overline{F}(x)$, and my loss can be exactly this quantity (when $\sup_{y\in Y} F(x,y)$ is achieved), or a smaller quantity arbitrarily closer to it. As a result, if I am risk-averse, my selection should minimize my worst-case loss $\overline{F}(x)$ over $x \in X$, that is, I should solve the optimization problem

$$\mathrm{Opt}(P) = \min_{x\in X}\left[\overline{F}(x) = \sup_{y\in Y} F(x,y)\right] \qquad (P)$$

$$\left[\mathrm{Opt}(P) = \inf_{x\in X}\sup_{y\in Y} F(x,y)\right]$$

(P) is called the *primal* optimization problem associated with the game in question.

- You make your choice $y \in Y$ first, and it becomes known to me when I make my choice $x \in X$.

  In this case, you should be ready to get for a choice $y \in Y$ the sum as small as $\underline{F}(y) := \inf_{x\in X} F(x,y)$: selecting y, you never will get less than $\underline{F}(x)$, and your gain can be exactly this quantity (when $\inf_{x\in X} F(x,y)$ is achieved), or a larger quantity arbitrarily closer to it. As a result, if you are risk-averse, your selection should maximize your worst-case gain $\underline{F}(y)$ over $y \in Y$, that is, you should solve the optimization problem

$$\mathrm{Opt}(D) = \max_{y\in Y}\left[\underline{F}(y) = \inf_{x\in X} F(x,y)\right] \qquad (D)$$

$$\left[\mathrm{Opt}(D) = \sup_{y\in Y}\inf_{x\in X} F(x,y)\right]$$

  (D) is called the *dual* optimization problem associated with the game.

Common sense says that the second case is better for me than the first one, so that $\mathrm{Opt}(D)$ should be $\leq \mathrm{Opt}(P)$. This indeed is the case:

Proposition 3.5 (Weak duality in antagonistic games). *In the situation in question, the primal objective $\overline{F}(x)$ is everywhere on X*

greater than or equal to the dual objective $\underline{F}(y)$ *everywhere on* Y:

$$\forall(x \in X, y \in Y) : \overline{F}(x) \geq \underline{F}(y),$$

implying that

$$\inf_{x\in X} \sup_{y\in Y} F(x,y) = \mathrm{Opt}(P) \geq \mathrm{Opt}(D) := \sup_{y\in Y} \inf_{x\in X} F(x,y).$$

Indeed, let $\bar{x} \in X$ and $\bar{y} \in Y$. Then $\overline{F}(\bar{x}) = \sup_{y\in Y} F(\bar{x}, y) \geq F(\bar{x}, \bar{y})$ and $\underline{F}(\bar{y}) = \inf_{x\in X} F(x, \bar{y}) \leq F(\bar{x}, \bar{y})$, whence $\underline{F}(\bar{y}) \leq \overline{F}(\bar{x})$. □

Saddle points: The challenging question is, of course, what should you and I do when making our choices simultaneously, so that neither you, nor I know the choice of the adversary when making our own choices? While this question has no meaningful answer in general, there is a case when it admits a good answer — this is the case when our game has *an equilibrium*, that is, there exists a *saddle point* of F on $X \times Y$ — a pair $(x_*, y_*) \in X \times Y$ such that $F(x, y_*)$ as a function of x admits its minimum over $x \in X$ at $x = x_*$, and $F(x_*, y)$ as a function of $y \in Y$ admits its maximum at y_*:

$$x_* \in X \;\&\; y_* \in Y \;\&\; F(x, y_*) \geq F(x_*, y_*) \geq F(x_*, y) \;\forall(x \in X, y \in Y). \tag{3.16}$$

A point (x_*, y_*) with the just indicated property indeed is an equilibrium: if you stick to your choice y_*, I have no incentive to deviate from my choice x_* — such a deviation can only increase my loss. Similarly, if I stick to my choice x_*, you have no incentive to deviate from your choice y_* — such a deviation can only decrease your gain. Note also that were I informed in advance that your choice is y_*, x_* still would be one of my optimal choices, and similarly for you.

Saddle points do not necessarily exist; the next proposition provides necessary and sufficient conditions for their existence and fully specifies their structure.

Proposition 3.6 (Existence and structure of saddle points). *The antagonistic game in question has saddle points if and only if both the primal problem* (P) *and the dual problem* (D) *are solvable with equal optimal values. Whenever this is the case, saddle points are exactly the primal–dual optimal pairs* (x_*, y_*), *that is, pairs with*

x_* being an optimal solution to (P), and y_* being an optimal solution to (D), and for every saddle point (x_*, y_*) one has

$$\inf_{x\in X}\sup_{y\in Y} F(x,y) = \mathrm{Opt}(P) = F(x_*, y_*) = \mathrm{Opt}(D) = \sup_{y\in Y}\inf_{x\in X} F(x,y). \tag{3.17}$$

Proof. In one direction: Let (P) and (D) be solvable with equal optimal values, let x_* be an optimal solution to (P), and y_* be an optimal solution to (D). Then

$$\begin{aligned}\mathrm{Opt}(D) = \underline{F}(y_*) = \inf_{x\in X} F(x, y_*) \le F(x_*, y_*) \le \sup_{y\in Y} F(x_*, y)\\ = \overline{F}(x_*) = \mathrm{Opt}(P),\end{aligned}$$

and since $\mathrm{Opt}(P) = \mathrm{Opt}(D)$, we conclude that $\mathrm{Opt}(P) = \mathrm{Opt}(D) = F(x_*, y_*)$, that is,

$$\mathrm{Opt}(P) = \overline{F}(x_*) = F(x_*, y_*) = \underline{F}(y_*) = \mathrm{Opt}(D).$$

Consequently,

$$\begin{aligned}F(x, y_*) \ge \underline{F}(y_*) = F(x_*, y_*)\,\forall x \in X\ \&\\ F(x_*, y) \le \overline{F}(x_*) = F(x_*, y_*)\,\forall y \in Y,\end{aligned}$$

that is, (x_*, y_*) is a saddle point of F on $X \times Y$.

In the opposite direction: Let (x_*, y_*) be a saddle point of F on $X \times Y$, so that (3.16) holds true. This relation says, in particular, that

$$\underline{F}(y_*) = F(x_*, y_*) = \overline{F}(x_*), \tag{3.18}$$

which combines with Weak duality (Proposition 3.5) to imply that

$$\mathrm{Opt}(P) = \overline{F}(x_*) = \underline{F}(y_*) = \mathrm{Opt}(D), \tag{3.19}$$

that is, (P) and (D) are solvable with equal optimal values, x_* and y_* being optimal solutions to the respective problems.

(3.17) is an immediate conclusion of (3.18) and (3.19). □

3.3.3.2 *Saddle points in bilinear polyhedral games*

As we have already mentioned, not every game has a saddle point. The standard sufficient condition for existence of such a point is as follows:

Theorem 3.4 (Sion–Kakutani). *Let $X \subset \mathbf{R}^m$ and $Y \subset \mathbf{R}^n$ be nonempty closed and bounded convex sets, and $F(x,y) : X \times Y \to \mathbf{R}$ be a continuous function convex in $x \in X$ for every $y \in Y$ and concave in $y \in Y$ for every $x \in X$. Then F has a saddle point on $X \times Y$.*

The Sion–Kakutani Theorem goes beyond our LO-oriented book. We are about to prove the polyhedral version of this theorem — the one where X, Y are polyhedral sets, and F is bi-affine:

$$\begin{aligned} X &= \{x \in \mathbf{R}^m : Px \le p\}, Y = \{y \in \mathbf{R}^n : Qy \le q\}, \\ F(x,y) &= a^T x + b^T y + x^T C y. \end{aligned} \tag{3.20}$$

Assume, similar to the above, that X and Y are nonempty and bounded. For the antagonistic game in question, we have

$$\begin{aligned} \overline{F}(x) &= \max_y \left\{a^T x + b^T y + x^T C y : Qy \le q\right\} \\ &= a^T x + \min_z \left\{q^T z : Q^T z = b + C^T x, z \ge 0\right\} \\ &\quad \text{[LO duality as applied to LO with bounded} \\ &\quad \text{nonempty feasible set]}, \\ \Rightarrow \mathrm{Opt}(P) &= \min_{x \in X} \overline{F}(x) \\ &= \min_{x,z} \left\{a^T x + q^T z : Q^T z - C^T x \right. \\ &\qquad \left. = b, Px \le p, z \ge 0\right\} \qquad (a) \\ \underline{F}(y) &= \min_x \left\{a^T x + b^T y + x^T C y : Px \le p\right\} \\ &= b^T y + \max_w \left\{-p^T w : P^T w + Cy = -a, w \ge 0\right\} \\ &\quad \text{[LO duality as applied to LO with bounded} \\ &\quad \text{nonempty feasible set]}, \\ \Rightarrow \mathrm{Opt}(D) &= \max_{y \in Y} \underline{F}(y) \\ &= \max_{y,w} \left\{b^T y - p^T w : P^T w + Cy \right. \\ &\qquad \left. = -a, Qy \le q, w \ge 0\right\}. \qquad (b) \end{aligned}$$

By their origin, the LO programs in (a) and (b) are solvable. Let us build the LO program dual to (a). Denoting by $y \in \mathbf{R}^n$, $-w \leq 0$, and $\theta \geq 0$ the Lagrange multipliers for the constraints in (a), the dual to (a) problem has the same optimal value as the (solvable!) problem (a) and reads

$$[\mathrm{Opt}(P) =]\mathrm{Opt} = \max_{y,w,\theta} \left\{ b^T y - p^T w : a = -Cy - P^T w, \right.$$
$$\left. q = Qy + \theta, w \geq 0, \theta \geq 0 \right\},$$

which is equivalent to (b). Thus, we are in the situation where (P) and (D) are solvable with equal optimal values. Applying Proposition 3.6, we arrive at the following result:

Proposition 3.7. *Let $X = \{x \in \mathbf{R}^m : Px \leq p\}$ and $Y = \{y \in \mathbf{R}^n : Qy \leq q\}$ be nonempty and bounded polyhedral sets, and $F(x,y) = a^T x + b^T y + x^T Cy : X \times Y \to \mathbf{R}$ be a bi-affine function. Then the antagonistic game given by X, Y, F has saddle points. The x-components of the saddle points are the x-components of optimal solutions to the solvable LO problem*

$$\min_{x,z} \left\{ a^T x + q^T z : Q^T z - C^T x = b, Px \leq p, z \geq 0 \right\}, \qquad \text{(a)}$$

and the y-components of these saddle points are the y-components of optimal solutions to the solvable LO problem

$$\max_{y,w} \left\{ b^T y - p^T w : P^T w + Cy = -a, Qy \leq q, w \geq 0 \right\}, \qquad \text{(b)}$$

which is the LO dual of (a). *The value of the game — the common value of the primal and dual optimization problems associated with the game — is equal to the common optimal value of the LO programs* (a), (b).

Example: Matrix game in mixed strategies. Consider the game where my choices, called my *pure strategies*, are selected in m-element set $X = \{1, 2, \ldots, m\}$, and your choices — your pure strategies — are selected in n-element set $Y = \{1, 2, \ldots, n\}$. In this case, a cost function can be identified with $m \times n$ matrix $F = [F_{ij}]$; I select a row i in this matrix, you select a column j in it, and my resulting loss, a.k.a. your gain, is F_{ij}; this is called *matrix game*. In this game, saddle points, if any, are pairs of indices of entries which simultaneously are among the smallest in their columns and among the largest in their rows; existence of entries of this type is a rare commodity. When saddle points do not exist, it is unclear what is a meaningful solution to the game. Of course, we could use as our choices the solutions to (P) and (D), thus optimizing our worst-case outcomes, but this approach does not make much sense. For example, when $F = \begin{bmatrix} 2 & 4 & 0 \\ 1 & 0 & 5 \end{bmatrix}$, the only optimal solution to (P) is $i_* = 1$, with $\mathrm{Opt}(P) = 4$, and the only optimal solution to (D) is $j_* = 1$ with $\mathrm{Opt}(D) = 1$. This being said, my choice $i = 1$ and your choice $j = 1$ result in loss/gain 2, which is strictly in-between $\mathrm{Opt}(D)$ and $\mathrm{Opt}(P)$; were you told in advance that my choice will be $i = 1$, you would choose $j = 2$ and gain 4 rather than 2, and were I told in advance that your choice is $j = 1$, I would choose $i = 2$ and lose 1 rather than 2.

To arrive at a meaningful notion of a solution to a matrix game, in the 1940s Neumann and Morgenstern proposed to pass from solving the matrix game with deterministic pure strategies to solving it with *randomized* solutions, as when I draw $i \in X$ at random from probability distribution x on X (so that $x \in \Delta_m := \{x \in \mathbf{R}^m_+ : \sum_i x_i = 1\}$ and you draw j at random, independently of me, from probability distribution $y \in \Delta_n$. As a result, my loss and your gain become random with expected value $\sum_{i,j} F_{ij} x_i y_j = x^T F y$, and we arrive at solving the matrix game with *mixed strategies*; our choices in this new game are our *mixed strategies* $x \in \Delta_m$ and $y \in \Delta_n$, and my loss (a.k.a. your gain) stemming from choices x, y is $x^T F y$. The resulting game, by Proposition 3.7, has a saddle point (x_*, y_*), which can be interpreted as follows: imagine that we are playing a matrix game not just once, but round by round, drawing our choices in subsequent rounds, independently of each other and across the rounds, from respective probability distributions $x \in \Delta_m$ and $y \in \Delta_n$. By the Law of Large Numbers, our average, over expanding time horizon, per-round loss/gain would approach $x^T F y$, so that a saddle point in mixed strategies corresponds to a behavior that would be meaningful in the long run in the repeated game.

In the above 2×3 matrix game, the values from mixed strategies is $2\frac{2}{9}$, and those from the saddle point mixed strategies are $x_* = [\frac{5}{9}; \frac{4}{8}]$ and $y_* = [0; \frac{5}{9}; \frac{4}{9}]$.

3.3.4 *Extending calculus of polyhedral representability

3.3.4.1 *Polyhedral representability of Legendre transform

Let f be a proper polyhedrally representable function on $\mathbf{R}^n$:

$$\emptyset \neq \{[x;\tau] : \tau \geq f(x)\} = \{[x;\tau] : \exists w : Px + \tau p + Qw \leq r\}. \quad (3.21)$$

The *Legendre* (a.k.a. *Fenchel*) *transform* of f is the function

$$f_*(x_*) = \sup_{x \in \mathrm{Dom}\, f} \{x^T x_* - f(x)\} \in \mathbf{R} \cup \{+\infty\}. \quad (3.22)$$

Proposition 3.8. *The Legendre transform of a proper p.r.f. is a proper p.r. function with polyhedral representation readily given by the one of f:*

$$\begin{aligned}
\{[x_*;\tau_*] : \tau_* \geq f_*(x_*)\} &= \{[x_*;\tau_*] : \exists w_* : P_* x_* + \tau_* p_* + Q_* w_* \leq r_*\}, \\
P_* &= [\ ;\ ;\ ; -I_n; I_n;\ ;\ ;\ ], \\
p_* &= [-1;\ ;\ ;\ ;\ ;\ ;\ ;\ ] \\
Q_* &= [r^T; p^T; -p^T; P^T; -P^T; Q^T; -Q^T; -I_m], \\
r_* &= [\ ; -1; 1;\ ;\ ;\ ;\ ;\ ] \qquad (3.23)
\end{aligned}$$

(blanks, as always stand for zero blocks, $n = \dim x$, and $m = \dim r$). Besides this, Legendre transform is involution: $(f_*)_* = f$.

Proof. We have

$$\begin{aligned}
&\{[x_*;\tau_*] : \tau_* \geq f_*(x_*)\} \\
&= \left\{[x_*;\tau_*] : \sup_x [x^T x_* - f(x)] \leq \tau_*\right\} \\
&= \left\{[x_*;\tau_*] : \sup_{x,\tau,w} \left[x^T x_* - \tau : Px + \tau p + Qw \leq r\right] \leq \tau_*\right\} \\
&= \left\{[x_*;\tau_*] : \min_{w_* \geq 0} \left[r^T w_* : P^T w_* = x_*, -p^T w_* = 1, Q^T w_* = 0\right] \leq \tau_*\right\}
\end{aligned}$$

[LO Duality; note that LO program

$$\sup_{x,\tau,w}\left[x^Tx_* - \tau : Px + \tau p + Qw \le r\right] \text{ is feasible}],$$

$$\begin{aligned}&= \big\{[x_*;\tau_*] : \exists w_* \ge 0 : P^Tw_* - x_* = 0, 1 + p^Tw_* = 0,\\ &\quad Q^Tw_* = 0, r^Tw_* \le \tau_*\big\} \ (!)\\ &= \left\{[x_*;\tau_*] : \exists w_* : \underbrace{[\,;\,;\,;-I_n;I_n;\,;\,;\,]}_{P_*}x_* + \tau_*\underbrace{[-1;\,;\,;\,;\,;\,;\,;\,]}_{p_*}\right.\\ &\quad + \underbrace{[r^T;p^T;-p^T;P^T;-P^T;Q^T;-Q^T;-I_m]}_{Q_*}\\ &\quad \left. w_* \le \underbrace{[\,;-1;1;\,;\,;\,;\,;\,]}_{r_*}\right\}.\end{aligned}$$

We see that $p_* \le 0$ and that (3.23) holds true. Next, f is proper; selecting $x \in \operatorname{Dom} f$, the LO program $\min_{\tau,w}\{\tau : Px + \tau p + Qw \le r\}$ is solvable, and therefore the dual problem is solvable as well, implying that there exists $\bar{w}_* \ge 0$ such that $Q^T\bar{w}_* = 0$ and $p^T\bar{w}_* + 1 = 0$. Setting $\bar{x}_* = P^T\bar{w}_*$, $\bar{\tau}_* = r^T\bar{w}_*$, we conclude from (!) that $x_* = \bar{x}_*, \tau_* = \bar{\tau}_*, w_* = \bar{w}_*$ satisfy $P_*\bar{x}_* + \bar{\tau}_*p_* + Q_*\bar{w}_* \le r_*$, whence $[\bar{x}_*;\bar{\tau}_*]$ belongs to the right-hand side set in (3.23), so that $f_*(\bar{x}_*) < \infty$ by (3.23). On the other hand, since f is proper, $f_*(x_*) > -\infty$ for every x_*. Thus, $f_*(\bar{x}_*)$ is finite. As a result, (3.23) says that the (a) relation $\tau_*p_* + Q_*w_* \le 0$ implies that $\tau_* \ge 0$;[5] further, (b) the relation $\tau_*p_* + Q_*w_* \le 0$ is satisfied when $\tau_* = 1$ and $w_* = 0$ (since $p_* \le 0$). Besides this, we have already verified that (c) the right-hand side set in (3.23) is nonempty. Taken together, by Proposition 2.19 (a), (b), (c) imply that the right-hand side of (3.23) is a polyhedral representation of a proper p.r.f. That is, f_* is a proper p.r.f. with polyhedral representation (3.23), as claimed.

[5]Indeed, assuming that $\widehat{\tau}_*p_* + Q_*\widehat{w}_* \le 0$ for some $\widehat{w}_*$ and $\widehat{\tau}_* < 0$, we would get $P_*\bar{x}_* + [\bar{\tau}_* + s\widehat{\tau}_*]p_* + Q_*[\bar{w}_* + s\widehat{w}_*] \le r_*$ for all $s \ge 0$, implying by (!) and (3.23) that $f_*(\bar{x}_*) \le \bar{\tau}_* + s\widehat{\tau}_*$ for all $s \ge 0$, which is impossible, since $\widehat{\tau}_* < 0$ and $f_*(\bar{x}_*) \in \mathbf{R}$.

Now, let us compute the Fenchel transform of the proper p.r.f. function f_*. We have

$$\{[x;\tau] : \tau \geq (f_*)_*(x)\}$$
$$= \left\{[x;\tau] : \sup_{x_*,\tau_*,w_*} \left[x^T x_* - \tau_* : P_* x_* + \tau_* p_* + Q_* w_* \leq r_*\right] \leq \tau\right\}$$
$$= \left\{[x;\tau] : \exists z \geq 0 : P_*^T z = x, p_*^T z = -1, Q_*^T z = 0, r_*^T z \leq \tau\right\}$$

$\Big[$LO Duality; note that LO program

$$\sup_{x_*,\tau_*,w_*} \left[x^T x_* - \tau_* : P_* x_* + \tau_* p_* + Q_* w_* \leq r_*\right] \text{ is feasible}\Big],$$

$$= \left\{ [x;\tau] : \exists z = [z_r; z_p^+; z_p^-; z_P^+; z_P^-, z_Q^+; z_Q^-; z_I] \geq 0 : z_P^- - z_P^+ \right.$$
$$= x,\ z_r = 1, \underbrace{z_p^- - z_p^+}_{=:\bar\tau} \leq \tau, z_r r + [z_p^+ - z_p^-]p$$
$$+ P[z_P^+ - z_P^-] + Q\underbrace{[z_Q^+ - z_Q^-]}_{=:-w} - z_I = 0,$$
$$\left. z_p^\pm \geq 0,\ z_P^\pm \geq 0,\ z_Q^\pm \geq 0,\ z_I \geq 0 \right\}$$
$$= \{[x;\tau] : \exists w, \bar\tau : Px + \bar\tau p + Qw \leq r, \bar\tau \leq \tau\}$$
$$= \{[x;\tau] : \exists w, \tau : Px + \tau p + Qw \leq r\}$$

where the concluding equality is due to the fact that $\{\tau : \exists w : \tau p + Qw \leq 0\} = \{\tau \geq 0\}$ by Proposition 2.19. Invoking (3.21), we conclude that

$$\{[x;\tau] : \tau \geq (f_*)_*(x)\} = \{[x;\tau] : \tau \geq f(x)\},$$

that is $(f_*)_* \equiv f$. □

We remark that the Legendre transform is naturally extended, the involutivity property being preserved, from proper p.r.f.'s to proper convex functions with closed epigraphs.

3.3.4.2 *Support functions of polyhedral sets

Let $X \subset \mathbf{R}^n$ be a nonempty polyhedral set. The *characteristic function* $\chi_X(x)$ of X is, by definition, the function which is equal to 0 on X and to $+\infty$ outside of X; needless to say, this is a proper p.r.f., see Proposition 1.1. The Legendre transform of this function — the function

$$\mathrm{Supp}_X(x_*) = \sup_x \left[x_*^T x - \chi_X(x)\right] = \sup_{x \in X} x_*^T x : \mathbf{R}^n \to \mathbf{R} \cup \{+\infty\}$$

— is called the *support function* of X. For example,

- The characteristic function of the entire space is identically zero, and the support function is zero at the origin and $+\infty$ outside of the origin.
- The characteristic function of the singleton $\{0\}$ is zero at the origin and $+\infty$ outside the origin, and the support function is identically zero.
- The support function of the unit box $\{x \in \mathbf{R}^n : \|x\|_\infty \leq 1\}$ is the ℓ_1-norm $\|\cdot\|_1$, and the support function of the unit $\|\cdot\|_1$-ball $\{x : \|x\|_1 \leq 1\}$ is the ℓ_∞ norm $\|\cdot\|_\infty$.

Here is the "executive summary" on support functions.

Observation 3.1. (i) A function ϕ is the support function of a nonempty polyhedral set in $\mathbf{R}^n$ if and only if ϕ is proper p.r.f. which is positively homogeneous of homogeneity degree 1:

$$(x \in \mathrm{Dom}\,\phi, t \geq 0) \Rightarrow tx \in \mathrm{Dom}\,\phi \;\&\; \phi(tx) = t\phi(x).$$

(ii) If X, Y are nonempty polyhedral sets in $\mathbf{R}^n$, then $X \subset Y$ iff $\mathrm{Supp}_X(\cdot) \leq \mathrm{Supp}_Y(\cdot)$. In particular, the support function Supp_X of X "remembers X" — support functions of two nonempty polyhedral sets coincide with each other if and only if the sets are the same.

Proof. (i) In one direction: the fact that the support function of a nonempty polyhedral set is a proper p.r.f. is readily given by Proposition 3.8, and positive homogeneity, of degree 1, of the support function is evident. In the opposite direction: let ϕ be a proper p.r. function that is positively homogeneous of degree 1. Taking into account Proposition 3.8, all we need in order to prove that ϕ is the support function of a nonempty polyhedral set is to verify that the Legendre transform ϕ_* of ϕ is the characteristic function of a nonempty polyhedral set. By Proposition 3.8, ϕ_* is a proper p.r.f., so that all we need to verify is that the only real value taken by ϕ_* is zero. Indeed, assuming that $x \in \mathrm{Dom}\,\phi_*$, we have

$$\begin{aligned}\phi_*(x) &= \sup_{x_*}\left[x^T x_* - \phi(x_*)\right] = \sup_{x_*\in\mathrm{Dom}\,\phi}\left[x^T x_* - \phi(x_*)\right] \\ &\underbrace{=}_{(a)} \sup_{y_*\in\mathrm{Dom}\,\phi}\left[x^T[2y_*] - \phi(2y_*)\right] \underbrace{=}_{(b)} 2\sup_{y_*\in\mathrm{Dom}\,\phi}\left[x^T y_* - \psi(y_*)\right] \\ &= 2\phi_*(x),\end{aligned}$$

where (a) and (b) are due to positive homogeneity of ϕ (which, in particular, says that when y_* runs through the entire $\mathrm{Dom}\,\phi$, $2y_*$ runs through the entire $\mathrm{Dom}\,\phi$ as well). The resulting relation $\phi(x_*) = 2\phi(x_*)$ implies that $\phi(x_*) = 0$. □

(ii) If f, g are proper p.r.f.'s on $\mathbf{R}^n$ and $f(\cdot) \geq g(\cdot)$, then, by definition of the Legendre transform, we have $f_* \leq g_*$. Taking into account that the transform is involution, we conclude that proper p.r.f.'s f, g are linked by the relation $f \geq g$ if and only if f_* and g_* are linked by $f_* \leq g_*$. In particular, for support functions Supp_X, Supp_Y of nonempty polyhedral sets X, Y one has $\mathrm{Supp}_X \leq \mathrm{Supp}_Y$ if and only if $\chi_X \geq \chi_Y$, and the latter relation clearly is the same as $X \subset Y$. □

We conclude this section with a "streamlined" polyhedral representation of the support function of a p.r. set; this should be considered as a new rule in the calculus of polyhedral sets and polyhedrally representable functions which we started in Section 1.3.3. The following numeration of calculus rules continues the one in Section 1.3.3.

F.6. *Taking the support function of a polyhedrally representable set:* a polyhedral representation

$$X = \{x \in \mathbf{R}^n : \exists w : Px + Qw \leq r\}$$

of a *nonempty* polyhedral set can be easily converted to a polyhedral representation of its support function:

$$\left\{[c;\tau] : \tau \geq \mathrm{Supp}_X(c) := \sup_{x\in X} c^T x\right\}$$
$$= \left\{[c;\tau] : \exists\lambda : \lambda \geq 0, P^T\lambda = c, Q^T\lambda = 0, r^T\lambda \leq \tau\right\}.$$

To prove the latter relation, note that from the definition of $\mathrm{Supp}_X(c)$ and from the polyhedral representation of X in question, it follows immediately that $\mathrm{Supp}_X(c)$ is the optimal value in the (feasible, since X is nonempty) LO program

$$\max_{[x;w]} \left\{c^T x : Px + Qw \leq r\right\},$$

and it remains to apply the Linear Programming Duality Theorem.

The following two calculus rules are immediate corollaries of F.6:

S.6. A polyhedral representation

$$X = \{x \in \mathbf{R}^n : \exists w : Px + Qw \leq r\}$$

of a polyhedral set containing the origin can be straightforwardly converted into a p.r. of its polar as follows:

$$\begin{aligned}\mathrm{Polar}\,(X) &:= \{\xi : \xi^T x \leq 1\,\forall x \in X\}\\ &= \{\xi : \mathrm{Supp}_X(\xi) \leq 1\}\\ &= \{\xi : \exists\lambda : \lambda \geq 0, P^T\lambda = \xi, \quad Q^T\lambda = 0, r^T\lambda \leq 1\}.\end{aligned}$$

S.7. Let X be a polyhedral cone given by a p.r.

$$X = \{x : \exists w : Px + Qw \leq r\}.$$

Then the cone X_* dual to X admits the p.r.

$$\begin{aligned}X_* &:= \{\xi : \xi^T x \geq 0\,\forall x \in X\} = \{\xi : \mathrm{Supp}_X(-\xi) \leq 0\}\\ &= \left\{\xi : \exists\lambda : \lambda \geq 0, P^T\lambda + \xi = 0, Q^T\lambda = 0, r^T\lambda \leq 0\right\}.\end{aligned}$$

3.3.5 *Polyhedral representability of the cost function of an LO program, a.k.a. sensitivity analysis*

Consider an LO program in the form

$$\mathrm{Opt}(c,b) = \max_x \left\{c^T x : Ax \le b\right\} \quad [A : m \times n], \tag{3.24}$$

which we will denote also as $(P[c,b])$. There are situations when either b or c, instead of being fixed components of the data, are themselves decision variables in a certain "master program" which involves, as a "variable," the optimal value $\mathrm{Opt}(c,b)$ of (3.24) as well.

To give an example, let (3.24) be a production planning problem, where b is the vector of available resources, x is a production plan, and $\mathrm{Opt}(c,b)$ is the maximum, given the resources, volume of sales. Now, imagine that we can buy the resources at certain prices, perhaps under additional restrictions on what can and what cannot be bought (like upper bound on the total cost of ordering the resources, lower and upper bounds on the amounts of every one of the resources we can buy, etc.). In this situation, the master problem might be to maximize our net profit (volume of sales minus the cost of resources) by choosing both the resources to be used (b) and how to produce (x).

A highly instructive example of a situation where c is varying will be given later ("Robust Optimization").

In such a situation, in order to handle the master program, we need a polyhedral representation of $\mathrm{Opt}(c,b)$ as a function of the "varying parameter" (b or c) in question.

3.3.5.1 *Opt(c,b) as a function of b*

Let us fix c and write $(P_c[b])$ instead of $(P[c,b])$ and $\mathrm{Opt}_c(b)$ instead of $\mathrm{Opt}(c,b)$ to stress the fact that c is fixed and we treat the optimal value of (3.24) as a function of the right-hand side vector b. Let us make the following assumption:

(!) *For some value $\bar{b}$ of b, program $(P_c[b])$ is feasible and bounded.*

Then the problem dual to $(P[b])$, that is,

$$\min_\lambda \left\{b^T\lambda : \lambda \ge 0, A^T\lambda = c\right\} \tag{$D_c[b]$}$$

is feasible when $b = \bar{b}$. But the feasible set of the latter problem does not depend on b, so that, by Weak duality, $(P_c[b])$ is bounded for all b, and thus $\mathrm{Opt}_c(b)$ is a function taking real values and, perhaps, the value $-\infty$ (the latter happens at those b for which $(P_c[b])$ is infeasible). Taking into account that a feasible and bounded LO program is solvable, we have the equivalence

$$\mathrm{Opt}_c(b) \geq t \Leftrightarrow \exists x : Ax \leq b \ \& \ c^T x \geq t, \tag{3.25}$$

which is a polyhedral representation of the *hypograph* $\{[b;t] : t \leq \mathrm{Opt}_c(b)\}$ of $\mathrm{Opt}_c(\cdot)$. As a byproduct, we see that $\mathrm{Opt}_c(\cdot)$ is a *concave* function.

Now, the domain D of $\mathrm{Opt}_c(\cdot)$ — the set of values of b where $\mathrm{Opt}_c(b)$ a real number — clearly is the set of those b's for which $(P_c[b])$ is feasible (again: under our assumption that program $(P_c[b])$ is feasible and bounded for some $b = \bar{b}$, it follows that the program is solvable whenever it is feasible). Let $\bar{b} \in D$, and let $\bar{\lambda}$ be the optimal solution to the program $(D_c[\bar{b}])$. Then $\bar{\lambda}$ is dual feasible for *every* program $(P_c[b])$, whence by Weak duality

$$\mathrm{Opt}_c(b) \leq \bar{\lambda}^T b = \bar{\lambda}^T \bar{b} + \bar{\lambda}^T (b - \bar{b}) = \mathrm{Opt}_c(\bar{b}) + \bar{\lambda}^T (b - \bar{b}).$$

The resulting inequality

$$\forall b : \mathrm{Opt}_c(b) \leq \Phi_{\bar{b}}(b) := \mathrm{Opt}_c(\bar{b}) + \bar{\lambda}^T (b - \bar{b}) \tag{3.26}$$

resembles the Gradient inequality $f(y) \leq f(x) + (y - x)^T \nabla f(x)$ for smooth concave functions f; geometrically, it says that the graph of $\mathrm{Opt}_c(b)$ never goes above the graph of the affine function $\Phi_{\bar{b}}(b)$ and touches the latter graph at the point $[\bar{b}; \mathrm{Opt}_c(\bar{b})]$. Now, recall that $\mathrm{Opt}_c(b)$ is a polyhedrally representable concave function and thus is the restriction on D of the minimum of finitely many affine functions of b as follows:

$$b \in D \Rightarrow \mathrm{Opt}_c(b) = \min_{1 \leq i \leq I} \phi_i(x), \ \ \phi_i(x) = \alpha_i^T b + \beta_i.$$

Assuming w.l.o.g. that all these functions are distinct from each other and taking into account that D clearly is full-dimensional (since whenever $b' \geq b \in D$, we have $b' \in D$ as well), we see that $\mathrm{Opt}_c(b)$ is differentiable almost everywhere on D (specifically, everywhere in $\mathrm{int}\, D$, except for the union of finitely many hyperplanes given by the solvable equations of the form $\phi_i(x) = \phi_j(x)$ with $i \neq j$). At every point $\bar{b} \in \mathrm{int}\, D$ where $\mathrm{Opt}_c(b)$ is differentiable, (3.26) is possible only when $\bar{\lambda} = \nabla \mathrm{Opt}_c(\bar{b})$, and we arrive at the following conclusion:

Let $(P_c[b])$ be feasible and bounded for some b. Then the set $D = \mathrm{Dom}\,\mathrm{Opt}_c(\cdot)$ is a polyhedral cone with a nonempty interior, and at every point $\bar{b} \in \mathrm{int}\ D$ where the function $\mathrm{Opt}_c(\cdot)$ is differentiable (and this is so everywhere on int D except for the union of finitely many hyperplanes), *the problem $(D_c[\bar{b}])$ has a unique optimal solution which is the gradient of $\mathrm{Opt}_c(\cdot)$ at $\bar{b}$.*

Of course, we can immediately recognize where the domain D of $\mathrm{Opt}_c(\cdot)$ and the "pieces" $\phi_i(\cdot)$ come from. By Linear Programming Duality Theorem, we have

$$\mathrm{Opt}_c(b) = \min_{\lambda} \left\{ b^T\lambda : \lambda \geq 0, A^T\lambda = c \right\}.$$

We are in the situation where the feasible domain Λ of the latter problem is nonempty; besides this, it clearly does not contain lines. By Theorem 2.12, this feasible domain is

$$\mathrm{Conv}\{\lambda_1, \ldots, \lambda_S\} + \mathrm{Cone}\,\{\rho_1, \ldots, \rho_T\},$$

where λ_i are the vertices of Λ, and ρ_j are the (directions of the) extreme rays of $\mathrm{Rec}(\Lambda)$. We clearly have

$$D = \{b : \rho_j^T b \geq 0,\ 1 \leq j \leq T\}, b \in D \Rightarrow \mathrm{Opt}_c(b) = \min_{1 \leq i \leq S} \lambda_i^T b,$$

so that D is exactly the cone dual to $\mathrm{Rec}(\Lambda)$, and we can take $I = S$ and $\phi_i(b) = \lambda_i^T b$, $1 \leq i \leq S$. Here, again we see how powerful polyhedral representations of functions are as compared to their straightforward representations as maxima of affine pieces: the number S of pieces in $\mathrm{Opt}_c(\cdot)$ and the number of linear inequalities specifying the domain of this function can be — and typically are — astronomical, while the polyhedral representation (3.25) of the function is fully tractable.

3.3.5.2 *Law of diminishing marginal returns*

The concavity of $\mathrm{Opt}_c(b)$ as a function of b, however simple this fact be, has important "real life" consequences. Assume, as in the motivating example above, that b is the vector of resources available for a production process, and we should buy these resources at a market at certain prices p_i forming price vector p. Interpreting the objective in (3.24) as our income (the total dollar value of our sales), and denoting by M our investments in the resources, the problem of optimizing the income becomes the LO program

$$\mathrm{Opt}(M) = \max_{x,b} \left\{ c^T x : Ax \leq b, Qb \leq q, p^T b \leq M \right\},$$

where $Qb \leq q$ are "side constraints" on b (like nonnegativity, upper bounds on the maximal available amounts of various resources, etc.) In this problem, the right-hand side vector is $[q; M]$; we fix once for ever q and treat M as a varying parameter, as expressed by the notation $\mathrm{Opt}(M)$. Note that the feasible set of the problem extends with M; thus, assuming that the problem is feasible and bounded for certain $M = M_0$, it remains feasible and bounded for $M \geq M_0$. We conclude that $\mathrm{Opt}(M)$ is a finite, nondecreasing and, as we have shown, *concave* function of $M \geq M_0$. Concavity implies that $\mathrm{Opt}(M + \delta) - \mathrm{Opt}(M)$, where $\delta > 0$ is fixed, is decreasing (perhaps, nonstrictly) as M grows. In other words, *the reward for an extra $1 in the investment can only decrease as the investment grows.* In Economics, this is called *the law of diminishing marginal returns.*

3.3.5.3 *Opt(c, b) as a function of c*

Now, let us treat b as once forever fixed, and c — as a varying parameter in (3.24), and write $(P^b[c])$ instead of (3.24) and $\mathrm{Opt}^b(c)$ instead of $\mathrm{Opt}(c, b)$. Assume from now on that $(P^b[c])$ is feasible (this fact is independent of a particular value of c). Then the relation

$$\mathrm{Opt}^b(c) \leq \tau \tag{$*$}$$

is equivalent to the fact that the problem $(D_c[b])$ is solvable with optimal value $\leq \tau$. Applying the Linear Programming Duality Theorem, we arrive at the equivalence

$$\mathrm{Opt}^b(c) \leq \tau \Leftrightarrow \exists \lambda : A^T \lambda = c, \lambda \geq 0, b^T \lambda \geq \tau; \tag{3.27}$$

this equivalence is a polyhedral representation of $\mathrm{Opt}^b(c)$ as a *convex* function. As a byproduct, we see that $\mathrm{Opt}^b(c)$ is convex. The latter could be easily seen from the very beginning: $\mathrm{Opt}^b(c)$ is the supremum of the family $c^T x$ of linear (and thus convex) functions of c over the set $\{x : Ax \leq b\}$ of values of the "parameter" x and thus is convex (see calculus of convex functions).

In the above construction it was assumed that (3.24) is feasible. When it is not the case, the situation is trivial: $\mathrm{Opt}^b(c)$ is $-\infty$ identically, and we do *not* treat such a function as convex (convex function must take finite values and the value $+\infty$ only). Thus, when (3.24) is infeasible, we cannot say that $\mathrm{Opt}^b(c)$ is a "polyhedrally representable convex function," in spite of the fact that the set of all pairs (c,τ) satisfying $(*)$ — the entire (c,τ)-space! — clearly is polyhedral.

Now, let $\bar{c} \in \mathrm{Dom}\,\mathrm{Opt}^b(\cdot)$. Then the problem $(P^b[\bar{c}])$ is feasible and bounded, and thus is solvable. Denoting by $\bar{x}$ an optimal solution to $(P^b[\bar{c}])$ and observing that $\bar{x}$ is feasible for every program $(P^b[c])$, we have

$$\mathrm{Opt}^b(c) \geq c^T\bar{x} = \bar{c}^T\bar{x} + (c-\bar{c})^T\bar{x} = \mathrm{Opt}^b(\bar{c}) + \bar{x}^T(c-\bar{c}),$$

that is,

$$\forall c : \mathrm{Opt}^b(c) \geq \mathrm{Opt}^b(\bar{c}) + \bar{x}^T(c-\bar{c}), \tag{3.28}$$

which looks as the Gradient inequality for a smooth convex function. To simplify our analysis, note that if A has a nontrivial kernel, then $\mathrm{Opt}^b(c)$ clearly is $+\infty$ when c is not orthogonal to $\mathrm{Ker}\,A$; if c is orthogonal to $\mathrm{Ker}\,A$, then $\mathrm{Opt}^b(c)$ remains intact when we augment the constraints of $(P^b[c])$ with the linear equality constraints $x \in L = (\mathrm{Ker}\,A)^{\perp}$. Replacing, if necessary, $\mathbf{R}^n$ with L, let us assume that $\mathrm{Ker}\,A = 0$, so that the feasible set Π of $(P^b[\cdot])$ does not contain lines. Since it is nonempty, we have

$$\Pi = \mathrm{Conv}\{x_1,\ldots,x_S\} + \mathrm{Cone}\,\{r_1,\ldots,r_T\},$$

where x_i are the vertices of Π, and r_j are the (directions of the) extreme rays of $\mathrm{Rec}(\Pi)$. We now clearly have

$$\mathrm{Dom}\,\mathrm{Opt}^b(\cdot) = -(\mathrm{Rec}(\Pi))^*,\ c \in \mathrm{Dom}\,\mathrm{Opt}^b(\cdot) \Rightarrow \mathrm{Opt}^b(c) = \max_i x_i^T c.$$

Since the cone $\mathrm{Rec}(\Pi)$ is pointed due to $\mathrm{Ker}\,A = \{0\}$, the domain of $\mathrm{Opt}^b(\cdot)$ possesses a nonempty interior (see the proof of Proposition 2.14), and $\mathrm{Opt}^b(c)$ is differentiable everywhere in this interior except for the union of finitely many hyperplanes. This combines with (3.28) to imply the following result, completely symmetric to the one we got for $\mathrm{Opt}_c(\cdot)$:

Let $\mathrm{Ker}\, A = \{0\}$ *and* $(P^b[\cdot])$ *be feasible. Then* $P = \mathrm{Dom}\,\mathrm{Opt}^b(\cdot)$ *is a polyhedral cone with a nonempty interior, and at every point* $\bar{c} \in \mathrm{int}\, P$ *where the function* $\mathrm{Opt}^b(\cdot)$ *is differentiable* (and this is so everywhere on int P except for the union of finitely many hyperplanes), *the problem* $(P^b[\bar{c}])$ *has a unique optimal solution which is the gradient of* $\mathrm{Opt}^b(\cdot)$ *at* $\bar{c}$.

3.3.6 *Applications in robust LO*

Polyhedral representability of $\mathrm{Opt}^b(c)$ plays a crucial role in *Robust Linear Optimization* — a (reasonably novel) methodology for handling LO problems with *uncertain* data. Here is the story.

3.3.6.1 *Data uncertainty in LO: Sources*

Typically, the data of real world LOs

$$\max_x \left\{c^T x : Ax \le b\right\} \quad [A = [a_{ij}] : m \times n] \tag{LO}$$

are not known exactly when the problem is being solved. The most common reasons for data uncertainty are as follows:

- some data entries (future demands, returns, etc.) do not exist when the problem is solved and hence are replaced with their forecasts. These data entries are thus subject to *prediction errors*;
- some of the data (parameters of technological devices/processes, contents associated with raw materials, etc.) cannot be measured exactly, and their true values drift around the measured "nominal" values; these data are subject to *measurement errors*;
- some of the decision variables (intensities with which we intend to use various technological processes, parameters of physical devices we are designing, etc.) cannot be implemented exactly as computed. The resulting *implementation errors* are equivalent to appropriate artificial data uncertainties.

  Indeed, the contribution of a particular decision variable x_j to the left-hand side of constraint i is the product $a_{ij}x_j$. A typical implementation error can be modelled as $x_j \mapsto (1 + \xi_j)x_j + \eta_j$, where ξ_j is the multiplicative, and η_j is the additive component of the error. The effect of this error is *as if* there were no implementation error at all, but the coefficient a_{ij} got the multiplicative perturbation: $a_{ij} \mapsto a_{ij}(1 + \xi_j)$, and the right-hand side b_i of the constraint got the additive perturbation $b_i \mapsto b_i - \eta_j a_{ij}$.

3.3.6.2 *Data uncertainty: Dangers*

In the traditional LO methodology, a small data uncertainty (say, 0.1% or less) is just ignored; the problem is solved *as if* the given ("nominal") data were exact, and the resulting *nominal* optimal solution is what is recommended for use, in the hope that small data uncertainties will not affect significantly the feasibility and optimality properties of this solution, or that small adjustments of the nominal solution will be sufficient to make it feasible. In fact, these hopes are not necessarily justified, and sometimes even a small data uncertainty deserves significant attention. We are about to present two instructive examples of this type.

Motivating example I: Synthesis of antenna arrays. Consider a monochromatic transmitting antenna placed at the origin. Physics says that

(1) The directional distribution of energy sent by the antenna can be described in terms of the *antenna's diagram* which is a complex-valued function $D(\delta)$ of a 3D direction δ. The directional distribution of energy sent by the antenna is proportional to $|D(\delta)|^2$.
(2) When the antenna comprises several antenna elements with diagrams $D_1(\delta), \ldots, D_k(\delta)$, the diagram of the antenna is just the sum of the diagrams of the elements.

In a typical Antenna Design problem, we are given several antenna elements with diagrams $D_1(\delta), \ldots, D_n(\delta)$ and are allowed to multiply these diagrams by complex *weights* x_i (which in reality corresponds to modifying the output powers and shifting the phases of the elements). As a result, we can obtain, as a diagram of the array, any function of the form

$$D(\delta) = \sum_{j=1}^{n} x_j D_j(\delta),$$

and our goal is to find the weights x_j which result in a diagram as close as possible, in a prescribed sense, to a given "target diagram" $D_*(\delta)$.

Example: Antenna design: Consider that a planar antenna comprises a central circle and 9 concentric rings of the same area as the circle (Fig. 3.2(a)) in the XY-plane ("Earth's surface"). Let the wavelength be $\lambda = 50$ cm, and the outer radius of the outer ring be 1 m (twice the wavelength).

One can easily see that the diagram of a ring $\{a \le r \le b\}$ in the plane XY (r is the distance from a point to the origin) as a function of a 3-dimensional direction δ depends on the altitude (the angle θ between the direction and the plane) only. The resulting function of θ turns out to be *real-valued*, and its analytic expression is

$$D_{a,b}(\theta) = \frac{1}{2}\int_a^b \left[\int_0^{2\pi} r\cos\left(2\pi r\lambda^{-1}\cos(\theta)\cos(\phi)\right) d\phi\right] dr.$$

Figure 3.2(b) represents the diagrams of our 10 rings for $\lambda = 50$cm.

Assume that our goal is to design an array with a real-valued diagram which should be axial symmetric with respect to the Z-axis and should be "concentrated" in the cone $\pi/2 \ge \theta \ge \pi/2 - \pi/12$. In other words, our target diagram is a real-valued function $D_*(\theta)$ of the altitude θ with $D_*(\theta) = 0$ for $0 \le \theta \le \pi/2 - \pi/12$ and $D_*(\theta)$ somehow approaching 1 as θ approaches $\pi/2$. The target diagram $D_*(\theta)$ used in this example is given in Fig. 3.2(c) (bold curve).

Let us measure the discrepancy between a synthesized diagram and the target one by the Tschebyshev distance, taken along the equidistant 240-point grid of altitudes, i.e., by the quantity

$$\tau = \max_{i=1,\ldots,240}\left|D_*(\theta_i) - \sum_{j=1}^{10} x_j \underbrace{D_{r_{j-1},r_j}(\theta_i)}_{D_j(\theta_i)}\right|, \quad \theta_i = \frac{i\pi}{480}.$$

Our design problem is simplified considerably by the fact that the diagrams of our "building blocks" and the target diagram are real-valued; thus, we need no complex numbers, and the problem we should finally solve is

$$\min_{\tau\in\mathbf{R},x\in\mathbf{R}^{10}}\left\{\tau : -\tau \le D_*(\theta_i) - \sum_{j=1}^{10} x_j D_j(\theta_i) \le \tau,\ i = 1,\ldots,240\right\}. \tag{3.29}$$

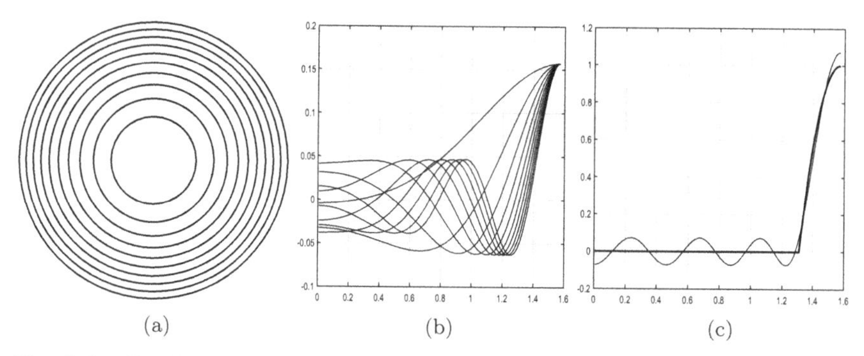

Fig. 3.2. Synthesis of antennae array. (a) 10 array elements of equal areas in the XY-plane; the outer radius of the largest ring is 1m, the wavelength is 50cm. (b) "Building blocks" — the diagrams of the rings as functions of the altitude angle θ. (c) The target diagram (monotone) and the synthesized diagram (oscillating).

This is a simple LP program; its optimal solution x^* results in the diagram depicted at Fig. 3.2(c) (thin curve). The uniform distance between the actual and the target diagrams is ≈ 0.0726 (recall that the target diagram varies from 0 to 1).

Now, recall that our design variables are characteristics of certain physical devices. In reality, of course, we cannot tune the devices to have precisely the optimal characteristics x_j^*; the best we may hope for is that the actual characteristics $x_j^{\rm fct}$ will coincide with the desired values x_j^* within a small margin ρ, say, $\rho = 0.1\%$ (this is a fairly high accuracy for a physical device):

$$x_j^{\rm fct} = (1+\xi_j)x_j^*,\ |\xi_j| \le \rho = 0.001.$$

It is natural to assume that the *actuation errors* ξ_j are random with the mean value equal to 0; it is perhaps not a great sin to assume that these errors are independent of each other. Note that as it was already explained, the consequences of our actuation errors are *as if* there were no actuation errors at all, but the coefficients $D_j(\theta_i)$ of variables x_j in (3.29) were subject to perturbations $D_j(\theta_i) \mapsto (1+\xi_j)D_j(\theta_i)$.

Since the actual weights differ from their desired values x_j^*, the actual (random) diagram of our array of antennae will differ from the "nominal" one we see in Fig. 3.2(c). How large could the difference be? Looking at Fig. 3.3, we see that the difference can be dramatic. The diagrams corresponding to $\rho > 0$ are not even the worst case:

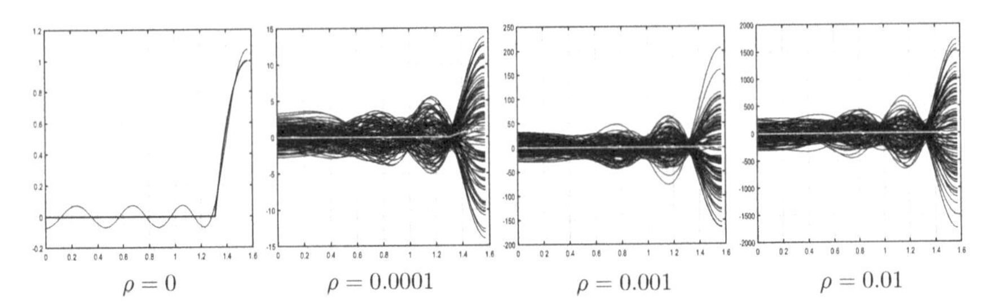

Fig. 3.3. "Dream and reality," nominal optimal design: samples of 100 actual diagrams for different uncertainty levels. White line: the target diagram.

Table 3.1. Quality of nominal antenna design: dream and reality. Data from 100 samples of actuation errors per each uncertainty level ρ.

	Dream	Reality					
	$\rho = 0$	$\rho = 0.0001$			$\rho = 0.01$		
	Value	Min	Mean	Max	Min	Mean	Max
$\|\cdot\|_\infty$-distance to target	0.0726	0.512	5.696	14.987	76.9	635.0	1748
Energy concentration	81.42%	1.15%	16.37%	47.28%	0.54%	14.97%	39.34%

given ρ, we just have taken as $\{\xi_j\}_{j=1}^{10}$ 100 samples of 10 independent numbers distributed uniformly in $[-\rho, \rho]$ and have plotted the diagrams corresponding to $x_j = (1 + \xi_j)x_j^*$. Pay attention not only to the shape, but also to the scale (Table 3.1): the target diagram varies from 0 to 1, and the nominal diagram (the one corresponding to the exact optimal x_j) differs from the target by no more than 0.0726 (this is the optimal value in the "nominal" problem (3.29)). The data in Table 3.1 show that when $\rho = 0.001$, the typical $\|\cdot\|_\infty$ distance between the actual diagram and the target one is larger by 3 (!) orders of magnitude. Another meaningful way, also presented in Table 3.1, to understand what is the quality of our design is via *energy concentration* — the fraction of the total emitted energy which "goes up," that is, is emitted along the spatial angle of directions forming an angle at most $\pi/12$ with the Z-axis. For the nominal design, *the dream* (i.e., with no actuation errors) energy concentration is as high as 81% — quite respectable, given that the spatial angle in question forms just 3.41% of the entire hemisphere. This high concentration, however, exists only in our imagination, since actuation errors of magnitude ρ as low as 0.01% reduce the average energy concentration (which, same as the diagram itself, now becomes random) to just 16%; the lower 10% quantile of this random quantity is as small as 2.2% — 1.5 times less than the fraction (3.4%) which the "going up" directions form among all directions. The bottom line is that "in reality" our nominal optimal design is completely meaningless.

Motivating example II: NETLIB Case Study: NETLIB includes about 100 not very large LOs, mostly of real-world origin, used as the standard benchmark for LO solvers. In the study to be described, we used this collection in order to understand how "stable" the feasibility properties of the standard — "nominal" — optimal solutions are with respect to small uncertainty in the data. To motivate the methodology of this "case study," here is the constraint # 372 of the problem `PILOT4` from NETLIB:

$$\begin{aligned}
a^T x \equiv & -15.79081x_{826} - 8.598819x_{827} - 1.88789x_{828} - 1.362417x_{829} \\
& -1.526049x_{830} - 0.031883x_{849} - 28.725555x_{850} \\
& -10.792065x_{851} - 0.19004x_{852} - 2.757176x_{853} \\
& -12.290832x_{854} + 717.562256x_{855} - 0.057865x_{856} \\
& -3.785417x_{857} - 78.30661x_{858} - 122.163055x_{859} \\
& -6.46609x_{860} - 0.48371x_{861} - 0.615264x_{862} \\
& -1.353783x_{863} - 84.644257x_{864} - 122.459045x_{865} \\
& -43.15593x_{866} - 1.712592x_{870} - 0.401597x_{871} \\
& +x_{880} - 0.946049x_{898} - 0.946049x_{916} \\
\geq & \; b \equiv 23.387405
\end{aligned} \tag{C}$$

The related *nonzero* coordinates in the optimal solution x^* of the problem, as reported by `CPLEX` (one of the best commercial LP solvers), are as follows:

$$\begin{array}{ll}
x^*_{826} = 255.6112787181108 & x^*_{827} = 6240.488912232100 \\
x^*_{828} = 3624.613324098961 & x^*_{829} = 18.20205065283259 \\
x^*_{849} = 174397.0389573037 & x^*_{870} = 14250.00176680900 \\
x^*_{871} = 25910.00731692178 & x^*_{880} = 104958.3199274139
\end{array}$$

The indicated optimal solution makes (C) an equality within machine precision.

Observe that most of the coefficients in (C) are "ugly reals" like -15.79081 or -84.644257. We have all reasons to believe that coefficients of this type characterize certain technological devices/processes, and as such *they could hardly be known to high accuracy.* It is quite natural to assume that the "ugly coefficients" are in fact uncertain — they coincide with the "true" values of the corresponding data within accuracy of 3–4 digits, not more. The only exception is the coefficient 1 of x_{880} — it perhaps reflects the structure of the underlying model and is therefore exact — "certain."

Assuming that the uncertain entries of a are, say, 0.1%-accurate approximations of unknown entries of the "true" vector of coefficients $\tilde{a}$, we looked at what would be the effect of this uncertainty on the validity of the "true" constraint $\tilde{a}^T x \geq b$ at x^*. Here is what we found:

- The minimum (over all vectors of coefficients $\tilde{a}$ compatible with our "0.1%-uncertainty hypothesis") value of $\tilde{a}^T x^* - b$ is < -104.9; in other words, the violation of the constraint can be as large as 450% of the right-hand side!
- Treating the above worst-case violation as "too pessimistic" (why should the true values of all uncertain coefficients differ from the values indicated in (C) in the "most dangerous" way?), consider a more realistic measure of violation. Specifically, assume that the true values of the uncertain coefficients in (C) are obtained from the "nominal values" (those shown in (C)) by random perturbations $a_j \mapsto \tilde{a}_j = (1 + \xi_j)a_j$ with independent and, say, uniformly distributed on $[-0.001, 0.001]$ "relative perturbations" ξ_j. What will be a "typical" relative violation

$$V = \frac{\max[b - \tilde{a}^T x^*, 0]}{b} \times 100\%$$

of the "true" (now random) constraint $\tilde{a}^T x \geq b$ at x^*? The answer is nearly as bad as for the worst scenario:

Prob$\{V > 0\}$	Prob$\{V > 150\%\}$	Mean(V)
0.50	0.18	125%

Note: Relative violation of constraint # 372 in PILOT4 (1,000-element sample of 0.1% perturbations of the uncertain data).

We see that *quite small (just 0.1%) perturbations of "clearly uncertain" data coefficients can make the "nominal" optimal solution x^* heavily infeasible, and thus, practically meaningless.*

A "case study" reported in Ben-Tal and Nemirovski (2000) shows that the phenomenon we have just described is not an exception — in 13 of 90 *NETLIB* Linear Programming problems considered in this study, already 0.01%-perturbations of "ugly" coefficients result in violations of some constraints as evaluated at the nominal optimal solutions by more than 50%. In 6 of these 13 problems, the magnitude of constraint violations was over 100%, and in `PILOT4` — "the champion" — it was as large as 210,000%, that is, 7 orders of magnitude larger than the relative perturbations in the data.

The conclusion is as follows:

> *In applications of LO, there exists a real need of a technique capable of detecting cases when data uncertainty can heavily affect the quality of the nominal solution, and in these cases to generate a "reliable" solution, one that is immunized against uncertainty.*

3.3.6.3 *Uncertain linear problems and their robust counterparts*

We are about to introduce the *Robust Counterpart* approach to uncertain LO problems aimed at coping with data uncertainty.

Uncertain LO problem: We start with

Definition 3.1. An uncertain Linear Optimization problem is a collection

$$\left\{\max_x \left\{c^T x + d : Ax \leq b\right\}\right\}_{(c,d,A,b)\in\mathcal{U}} \qquad (\mathrm{LO}_{\mathcal{U}})$$

of LO problems (instances) $\min_x \left\{c^T x + d : Ax \leq b\right\}$ of common structure (i.e., with common numbers m of constraints and n of variables) with the data varying in a given uncertainty set $\mathcal{U} \subset \mathbf{R}^{(m+1)\times(n+1)}$.

We always assume that the uncertainty set is parameterized, in an affine fashion, by *perturbation vector* ζ varying in a given *perturbation set* $\mathcal{Z}$:

$$\mathcal{U} = \left\{ \left[\begin{array}{c|c} c^T & d \\ \hline A & b \end{array}\right] = \underbrace{\left[\begin{array}{c|c} c_0^T & d_0 \\ \hline A_0 & b_0 \end{array}\right]}_{\substack{\text{nominal} \\ \text{data } D_0}} + \sum_{\ell=1}^{L} \zeta_\ell \underbrace{\left[\begin{array}{c|c} c_\ell^T & d_\ell \\ \hline A_\ell & b_\ell \end{array}\right]}_{\substack{\text{basic} \\ \text{shifts } D_\ell}} : \zeta \in \mathcal{Z} \subset \mathbf{R}^L \right\}. \tag{3.30}$$

For example, when speaking about **PILOT4**, we, for the sake of simplicity, tacitly assumed uncertainty only in the constraint matrix, specifically, as follows: every coefficient a_{ij} is allowed to vary, independently of all other coefficients, in the interval $[a^{\rm n}_{ij} - \rho_{ij}|a^{\rm n}_{ij}|, a^{\rm n}_{ij} + \rho_{ij}|a^{\rm n}_{ij}|]$, where $a^{\rm n}_{ij}$ is the nominal value of the coefficient — the one in the data file of the problem as presented in NETLIB, and ρ_{ij} is the perturbation level, which in our experiment was set to 0.001 for all "ugly" coefficients $a^{\rm n}_{ij}$ and was set to 0 for "nice" coefficients, like the coefficient 1 at x_{880}. Geometrically, the corresponding perturbation set is just a box

$$\zeta \in \mathcal{Z} = \left\{ \zeta = \{\zeta_{ij} \in [-1,1]\}_{i,j:a^{\rm n}_{ij} \text{ is ugly}} \right\},$$

and the parameterization of the a_{ij}-data by the perturbation vector is

$$a_{ij} = \begin{cases} a^{\rm n}_{ij}(1+\zeta_{ij}), & a^{\rm n}_{ij} \text{ is ugly,} \\ a^{\rm n}_{ij}, & \text{otherwise.} \end{cases}$$

3.3.6.4 *Robust counterpart of uncertain LO*

Note that a *family* of optimization problems like (LO$_\mathcal{U}$), in contrast to a single optimization problem, is not associated by itself with the concepts of feasible/optimal solution and optimal value. How to define these concepts depends on the underlying "decision environment." Here, we focus on an environment with the following characteristics:

A.1. All decision variables in (LO$_\mathcal{U}$) represent "here and now" decisions; they should be assigned specific numerical values as a result of solving the problem *before* the actual data "reveals itself."

A.2. The decision maker is fully responsible for consequences of the decisions to be made when, and only when, the actual data are within the prespecified uncertainty set $\mathcal{U}$ given by (3.30).

A.3. The constraints in (LO$_\mathcal{U}$) are "hard" — we cannot tolerate violations of constraints, even small ones, when the data are in $\mathcal{U}$.

Note that A.1–A.3 are *assumptions* on our decision environment (in fact, the strongest ones within the methodology we are presenting); while being meaningful, these assumptions are in no sense automatically valid.[6]

Assumptions A.1–A.3 determine, essentially in a unique fashion, what the meaningful, "immunized against uncertainty," feasible solutions are to the uncertain problem ($\mathrm{LO}_{\mathcal{U}}$). By A.1, these should be fixed vectors; by A.2 and A.3, they should be *robust feasible*, that is, they should satisfy all the constraints, whatever the realization of the data from the uncertainty set. We have arrived at the following definition.

Definition 3.2. A vector $x \in \mathbf{R}^n$ is a "robust feasible" solution to ($\mathrm{LO}_{\mathcal{U}}$) if it satisfies all realizations of the constraints from the uncertainty set, that is,

$$Ax \leq b \quad \forall (c, d, A, b) \in \mathcal{U}. \tag{3.31}$$

As for the objective value to be associated with a robust feasible solution, assumptions A.1–A.3 do not prescribe it in a unique fashion. However, "the spirit" of these worst-case-oriented assumptions leads naturally to the following definition:

Definition 3.3. Given a candidate solution x, the "robust" value $\widehat{c}(x)$ of the objective in ($\mathrm{LO}_{\mathcal{U}}$) at x is the smallest value of the "true" objective $c^T x + d$ over all realizations of the data from the uncertainty set:

$$\widehat{c}(x) = \inf_{(c,d,A,b)\in\mathcal{U}} [c^T x + d]. \tag{3.32}$$

After we agree what are meaningful candidate solutions to the uncertain problem ($\mathrm{LO}_{\mathcal{U}}$) and how to quantify their quality, we can seek for the best robust value of the objective among all robust feasible solutions to the problem. This brings us to the central concept of

---

[6]By these reasons, Robust Optimization addresses, along with these assumptions, their relaxed versions. What are these relaxations and what is their "computational price" — these issues go far beyond the scope of this book; the interested reader is addressed to Ben-Tal *et al.* (2009); Bertsimas and Den Hertog (2022) and references therein.

the RO methodology, *Robust Counterpart* of an uncertain optimization problem, which is defined as follows:

Definition 3.4. The Robust Counterpart of the uncertain LO problem ($\mathrm{LO}_{\mathcal{U}}$) is the optimization problem

$$\max_x \left\{ \widehat{c}(x) - \inf_{(c,d,A,b)\in\mathcal{U}} [c^T x + d] : Ax \leq b \ \forall (c,d,A,b) \in \mathcal{U} \right\} \quad (3.33)$$

of maximizing the robust value of the objective over all robust feasible solutions to the uncertain problem.

An optimal solution to the Robust Counterpart is called a robust optimal solution to ($\mathrm{LO}_{\mathcal{U}}$), and the optimal value of the Robust Counterpart is called the robust optimal value of ($\mathrm{LO}_{\mathcal{U}}$).

In a nutshell, the robust optimal solution is simply "the best uncertainty-immunized" solution we can associate with our uncertain problem.

3.3.6.5 *Tractability of the RC*

For the outlined methodology to be a practical tool rather than a wishful thinking, the RO of an uncertain LO problem should be efficiently solvable. We believe (and this belief will be justified in the "algorithmic" part of our book) that LO programs are efficiently solvable; but the RO of uncertain LO problem, as it appears in (3.33), is *not* an LO program — when the uncertainty set $\mathcal{U}$ is infinite (which typically is the case), (3.33) is an optimization program with *infinitely many* linear inequality constraints, parameterized by the uncertain data! And programs of this type (the so-called *semi-infinite* LO's) are not necessarily tractable...

The situation, however, is not that bad. We are about to demonstrate — and this is where the equivalence (3.27) is instrumental — that *the Robust Counterpart of an uncertain LO problem with a nonempty polyhedral uncertainty set* $\mathcal{U}$ (given by its polyhedral representation) *is equivalent to an explicit LO program* (and thus is computationally tractable).

The reasoning goes as follows. Introducing slack variable t, we can rewrite (3.33) as a problem with a "certain" linear objective and semi-infinite constraints:

$$\max_{y=[x;t]}\left\{t:\begin{array}{l}t-c^Tx-d\leq 0\\ \sum_j a_{ij}x_j-b_j\leq 0,\ 1\leq j\leq m\end{array}\right\}\ \forall(c,d,A,b)\in\mathcal{U}\Bigg\}. \qquad (!)$$

To save the notation, let η stand for the data (c, d, A, b) which we can treat as a vector of certain dimension N (in this vector, the first n entries are those of c, the next entry is d, the next mn entries are the coefficients of the constraint matrix, written, say, column by column, and the last m entries are those of b). Note that (!) is of the generic form

$$\max_y\left\{h^Ty:\forall\eta\in\mathcal{U}:p_\ell^T[y]\eta\leq q_\ell[y],\ 1\leq\ell\leq L\right\}, \qquad (!!)$$

where $L = m + 1$, and $p_\ell[y]$, $q_\ell[y]$ are known *affine* vector- and real-valued functions of y. All we need in order to convert (!!) into an ordinary LO is to represent every one of the semi-infinite constraints

$$p_\ell^T[y]\eta\leq q_\ell[y]\ \forall\eta\in\mathcal{U} \qquad (*_\ell)$$

by a finite system of linear inequalities on y and on appropriate slack variables. Now, for every ℓ and y fixed, $(*_\ell)$ says that

$$\sup_{\eta\in\mathcal{U}}p_\ell^T[y]\eta\leq q_\ell[y]; \qquad (**)$$

Recalling that $\mathcal{U}$ is polyhedral and we are given a polyhedral representation of this set, let it be

$$\mathcal{U}=\{\eta:\exists u:P\eta+Qu\leq r\},$$

the supremum in η in the left-hand side of $(**)$ can be represented as the optimal value in an LO program with variables η, u, specifically, the program

$$\mathrm{Opt}(y)=\max_{\eta,u}\left\{p_\ell^T[y]\eta:P\eta+Qu\leq r\right\}.$$

This program is feasible (since $\mathcal{U}$ is nonempty), and we can invoke the equivalence (3.27) to conclude that

$$\mathrm{Opt}(y)\leq q_\ell(y)\Leftrightarrow\exists w:w\geq 0,[P;Q]^Tw=[p_\ell[y];\underbrace{0;...;0}_{\dim u}],r^Tw\leq q_\ell[y].$$

The bottom line is that y *satisfies* $(*_\ell)$ *if and only if it can be extended, by a properly chosen* $w = w^\ell$, *to a feasible solution to the* *system of linear inequalities*

$$w^\ell \geq 0,\ P^T w^\ell = p_\ell[y],\ Q^T w^\ell = 0,\ r^T w^\ell \leq q_\ell[y] \qquad (S_\ell)$$

in variables y, w^ℓ (to see that the inequalities indeed are linear, note that $p_\ell[y]$ and $q_\ell[y]$ are affine in y).

We are done: replacing every one of the L semi-infinite constraints in (!!) with the corresponding systems (S_ℓ), $\ell = 1, \ldots, L$, we end up with an equivalent reformulation of (!!) as an LO program, let it be called (L), with variables $y, w^1, \ldots, w^L$, equivalence meaning that y is feasible for (!!) if and only if y can be augmented by properly chosen $w^1, \ldots, w^L$ to yield a feasible solution to (L), while the objective in (L) is the same as in (!). Note that given a polyhedral representation of $\mathcal{U}$, building the resulting LO is a purely algorithmic and efficient process, and that the sizes of this LO are polynomial in the sizes of the instances of the original uncertain LO problem and the sizes of the polyhedral description of the uncertainty set.

How it works: Robust Antenna Design. In the situation of the Antenna Design problem (3.29), the "physical" uncertainty comes from the actuation errors $x_j \mapsto (1 + \xi_j)x_j$; as we have already explained, these errors can be modelled equivalently by the perturbations $D_j(\theta_i) \mapsto D_{ij} = (1 + \xi_j)D_j(\theta_i)$ in the coefficients of x_j. Assuming that the errors ξ_j are bounded by a given *uncertainty level* ρ, and that this is the only *a priori* information on the actuation errors, we end up with the uncertain LO problem

$$\left\{ \min_{x,\tau} \left\{ \tau : -\tau \leq \sum_{j=1}^{J=10} D_{ij}x_j - D_*(\theta_i) \leq \tau, 1 \leq i \leq I = 240 \right\} : \right.$$
$$\left. |D_{ij} - D_j(\theta_i)| \leq \rho |D_j(\theta_i)| \right\}.$$

The Robust Counterpart of the problem is the semi-infinite LO program

$$\min_{x,\tau} \left\{ \tau : -\tau \leq \sum_j D_{ij}x_j \leq \tau, 1 \leq i \leq I\ \forall D_{ij} \in [\underline{G}_{ij}, \overline{G}_{ij}] \right\} \qquad (3.34)$$

with $\underline{G}_{ij} = G_j(\theta_i) - \rho|G_j(\theta_i)|$, $\overline{G}_{ij} = G_j(\theta_i) + \rho|G_j(\theta_i)|$. The generic form of this semi-infinite LO is

$$\min_y \left\{ c^T y : Ay \leq b \forall [A, b] : [\underline{A}, \underline{b}] \leq [A, b] \leq [\overline{A}, \overline{b}] \right\}, \qquad (3.35)$$

where $\leq$ for matrices is understood entrywise and $[\underline{A}, \underline{b}] \leq [\overline{A}, \overline{b}]$ are two given matrices. This is a very special case of a polyhedral uncertainty set, so that our theory says that the RC is equivalent to an explicit LO program. In fact, we can point out an LO reformulation of the Robust Counterpart without reference to any theory: it is easy to see (check it!) that (3.35) is equivalent to the LO program

$$\min_{y,z}\left\{c^T y : \frac{1}{2}[\overline{A}+\underline{A}]y + \frac{1}{2}[\overline{A}-\underline{A}]z \leq \underline{b}, -z \leq y \leq z\right\}. \quad (3.36)$$

Solving (3.34) for the uncertainty level $\rho = 0.01$, we end up with the robust optimal value 0.0910, which, while being 15% worse than the nominal optimal value 0.0726 (the latter, as we have seen, exists only in our imagination and says nothing about the actual performance of the nominal optimal design), still is reasonably small. Note that the robust optimal value, in sharp contrast with the nominally optimal one, does say something meaningful about the actual performance of the underlying *robust* design. In our experiments, we have tested the robust optimal design associated with the uncertainty level $\rho = 0.01$ versus actuation errors of this and larger magnitudes. The results are presented in Fig. 3.4 and in Table 3.3. Comparing this figure and table with their "nominal design" counterparts, we see that the robust design is incomparably better than the nominal one.

To conclude the antenna story, it makes sense to explain why the nominal design is that unstable with respect to small implementation errors. The target diagram D_*, same as the diagrams D_i of rings, are functions with values of order of one; as a matter of fact, D_i's are nearly linearly dependent, so that the problem of best uniform approximation of D_* by linear combination of D_i's is ill-posed, which, as we see from Table 3.2, makes some of the entries in the nominal optimal solution as large as $10^4 - 10^5$. Combination of functions of magnitude of order of one with that of large coefficients may happen

Table 3.2. Nominal and robust designs.

	x_1	x_2	x_3	x_4	x_5	x_6	x_7	x_8	x_9	x_{10}
Nominal	−2.8e2	0.00	1.6e4	−9.5e4	2.7e5	−4.6e5	4.9e5	−3.2e5	1.2e5	−2.0e4
Robust	0.67	1.11	0.00	0.00	2.00	0.60	0.00	−1.91	0.00	4.37

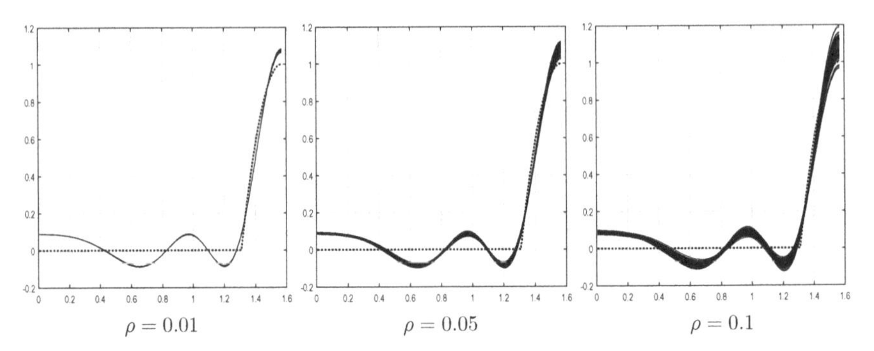

Fig. 3.4. "Dream and reality," robust optimal design: samples of 100 actual diagrams for different uncertainty levels. Dotted line: the target diagram.

Table 3.3. Quality of robust antenna design. Data of over 100 samples of actuation errors per each uncertainty level ρ. For comparison: for nominal design, with the uncertainty level as small as $\rho = 0.001$, the average $\|\cdot\|_\infty$-distance of the actual diagram to target is as large as 61.3, and the expected energy concentration is as low as 16.4%.

	$\rho = 0.01$			$\rho = 0.1$		
	Min	Mean	Max	Min	Mean	Max
$\|\cdot\|_\infty$-distance to target	0.0879	0.0892	0.0904	0.0892	0.1147	0.1796
Energy concentration	72.15%	72.70%	73.21%	66.58%	72.74%	78.38%

to be of order of one only when combining D_i with the coefficients in question results in significant cancellations. However, implementation errors of small *relative* magnitude, even as small as 0.01%, result in perturbations of magnitude $1 - 10$ in some of the coefficients. These perturbations have, of course, no reason to cancel each other and well may result in perturbations of magnitude $1 - 10$ in the resulting diagram; this is exactly what we observe in Fig. 3.3. In contrast to the nominal design, the robust one accounts for implementation errors in advance and is therefore enforced to take care of magnitudes of the entries in the robust solution. As we see from Table 3.2, these latter entries are of order of one; as a result, small in relative scale implementation errors yield small perturbations in the coefficients and, consequently, small perturbations in the synthesized diagram.

How it works: NETLIB Case Study: The corresponding uncertainty model ("ugly coefficients a_{ij} in the constraint matrix independently of each other run through the segments $[a^{\rm n}_{ij}-\rho|a^{\rm n}_{ij}|, a^{\rm n}_{ij}+\rho|a^{\rm n}_{ij}|]$, $\rho > 0$ being the uncertainty level) clearly yields the RCs of the generic form (3.35). As explained above, these RCs can be straightforwardly converted to explicit LO programs which are of nearly the same sizes and sparsity as the instances of the uncertain LOs in question. It turns out that at the uncertainty level 0.1% ($\rho = 0.001$), all these RCs are feasible, that is, we can immunize the solutions against this uncertainty. Surprisingly, this immunization is "nearly costless" — the robust optimal values of all 90 NETLIB LOs considered in Ben-Tal and Nemirovski (2000) remain within 1% margin of the nominal optimal values. For further details, including what happens at larger uncertainty levels, see Ben-Tal and Nemirovski (2000).

3.3.7 **Application: Synthesis of linear controllers*

3.3.7.1 **Discrete time linear dynamical systems*

The most basic and well-studied entity in control is a *linear dynamical system* (LDS). In the sequel, we focus on *discrete time LDS* modelled as

$$\begin{aligned} x_0 &= z && \text{[initial condition]},\\ x_{t+1} &= A_t x_t + B_t u_t + R_t d_t, && \text{[state equations]},\\ y_t &= C_t x_t + D_t d_t && \text{[outputs]}. \end{aligned} \tag{3.37}$$

In this description,

- $t = 0, 1, 2, \ldots$ are time instants,
- $x_t \in \mathbf{R}^{n_x}$ is the *state* of the system at instant t,
- $u_t \in \mathbf{R}^{n_u}$ is the *control* generated by the system's controller at instant t,
- $d_t \in \mathbf{R}^{n_d}$ is the *external disturbance* coming from the system's environment at instant t,
- $y_t \in \mathbf{R}^{n_y}$ is the *observed output* at instant t,
- $A_t, B_t, \ldots, D_t$ are (perhaps depending on t) matrices of appropriate sizes specifying the system's dynamics and the relations between states, controls, external disturbances and outputs.

What we have described so far is called an *open-loop* system (or open loop plant). This plant should be augmented by a *controller* which generates subsequent controls. The standard assumption is that the control u_t is generated in a deterministic fashion and depends on the outputs $y_0, y_1, \dots, y_t$ observed prior to instant t and at this very instant (*nonanticipative*, or *causal* control):

$$u_t = U_t(y_0, \dots, y_t); \tag{3.38}$$

here U_t are arbitrary everywhere defined functions of their arguments taking values in $\mathbf{R}^{n_u}$. Plant augmented by a controller is called a *closed-loop* system; its behavior clearly depends on the initial state and external disturbances only.

3.3.7.2 **Affine control*

The simplest (and extremely widely used) form of control law is *affine* control, where u_t are affine functions of the outputs:

$$u_t = \xi_t + \Xi_0^t y_0 + \Xi_1^t y_1 + \cdots + \Xi_t^t y_t, \tag{3.39}$$

where ξ_t are vectors, and Ξ_τ^t, $0 \le \tau \le t$, are matrices of appropriate sizes.

Augmenting linear open-loop system (3.37) with affine controller (3.39), we get a well-defined closed-loop system in which states, controls and outputs are *affine* functions of the initial state and external disturbances; moreover, x_t depends solely on the initial state z and the collection $d^{t-1} = [d_0; \dots; d_{t-1}]$ of disturbances prior to instant t, while u_t and y_t may depend on z and the collection $d^t = [d_0; \dots; d_t]$ of disturbances including the one at instant t.

3.3.7.3 **Design specifications and the analysis problem*

The entities of primary interest in control are states and controls; we can arrange states and controls into a long vector — the *state–control trajectory*

$$w^N = [x_1; x_2; \dots; x_N; u_0; u_1, \dots; u_{N-1}];$$

here N is the *time horizon* on which we are interested in the system's behavior. With affine control law (3.39), this trajectory is an *affine* function of z and d^{N-1}:

$$w^N = w^N(z, d^{N-1}) = \omega_N + \Omega_N[z; d^{N-1}]$$

with vector ω_N and matrix Ω_N readily given by the matrices $A_t, \ldots, D_t$, $0 \leq t < N$, from the description of the open-loop system and by the collection $\vec{\xi}^N = \{\xi_t, \Xi^t_\tau : 0 \leq \tau \leq t < N\}$ of the parameters of the affine control law (3.39).

Imagine that the "desired behavior" of the closed-loop system on the time horizon in question is given by a system of *linear* inequalities

$$\mathcal{B}w^N \leq b(\zeta) \qquad [b(\zeta)\text{: affine in } \zeta := [z; d^{N-1}]], \tag{3.40}$$

which should be satisfied by the state–control trajectory provided that the initial state z and the disturbances d^{N-1} vary in their "normal ranges" $\mathcal{Z}$ and $\mathcal{D}^{N-1}$, respectively; this is a pretty general form of design specifications. The fact that w^N depends *affinely* on $\zeta := [z; d^{N-1}]$ makes it easy to solve the *analysis problem*: to check whether a given control law (3.39) ensures the validity of design specifications (3.40). Indeed, to this end we should check whether the functions $[\mathcal{B}w^N - b(\zeta)]_i$, $1 \leq i \leq I$ (I is the number of linear inequalities in (3.40)) remain nonpositive whenever

$$\zeta := [z; d^{N-1}] \in \mathcal{Z}\mathcal{D}^{N-1} := \mathcal{Z} \times \mathcal{D}^{N-1}.$$

For a given control law, the functions in question are explicitly given *affine* functions $\phi_i(\zeta)$ of ζ, so that what we need to verify is that

$$\max_{\zeta} \left\{\phi_i(\zeta) : \zeta \in \mathcal{Z}\mathcal{D}^{N-1}\right\} \leq 0, \; i = 1, \ldots, I.$$

Whenever $\mathcal{D}^{N-1}$ and $\mathcal{Z}$ are explicitly given convex sets, the latter problems are convex and thus easy to solve. Moreover, if $\mathcal{Z}\mathcal{D}^{N-1}$ is given by polyhedral representation:

$$\mathcal{Z}\mathcal{D}^{N-1} = \{\zeta : \exists v : P\zeta + Qv \leq r\} \tag{3.41}$$

(this is a pretty flexible and general enough way to describe typical ranges of disturbances and initial states), the analysis problem reduces to a bunch of explicit LPs

$$\max_{\zeta, v} \left\{\phi_i(\zeta) : P\zeta + Qv \leq r\right\}, \; 1 \leq i \leq I;$$

the answer in the analysis problem is positive if and only if the optimal values in all these LPs are nonpositive.

3.3.7.4 *Synthesis problem

As we have seen, with affine control and affine design specifications, it is easy to check whether a given control law meets the specifications. The basic problem in linear control is, however, somehow different: usually we need to *build* an affine control which meets the design specifications (or to detect that no such control exists). And here we run into a difficult problem: while state–control trajectory for a given affine control is an easy-to-describe *affine* function of $[z; d^{N-1}]$, its dependence on the collection $\vec{\xi}^N$ of the parameters of the control law is highly nonlinear, even for a time-invariant system (the matrices $A_t, \ldots, D_t$ are independent of t) and a control law as simple as the time-invariant linear feedback: $u_t = Ky_t$. Indeed, due to the dynamic nature of the system, in the expressions for states and controls *powers* of the matrix K will be present. Highly nonlinear dependence of states and controls on $\vec{\xi}^N$ makes it impossible to optimize efficiently w.r.t. the parameters of the control law and thus makes the synthesis problem extremely difficult.

The situation, however, is far from being hopeless. We are about to demonstrate that *one can re-parameterize affine control law in such a way that with the new parameterization both analysis and synthesis problems become tractable.*

Illustration: As a simple illustration, consider the situation as follows:

> Water supply in a village comes from a tank of capacity V which is filled by pumps taking water from a source of unlimited capacity. Denoting by x_t the amount of water ("level") in the tank at the beginning of hour t, $0 \leq t < N = 24$, the dynamics of this level is given by
>
> $$x_{t+1} = x_t + u_t - d_t,\ t = 0, 1, \ldots, 23$$
>
> where $x_0 = z$ is the level of tank at midnight, u_t, $0 \leq t \leq 23$, is the amount of water pumped into the tank during hour t, and d_t is the demand — the amount of water consumed during the same hour by the villagers. Pumping a unit of water into the tank during hour t costs $c_t \geq 0$.[7] We assume that x_t is observed at the beginning of hour t, when

[7]The dependence of pumping price on time is natural; say, the price of electricity at night is smaller than during the day hours. Note that typical per-hour demand for water exhibits similar behavior: During the day hours, it is essentially higher

the decision on u_t should be made, while d_t is not known at this time instant.

Given the range $\mathcal{Z} = [0, \overline{z}]$ of the initial level of water and upper and lower bounds $\overline{d}_t \geq \underline{d}_t \geq 0$ on the demand d_t, $0 \leq t \leq 23$, we want to design, in a nonanticipative fashion, *nonnegative* controls u_t, $0 \leq t \leq 23$, in such a way that the levels x_t remain nonnegative and not exceeding tank's capacity V whenever the initial state z and the demand trajectory $d^{23} = [d_0; \ldots; d_{23}]$ stay in their normal range:

$$\zeta := [z; d_0; d_1; \ldots; d_{23}] \in \mathcal{Z}\mathcal{D}^{23} := \{[z; d_0; \ldots; d_{23}] : 0 \leq z \leq \overline{z}, \underline{d}_t \leq d_t \leq \overline{d}_t\}$$

The dynamical system modeling the above story is

$$\begin{aligned} x_0 &= z \\ x_{t+1} &= x_t + u_t - d_t\,, \quad 0 \leq t < N = 24 \\ y_t &= x_t \end{aligned} \tag{3.42}$$

and the design specifications are given by the system of linear inequalities

$$0 \leq x_t \leq V,\ 1 \leq t \leq N = 24\ \&\ 0 \leq u_t,\ 0 \leq t \leq N - 1 = 23. \tag{3.43}$$

The total price of a given control policy on our 24-hour time horizon depends on the actual realization of ζ. In our illustration, we are interested in minimizing the maximum, over $\zeta \in \mathcal{Z}\mathcal{D}^{23}$, total price of pumping by seeking an *affine* controller satisfying the design specifications (3.43) for all realizations of $\zeta \in \mathcal{Z}\mathcal{D}^{23}$.

3.3.7.5 **Purified outputs and purified-output-based control laws*

Imagine that we "close" the open-loop system with a whatever (affine or nonaffine) control law (3.38) and in parallel with running the

---

than at night. These dependencies make utilizing a tank more attractive than pumping water to customers directly from the source: we can use low-cost night pumping to fill the tank and use this water at the largest extent possible during the day hours.

closed-loop system, run its model as follows:

$$
\begin{array}{|l|}
\hline
\text{System:} \\
\quad x_0 = z \\
x_{t+1} = A_t x_i + B_t u_t + R_t d_t, \\
\quad y_t = C_t x_t + D_t d_t \\
\hline
\text{Model:} \\
\quad \widehat{x}_0 = 0 \\
\widehat{x}_{t+1} = A_t \widehat{x}_i + B_t u_t, \\
\quad \widehat{y}_t = C_t \widehat{x}_t \\
\hline
\text{Controller:} \\
\quad u_t = U_t(y_0, \ldots, y_t) \qquad (!) \\
\hline
\end{array} \tag{3.44}
$$

Assuming that we know the matrices $A_t, \ldots, D_t$, we can run the model in an online fashion, so that at instant t, when the control u_t should be specified, we have at our disposal both the actual outputs $y_0, \ldots, y_t$ and the model outputs $\widehat{y}_0, \ldots, \widehat{y}_t$, and thus have at our disposal the *purified outputs*

$$v_\tau = y_\tau - \widehat{y}_\tau,\ 0 \leq \tau \leq t.$$

Now, let us ask ourselves what will happen if we, instead of building the controls u_t on the basis of actual outputs y_τ, $0 \leq \tau \leq t$, pass to controls u_t built on the basis of purified outputs v_τ, $0 \leq \tau \leq t$, i.e., replace the control law (!) with the control law of the form

$$u_t = V_t(v_0, \ldots, v_t). \tag{!!}$$

It is easily seen that nothing will happen:

Proposition 3.9. *For every control law $\{U_t(y_0, \ldots, y_t)\}_{t=0}^\infty$ of the form (!) there exists a control law $\{V_t(v_0, \ldots, v_t)\}_{t=0}^\infty$ of the form (!!) (and vice versa, for every control law of the form (!!) there exists a control law of the form (!)) such that the dependencies of actual states, outputs and controls on the disturbances and the initial state for both control laws in question are exactly the same.*

Moreover, the above "equivalence claim" remains valid when we restrict controls (!), (!!) to be affine in their arguments.

Proof of proposition is postponed till the end of this section.

The bottom line is that *every behavior of the closed-loop system which can be obtained with affine nonanticipative control* (3.39) *based on actual outputs, can also be obtained with the affine nonanticipative control law*

$$u_t = \eta_t + H_0^t v_0 + H_1^t v_1 + \cdots + H_t^t v_t, \tag{3.45}$$

based on purified outputs (and vice versa).

We have said that *as far as achievable behaviors of the closed-loop system are concerned, we loose* (and gain) *nothing when passing from affine output-based control laws* (3.39) *to affine purified output-based control laws* (3.45). At the same time, when passing to purified outputs, we get a huge bonus as follows:

> (#) *With control* (3.45), *the trajectory* w^N *of the closed-loop system turns out to be bi-affine: it is affine in* $\zeta = [z; d^{N-1}]$, *the parameters* $\vec{\eta}^N = \{\eta_t, H_\tau^t, 0 \le \tau \le t < N\}$ *of the control law being fixed, and it is affine in the parameters of the control law* $\vec{\eta}^N$, $[z; d^{N-1}]$ *being fixed,* and this bi-affinity, as we shall see in a while, is the key to efficient solvability of the synthesis problem.

The reason for bi-affinity is as follows (after this reason is explained, verification of bi-affinity itself becomes immediate): *purified output* v_t *is completely independent of the controls and is a known-in-advance affine function of* d^t, z. Indeed, from (3.44) it follows that

$$v_t = C_t\underbrace{(x_t - \widehat{x}_t)}_{\delta_t} + D_t d_t,$$

and that the evolution of δ_t is given by

$$\delta_0 = z,\ \delta_{t+1} = A_t\delta_t + R_t d_t,$$

and thus it is completely independent of the controls, meaning that δ_t and v_t indeed are known-in-advance (provided the matrices $A_t, \ldots, D_t$ are known-in-advance) affine functions of d^t, z.

Note that in contrast to the just outlined "control-independent" nature of purified outputs, the actual outputs are heavily control-dependent (indeed, u_t is affected by y_t, y_t is affected by x_t, and this state, in turn, is affected by past controls $u_0, \ldots, u_{t-1}$). This is why with the usual output-based affine control, the states and the controls are highly nonlinear in the parameters of the control law — to build u_i, we multiply matrices Ξ_τ^t (which are parameters of the control law) by outputs y_τ which by themselves already depend on the "past" parameters of the control law.

3.3.7.6 *Tractability of the synthesis problem

Assume that the normal range $\mathcal{ZD}^{N-1}$ of $\zeta = [z; d^{N-1}]$ is a nonempty and bounded set given by polyhedral representation (3.41), and let us prove that in this case the design specifications (3.40) reduce to a system of explicit linear inequalities in variables $\vec{\eta}^N$ — the parameters of the purified-output-based affine control law used to close the open-loop system — and appropriate slack variables. Thus, (parameters of) purified-output-based affine control laws meeting the design specifications (3.40) form a polyhedrally representable, and thus easy to work with, set.

The reasoning goes as follows. As stated by (#), the state-control trajectory w^N associated with affine purified-output-based control with parameters $\vec{\eta}^N$ is bi-affine in $\zeta = [z; d^{N-1}]$ and in $\vec{\eta}^N$, so it can be represented in the form

$$w^N = w[\vec{\eta}^N] + W[\vec{\eta}^N]\zeta,$$

where the vector-valued and the matrix-valued functions $w[\cdot]$, $W[\cdot]$ are affine and are readily given by the matrices $A_t, \ldots, D_t$, $0 \leq t \leq N-1$. Plugging the representation of w^N into the design specifications (3.40), we get a system of scalar constraints of the form

$$\alpha_i^T(\vec{\eta}^N)\zeta \leq \beta_i(\vec{\eta}^N),\ 1 \leq i \leq I, \tag{\&}$$

where the vector-valued functions $\alpha_i(\cdot)$ and the scalar functions $\beta_i(\cdot)$ are affine and readily given by the description of the open-loop system and by the data $\mathcal{B}$, b in the design specifications. What we want from $\vec{\eta}^N$ is to ensure the validity of every one of the constraints (&) for all ζ from $\mathcal{ZD}^{N-1}$, or, which is the same in view of (3.41), we want the optimal values in the LPs

$$\max_{\zeta, v} \left\{ \alpha_i^T(\vec{\eta}^N)\zeta : P\zeta + Qv \leq r \right\}$$

to be $\leq \beta_i(\vec{\eta}^N)$ for $1 \leq i \leq I$. Now, the LPs in question are feasible; passing to their duals, what we want become exactly the relations

$$\min_{s_i} \left\{ r^T s_i : P^T s_i = \alpha_i(\vec{\eta}^N),\ Q^T s_i = 0,\ s_i \geq 0 \right\} \leq \beta_i(\vec{\eta}^N),\ 1 \leq i \leq I.$$

The bottom line is that

A purified-output-based control law meets the design specifications (3.40) *if and only if the corresponding collection $\vec{\eta}^N$ of parameters can be augmented by properly chosen slack vector variables s_i to give a solution to the system of linear inequalities in variables $s_1, \ldots, s_I, \vec{\eta}^N$, specifically, the system*

$$P^T s_i = \alpha_i(\vec{\eta}^N),\ Q^T s_i = 0,\ r^T s_i \le \beta_i(\vec{\eta}^N),\ s_i \ge 0,\ 1 \le i \le I. \quad (3.46)$$

Remark 3.3. We can say that when passing from affine output-based control laws to affine *purified*-output-based ones we are all the time dealing with the same entities (affine nonanticipative control laws), but switch from one parameterization of these laws (the one by $\vec{\xi}^N$-parameters) to another parameterization (the one by $\vec{\eta}^N$-parameters). This re-parameterization is nonlinear, so that in principle there is no surprise that what is difficult in one of them (the synthesis problem) is easy in another one. This being said, note that with our re-parameterization we neither lose nor gain only as far as the *entire family* of linear controllers is concerned. Specific sub-families of controllers can be "simple-looking" in one of the parameterizations and be extremely difficult to describe in the other one. For example, time-invariant linear feedback $u_t = Ky_t$ looks pretty simple (just a linear subspace) in the $\vec{\xi}^N$-parameterization and forms a highly nonlinear manifold in the $\vec{\eta}^N$-one. Similarly, the purified-output-based control $u_t = Kv_t$ looks simple in the $\vec{\eta}^N$-parameterization and is difficult to describe in the $\vec{\xi}^N$-one. We can say that there is no such thing as "the best" parameterization of affine control laws — everything depends on what our goals are. For example, the $\vec{\eta}^N$-parameterization is well suited for synthesis of general-type affine controllers and becomes nearly useless when a linear feedback is sought.

Illustration (Continued): We are about to demonstrate how the outlined methodology works in our toy "water supply" illustration. In our experiment, we used tank capacity $V = 50$; the pumping costs c_t and the upper bounds on hourly demands $\overline{d}_t$ are shown on the left plots of Fig. 3.5; the lower bounds on hourly demands were set to 0. The upper bound $\overline{z}$ on the tank's level at the beginning of hour 0 was set to 10.

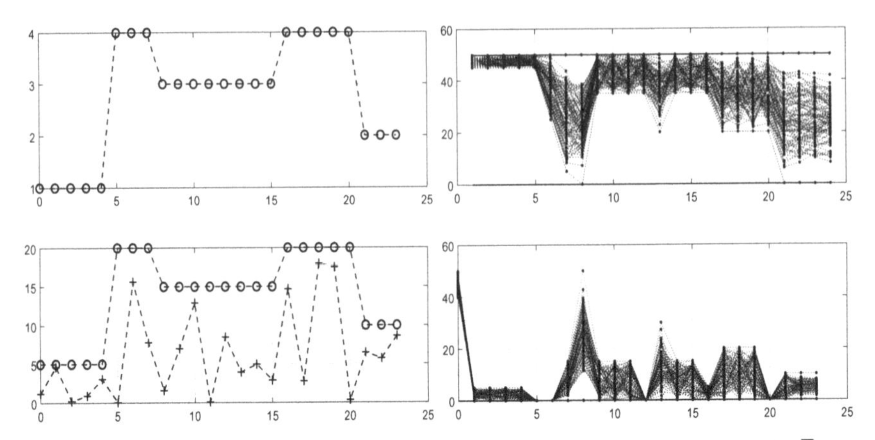

Fig. 3.5. Left, top: hourly pumping cost. Left, bottom: upper bounds $\overline{d}_t$ on hourly demands (o) and sample demand trajectory (+). Right: 100 sample level (top) and control (bottom) trajectories. Pay attention to how close to the tank's capacity are the tank levels in the first five ($1 \leq t \leq 5$) "cheap pumping — low demand" hours and to how intensive the pumping is in the beginning ($t = 0$).

The performance of the purified-output-based controller for (3.42) yielded by minimizing the worst-case total pumping cost over affine purified-output-based controllers satisfying the design specifications (3.43) for all $\zeta \in \mathcal{ZD}^{23}$ is illustrated by the right-hand side of Fig. 3.5, where we plot 100 state and control trajectories corresponding to 100 initial tank levels z and demand trajectories selected at random from the "uncertainty set" $\mathcal{ZD}^{23} = \{[z; d_0; ...; d_{23}] : 0 \leq z \leq \overline{z}, 0 \leq d_t \leq \overline{d}_t, 0 \leq t \leq 23\}$. The pumping costs in these simulations were as follows:

min	mean	median	max
50.00	472.13	473.31	905.00

Note: Total pumping costs in 100 simulations

The worst-case, w.r.t. $\zeta \in \mathcal{ZD}^{23}$, pumping cost for our controller is 915.00. To put this result into proper perspective, note that with ζ corresponding to the zero initial tank level and the largest possible hourly demands $d_t = \overline{d}_t$, $0 \leq t \leq 23$, the minimal possible pumping cost is 885.00 — just by 3.3% less than the cost guaranteed in the worst case by our controller. Thus, in our example, passing from nonanticipative affine controllers to "utopian" control policies utilizing *a priori* knowledge of ζ cannot reduce the worst-case, over

$\zeta \in \mathcal{ZD}^{23}$, pumping cost by more than 3.3%. In fact, the situation is even better: applying to our simple model Dynamic Programming, one can minimize the worst-case cost of pumping w.r.t. *all* nonanticipative output-based control laws, linear and nonlinear alike, and this minimum happens to be 915.0 — exactly what is yielded by our optimal purified-output-based controller, cf. Bertsimas *et al.* (2010).

3.3.7.7 **Clearing debts: Justification of Proposition* 3.9

In one direction: assume that the controller is given by $u_t = U_t(y_0, \ldots, y_t)$, $t = 0, 1, \ldots$, and let us prove that the same system's behavior can be obtained by purified-output-based controller $u_t = \widehat{U}_t(v_0, v_1, \ldots, v_t)$, $t = 0, 1, \ldots$. To this end, it suffices to prove by induction in t that with the output-based control, the outputs y_t are deterministic functions of purified outputs: $y_t = Y_t(v_0, v_1, \ldots, v_t)$ stemming from our control, $t = 0, 1, \ldots$. Base $t = 0$ is trivial, since by construction $y_0 = v_0$. Assuming that $y_\tau = Y_\tau(v_0, \ldots, v_\tau)$ for $0 \leq \tau \leq t$, we conclude that $u_\tau = \widehat{U}_\tau(v_0, \ldots, v_\tau) := U_\tau(Y_0(v_0), Y_1(v_0, v_1), \ldots, Y_\tau(v_0, \ldots, v_\tau))$. Looking at the model, we conclude that $\widehat{x}_{\tau+1}$, $0 \leq \tau \leq t$, are deterministic functions of $v_0, \ldots, v_\tau$: $\widehat{x}_{\tau+1} = \widehat{X}_{\tau+1}(v_0, \ldots, v_\tau)$. Consequently, $\widehat{y}_{t+1} = C_{t+1}\widehat{x}_{t+1}$ is a deterministic function of $v_0, \ldots, v_t$: $\widehat{y}_{t+1} = \widehat{Y}_{t+1}(v_0, \ldots, v_t)$, implying that $y_{t+1} = \widehat{y}_{t+1} + v_{t+1} = \widehat{Y}_{t+1}(v_0, \ldots, v_t) + v_{t+1}$, that is, y_{t+1} is a deterministic function of $v_0, \ldots, v_{t+1}$. The inductive step is complete. And since $y_t = Y_t(v_0, \ldots, v_t)$, we conclude that the output-based controls $u_t = U_t(y_0, \ldots, y_t)$ can be represented as deterministic functions of the purified outputs: $u_t = \widehat{U}_t(v_0, \ldots, v_t)$, as claimed. It is immediately seen that with affine in $[y_0; ...; y_t]$ functions $U_t(y_0, \ldots, y_t)$, the above construction results in affine in $[v_0; ...; v_t]$ functions $\widehat{U}_t(v_0, \ldots, v_t)$.

In the opposite direction: assume that the controller is given by $u_t = \widehat{U}_t(v_0, \ldots, v_t)$, $t = 0, 1, \ldots$, and let us prove that the same system's behavior can be obtained by the control policy of the form $u_t = U_t(y_0, y_1, \ldots, y_t)$, $t = 0, 1, \ldots$. To this end, it suffices to prove by induction in t that with the purified-output-based control, the purified outputs v_t are deterministic functions of actual outputs y_t: $v_t = \widehat{V}_t(y_0, \ldots, y_t)$ stemming from our control, $t = 0, 1, \ldots$. Base $t = 0$ is trivial, since by construction $v_0 = y_0$. Assuming that $v_\tau = \widehat{V}_\tau(y_0,\ldots,y_\tau)$ for $0 \le \tau \le t$, we conclude that for $0 \le \tau \le t$ the controls u_τ are deterministic functions of $y_0,\ldots,y_\tau$: $u_\tau = U_\tau(y_0,\ldots,y_\tau) := \widehat{U}_\tau(\widehat{V}_0(y_0),\ldots,\widehat{V}_\tau(y_0,\ldots,y_\tau))$, $\tau \le t$. As a result, from the model's dynamics $\widehat{x}_{t+1}$ is a deterministic function of $y_0,\ldots,y_t$: $\widehat{x}_{t+1} = \widehat{X}_{t+1}(y_0,\ldots,y_t)$, whence, again by the model's dynamics, $\widehat{y}_{t+1}$ is a deterministic function of $y_0,\ldots,y_t$, so that $v_{t+1} = y_{t+1} - \widehat{y}_{t+1}$ is a deterministic function of $y_0,\ldots,y_{t+1}$. The induction is complete. It remains to note that with v_t being deterministic functions of $y_0,\ldots,y_t$, $t = 1,2,\ldots$, the purified-output-based controls $u_t = \widehat{U}_t(v_0,\ldots,v_t)$ become deterministic functions of $y_0,\ldots,y_t$: $u_u = U_t(y_0,\ldots,y_t)$, and the resulting output-based control yields the same system's behavior as the purified-output-based control we started with. And here again, looking at the construction, we immediately conclude that when the initial purified-output-based controller is affine, so is the resulting output-based controller. □

3.3.8 ***Extending calculus of polyhedral representability: Majorization**

3.3.8.1 **Preliminaries*

We start with introducing two useful functions on $\mathbf{R}^n$. Given an integer k, $1 \le k \le n$, let us set

$$s_k(x) = \text{sum of the } k \text{ largest entries in } x$$

and

$$\|x\|_{k,1} = \text{sum of the } k \text{ largest magnitudes of entries in } x$$

(we already met with the latter function, see Section 1.2.2.3 and Lemma 1.1). For example,

$$s_2([3;1;2]) = 3+2 = 5,\ s_2([3,3,1]) = 3+3 = 6,\ \|[5,-1,-7]\|_{2,1} = 7+5 = 12.$$

We intend to demonstrate that these functions are polyhedrally representable, and to build their p.r.'s.

Function $s_k(\cdot)$: Let S_k be the set of all Boolean vectors (i.e., those with coordinates 0 and 1) in $\mathbf{R}^n$ which have exactly k coordinates equal to 1. We clearly have

$$s_k(x) = \max_{y \in S_k} y^T x. \tag{$*$}$$

Now, by the result stated in Task 2.4, S_k is exactly the set of vertices of the polyhedral set

$$Y = \left\{ y \in \mathbf{R}^n : 0 \le y_i \le 1, 1 \le i \le n, \sum_i y_i = k \right\};$$

since X is bounded and thus is the convex hull of its vertices, $(*)$ says that $s_k(x)$ is the support function of Y:

$$s_k(x) = \max_y \left\{ x^T y : 0 \le y_i \le 1, 1 \le i \le n, \sum_i y_i = k \right\},$$

so that a polyhedral representation of $s_k(\cdot)$ is readily given by the results of Section 3.3.4.2. Applying (3.27) and denoting $\lambda^-, \lambda^+, \mu$ the (vectors of) Lagrange multipliers associated with the constraints $y \ge 0$, $y \le [1; ...; 1]$ and $\sum_i y_i = k$, respectively, we get

$$\begin{aligned} &\{[x;\tau] : \tau \ge s_k(x)\} \\ &\quad = \left\{ [x;\tau] : \exists (\lambda^- \le 0, \lambda^+ \ge 0, \mu) : \lambda^- + \lambda^+ + \mu[1; ...; 1] \right. \\ &\qquad \left. = x, \sum_i \lambda_i^+ + k\mu \le \tau \right\}, \end{aligned}$$

which clearly simplifies to

$$\begin{aligned} &\{[x;\tau] : \tau \ge s_k(x)\} \\ &\quad = \left\{ [x;\tau] : \exists \lambda, \mu : x_i \le \lambda_i + \mu, 1 \le i \le n, \right. \\ &\qquad \left. \tau \ge \sum_{i=1}^n \lambda_i + k\mu, \lambda_i \ge 0 \, \forall i \right\}. \end{aligned} \tag{3.47}$$

We have built a p.r. for $s_k(\cdot)$ (cf. Lemma 1.1).

Remark. Since the functions $s_k(x)$ are p.r.f., so are the functions $s_{n-k}(x) - \sum_{i=1}^n x_i$, meaning that *the sum* $\underline{s}_k(x)$ *of* k *smallest entries of* $x \in \mathbf{R}^n$ *is a concave polyhedrally representable function of* x (indeed, $\underline{s}_k(x) = \sum_{i=1}^n x_i - s_{n-k}(x)$). What is important in these convexity/concavity results is that we speak about *sums* of k largest/smallest entries in x, not about the kth largest (or kth smallest) entry in x. One can demonstrate by examples that the kth largest entry $x^k(x)$ of a vector $x \in \mathbf{R}^n$ is neither a concave, nor a convex function of x, unless $k = 1$ ($x^1(x) = s_1(x)$ is convex) or $k = n$ ($x^n(x) = \underline{s}_1(x)$ is concave).

Function $\|x\|_{k,1}$**:** Denoting $|x| = [|x_1|; ...; |x_n|]$, we clearly have $\|x\|_{k,1} = s_k(|x|)$, which combines with (3.47) to imply that

$$\{[x;\tau] : \tau \geq \|x\|_{k,1}\} = \Bigg\{[x;\tau] : \exists \lambda, \mu : \\ \pm x_i \leq \lambda_i + \mu, 1 \leq i \leq n, \tau \\ \geq \sum_{i=1}^n \lambda_i + k\mu, \lambda_i \geq 0\,\forall i\Bigg\}, \tag{3.48}$$

which is a p.r. for $\|x\|_{k,1}$. We have built this p.r. in Lemma 1.1.

3.3.8.2 **Majorization*

Postponing for a moment functions $s_k(\cdot)$, let us look at something seemingly completely different — at the set Π_n of doubly stochastic $n \times n$ matrices (see Section 2.3.2.3). Recall that this is the polyhedral set in the space $\mathbf{R}^{n\times n} = \mathbf{R}^{n^2}$ of $n \times n$ matrices P given by the constraints

$$\Pi_n = \left\{P \in \mathbf{R}^{n\times n} : P_{ij} \geq 0\,\forall i,j, \sum_i P_{ij} = 1\,\forall j, \sum_j P_{ij} = 1\,\forall i\right\}.$$

Birkhoff's Theorem (Theorem 2.11) states that the vertices of Π_n are exactly the permutation $n \times n$ matrices. Since Π_n clearly is bounded, it is the convex hull of the set of its vertices, and we arrive at the following useful result:

An $n \times n$ matrix is doubly stochastic iff it is a convex combination of $n \times n$ permutation matrices.

We now make the following.

Observation 3.2. Let f be a convex and symmetric function on $\mathbf{R}^n$, the symmetry meaning that whenever x' is obtained from x by permuting the entries, one has $f(x') = f(x)$. Then for every x and every doubly stochastic matrix P, one has

$$f(Px) \leq f(x). \tag{3.49}$$

Verification is immediate: if P is doubly stochastic, then $P = \sum_i \lambda_i P^i$, where P^i are permutation matrices and λ_i are nonnegative weights summing up to 1. It follows that

$$f(Px) = f\left(\sum_i \lambda_i P^i x\right) \underbrace{\leq}_{(a)} \sum_i \lambda_i f(P^i x) \underbrace{=}_{(b)} \sum_i \lambda_i f(x) = f(x),$$

where (a) is given by Jensen's inequality and (b) is due to the symmetry of f. □

Observation 3.2 is the source of numerous useful inequalities. For starters, here is the derivation of the famous inequality between the arithmetic and the geometric means:

For nonnegative reals $x_1, \ldots, x_n$, it always holds that

$$\frac{x_1 + \cdots + x_n}{n} \geq (x_1 x_2 \ldots x_n)^{1/n}. \tag{$*$}$$

Indeed, it is not difficult to prove that the function $g(x) = (x_1, \ldots, x_n)^{1/n}$ is concave in the nonnegative orthant, and of course it is symmetric. Given an $x \geq 0$, specifying P as the $n \times n$ matrix with all entries equal to $1/n$ (this clearly is a doubly stochastic matrix) and applying (3.49) to x, P and the convex symmetric function $f = -g$, we get

$$g(Px) \geq g(x);$$

but $g(Px)$ clearly is the arithmetic mean of $x_1, \ldots, x_n$, and we are done.

We can get in the same fashion the inequality between the arithmetic and the harmonic means:

For positive reals $x_1, \ldots, x_n$*, it always holds that*

$$\frac{n}{\frac{1}{x_1} + \frac{1}{x_2} + \cdots + \frac{1}{x_n}} \leq \frac{x_1 + \cdots + x_n}{n}. \qquad (**)$$

Indeed, it is easy to see that the function $f(x) = \frac{1}{x_1} + \cdots + \frac{1}{x_n}$, regarded as a function of a positive vector x (i.e., extended by the value $+\infty$ outside the set of positive vectors), is convex (and of course symmetric). Given $x > 0$ and using the same P as above, we get from (3.49) as applied to x, P, f that

$$\frac{n^2}{x_1 + \cdots + x_n} \leq \frac{1}{x_1} + \cdots + \frac{1}{x_n},$$

which is nothing but (**).

In fact, both inequalities (*), (**) can be easily obtained directly from the Jensen inequality and do not need "a cannon" like Birkhoff's Theorem.[8] This is not so in our next example:

(!) *Let* $\mathbf{S}^n$ *be the space of symmetric* $n \times n$ *matrices; for a matrix* $X \in \mathbf{S}^n$*, let* $\lambda(X) \in \mathbf{R}^n$ *be the vector of eigenvalues* $\lambda_i(X)$ *of* X *(taken with their multiplicities in the nonascending order). For every convex and symmetric function* f *on* $\mathbf{R}^n$*, the function* $F(X) = f(\lambda(X))$ *on* $\mathbf{S}^n$ *is convex.*

To justify this claim, it suffices to verify the following, important by its own right, relation:

$$F(X) = \sup_{U \in \mathcal{O}_n} f(\mathrm{diag}(UXU^T)), \qquad (!!)$$

where $\mathcal{O}_n$ is the set of all orthogonal $n \times n$ matrices, and $\mathrm{diag}(Y)$ is the vector comprising the diagonal entries of a matrix Y. Indeed, taking (!!) for granted, we observe that $f(\mathrm{diag}(UXU^T))$ is convex along with f (calculus of convex functions, rule on affine substitution of argument). It remains to recall that the supremum of a whatever family of convex functions is convex as well.

To prove (!!), note that by the eigenvalue decomposition, a symmetric matrix X can be represented as $X = V \Lambda V^T$, where V is

---

[8]Indeed, when proving the arithmetic–geometric means inequality, we lose nothing by assuming that $x > 0$ and $\sum_i x_i = n$. Applying Jensen's inequality to the (clearly convex) function $-\ln(s)$ on the positive ray, we get $0 = -\ln(\frac{x_1 + \cdots + x_n}{n}) \leq \frac{1}{n}[-\ln(x_1) - \cdots - \ln(x_n)] = -\ln([x_1, \ldots, x_n]^{1/n})$, which is the same as the inequality we need.

orthogonal and Λ is the diagonal matrix with the eigenvalues of X on the diagonal. Denoting temporarily the right-hand side in (!!) by $G(X)$, we clearly have $G(X) \geq F(X) \equiv f(\lambda(X))$ (take $U = V^T$). To prove the opposite inequality, note that when U is orthogonal, we have $UXU^T = U(V\Lambda V^T)U^T = W\Lambda W^T$, where $W = UV$ is orthogonal along with U, V. It follows that

$$(UXU^T)_{ii} = \sum_{j,\ell} W_{ij}\Lambda_{j\ell}[W^T]_{\ell i} = \sum_{j=1}^{n} W_{ij}^2 \lambda_j(X).$$

Since W is orthogonal, the matrix $P = [W_{ij}^2]_{1\leq i,j\leq n}$ is doubly stochastic. We see that *the diagonal of UXU^T is the product of a double stochastic matrix and the vector of eigenvalues of X*, whence, by (3.49), $f(\mathrm{diag}(UXU^T)) \leq f(\lambda(X))$ (recall that f is symmetric and convex). The latter inequality holds true for every orthogonal U, whence $G(X) = \sup_{U\in\mathcal{O}_n} f(\mathrm{diag}(UXU^T)) \leq f(\lambda(X)) = F(x)$. □

The results above imply numerous useful facts that are nonevident at first glance, like

- the sum of $k \leq n$ largest eigenvalues of a symmetric $n\times n$ matrix X is a convex function of X, and its value at X is $\geq$ the sum of the k largest diagonal entries in X (use (!) — (!!) with $f = s_k$). Similar result holds for the sums of k largest magnitudes of eigenvalues and k largest magnitudes of the diagonal entries of X;
- the functions $\mathrm{Det}^{1/n}(X)$ and $\ln(\mathrm{Det}(X))$, regarded as functions of positive semidefinite (all eigenvalues are nonnegative, or, equivalently, $\xi^T X\xi \geq 0$ for all ξ) symmetric $n \times n$ matrix X are concave, and $\mathrm{Det}(X)$ is $\leq$ the product of the diagonal entries of X (use (!) — (!!) with the functions $f(s) = (s_1...s_n)^{1/n}$ and $f(s) = \sum_i \ln s_i$, $s \geq 0$);
- the function $\mathrm{Det}^{-1}(X)$, regarded as a function of positive definite (all eigenvalues positive) symmetric matrix X is convex;

to name just a few.

3.3.8.3 *Majorization principle

Observation 3.2 attracts our attention to the following question:

> *Given $x, y \in \mathbf{R}^n$, when can y be represented as Px with a doubly stochastic matrix P?*

The answer is given by

Theorem 3.5 (Majorization Principle). *A vector $y \in \mathbf{R}^n$ is the image of a vector $x \in \mathbf{R}^n$ under multiplication by a doubly stochastic matrix iff*

$$s_k(y) \leq s_k(x),\ 1 \leq k < n\ \&\ s_n(y) = s_n(x). \tag{3.50}$$

Proof. *Necessity*: let $y = Px$ for a doubly stochastic P. Then $s_k(y) \leq s_k(x)$, $1 \leq k \leq n$, by Observation 3.2, since $s_k(\cdot)$ is convex (we have seen it) and is clearly symmetric. And of course, multiplication by a doubly stochastic matrix preserves the sum of entries in a vector:

$$\sum_i (Px)_i = [1;...;1]^T Px = (P^T[1;...;1])^T x = [1;...;1]^T x;$$ [9]

so that $s_n(Px) = s_n(x)$.

Sufficiency: Assume that (3.50) holds true, and let us prove that $y = Px$ for some doubly stochastic matrix P. Both the existence of the representation in question and the validity of (3.50) are preserved when we permute entries in x and permute, perhaps differently, entries in y. Thus, in addition to (3.50), we can assume w.l.o.g. that

$$x_1 \geq x_2 \geq \cdots \geq x_n,\ y_1 \geq y_2 \geq \cdots \geq y_n.$$

Now, suppose that the representation we are looking for does not exist: $y \notin X = \{Px : P \in \Pi_n\}$, and let us lead this assumption to a contradiction. Since Π_n is polyhedral, so is X (as the image of Π_n under the linear mapping $P \mapsto Px$). Since $y \notin X$ and X is polyhedral, there exists a nonstrict linear inequality which is valid on X and is violated at y, or, equivalently, there exists $\xi \in \mathbf{R}^n$ such that

$$\xi^T y > \max_{x' \in X} \xi^T x'. \tag{!}$$

Now, if $\xi_i < \xi_j$ for i, j such that $i < j$, then, permuting ith and jth entry in ξ, the right-hand side in (!) remains intact (since X clearly

[9] To understand the equalities, note that if P is doubly stochastic, then so is P^T, and the product of a doubly stochastic matrix by the all-1 vector is this very vector.

is closed w.r.t. permutations of entries in a vector), and the left-hand side does not decrease (check it, keeping in mind that $\xi_i < \xi_j$ and $y_i \geq y_j$ due to $i < j$). It follows that arranging the entries in ξ in the nonascending order, we keep intact the right-hand side in (!) and can only increase the left-hand side, that is, (!) remains valid. The bottom line is that we can assume that $\xi_1 \geq \xi_2 \geq \cdots \geq \xi_n$. Now comes the punch line: Since y_i are in the nonascending order, we have $s_k(y) = y_1 + \cdots + y_k$, whence

$$y_k = s_k(y) - s_{k-1}(y), \quad 2 \leq k \leq n,$$

so that

$$\begin{aligned}
\xi^T y &= \xi_1 s_1(y) + \xi_2[s_2(y) - s_1(y)] + \xi_3[s_3(y) - s_2(y)] \\
&\quad + \cdots + \xi_n[s_n(y) - s_{n-1}(y)] \\
&= [\xi_1 - \xi_2]s_1(y) + [\xi_2 - \xi_3]s_2(y) \\
&\quad + \cdots + [\xi_{n-1} - \xi_n]s_{n-1}(y) + \xi_n s_n(y).
\end{aligned}$$

(What we have used is the identity $\sum_{i=1}^n a_i b_i = \sum_{i=1}^{n-1} [a_i - a_{i+1}] \sum_{j=1}^i b_j + a_n \sum_{j=1}^n b_j$; this discrete analogy of integration by parts is called *Abel transformation*). Similarly, $s_k(x) = x_1 + \cdots + x_k$, whence

$$\xi^T x = [\xi_1 - \xi_2]s_1(x) + [\xi_2 - \xi_3]s_2(x) + \cdots + [\xi_{n-1} - \xi_n]s_{n-1}(x) + \xi_n s_n(x).$$

Comparing the resulting expressions for $\xi^T y$ and $\xi^T x$ and taking into account that $\xi_k - \xi_{k+1} \geq 0$, $s_k(y) \leq s_k(x)$ and $s_n(y) = s_n(x)$, we conclude that $\xi^T y \leq \xi^T x$. Since $x \in X$, the latter inequality contradicts (!). We have arrived at a desired contradiction. □

3.4 Exercises

Exercise 3.1.* (1) Prove Gordan Theorem on Alternative:

> *A system of strict homogeneous linear inequalities $Ax < 0$ in variables x has a solution iff the system $A^T\lambda = 0, \lambda \geq 0$ in variables λ has only the trivial solution $\lambda = 0$.*

(2) Prove Motzkin Theorem on Alternative:

A system $Ax < 0$, $Bx \leq 0$ of strict and nonstrict homogeneous linear inequalities has a solution iff the system $A^T\lambda + B^T\mu = 0, \lambda \geq 0, \mu \geq 0$ in variables λ, μ has no solution with $\lambda \neq 0$.

Exercise 3.2 (Do we need strict inequalities?).† Given the system

$$\begin{aligned} Ax &< p \\ Bx &\leq q \end{aligned} \qquad (S)$$

of finitely many strict and nonstrict linear inequalities in variables $x \in \mathbf{R}^n$, build a system of nonstrict linear inequalities (S') which is solvable if and only if (S) is, with feasible solutions to (S), easily convertible into feasible solutions to (S'), and vice versa.

Exercise 3.3.* For the systems of constraints to follow, write them down equivalently in the standard form $Ax < b, Cx \leq d$ and point out their solvability status ("solvable — unsolvable") along with the corresponding certificates.

(1) $x \leq 0$ $(x \in \mathbf{R}^n)$,
(2) $x \leq 0$ & $\sum_{i=1}^n x_i > 0$ $(x \in \mathbf{R}^n)$,
(3) $-1 \leq x_i \leq 1, 1 \leq i \leq n, \sum_{i=1}^n x_i \geq n$ $(x \in \mathbf{R}^n)$,
(4) $-1 \leq x_i \leq 1, 1 \leq i \leq n, \sum_{i=1}^n x_i > n$ $(x \in \mathbf{R}^n)$,
(5) $-1 \leq x_i \leq 1, 1 \leq i \leq n, \sum_{i=1}^n ix_i \geq \frac{n(n+1)}{2}$ $(x \in \mathbf{R}^n)$,
(6) $-1 \leq x_i \leq 1, 1 \leq i \leq n, \sum_{i=1}^n ix_i > \frac{n(n+1)}{2}$ $(x \in \mathbf{R}^n)$,
(7) $x \in \mathbf{R}^2$, $|x_1| + x_2 \leq 1, x_2 \geq 0, x_1 + x_2 = 1$,
(8) $x \in \mathbf{R}^2$, $|x_1| + x_2 \leq 1, x_2 \geq 0, x_1 + x_2 > 1$,
(9) $x \in \mathbf{R}^4$, $x \geq 0$, sum of two largest entries in x does not exceed 2, $x_1 + x_2 + x_3 \geq 3$,
(10) $x \in \mathbf{R}^4$, $x \geq 0$, sum of two largest entries in x does not exceed 2, $x_1 + x_2 + x_3 > 3$.

Exercise 3.4.* Let $(\mathcal{S})$ be the following system of linear inequalities in variables $x \in \mathbf{R}^3$:

$$x_1 \leq 1, \quad x_1 + x_2 \leq 1, \quad x_1 + x_2 + x_3 \leq 1. \qquad (\mathcal{S})$$

In the following list, point out which inequalities are, and which are not, consequences of the system, and certify your claims as explained in examples in items 1 and 2.

(1) $3x_1 + 2x_2 + x_3 \leq 4$, – consequence (sum up the original inequalities)
(2) $3x_1+2x_2+x_3 \leq 2$, – not a consequence (set $x_1 = 1, x_2 = x_3 = 0$)
(3) $3x_1 + 2x_2 \leq 3$,
(4) $3x_1 + 2x_2 \leq 2$,
(5) $3x_1 + 3x_2 + x_3 \leq 3$,
(6) $3x_1 + 3x_2 + x_3 \leq 2$.

Make a generalization: prove that a linear inequality $px_1+qx_2+rx_3 \leq s$ is a consequence of $(\mathcal{S})$ if and only if $s \geq p \geq q \geq r \geq 0$.

Exercise 3.5.* Is the inequality $x_1+x_2 \leq 1$ a consequence of the system $x_1 \leq 1, x_1 \geq 2$? If yes, can it be obtained by taking a legitimate weighted sum of inequalities from the system and the identically true inequality $0^T x \leq 1$, as it is suggested by the Inhomogeneous Farkas Lemma?

Exercise 3.6.* Certify the correct statements in the following list:

(1) The polyhedral set $X = \{x \in \mathbf{R}^3 : x \geq [1/3; 1/3; 1/3], \sum_{i=1}^3 x_1 \leq 1\}$ is nonempty.
(2) The polyhedral set $X = \{x \in \mathbf{R}^3 : x \geq [1/3; 1/3; 1/3], \sum_{i=1}^3 x_1 \leq 0.99\}$ is empty.
(3) The linear inequality $x_1 + x_2 + x_3 \geq 2$ is violated somewhere on the polyhedral set $X = \{x \in \mathbf{R}^3 : x \geq [1/3; 1/3; 1/3], \sum_{i=1}^3 x_i \leq 1\}$.
(4) The linear inequality $x_1 + x_2 + x_3 \geq 2$ is violated somewhere on the polyhedral set $X = \{x \in \mathbf{R}^3 : x \geq [1/3; 1/3; 1/3], \sum_{i=1}^3 x_i \leq 0.99\}$.
(5) The linear inequality $x_1 + x_2 \leq 3/4$ is satisfied everywhere on the polyhedral set $X = \{x \in \mathbf{R}^3 : x \geq [1/3; 1/3; 1/3], \sum_{i=1}^3 x_i \leq 1.05\}$.
(6) The polyhedral set $Y = \{x \in \mathbf{R}^3 : x_1 \geq 1/3, x_2 \geq 1/3, x_3 \geq 1/3\}$ is not contained in the polyhedral set $X = \{x \in \mathbf{R}^3 : x \geq [1/3; 1/3; 1/3], \sum_{i=1}^3 x_i \leq 1\}$.
(7) The polyhedral set $Y = \{x \in \mathbf{R}^3 : x \geq [1/3; 1/3; 1/3], \sum_{i=1}^3 x_i \leq 1\}$ is contained in the polyhedral set $X = \{x \in \mathbf{R}^3 : x_1 + x_2 \leq 2/3, x_2 + x_3 \leq 2/3, x_1 + x_3 \leq 2/3\}$.
(8) The polyhedral set $X = \{x \in \mathbf{R}^3 : x \geq [1/3; 1/3; 1/3], \sum_{i=1}^3 x_i \leq 1\}$ is bounded.

(9) The polyhedral set $X = \{x \in \mathbf{R}^3 : x_1 \geq 1/3, x_2 \geq 1/3, \sum_{i=1}^3 x_i \leq 1\}$ is unbounded.

Exercise 3.7.* Consider the LO program

$$\text{Opt} = \max_x \{x_1 : x_1 \geq 0, x_2 \geq 0, ax_1 + bx_2 \leq c\}, \qquad (P)$$

where a, b, c are parameters. Answer the following questions and certify your answers:

(1) Let $c = 1$. Is the problem feasible?
(2) Let $a = b = 1, c = -1$. Is the problem feasible?
(3) Let $a = b = 1, c = -1$. Is the problem bounded?
(4) Let $a = b = c = 1$. Is the problem bounded?
(5) Let $a = 1, b = -1, c = 1$. Is the problem bounded?
(6) Let $a = b = c = 1$. Is it true that $\text{Opt} \geq 0.5$?
(7) Let $a = b = 1, c = -1$. Is it true that $\text{Opt} \leq 1$?
(8) Let $a = b = c = 1$. Is it true that $\text{Opt} \leq 1$?
(9) Let $a = b = c = 1$. Is it true that $x_* = [1; 1]$ is an optimal solution of (P)?
(10) Let $a = b = c = 1$. Is it true that $x_* = [1/2; 1/2]$ is an optimal solution of (P)?
(11) Let $a = b = c = 1$. Is it true that $x_* = [1; 0]$ is an optimal solution of (P)?

Exercise 3.8.* Write down problems dual to the following LO programs:

(1) $$\max_{x \in \mathbf{R}^3} \left\{ x_1 + 2x_2 + 3x_3 : \begin{array}{l} x_1 - x_2 + x_3 = 0 \\ x_1 + x_2 - x_3 \geq 100 \\ x_1 \leq 0 \\ x_2 \geq 0 \\ x_3 \geq 0 \end{array} \right\},$$

(2) $\max_{x \in \mathbf{R}^n} \left\{ c^T x : Ax = b, x \geq 0 \right\}$,

(3) $\max_{x \in \mathbf{R}^n} \left\{ c^T x : Ax = b, \underline{u} \leq x \leq \overline{u} \right\}$,

(4) $\max_{x,y} \left\{ c^T x : Ax + By \leq b, x \leq 0, y \geq 0 \right\}$.

Exercise 3.9.† Consider a primal–dual pair of LO programs

$$\max_x \left\{ c^T x : \begin{array}{c} Px \le p \\ Qx \ge q \\ Rx = r \end{array} \right\} \tag{P}$$

$$\min_{\lambda=[\lambda_\ell,\lambda_g,\lambda_e]} \left\{ p^T\lambda_\ell + q^T\lambda_g + r^T\lambda_e : \begin{array}{l} \lambda_\ell \ge 0 \\ \lambda_g \le 0 \\ P^T\lambda_\ell + Q^T\lambda_g + R^T\lambda_e = c \end{array} \right\}. \tag{D}$$

Assume that both problems are feasible, and that the primal problem does contain inequality constraints. Prove that the feasible set of at least one of these problems is unbounded.

Exercise 3.10.† For positive integers $k \le n$, let $s_k(x)$ be the sum of the k largest entries in a vector $x \in \mathbf{R}^n$, e.g., $s_2([1;1;1]) = 1+1 = 2$, $s_2([1;2;3]) = 2+3 = 5$. Find a polyhedral representation of $s_k(x)$. *Hint:* Take into account that the extreme points of the set $\{x \in \mathbf{R}^n : 0 \le x_i \le 1, \sum_i x_i = k\}$ are exactly the 0/1 vectors from this set, and derive from this that

$$s_k(x) = \max_y \left\{ y^T x : 0 \le y_i \le 1 \,\forall i, \sum_i y_i = k \right\}.$$

Exercise 3.11.† Consider the scalar linear constraint

$$a^T x \le b \tag{1}$$

with uncertain data $a \in \mathbf{R}^n$ (b is certain) varying in the set

$$\mathcal{U} = \left\{ a : |a_i - a_i^*|/\delta_i \le 1, 1 \le i \le n, \sum_{i=1}^n |a_i - a_i^*|/\delta_i \le k \right\}, \tag{2}$$

where a_i^* are given "nominal data," $\delta_i > 0$ are given quantities and $k \le n$ is an integer (in the literature, this is called "budgeted uncertainty"). Rewrite the Robust Counterpart

$$a^T x \le b \,\forall a \in \mathcal{U} \tag{RC}$$

in a tractable LO form (that is, write down an explicit system (S) of linear inequalities in variables x and additional variables such that x satisfies (RC) if and only if x can be extended to a feasible solution of (S)).

Part II

Classical Algorithms of Linear Optimization: The Simplex Method

Chapter 4

Simplex Method

In this chapter, we focus on the historically first algorithm for solving LO programs — the famous Simplex method invented by George Dantzig in late 1940s. The importance of this invention can hardly be overestimated: the algorithm turned out to be extremely successful in actual computations and was *the* working horse of LO for over more than 4 decades. Today we have at our disposal also other, theoretically more advantageous (as well as better suited for many practical applications) LO algorithms; nevertheless, the Simplex method still remains indispensable in numerous applications.

As is the case with nearly all computational methods, the Simplex method is not a *single* well-defined algorithm; it is rather a common name for a *family* of algorithms of common structure. In our book, we will focus on the most basic members of this family — the *Primal* and the *Dual Simplex methods.* These methods heavily exploit the specific geometry of LO. The informal "high level" description of the Simplex method is quite transparent and natural. Assume that the feasible set of an LO program is nonempty and does not contain lines. From the theory developed in Chapter 2 it then follows that if the program is feasible, its optimal solution, if any, can be found among the finitely many candidates — the vertices of the feasible set. The Simplex method moves along the vertices according to the following scheme (see Fig. 4.1):

- staying at a current vertex v, the method checks whether there is an edge — a one-dimensional face of the feasible set — which contains v and is an *improving* one, that is, moving from v along

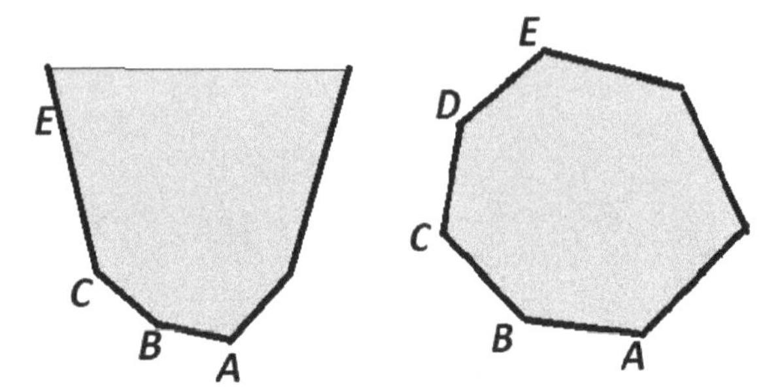

Fig. 4.1. Geometry of Simplex Method. The objective to be maximized is the ordinate ("height"). Left: method starts from vertex A and ascends to vertex C where an improving ray CE is discovered, meaning that the problem is unbounded. Right: method starts from vertex A and ascends to the optimal vertex E.

this edge, the objective improves (increases when speaking of a maximization problem and decreases when the problem is a minimization one). There are three possibilities:

- an improving edge does not exist. It can be proved that in this case the vertex is the optimal solution;
- there exists an improving edge, which, geometrically, is a ray, that is, moving from v along this edge, we never leave the feasible set. In this case, the problem clearly is unbounded;
- there exists an improving edge which is a nontrivial segment and thus has two vertices which, being vertices of a face of the feasible set, are vertices of this set itself (Proposition 2.10). One of these vertices is v; denoting the other vertex by v^+ and taking into account that the edge is an improving one, the value of the objective at v^+ is better than its value at v. The Simplex method moves from v to v^+, and proceeds from this point as from v.

Since there are only finitely many vertices and the method before termination strictly improves the objective and thus cannot visit the same vertex twice, it in finitely many steps either finds an improving *ray* and terminates with the (correct) claim that the program is unbounded, or arrives at a vertex which does not admit an improving edge and thus is optimal. The essence of the matter is in the fact that there are relatively simple, quite transparent and fully algorithmic algebraic tools which, modulo "degenerate cases" (which need a special treatment and never occur when the problem is "in general position"), make it easy to implement the above strategy, that is, to find, given a vertex, an improving edge, or to detect correctly that no one exists.

Note that the outlined "finiteness" of the Simplex method by itself does not promise much — if finiteness were the only goal, why not use the Fourier–Motzkin elimination scheme (see p. 42)? The only upper bound on the number of steps of the Simplex method (i.e., the number of vertices visited before termination) given by the proof of the method's finiteness is that the total number of steps is bounded by the total number of vertices, and the latter can be astronomically large: the polytope as simple as the n-dimensional box has 2^n vertices! It is a kind of miracle that "in reality" the method visits a negligible part of all vertices (empirically speaking, just a moderate multiple of the number of equality constraints in the standard form of the program) and is a surprisingly successful algorithm capable of solving routinely "real world" LO programs with tens and hundreds of thousands of variables and constraints in reasonable time.

The constructions and results to follow are quite standard; for their origin, see [Bertsimas and Tsitsiklis (1997), Section 3.10].

4.1 Simplex method: Preliminaries

Primal and Dual Simplex methods (PSM and DSM for short) are directly applicable to an LO program in the standard form

$$\begin{aligned} \mathrm{Opt}(P) &= \max_{x\in\mathbf{R}^n}\left\{c^Tx : x\in X=\{x: Ax=b,\ x\geq 0\}\right\},\\ A &= [a_1^T;\ldots;a_m^T]; \end{aligned} \tag{4.1}$$

as we remember from Chapter 1, every LO program can be straightforwardly converted to this form.

Assumption: From now on, we make the following

> **Assumption:** *The system of linear constraints $Ax = b$ in (4.1) is feasible, and the equations in this system are linearly independent, so that the rank of A is equal to the number m of rows in A.*

Note that checking the solvability of a system of linear equations $Ax = b$ is a simple Linear Algebra task; if the system is not feasible, so is the LO program (4.1), and its processing is therefore complete.

If the system $Ax = b$ is solvable, then the solution set remains intact when we eliminate from the system, one at a time, equations which are linear combinations of the remaining equations until a system with linearly independent equations is built; this reduction of redundant linear equations again is a simple Linear Algebra task. We see that our Assumption (which by default acts everywhere in the sequel) in fact does not restrict generality.

4.2 Geometry of an LO program in the standard form

Our next step is to understand how our results on the geometry of a general type LO program should be specialized when the program is in the standard form. We start with building up the problem dual to (4.1).

4.2.1 *The dual of an LO program in the standard form*

Written in the form of (3.8), problem (4.1) reads

$$\mathrm{Opt} = \max_x \left\{ c^T x : \left\{ \begin{array}{c} x \geq 0 \ (g) \\ Ax = b \ (e) \end{array} \right\} \right\}$$

and its dual, as built in Section 3.2.1, reads

$$\mathrm{Opt}(D) = \min_{\lambda=[\lambda_g;\lambda_e]} \left\{ d^T\lambda := 0^T\lambda_g + b^T\lambda_e : \left\{ \begin{array}{r} \lambda_g \leq 0 \\ \lambda_g + A^T\lambda_e = c \end{array} \right\} \right\}. \tag{4.2}$$

By eliminating λ_g and renaming λ_e as y, we pass to the equivalent form of the dual problem, namely,

$$\mathrm{Opt}(D) = \min_y \left\{ b^T y : c - A^T y \leq 0 \right\}. \tag{4.3}$$

(!) *From now on we call* (4.3) *the dual of the LO program in the standard form* (4.1).

The optimality conditions read (see Proposition 3.4) as follows:

Proposition 4.1. *A feasible solution x to (4.1) is optimal iff there exists a vector of Lagrange multipliers $y \in \mathbf{R}^m$ such that $c - A^T y \leq 0$ and*

$$x_j(c - A^T y)_j = 0,\ 1 \leq j \leq n \qquad \text{[complementary slackness]}.$$

If y satisfies these requirements w.r.t. some feasible solution x to (4.1), then y is an optimal solution to the dual problem (4.3).

Note that when replacing the original objective $c^T x$ in (4.1) with $[c - A^T y]^T x$, on the feasible set of the program the objective is changed by a constant (specifically, by $-y^T b$), and the primal problem remains intact. In LO terminology, such an equivalent modification $c \mapsto c - A^T y$ of the primal objective is called *passing to reduced costs.*[1] Proposition 4.3 says that a primal feasible solution x is optimal iff there exist reduced costs $c - A^T y$ which "make optimality evident" — all the reduced costs are nonpositive and complementary slackness w.r.t. x takes place, so that the new objective is nonpositive everywhere on the nonnegative orthant, while being zero at x due to the complementary slackness; thus, x maximizes the new objective on the *entire nonnegative orthant*, not to speak of maximizing the new (and thus the old) objective on the primal feasible set.

4.2.2 *Bases and basic feasible solutions*

Now it is time to investigate the vertices of the primal feasible set $X = \{x \in \mathbf{R}^n : x \geq 0 : Ax = b\}$ and the dual feasible set $Y = \{y \in \mathbf{R}^m : c - A^T y \leq 0\}$. Note that X does not contain lines (as a part of the nonnegative orthant $\mathbf{R}^n_+$ which does not contain lines), and Y does not contain lines due to the fact that A is of rank m, that is, A^T has a trivial kernel.

We call a collection $I = \{i_1, \ldots, i_m\}$ of m distinct indices from the index set $\{1, \ldots, n\}$ a *basis* of (4.1) if the corresponding columns

---

[1]This terminology fits a *minimization* LO in the standard form, where c_i can be interpreted as costs. In our maximization setting of the primal problem, the name "reduced profits" would be more to the point, but we prefer to stick to the standard terminology.

$A_{i_1}, \ldots, A_{i_m}$ of A are linearly independent (and thus form a linear basis in the space $\mathbf{R}^m$ to which they belong). Since A is of rank m, bases do exist. If I is a basis, we can partition every vector $x \in \mathbf{R}^n$ into its m-dimensional *basic part* x_I comprising of entries with indices from I and its *nonbasic part* $x_{\overline{I}}$ comprising all remaining entries; we shall denote this partition as $x = (x_I, x_{\overline{I}})$. We can similarly partition the matrix A: $A = (A_I, A_{\overline{I}})$, where A_I is the $m \times m$ matrix comprising the columns of A with indices from I,[2] and $A_{\overline{I}}$ comprises all remaining columns of A. With these partitions, A_I is an $m \times m$ *nonsingular* matrix, and the system of linear constraints in (4.1) can be written down as

$$\begin{aligned} & A_I x_I + A_{\overline{I}} x_{\overline{I}} = b \\ \Leftrightarrow\ & x_I + [A_I]^{-1} A_{\overline{I}} x_{\overline{I}} = [A_I]^{-1} b. \end{aligned} \tag{4.4}$$

We can now satisfy the primal equality constraints by setting $x_{\overline{I}} = 0$ and $x_I = [A_I]^{-1}b$, thus arriving at a *basic primal solution*

$$x^I = (x_I = [A_I]^{-1}b, x_{\overline{I}} = 0)$$

associated with basis I; this solution satisfies the primal equality constraints, but is not necessarily feasible for the primal problem; it is primal feasible iff $x_I \geq 0$.

Similarly, given a basis I, we can try to find a dual solution y which makes the dual inequality constraints with indices from I active, that is, such that

$$(A^T y - c)_i = 0,\ i \in I.$$

The matrix of coefficients in this system of m linear equations in m variables y is A_I^T, so that the system has a unique solution

$$y^I = [A_I]^{-T} c_I; \qquad [B^{-T} = [B^{-1}]^T]$$

y^I is called the *basic dual solution*, and the vector $c^I = c - A^T y^I$ — the *reduced costs associated with the basis I*. The basic dual solution

[2]These indices are somehow assigned serial numbers, so that $I = \{i_1, \ldots, i_m\}$ and $A_I = [A_{i_1}, \ldots, A_{i_m}]$, where A_i is ith column in A.

y^I is not necessarily feasible for the dual problem; it is feasible iff the corresponding vector of reduced costs is nonpositive:

$$c^I := c - A^T[A_I]^{-T}c_I \leq 0.$$

Note that by construction all basic entries — those with indices from I — in this vector are zeros.

The role of bases in our context stems from the following simple.

Proposition 4.2.

(i) *Let v be a feasible solution to the primal program (4.1). v is a vertex of the primal feasible set X iff $v = x^I$ for some basis I. Equivalently: vertices of X are exactly the feasible primal basic solutions.*

(ii) *Let y be a feasible solution to the dual program (4.3). y is a vertex of the dual feasible set Y iff $y = y^I$ for some basis I. Equivalently: vertices of Y are exactly the feasible dual basic solutions.*

Proof. (i) In one direction: let v be a vertex of X, and let $J = \{i : v_i > 0\}$. Assume, first, that $J \neq \emptyset$ (that is, $v \neq 0$). We claim that the columns of A with indexes from J are linearly independent.

> Indeed, assuming the opposite, we could find a vector $h \in \mathbf{R}^n$ which is nonzero, has zero entries with indexes not in J, and satisfies $Ah = 0$. For small positive t, the vectors $v + th$ and $v - th$ belong to X, which is impossible, since v is a vertex of X and $h \neq 0$.

Since the columns $A_j, j \in J$ of A are linearly independent and the rank of A is m, we can extend J to m-element set of indexes which form a basis; clearly, v is the corresponding basic feasible solution. And if $J = \emptyset$, then $v = 0$ and $b = Av = 0$, implying that $v = 0$ is the basic feasible solution for *whatever* basis (and bases do exist!).

In the opposite direction: Let v be a basic feasible solution associated with basis I, and let us prove that v is a vertex of X, that is, the only h resulting in feasible $v \pm h$ is $h = 0$. Indeed, for such an h, the entries with indexes j outside of I should be zero (since $x_j \pm h_j = 0 \pm h_j$ should be nonnegative). Next, $Ah = A(v+h) - Av = b - b = 0$; since $h_j = 0$ for $j \notin I$ we get $\sum_{j \in I} h_j A_j = 0$, and since A_j, $j \in I$, are linearly independent, we get $h_j = 0, j \in I$. Thus, $h = 0$, as claimed. (i) is proved. $\square$

(ii) Let $\bar{y}$ be the dual basic feasible solution associated with basis I. This is a feasible solution to the system of inequalities $A^T y \geq c$ which makes active $m = \dim y$ inequalities of the system, namely, $A_j^T \bar{y} = c_j$, $j \in I$, and the vectors of coefficients of these inequalities are linearly independent, since I is a basis. By algebraic characterization of extreme points of a polyhedral set, $\bar{y}$ is a vertex of the dual feasible set. In the opposite direction: if $\bar{y}$ is an extreme point of the dual feasible set, then, by the same algebraic characterization of extreme points, among the inequalities of the system $Ay \geq c$ which are active at $\bar{y}$ there should be $m = \dim y$ inequalities with linearly independent vectors of coefficient, that is, there exists a basis I such that $A_j^T \bar{y} = c_j$ for $j \in I$; besides this, $\bar{y}$ is dual feasible. We conclude that $\bar{y}$ is the (unique!) basic dual solution associated with basis I, and this basic solution is dual feasible. □

Some remarks are in order.

A. Proposition 4.2 suggests the following conceptual scheme for enumerating the vertices of the feasible set X of the program (4.1): we look one by one at all m-element subsets I of the index set $\{1, \dots, n\}$ and skip those which are not bases. When I is a basis, we find the (unique) solution to the system $Ax = b$ with the zero nonbasic part ("primal basic solution" in the LO terminology), that is, the solution $x^I = (x_I = [A_I]^{-1}b, x_{\bar{I}} = 0)$. If x_I is nonnegative, we get a vertex of the primal feasible set ("primal basic feasible solution" in the LO terminology). In this fashion, we get all the vertices of X, if any. A similar process can be used to enumerate the vertices of the dual feasible set; now these are exactly the dual basic solutions y^I which happen to be dual feasible.

B. Pay attention to the fact that while every vertex v of X is a primal basic feasible solution, the corresponding basis is not always uniquely defined by the vertex. This basis definitely is unique, if the vertex is *nondegenerate*, that is, possesses exactly m nonzero (and thus positive) entries. In this case, the basis which makes the vertex a primal basic feasible solution comprises indices of the m nonzero entries in the vertex. A degenerate — with less than m positive entries — vertex v can be defined by many different bases. To give an extreme example, consider the case of $b = 0$. In this case, X is a pointed cone and as such has exactly one vertex — the origin;

but when $b = 0$, the origin is the primal basic feasible solution for *every* basis of the program, and there could be astronomically many of them.

Similarly, every vertex of the dual feasible set is a dual basic feasible solution, but a particular vertex y of Y can be associated with more than one basis. It may happen only when the vertex y is *degenerate*, meaning that the number of zero entries in the corresponding vector of reduced costs $c - A^T y$ is $> m$. If the vertex is nondegenerate — there are exactly m zero entries in the associated y vector of reduced costs — there exists exactly one basis I such that $y = y^I$, namely, the basis comprising indices of the m zero entries in $c - A^T y$.

Potential degeneracy (presence of degenerate vertices) of an LO program needs a special (as we shall see, not too difficult) treatment. Speaking about "general" LO programs of given sizes m, n, degeneracy (presence of at least one degenerate primal and/or dual basic feasible solution) is a "rare phenomenon" — the data of degenerate programs form a set of Lebesgue measure zero in the space $\mathbf{R}^{m+n+mn}$ of data of all programs of these sizes. In "general position," an LO program is nondegenerate, and every subset I of the index set $\{1, \ldots, n\}$ is a basis. Nevertheless, there are important special classes of LOs where a significant part of the data are "hard zeros" (coming from the problem's structure), and in these cases degeneracy can be typical.

C. An important point is that given a primal basic feasible solution x^I, we can try to certify its optimality by building the dual basic solution y^I associated with the same basis. Observing that by construction x^I and y^I satisfy the complementary slackness condition $x^I_j[c - A^T y^I]_j = 0$ for all j (indeed, by construction the first factor can be nonzero only for $j \in I$, and the second — only for $j \notin I$), we see that *if y^I happens to be dual feasible, x^I and y^I are optimal solutions to the respective problems.* By exactly the same token, given a dual basic feasible solution y^I, we can try to certify its dual optimality by building the primal basic solution x^I; *if it happens to be primal feasible, x^I and y^I again are optimal solutions to the respective problems.*

From the above discussion it follows that a *sufficient* condition for a feasible basic primal solution $\bar{x}$ associated with some basis I

to be primal optimal is dual feasibility of the dual basic solution $\bar{y}$ associated with basis I. This sufficient condition is also necessary, provided that $\bar{x}$ is nondegenerate; indeed, when $\bar{x}$ is primal optimal, there should be a dual feasible solution $\widehat{y}$ certifying optimality, that is, such that $[c - A^T\widehat{y}]_j[\bar{x}]_j = 0$ for all j, implying that $[c - A^T\widehat{y}]_j = 0$ for all j with $\bar{x}_j > 0$; for a *nondegenerate* feasible basic solution $\bar{x}$ associated with basis I, this says that $\widehat{y}$ is (the unique!) dual feasible basic solution associated with the same basis. However, when $\bar{x}$ is a primal optimal *degenerate* basic solution associated with some basis I, the basic dual solution associated with I is not necessarily dual feasible, so that the above sufficient condition for optimality of a primal feasible basic solution is, in general, not necessary. An "extreme example" here is the problem $\max_x\{\sum_{i=1}^n ix_i : x \geq 0, \sum_i x_i = 0\}$. In this problem, $x = 0$ is the only feasible, and thus, optimal solution to the primal problem, and it is basic for *every* basis, which under the circumstances is every single-element subset of $\{1, \ldots, n\}$. The basic dual solution y associated with basis $I = \{i\}$ is $y = i$, and this solution is dual feasible only when $i = n$. We see that when the basic primal optimal solution is degenerate, and thus is basic for perhaps more than one basis, not all dual basic solutions associated with these bases certify optimality.

D. The strategy implemented in Simplex method follows the recommendations of the previous item. Specifically,

- the *Primal Simplex method* generates subsequent primal basic feasible solutions, improving at every step (strictly if the current solution is nondegenerate) the primal objective, and augments this process by building associated dual basic solutions until either unboundedness of the primal problem is detected, or a *feasible* dual basic solution is met, thus certifying optimality of the current primal and dual basic feasible solutions;
- the *Dual Simplex method* generates subsequent dual basic feasible solutions, improving at every step (strictly, if the current solution is nondegenerate) the dual objective, and augments this process by building associated primal basic solutions until either unboundedness of the dual problem is detected, or a *feasible* primal basic solution is met, thus certifying optimality of the current primal and dual basic feasible solutions.

In both cases, the finiteness of the method in the nondegenerate case[3] follows from the fact that the corresponding objective strictly improves from step to step, which makes it impossible to visit twice the same primal (in the Primal Simplex Method) or dual (in the Dual Simplex Method) vertex; since the number of vertices is finite, the method *must* terminate after finitely many steps.

E. Finally, we make the following observation closely related to detecting unboundedness:

Observation 4.1. Let I be a basis.

(i) Assume that $\jmath \notin I$ and $w \in \mathbf{R}^n$ is a nonzero vector such that $w \geq 0$, $w_j = 0$ when $j \notin I \cup \{\jmath\}$ and $Aw = 0$, and that the feasible set $X = \{x \in \mathbf{R}^n : x \geq 0, Ax = b\}$ of the primal problem (4.1) is nonempty. Then w is the direction of an extreme ray of the recessive cone $\mathrm{Rec}(X)$ of X.
(ii) Assume that $\mu \in \mathbf{R}^m$ is a nonzero vector such that $A^T\mu \geq 0$ and $(A^T\mu)_j = 0$ for all but one indices $j \in I$, and that the feasible set $Y = \{y : c - A^T y \leq 0\}$ of the dual problem (4.3) is nonempty. Then μ is the direction of an extreme ray of the recessive cone $\mathrm{Rec}(Y)$ of Y.

Indeed, the recessive cone of X is $\mathrm{Rec}(X) = \{x \in \mathbf{R}^n : x \geq 0, Ax = 0\}$. w clearly belongs to this cone and makes equalities $n - 1$ of the homogeneous constraints defining $\mathrm{Rec}(X)$, specifically, m constraints $a_i^T x = 0$, $i = 1, \ldots, m$ and $n-m-1$ constraints $e_j^T x = 0$, $j \notin I \cup \{\jmath\}$. Since the rows of A_I are linearly independent, the $n - 1$ vectors $a_1, \ldots, a_m, \{e_j\}_{j \notin I \cup \{\jmath\}}$ are linearly independent (why?), and (i) follows from Proposition 2.16(i).

Similarly, the recessive cone of Y is $\mathrm{Rec}(Y) = \{y \in \mathbf{R}^m : A^T y \geq 0\}$; μ clearly belongs to this cone and makes equalities $m - 1$ among the m homogeneous linear equations $A_j^T y \geq 0$, $j \in I$, participating in the description of the cone. Since the vectors A_j, $j \in I$, are linearly independent, (ii) follows from Proposition 2.16(i). □

---

[3]When speaking about PSM, the nondegenerate case is defined as the case when all basic feasible primal solutions are nondegenerate; for DSM, nondegeneracy means that all dual feasible basic solutions are nondegenerate.

4.3 Simplex method

4.3.1 *Primal simplex method*

We are ready to present the PSM. In what follows, paragraphs in italics constitute the description of the method, while the usual text between these paragraphs contains explanations.

In the description to follow, we assume that the program (4.1) is feasible and, moreover, we have at our disposal a starting point which is a basis associated with a primal basic feasible solution to the program.

At step t, the current basic feasible solution x^I associated with the current basis I and this basis are updated according to the following rules.

A. *We compute the vector* $c^I = c - A^T y^I$, $y^I = [A_I]^{-T} c_I$, *of the reduced costs associated with the basis* I. *If* c^I *is nonpositive, we terminate with the claim that* x^I *is an optimal solution to the primal program, and* y^I *is the optimal solution to the dual program, otherwise we pass to item* B.

B. *We pick an index* j *— a pivot — such that the reduced cost* c^I_j *is positive* (such an index does exist, otherwise we were not invoking **B**; the index does not belong to I, since by construction the basic reduced costs c^I_i (those with $i \in I$) are zeros). *We then try to increase the variable* x_j *(which was zero in the solution* x^I*), allowing for updating the basic entries* x_i, $i \in I$, *in a feasible solution, and keeping the entries with indices outside of* $I \cup \{j\}$ *zeros. Specifically, let* $x(t)$, $t \geq 0$, *be given by*

$$x_i(t) = \begin{cases} 0, & i \notin I \cup \{j\} \\ t, & i = j \\ x^I_i - t([A_I]^{-1} A_j)_i, & i \in I. \end{cases} \tag{4.5}$$

Comment: The origin of (4.5) is as follows. $x(t)$ is the feasible solution of the system $Ax = b$ such that $x_j(t) = t$ and $x_i(t) = 0$ when $i \notin I \cup \{j\}$. As is seen from the second relation in (4.4), the basic entries $x_i(t)$, $i \in I$, in $x(t)$ should be exactly as stated in (4.5). Note that while $x(t)$ satisfies the constraints $Ax = b$ for all $t \geq 0$, the feasibility of $x(t)$ for the program (which amounts to nonnegativity of all entries in $x(t)$) depends on what happens with the basic entries $x_i(t)$. For every $i \in I$, there are just two possibilities:

- (A) the associated quantity $([A_I]^{-1}A_j)_i$ is nonpositive. In this case, the variable $x_i(t)$ is nonnegative for every $t \geq 0$.
- (B) the associated quantity $([A_I]^{-1}A_j)_i$ is positive, in which case $x_i(t)$ is nonnegative iff $t \leq t_i := [x_i^I]\left[([A_I]^{-1}A_j)_i\right]^{-1}$ and becomes negative when $t > t_i$. Note that t_i is nonnegative since x_i^I is so.

We check whether (A) *takes place for all basic indices i (i.e., indices from the current basis I). If it is the case, we terminate and claim that the program is unbounded. If there are basic indices for which* (B) *takes place, we define t_* as the minimum, over these basic indices, of the corresponding t_i's, and set i_* equal to the basic index corresponding to this minimum, as follows:*

$$t_* = \min_i \left\{ t_i := [x_i^I]\left[([A_I]^{-1}A_j)_i\right]^{-1} : ([A_I]^{-1}A_j)_i > 0 \right\},$$
$$i_* : ([A_I]^{-1}A_j)_{i_*} > 0 \ \& \ t_* = t_{i_*}.$$

We specify our new basis as $I^+ = [I \backslash \{i_*\}] \cup \{j\}$ (in the LO terminology, the variable x_j enters the basis, the variable x_{i_*} leaves it), *our new basic feasible solution as* $x^{I^+} = x(t_*)$, *and pass to the next step of the method.*

Several justifications and comments are in order.

I. According to the above description, a step of the PSM can lead to three outcomes as follows:

(1) Termination with the claim that the current primal basic feasible solution x^I is optimal; this happens when all the reduced costs c_i^I are nonpositive.

In this situation, the claim is correct due to the discussion in item **C** of Section 4.2.2, see p. 289.

(2) Termination with the claim that the problem is unbounded; this happens when for all basic indices, (A) is the case.

In the situation in question, the claim is correct. Indeed, in this situation $x(t)$ is feasible for the primal program for all $t \geq 0$. As we remember, replacing the original objective $c^T x$ with the objective $[c^I]^T x$ reduces to adding a constant to the restriction of the objective on the primal feasible plane. Since $c_i^I = 0$ for $i \in I$, $c_j^I > 0$ (due to the origin of j) and all entries in $x(t)$ with indices not in $I \cup \{j\}$ are zeros, we have

$$[c^I]^T x(t) = \sum_{i \in I} c_i^I x_i(t) + c_j^I x_j(t) = c_j^I x_j(t) = c_j^I t \to +\infty,\ t \to \infty,$$

whence also $c^T x(t) \to +\infty$ as $t \to \infty$, thus certifying that the problem is unbounded. Note that in the case in question the improving direction $\frac{d}{dt}x(t)$ is the direction of an extreme ray of $\mathrm{Rec}(X)$, $X = \{x : x \geq 0, Ax = b\}$ being the feasible domain of (4.1), see Observation 4.1.

(3) Passing to a new step with the claims that I^+ is a basis, and $x(t_*)$ is the corresponding primal basic feasible solution.

In this situation, the claims are also correct, which can be seen as follows. From the origin of t_* and i_* it is clear, first, that $x(t_*)$ satisfies the equality constraints of the problem and is nonnegative (i.e., is a feasible solution), and, second, that $(x(t_*))_i = 0$ when $i = i_*$ and when $i \notin I \cup \{j\}$, that is, $(x(t_*))_i$ can be positive only when $i \in I^+$. Further, by construction I^+ is an m-element subset of $\{1, \ldots, n\}$. Thus, all which remains to be verified in order to support the claims in question is that I^+ is a basis, that is, that the m columns A_i, $i \in I^+$, of A are linearly independent. This is immediate: assuming the opposite and taking into account that the columns A_i, $i \in I \backslash \{i_*\}$ of A "inherited" from I are linearly independent, the only possibility for A_i, $i \in I^+$, to be linearly dependent is that A_j is a linear combination of these inherited columns, which amounts to the existence of a representation

$$A_j = \sum_{i=1}^{n} \mu_i A_i$$

with $\mu_i = 0$ whenever $i \notin I$ or $i = i_*$. But then the vector

$$y(t) = x^I - t\mu + te_j$$

satisfies exactly the same requirements as the vector $x(t)$: $Ay(t) = Ax^I = b$, all coordinates in $y(t)$, except for those with indices from $I \cup \{j\}$, are zero, and the jth coordinate equals to t. From the explanation where $x(t)$ comes from it follows that $x(t) = y(t)$, which, in particular, implies that $(x(t))_{i_*}$ is independent of t (recall that $\mu_{i_*} = 0$). But the latter is impossible due to

$$x_{i_*}(t) = x^I_{i_*} - t([A_I]^{-1}A_j)_{i_*}$$

and the fact that $([A_I]^{-1}A_j)_{i_*} > 0$.

The bottom line is that *PSM is a well-defined procedure* — when running it, at every step we either terminate with correct claims "the current basic feasible solution is optimal" or "the problem is unbounded," or have a possibility to make the next step, since the updated I and x^I are what they should be (a basis and the associated primal basic feasible solution) in order make the next step well defined.

II. The correctness of PSM is good news, but by itself this news is not sufficient: we want to get an optimal solution, and not just "to run." What is crucial in this respect is the following

Observation 4.2. The PSM is a monotone process: if x^I and x^{I^+} are two consecutive primal basic feasible solutions generated by the method, then $c^Tx^I \leq c^Tx^{I^+}$, with the inequality being strict unless $x^I = x^{I^+}$. The latter never happens when the basic feasible solution x^I is nondegenerate.

The verification is immediate. In the notation from the description of the PSM and due to the explanations above, we have $c^Tx^{I^+} - c^Tx^I = [c^I]^Tx^{I^+} - [c^I]^Tx^I = c^I_j t_* \geq 0$, the equality being possible iff $t_* = 0$. The latter can happen only when x^I is degenerate, since otherwise $x^I_i > 0$ for all $i \in I$ and thus $t_i > 0$ for all $i \in I$ for which t_i are well defined. □

We arrive at the following.

Corollary 4.1. *The Primal Simplex method, initiated at a primal basic feasible solution of* (4.1), *possesses the following property: if the*

method terminates at all, the result upon termination is either a primal basic feasible solution which is an optimal solution to the program, or is a correct claim that the problem is unbounded. In the first case, the method produces not only the primal optimal solution, but also the corresponding optimality certificate — an optimal basic solution to the dual problem. In the second case, the method produces a ray which is contained in the feasible set of the problem and is a ray along which the objective increases, so that this ray is a certificate of unboundedness. In fact, this ray is an extreme ray of the recessive cone of the primal feasible set.

The method definitely terminates in finitely many steps, provided that the program is primal nondegenerate (i.e., all primal basic feasible solutions are so).

This result is an immediate corollary of our preceding observations. The only claim that indeed needs a comment is the one of finite termination on a nondegenerate problem. On a closer inspection, this property appears immediate as well: when the problem is nondegenerate, every step before termination increases the value of the objective, implying that the method cannot visit the same vertex more than once. Since the number of vertices is finite, and the method before termination moves from vertex to vertex, the number of steps before termination is finite as well. □

4.3.2 *Tableau implementation and example*

To illustrate the Primal Simplex Method, let us work out an example. In this example, we will use the so-called *full tableau form* of the algorithm well suited for both thinking about the method and solving LO programs by hand (while now the latter process takes place in classrooms only, once upon a time this was how the LO programs, of course, toy ones as per today's scale, were actually solved). The idea is that if I is a basis of the program of interest, then the original program is equivalent to the program

$$\max_x \left\{ [c^I]^T x : [A_I]^{-1} A x = [A_I]^{-1} b \right\}, \; c^I = c - A^T [A_I]^{-T} c_I \qquad (*)$$

and we can keep the data of this equivalent problem, along with the current basic feasible solution x^I, in a tableau of the form

$$\begin{array}{|c|c|}\hline -c_I^T x_I^I & [c^I]^T \\ \hline [A_I]^{-1}b & [A_I]^{-1}A \\ \hline \end{array}$$

or, in more detailed form,

$$\begin{array}{|c|c|}\hline -c_I^T x_I^I & c_1^I \dots c_n^I \\ \hline x_{i_1}^I & [[A_I]^{-1}A]^1 \\ \vdots & \dots \\ x_{i_m}^I & [[A_I]^{-1}A]^m \\ \hline \end{array}$$

where $I = \{i_1, \dots, i_m\}$ and $[B]^s$ stands for s-th row of a matrix B. It should be stressed that what we keep in the tableau are the *values* of the corresponding expressions, not the expressions themselves. It is convenient to count the rows and the columns of a tableau starting from 0; thus, the zeroth row contains the *minus* value of the objective at the current basic feasible solution augmented by the reduced costs associated with this basis; rows $1, 2, \dots$ are labeled by the current basic variables and contain the entries of the vector $[A_I]^{-1}b$ augmented by the entries of the corresponding rows in $[A_I]^{-1}A$. Since $(*)$ is equivalent to the original problem, we can think about every iteration as about *the very first iteration of the method as applied to* $(*)$, and our goal is to update the tableau representing the current equivalent reformulation $(*)$ of the problem of interest into the tableau representing the next equivalent reformulation of the problem of interest.

Now let us work out a numerical example. The initial program is

$$\begin{array}{ll} \max & 5x_1 + 3x_2 + 6x_3 \\ \text{subject to} & \\ & \begin{array}{r} x_1 + x_2 + 2x_3 \leq 30 \\ 4x_1 + x_2 + 4x_3 \leq 60 \\ 2x_1 + x_2 + \ \ x_3 \leq 30 \\ x_1, x_2, x_3 \geq 0 \end{array} \end{array}$$

We introduce slack variables to convert the problem to the standard form, thus arriving at the program

$$\begin{array}{ll} \max & 5x_1 + 3x_2 + 6x_3 \\ \text{subject to} & \\ & x_1 + x_2 + 2x_3 + x_4 \qquad\qquad = 30 \\ & 4x_1 + x_2 + 4x_3 \qquad + x_5 \qquad = 60 \\ & 2x_1 + x_2 + \ x_3 \qquad\qquad + x_6 = 30 \\ & \quad x_1, ..., x_6 \geq 0 \end{array}$$

which allows us to point out a starting basis $I = \{4, 5, 6\}$ and the starting basic feasible solution x^I with nonzero entries $x_4 = 30$, $x_5 = 60$, $x_6 = 30$. The first tableau is

	x_1	x_2	x_3	x_4	x_5	x_6
0	5	3	6	0	0	0
$x_4 = 30$	1	1	2	1	0	0
$x_5 = 60$	$\underline{4}$	1	4	0	1	0
$x_6 = 30$	2	1	1	0	0	1

Note: the top – incomplete – row of the tableau contains labels of the columns (the seroth column haz no label); these labels go from tableau to tableau.

The reduced cost of x_1 is positive; let this be the variable entering the basis. When trying to replace $x_1 = 0$ with $x_1 = t \geq 0$, keeping x_2 and x_3 zeros, the basic variables x_4, x_5, x_6 start to change: $x_4(t) = 30 - 1 \cdot t$, $x_5(t) = 60 - 4 \cdot t$, $x_6(t) = 30 - 2 \cdot t$. The largest t which keeps all these variables nonnegative is $t_* = 60/4 = 30/2 = 15$, and it is clear where this value comes from: this is just the minimum, over all labelled rows in the tableau *containing positive entries in the pivoting column*[4] (in our case, the column of x_1), of the ratios "value of the basic variable labelling the row" (shown in the zeroth column of the tableau) to the entry in the intersection of the row and the pivoting column:

$$t_* = \min \left[\frac{30}{1}, \frac{60}{4}, \frac{30}{2} \right]$$

In fact, the minimum of this ratio is achieved simultaneously in two rows, and this is in our will to decide which one of the current basic

[4]Were there no positive entries in the cells in question (which is not the case in our example), we would terminate with the claim that the problem is unbounded.

variables, x_5 or x_6, is leaving the basis. Let us say this is x_5; thus, at the first iteration x_1 enters the basis, and x_5 leaves it. so that the new basis is $I^+ = \{4, 1, 6\}$. It remains to update the tableau, that is, to replace the rows in the tableau representing the equality constraints by their linear combinations in order to get the unit matrix (which in the initial tableau is in the columns 4,5,6) in the columns 4, 1, 6 corresponding to our new basis, and augment this transformation by updating the column of basic variables and the "zeroth row" containing the minus value of the objective and the reduced costs. The rules for this updating are as follows (check their validity!):

- we mark somehow the *pivoting element* — the element in the pivoting column and the row of the basic variable which leaves the basis (in our tableau, the pivoting element is underlined)
- we divide the entire *pivoting row* (i.e., the row of the pivoting element) by the pivoting element, thus making the pivoting element equal to 1; note that in the "basic variables" part of the pivoting row we get exactly t_*, that is, the would-be value of the variable (x_1) which enters the basis; we update accordingly the "basic variable label" of this row, thus arriving at the intermediate tableau

	x_1	x_2	x_3	x_4	x_5	x_6
0	5	3	6	0	0	0
$x_4 = 30$	1	1	2	1	0	0
$x_1 = 15$	$\underline{1}$	0.25	1	0	0.25	0
$x_6 = 30$	2	1	1	0	0	1

- finally, we subtract from every non-pivoting row of the intermediate tableau a multiple of the updated pivoting row, the coefficient of the multiple being the entry in the pivoting column of the non-pivoting row we are processing; as a result, all entries in the pivoting column, aside of the one of the pivoting row, become zeros. What we get is nothing but our new tableau:

	x_1	x_2	x_3	x_4	x_5	x_6
-75	0	1.75	1	0	-1.25	0
$x_4 = 15$	0	0.75	$\underline{1}$	1	-0.25	0
$x_1 = 15$	1	0.25	1	0	0.25	0
$x_6 = 0$	0	0.5	-1	0	-0.5	1

Now we process the new tableau exactly in the same fashion as the initial one:

— there are two variables with positive reduced cost — x_2 and x_3; we choose one of them, say, x_3, as the variable entering the basis, so that the pivoting column is the column of x_3
— we divide the values of basic variables shown in the tableau by the corresponding values in the pivoting column, *skipping the divisions by nonpositive entries of the latter column*; the results are $\frac{15}{1} = 15$ in the first row and $\frac{15}{1} = 15$ in the second row. We then call the row corresponding to the smallest ratio our new pivoting row; the element in intersection of this row and pivoting column is our new pivoting element. In our example, we can choose both the first and the second row as the pivoting one; let our choice be the first row
— we now mark the pivoting element, divide by it the pivoting row (with our numbers, this does not change the pivoting row) and replace the basic variable label x_4 in this row (x_4 is the variable which leaves the basis) with the label x_3 (this is the variable which enters the basis, so that the new basis is $\{3, 1, 6\}$)
— finally, we subtract from all non-pivoting rows of the tableau, including the zeroth row, the multiples of the (already updated) pivoting row to zero out entries in the pivoting column, thus arriving at the new tableau

	x_1	x_2	x_3	x_4	x_5	x_6
-90	0	1	0	-1	-1	0
$x_3 = 15$	0	0.75	1	1	-2	0
$x_1 = 0$	1	-1	0	-1	1	0
$x_6 = 15$	0	$\underline{1.25}$	0	1	-0.75	1

In the zeroth row of the new tableau, there is only one positive reduced cost, the one of x_2, meaning that x_2 should enter the basis. The pivoting row corresponds now to the minimum of the ratios $\frac{15}{0.75}$ (row of x_3) and $\frac{15}{1.25}$ (row of x_6), that is, the pivoting row is the one of x_6, the pivoting element is 1.25, and the variable leaving the basis is x_6 (so that the new basis is $\{3, 1, 2\}$). We proceed by

— dividing the pivoting row by the pivoting element and updating the basic variable label in the row from x_6 (this variable leaves the basis) to x_2 (this is the variable which enters the basis). The updated pivoting row is

$x_2 = 12$	0	$\underline{1}$	0	0.8	−0.6	0.8

We then subtract from all non-pivoting rows of the tableau multiples of the just transformed pivoting row to zero out the corresponding entries in the pivoting column. The result is the new tableau

	x_1	x_2	x_3	x_4	x_5	x_6
−102	0	0	0	−1.8	−0.4	−0.8
$x_3 = 6$	0	0	1	0.4	−0.4	−0.8
$x_1 = 6$	1	0	0	−0.6	0.2	−0.6
$x_2 = 12$	0	1	0	0.8	0.2	0.4

In this tableau, all reduced costs are nonpositive, meaning that we have built the optimal solution to the problem, This optimal solution is

$$x_1 = 6, \quad x_2 = 12, \quad x_3 = 6, \quad x_5 = x_6 = x_7 = 0$$

(this is said by the column of the values of basic variables), the optimal value is 102 (*minus* the number in very first cell of the zeroth row of the tableau) and the vector of the reduced costs $c - A^T y_*$ (that is, λ_g in (4.2)) corresponding to the dual optimal solution y_* is the vector $[0; 0; 0; -1.8; -0.4; -0.8]$ the transpose of which we see in the "reduced costs" part of the zeroth row of the final tableau.

4.3.3 *Preventing cycling*

We have seen that the Primal Simplex method, *if it terminates*, solves the problem it is applied to, and that the method definitely terminates when the program is primal nondegenerate (i.e., all primal basic feasible solutions have exactly m nonzero entries). The degenerate case needs special treatment, since here, without additional precautions, the method indeed can "loop forever," staying at the same degenerate and nonoptimal basic feasible solution; all that changes from step to step is the bases specifying this solution, but not the solution itself. Fortunately, it is easy to prevent cycling by applying the *lexicographic pivoting rule*, which we are about to explain.

Lexicographic order: We start with the notion of the *lexicographic order* on $\mathbf{R}^n$. Specifically, we say that vector $u \in \mathbf{R}^n$ *lexicographically dominates* vector $v \in \mathbf{R}^n$ (notation: $u \geq_L v$) if either $u = v$, or the first nonzero entry in the difference $u - v$ is positive. We write $u >_L v$, when $u \geq_L v$ and $u \neq v$, same as writing $v \leq_L u$ and $v <_L u$ as equivalents of $u \geq_L v$, $u >_L v$, respectively. For example, here is a sample of valid lexicographic inequalities:

$$[1;2;3] \leq_L [1;2;4]; [1;2;3] <_L [2;-1;100]; [2;-1;10] >_L [2;-1;9].$$

Note that the lexicographic inequalities follow the same arithmetic as the usual arithmetic inequalities $\leq$, $<$ between reals and the coordinate-wise inequalities $u \leq v$, $u < v$ for vectors, e.g.,

$$\begin{array}{l}
u \leq_L u \text{ [reflexivity]},\\
(u \geq_L v) \,\&\, (v \leq_L u) \Rightarrow u = v \text{ [anti-symmetry]},\\
(u \leq_L v) \,\&\, (v \leq_L w) \Rightarrow u \leq_L w \text{ [transitivity]},\\
\left.\begin{array}{l} u <_L v,\, \mathbf{R} \ni \lambda > 0 \Rightarrow \lambda u <_L \lambda v;\\ u \leq_L v, u' <_L v' \Rightarrow u + u' <_L v + v' \end{array}\right\} \text{[compatibility with linear operations]}.
\end{array}$$

Exactly as in the case of arithmetic inequalities with reals, and in contrast to what happens to the coordinate-wise vector inequality $\leq$ between vectors, the lexicographic order is *complete* — for every pair u, v of vectors of the same dimension, we have either $u <_L v$, or $u = v$, or $u >_L v$, and these three possibilities are mutually exclusive.

Lexicographic pivoting rule: This is a specific rule for choosing a pivoting element in the Primal Simplex method in the case when such a choice is necessary (that is, the current tableau contains positive reduced costs). The rule is as follows:

> (L) *Given the current tableau which contains positive reduced costs c_j^I, choose as the index of the pivoting column a j such that $c_j^I > 0$.*
>
> *Denote by u_i the entry of the pivoting column in row i, $i = 1, \ldots, m$, and let some of u_i be positive* (recall that otherwise the PSM terminates with the (correct) claim that the problem is unbounded). *Normalize every row i with $u_i > 0$ by dividing all its entries, including those in the zeroth column, by u_i, and choose among the resulting $(n+1)$-dimensional row vectors the smallest w.r.t the lexicographic order, let its index be i_*. The pivoting row is the one with index i_*, so that the basic variable (the one which labels the row) leaves the basis, while the variable x_j enters the basis.*

Observe that at a nonterminal iteration, the outlined rule defines i_* in a unique fashion, and that it is compatible with the construction of the Simplex method, that is, i_* corresponds to the minimal, over the rows with $u_i > 0$, ratio of the zeroth entry in the row to the element of the row in the pivoting column (in the description of an iteration of the PSM, these ratios were called t_i, and their minimum was called t_*).

Indeed, taking into account that the lexicographic order is complete, in order to prove that i_* is uniquely defined it suffices to verify that the normalizations of the rows $1, \ldots, m$ in the tableau are distinct from each other, or, which is the same, that before normalization, these rows were not proportional to each other. The latter is evident (look what happens in the columns indexed by basic variables). Now, the zeroth entries of the normalized rows i with $u_i > 0$ are exactly the quantities t_i from the description of a simplex iteration, and by definition of the lexicographic order, the lexicographically minimal among these normalized rows corresponds to the smallest value t_* of t_i's, as claimed.

Thus, in the context of PSM, the lexicographic pivoting rule is completely legitimate. Its role stems from the following.

Theorem 4.1. *Let the Primal Simplex method with pivoting rule (L) be initialized by a primal basic feasible solution, and, in addition, let all rows in the initial tableau, except for the zeroth row, be lexicographically positive. Then*

(i) *every row in the tableau, except for the zeroth one, remains lexicographically positive at all nonterminal iterations of the algorithm;*
(ii) *when passing from the current tableau to the next one, if any, the vector of reduced costs strictly lexicographically decreases;*
(iii) *the method terminates in finitely many iterations.*

Proof. (i) It suffices to prove that if all rows with indices $i = 1, 2, \ldots, m$ in the current tableau are lexicographically positive and the next tableau does exist (i.e., the method does not terminate at the iteration in question), then every row i, $i = 1, \ldots, m$, in the next tableau will also be lexicographically positive. Let the index of the pivoting row be ℓ, and the index of the pivoting column be j. Denoting $u_1, \ldots, u_m$ the entries in the pivoting column and ith row, and invoking the lexicographic pivoting rule, we have

$$u_\ell > 0 \,\&\, \frac{\ell\text{th row}}{u_\ell} <_L \frac{i\text{th row}}{u_i} \quad \forall (i : i \neq \ell \& u_i > 0).$$

Let r_i be ith row in the current tableau, and r_i^+ be the ith row in the next tableau, $1 \leq i \leq m$. Then by description of the method:

- when $i = \ell$, $r_i^+ = u_\ell^{-1} r_\ell$, so that $r_\ell^+ >_L 0$ due to $r_\ell >_L 0$ (recall that we are under the assumption that $r_i >_L 0$ for all $1 \leq i \leq m$);
- when $i \neq \ell$ and $u_i \leq 0$, we have $r_i^+ = r_i - \frac{u_i}{u_\ell} r_\ell >_L 0$ due to $r_i >_L 0$, $r_\ell >_L 0$ and $-u_i/u_\ell \geq 0$;
- when $i \neq \ell$ and $u_i > 0$, we have $r_i^+ = r_i - \frac{u_i}{u_\ell} r_\ell^+$, or

$$\frac{1}{u_i} r_i^+ = \frac{1}{u_i} r_i - \frac{1}{u_\ell} r_\ell.$$

  By the lexicographic pivoting rule, the ℓth normalized row $\frac{1}{u_\ell} r_\ell$ is strictly less lexicographically than the ith normalized row $\frac{1}{u_i} r_i$ (due to $u_i > 0$ and $i \neq \ell$), whence $\frac{1}{u_i} r_i^+ >_L 0$, whence $r_i^+ >_L 0$ as well.

(i) is proved.

(ii) By the description of the method, the reduced costs c^I and c^{I^+} in the current and the next tableaus are linked by the relation

$$c^{I^+} = c^I - \frac{c_j^I}{u_\ell} [r_\ell]^T.$$

Since $c_j^I > 0$, $u_\ell > 0$, and $r_\ell >_L 0$, we have $c^{I^+} <_L c^I$. (ii) is proved.

(iii) As (ii) implies that with the lexicographic pivoting rule in force, no vector of reduced costs can be generated twice, consequently, with this rule no basis can be visited twice (since the vector of reduced costs associated with a basis is uniquely defined by this basis). The latter combines with the fact that the number of bases is finite to imply that the method terminates after finitely many steps. □

In view of Theorem 4.1, all we need in order to guarantee finiteness of the Primal Simplex method, is to ensure lexicographic positivity of rows $1, 2, \ldots, m$ in the initial tableau. This can be achieved as follows. Recall that for the time being we have designed the PSM only for the case when we have at our disposal from the very beginning an initial basic feasible solution. Renaming the variables, we can assume w.l.o.g. that the initial basis is $I = \{1, \ldots, m\}$, so that the initial contents of the A-part of the tableau is $[A_I]^{-1}A = [I_m, [A_I]^{-1}A_{\bar{I}}]$. Thus, ith row in the initial tableau, $1 \leq i \leq m$, is of the form $[a, 0, \ldots, 0, 1, \ldots]$, where $a \geq 0$ is the value of a certain basic variable in the initial primal basic feasible solution; a row of this type clearly is lexicographically positive.

A simpler pivoting rule, due to Bland, which provably prevents cycling is as follows:

> **Smallest subscript pivoting rule:** *Given a current tableau containing positive reduced costs, find the smallest index j such that jth reduced cost is positive; take x_j as the variable to enter the basis. Assuming that this choice of the pivoting column does not lead to immediate termination due to the problem's unboundedness, choose among all legitimate (i.e., compatible with the description of the PSM) candidates for the role of the pivoting row the one with the smallest index and use it as the pivoting row.*

4.3.4 *How to start the PSM*

Our description of the PSM is still incomplete; we have assumed that we know in advance a primal basic feasible solution to the problem, and to find this solution is a nontrivial (and not always achievable — what if the problem is infeasible?) task. Luckily enough, this problem can be resolved by the same PSM as applied to a properly defined auxiliary problem. We are about to describe one of the implementations of this type — the *two-phase* PSM. As applied to the LO program in the standard form

$$\max_x \left\{ c^T x : Ax = b, x \geq 0 \right\} \tag{4.6}$$

with $m \times n$ matrix A of rank m, the method works as follows.

First phase: Multiplying, if necessary, some of the equality constraints by -1, we can enforce b to be ≥ 0. After this, we introduce

m additional variables $s_1, \ldots, s_m$ and build the auxiliary LO program

$$\text{Opt} = \max_{x,s} \left\{ -\sum_{i=1}^{m} s_i : Ax + s = b, x \geq 0, s \geq 0 \right\}. \tag{4.7}$$

Note that

(i) the original problem is feasible iff the optimal value in the auxiliary problem is 0, and
(ii) the auxiliary problem is feasible and bounded, and we can point out its primal basic feasible solution, specifically, $x = 0, s = b$, along with the associated basis $I = \{n+1, \ldots, n+m\}$.

Using $[0; \ldots; 0; b]$ as the starting point, we solve the (solvable!) auxiliary problem by the PSM, eventually arriving at its optimal solution $[x_*; s_*]$. If $s_* \neq 0$, the optimal value in the auxiliary problem is negative, meaning, by (i), that the original problem is infeasible.

In the case in question, the optimal solution y_* to the dual to (4.7), which we get as a byproduct of solving (4.7) by the PSM, satisfies

$$d - [A, I_m]^T y_* \leq 0, y_*^T b = \text{Opt} < 0,$$

where $d = [0; \ldots; 0; -1; \ldots; -1]$ is the objective of the auxiliary problem. The vector inequality here implies that $\mu := A^T y_* \geq 0$. Now, taking the weighted sum of the equality constraints in the original problem, the weights being $-[y_*]_i$, and adding the weighted sum of the inequalities $x_j \geq 0$, the weights being μ_j, we get the inequality

$$0^T x \equiv [\mu - A^T y_*]^T x \geq [-y_*]^T b;$$

since $y_*^T b = \text{Opt} < 0$, this inequality is contradictory, that is, we not only know that the original problem is infeasible, but have at our disposal a certificate for infeasibility.

Second phase: It remains to consider the case when $s_* = 0$. In this case, x_* is a feasible solution to the original problem, and the columns of A corresponding to positive entries in x_*, if any, are linearly independent (since $[x_*; s_*]$ is a basic solution to (4.7)). Since A is of rank m, we can include these columns in the collection of m linearly independent columns of A, thus arriving at a basis of A; clearly, x_* is the basic feasible solution to the problem of interest associated with this basis. Now, we have at our disposal both a primal basic feasible solution x_* and an associated basis for (4.6), and we solve this problem by the PSM.

4.3.5 *Dual simplex method*

Recall the geometric interpretation of the primal–dual pair of LO programs: we have two affine planes — the primal feasible plane $\mathcal{M}_P$ and the dual feasible plane $\mathcal{M}_D$ such that the corresponding linear subspaces are orthogonal complements to each other; our goal is to find in the intersections of the planes with the nonnegative orthant two vectors orthogonal to each other. When the primal problem is in the standard form (4.1), $\mathcal{M}_P$ is given as $\{x : Ax = b\}$, while $\mathcal{M}_D$ is the plane of *dual slacks* — vectors which can be represented as $A^T\lambda - c$. The geometry of the Primal Simplex method is as follows: at every step, we have at our disposal basis I. This basis specifies two vectors orthogonal to each other from $\mathcal{M}_P$ and $\mathcal{M}_D$:

- the primal solution $x^I = (x_I = [A_I]^{-1}b, x_{\overline{I}} = 0)$. This solution satisfies the necessary condition for being a vertex of the primal feasible set: among the constraints $Ax = b, x \geq 0$ which cut this set off $\mathbf{R}^n$, n constraints with linearly independent vectors of coefficients, specifically, the constraints $Ax = b$ and $x_j \geq 0$, $j \notin I$, are satisfied as equalities;
- the dual solution $\lambda^I = [A_I]^{-T}c_I$. This solution satisfies the necessary condition for being a vertex of the dual feasible set: among the constraints $A^T\lambda - c \geq 0$ which cut this set off $\mathbf{R}^m$, m constraints with linearly independent vectors of coefficients, specifically, the constraints $[A^T\lambda - c]_j \geq 0$ with $j \in I$, are satisfied as equalities.

By construction, the primal solution x^I belongs to $\mathcal{M}_P$, the dual slack $s^I = A^T\lambda^I - c = -c^I$ belongs to $\mathcal{M}_D$, and x^I, s^I satisfy the complementary slackness condition — $x^I_j s^I_j = 0$ for all j.

Now, in the Primal Simplex Method x^I is maintained to be primal feasible, while λ^I is not necessarily dual feasible. When λ^I happens to be dual feasible (that is, $s^I \geq 0$, or, equivalently, $c^I \leq 0$), we are done — we have built a pair of nonnegative vectors orthogonal to each other, $x^I \in \mathcal{M}_P$ and $s^I \in \mathcal{M}_D$. Until it happens, we update the basis in such a way that the primal objective either remains the same (which is possible only when the current primal basic feasible solution x^I is degenerate), or is improved.

An alternative, "symmetric" course of actions would be to maintain feasibility of the dual solution λ^I, sacrificing the primal feasibility of x^I. When x^I happens to be feasible, we are done, and while it does not happen, we update the basis in such a way that the dual objective improves. This is exactly what is going on in the Dual Simplex method. One could think that this method does not deserve a separate presentation, since geometrically this is nothing as the Primal Simplex method applied to the swapped pair of the problems — what used to be primal is now called dual, and vice versa. This conclusion is not completely correct, since an algorithm cannot directly access geometric entities; what it needs are algebraic descriptions of these entities, and in this respect primal–dual symmetry is not ideal — the algebraic representation of the primal feasible plane (for programs in the standard form we stick to — the representation by system of linear equations) is not identical to the one of the dual feasible plane (which is represented in the "parametric" form $\{s = A^T\lambda - c, \lambda \in \mathbf{R}^m\}$). By the outlined reasons, it makes sense to present an explicit algorithmic description of the Dual Simplex method. This is what we do next.

4.3.5.1 *A step of the dual Simplex method*

At a step of the Dual Simplex method (DSM) as applied to (4.1), we have ay our disposal current basis I (i.e., an m-element subset $\{i_1, \dots, i_m\}$ of $\{1, \dots, n\}$ such that the columns A_j of A with indices from I are linearly independent) along with the corresponding basic primal solution $x^I = (x_I = [A_I]^{-1}b, x_{\overline{I}} = 0)$, *which is not necessarily nonnegative*, and the basic dual solution $\lambda^I = [A_I]^{-T}c_I$, which is dual feasible: $c^I = c - A^T\lambda^I \leq 0$. We associate with the basis I a tableau, exactly as in the PSM:

$-c_I^T x_I^I$	$c_1^I \dots c_n^I$
$x_{i_1}^I$	$[[A_I]^{-1}A]^1$
$\vdots$	$\dots$
$x_{i_m}^I$	$[[A_I]^{-1}A]^m$

the difference from the PSM is that now all entries $c_1^I, \ldots, c_n^I$ in the zeroth row (except for the entry in the zeroth column) are nonpositive, while some of the entries $x_{i_1}^I, \ldots, x_{i_m}^I$ in the zeroth column can be negative.

A. It may happen that all $x_{i_\ell}^I$, $1 \leq \ell \leq m$, are nonnegative. In this case, we are done: λ^I is a feasible solution to the dual problem, x^I is a feasible solution to the primal problem, and x^I and the dual slack $-c^I$ satisfy the complementary slackness condition (since by construction of λ^I, $c_{i_\ell}^I = 0$ for all $\ell \leq m$), i.e., x^I and λ^I are optimal solutions to the respective problems.

B. Now, assume that some of $x_{i_\ell}^I$ are negative. Let us pick ℓ with this property and call the ℓth row of the tableau the *pivoting* row. We intend to eliminate i_ℓ from the basis. To this end, let us look at what happens when we update λ^I in such a way that i_ℓth reduced cost (which currently is zero) becomes $-t$, $t \geq 0$, and the reduced costs in the basis columns distinct from the i_ℓth column are kept zeros. In other words, we update λ^I according to $\lambda^I \mapsto \lambda(t) := \lambda^I + t[A_I]^{-T}e_\ell$. Observe that as a result of this shift, the dual objective reduces (i.e., improves); indeed,

$$b^T\lambda(t) = b^T\lambda^I + tb^T[A_I]^{-T}e_\ell = b^T\lambda^I + te_\ell^T[[A_I]^{-1}b] = b^T\lambda^I + tx_{i_\ell}^I,$$

and $x_{i_\ell}^I < 0$. Now, let us look at how passing from λ^I to $\lambda(t)$ affects the vector of reduced costs. These costs $c_j(t) = [c - A^T\lambda(t)]_j$ vary as follows:

- The reduced cost in the basic column i_ℓ becomes $-t$;
- The reduced costs in all other basic columns stay zero;
- The reduced cost in a nonbasic column j varies according to $c_j(t) = c_j^I - tA_j^T[A_I]^{-T}e_\ell = c_j^I - tp_j$, where p_j is the entry in the pivoting row of the tableau.

There are two possibilities:

B.1. All entries p_j, $j \geq 1$, in the pivoting row, except for those in the basic columns, are nonnegative. In this case, all reduced costs $c_j(t)$, in basic and nonbasic columns alike, are nonpositive for all $t \geq 0$, in other words, we have found an improving ray $\{\lambda(t) : t \geq 0\}$ in the dual problem; this ray certifies that the dual problem is unbounded, whence the primal problem is infeasible.

B.2. Among the entries p_j, $j \geq 1$, in the pivoting row, aside of those in the basic columns, there are negative ones. Then there exists the largest nonnegative value $\bar{t}$ of t for which $c_j(t)$ still are nonpositive for all j, specifically, the value

$$\bar{t} = \min_{j\geq 1: p_j<0} \frac{c_j}{p_j} = \frac{c_{j_*}}{p_{j_*}} \qquad [1 \leq j_* \notin I, p_{j_*} < 0],$$

(note that we can safely write $\min_{j:p_j<0} \frac{c_j}{p_j}$ instead of $\min_{j\notin I:p_j<0} \frac{c_j}{p_j}$, since the entries of the pivoting row in the basic columns are nonnegative, specifically, $m-1$ of them are zero, and the remaining one is equal to 1). Observe that when we exclude from the basis I the basic column A_{i_ℓ} and add to the basis the index $j_* \notin I$, we get a basis, and this is the new basis I^+ we deal with, with $\lambda^{I^+} = \lambda(\bar{t})$ being a new dual basic feasible solution.

Indeed, assuming that I^+ is not a basis, the column A_{j_*} is a linear combination of the $m-1$ columns A_{i_ν}, $1 \leq \nu \leq m$, $\nu \neq \ell$: $A_{j_*} = \sum_{\substack{\nu:1\leq\nu\leq m\\ \nu\neq\ell}} y_\nu A_{i_\nu}$. But then

$$p_{j_*} = A_{j_*}^T [A_I]^{-T} e_\ell = \sum_{\substack{\nu:1\leq\nu\leq m\\ \nu\neq\ell}} y_\nu [[A_I]^{-1} A_{i_\nu}]^T e_\ell = \sum_{\substack{\nu:1\leq\nu\leq m\\ \nu\neq\ell}} y_\nu e_\nu^T e_\ell = 0,$$

which is impossible, since p_{j_*} is negative.

We then update the information in the tableau, exactly as in the PSM, specifically,

- Call the column j_* the *pivoting column*, and the entry p_{j_*} in the intersection of the pivoting row and the pivoting column the pivoting entry;
- Replace the pivoting row with its normalization obtained by dividing all entries in the row, including the one in the zero column, by the pivoting entry, and replace the label x_{i_ℓ} of this row with x_{j_*};
- Update all other rows of the tableau, including the zeroth, by subtracting from them multiples of the normalized pivoting row, where

the multiples are chosen in such a way that the entry of the updated row in the pivoting column j_* becomes zero, and proceed to the next iteration of the DSM.

Numerical illustration: Let us illustrate the DSM rules with an example. The current tableau is

	x_1	x_2	x_3	x_4	x_5	x_6
8	-1	-1	-2	0	0	0
$x_4 = 20$	1	2	2	1	0	0
$x_5 = -20$	-2	-1	2	0	1	0
$x_6 = -1$	2	2	-1	0	0	1

which corresponds to $I = \{4, 5, 6\}$. The vector of reduced costs is nonpositive, and its basic entries are zero (a must for DSM).

1. Some of the entries in the basic primal solution (zeroth column) are negative. We select one of them, say, the entry in the row labeled by x_5, and call the corresponding row the pivoting row. Variable x_5 will leave the basis.
2. If all entries in the pivoting row, except for the one in the zeroth column, were nonnegative, we would terminate, claiming the dual problem unbounded, and the primal, infeasible. In our example, that does not happen, and we should pass to the next step of the DSM. To this end we

   (1) select the *negative* entries in the pivoting row outside of the zeroth column (all of them are nonbasic!) and divide by them the corresponding entries in the zeroth row, thus getting *nonnegative* ratios, in our example, the ratios $1/2 = (-1)/(-2)$ and $1 = (-1)/(-1)$;

   (2) pick the smallest of the computed ratios and call the corresponding column — the column of x_1 — the pivoting column. Variable x_1 which labels this column will enter the basis.

3. It remains to update the tableau, which is done exactly as in the PSM:

   (1) we normalize the pivoting row by dividing its entries by the *pivoting element* (the one in the intersection of the pivoting row and pivoting column) and change the label of this row

(the new label is the variable x_1 which enters the basis):

	x_1	x_2	x_3	x_4	x_5	x_6
8	-1	-1	-2	0	0	0
$x_4 = 20$	1	2	2	1	0	0
$x_1 = 10$	1	0.5	-1	0	-0.5	0
$x_6 = -1$	2	2	-1	0	0	1

(2) we subtract from all non-pivoting rows multiples of the (normalized) pivoting row to zero out the non-pivoting entries in the pivoting column, thus arriving at the tableau

	x_1	x_2	x_3	x_4	x_5	x_6
18	0	-0.5	-3	0	-0.5	0
$x_4 = 10$	0	1.5	3	1	0.5	0
$x_1 = 10$	1	0.5	-1	0	-0.5	0
$x_6 = -21$	0	1	1	0	1	1

The step of DSM is over. Our new basis is $I = \{4, 1, 6\}$, the vector of reduced costs ($[0, -0.5; -3; 0; -0.5; 0]$) is in the zeroth row part of nonzeroth columns, and the basic entries of the primal solution are in the zeroth column, so that this solution is $[10; 0; 0; 10; 0; -21]$.

To carry out the next step of the DSM, observe that in the new basic primal solution there is still a negative variable, specifically, x_6. Its row is the pivoting one. However, all entries in this row, except the one in the zeroth column, are nonnegative. Thus, the tableau we have ended up with witnesses that the dual problem is unbounded, and the primal is infeasible, and we terminate with this conclusion.

4.3.5.2 *Dual Simplex method: Convergence*

Observe that as a result of the iteration, we either

(a) conclude correctly that the current dual basic feasible solution is optimal and augment it by an optimal primal basic solution, or
(b) conclude correctly that the dual problem is unbounded (and therefore the primal problem is unsolvable), or

(c) pass from the current dual basic feasible solution to another solution of the same type, but with a strictly smaller value of the dual objective, or keep the dual solution intact and update the basis only. The second option is possible only when λ^I is a *degenerate* solution, meaning that not all nonbasic reduced costs corresponding to λ^I are strictly negative.

In view of these observations, exactly the same argument as in the PSM case leads to the following conclusion:

Corollary 4.2. *The Dual Simplex method, initiated at a dual basic feasible solution of* (4.3), *possesses the following property: if the method terminates at all, the result upon termination is either a pair of primal and dual basic feasible solutions which are optimal for the primal and dual programs* (4.1), (4.3), *or is a correct claim that the primal problem is infeasible. In the second case, the method produces a ray which is contained in the feasible set of the dual problem and is a ray along which the dual objective decreases, so that this ray is a certificate of primal infeasibility. In fact, this ray is an extreme ray of the recessive cone of the dual feasible set.*

The method definitely terminates in finitely many steps, provided that the program is dual nondegenerate (*i.e., all dual basic feasible solutions are so*).

Similar to the PSM, the Dual Simplex method admits pivoting rules which prevent cycling, various techniques for building the initial basic dual feasible solution, etc., see Bertsimas and Tsitsiklis (1997).

4.3.6 *Warm start*

Assume we have solved the LO program (4.1) to optimality by either Primal, or Dual Simplex method, and thus have at our disposal a basis I which gives rise to feasible primal and dual solutions satisfying the complementary slackness condition and thus are optimal for the respective programs. We are about to understand how this basis helps to solve a "close" LO program. This necessity arises when we intend to solve a series of relatively close to each other LO programs, as it is the case, e.g., in branch- and bound-type methods of Integer Programming.

Recall that the optimal basis is characterized by two requirements:

$$[A_I]^{-1}b \geq 0 \qquad \text{[feasibility]},$$
$$c - A^T[A_I]^{-T}c_I \leq 0 \quad \text{[optimality]}.$$

When these requirements are satisfied, $x^I = (x_I = [A_I]^{-1}b, x_{\bar{I}} = 0)$ is an optimal primal basic solution, and $\lambda_I = [A_I]^{-T}c_I$ is an optimal dual basic solution.

We now consider several typical updates of the problem.

4.3.6.1 *New variable is added*

Assume we extend x with a new decision variable x_{n+1}, A — with a new column A_{n+1} — and c — with a new entry c_{n+1} — keeping the rest of the data and the standard form of the problem intact. I still is a basis for the new problem, and we do not affect the feasibility condition. As far as the optimality condition is concerned, it remains satisfied when

$$c_{n+1} - A_{n+1}^T[A_I]^{-T}c_I \leq 0. \tag{!}$$

If this easy-to-verify condition holds true, x^I is an optimal solution to the new primal problem, and λ^I is an optimal solution to the new dual problem. If (!) is not satisfied, we can think of I as of the current basis for the PSM as applied to the new problem, and of x_{n+1} as of the variable entering this basis, and run PSM until the new problem is solved. This is called a "warm start," and usually with such a warm start the solution of the new problem takes significantly less time than if we were solving it from scratch.

4.3.6.2 *Changes in the cost vector c*

When we replace the cost vector c with $\bar{c} = c + t\delta c$, where δc is the "direction of change," and $t \in \mathbf{R}$ is the "step along this direction," the feasibility conditions remain intact; in order for optimality conditions to be satisfied, the perturbation should satisfy the requirement

$$\bar{c}^I = \bar{c} - A^T[A_I]^{-T}\bar{c}_I = c^I + t\left[\delta c - A^T[A_I]^{-T}\delta c_I\right] \leq 0.$$

Note that given δc and t, this fact is easy to verify; moreover, we can easily find the largest range T of values of t such that $\bar{c}^I$ has nonpositive nonbasic components. If the actual perturbation is "too large" to satisfy this requirement and some entries in the new vector of reduced costs are positive, we can run the PSM on the new problem starting with the basis I until the new problem is solved. This warm start usually saves a lot of computational effort as compared to solving the new problem from scratch.

Note that even when the shift $c \mapsto c + t\delta c$ keeps x^I primal optimal, the optimal dual solution can change — it becomes $[A_I]^{-T}[c_I + t\delta c_I]$, and thus changes, unless $t = 0$ of $\delta c_I = 0$.

4.3.6.3 *Changes in the right-hand side vector b*

Now, assume that we perturb the right-hand side vector b according to $b \mapsto b + t\delta b$. I still is a basis, and the optimality condition remains intact. In order for the feasibility condition to remain valid, we need

$$[A_I]^{-1}[b + t\delta b] \geq 0,$$

and here again we can easily find the largest range of values of t where the latter condition holds true. When t is in this range, we easily get a new primal optimal solution, while the optimal dual solution remains intact. If t is outside of the indicated range, we still have at our disposal a dual basic feasible solution for the new problem, and we can start the DSM at this solution to solve the new problem to optimality.

4.3.6.4 *Change in a nonbasic column of A*

When $j \notin I$, and we perturb jth column of A according to $A_j \mapsto A_j + t\delta a$, x^I remains a basic feasible solution to the new problem, A_I remains intact, and the only entity important for us, which can change is the jth entry in the vector of reduced costs, which changes according to

$$(c^I)_j \mapsto (c^I)_j - t(\delta a)^T [A_I]^{-T} c_I.$$

As above, we can easily find the largest range of values of t in which the perturbed reduced cost remains nonpositive, meaning that in this range both primal and dual optimal solutions remain intact. If perturbation in A_j runs out of this range, we can solve the new problem with the PSM started at the basis I, with the variable x_j entering the basis.

4.3.6.5 *New equality constraint is added to the primal problem*

Assume that A is augmented by a new row, a_{m+1}^T, and b, by the corresponding new entry b_{m+1}, so that the constraint matrix in the new problem is $A_+ = [A; a_{m+1}^T]$, and the right-hand side vector in the new problem is $b_+ = [b; b_{m+1}]$. In order to process the new problem, we check whether the rows in A_+ are linearly independent; if this is not the case, the new system of equality constraints is either infeasible, or the added equality constraint is redundant and does not affect the problem. When the rows of A_+ are linearly independent, we proceed by converting the optimal basic dual solution to the "old" problem into a feasible basic dual solution to the new problem, namely, as follows. We can easily find $\delta\lambda_i$, $1 \leq i \leq m$, such that $a_{m+1,j} = \sum_{i=1}^m \delta\lambda_i a_{ij}$ for all $j \in I$; indeed, $\delta\lambda = [A_I]^{-T}[a_{m+1}]_I$. Now, let $\bar{\lambda}$ be the optimal dual basic solution we end up with when solving the old problem, and let $\lambda(t) = [\bar{\lambda} + t\delta\lambda; -t]$. Then $[A_+]^T\lambda(t) = A^T\bar{\lambda} + tp$, $p = A^T\delta\lambda - a_{m+1}$. Observe that the entries of p with indices from I are zero, while p itself is nonzero, since the rows of A_+ are linearly independent. It follows that the reduced costs $c_j(t) = [c - A_+^T\lambda(t)]_j$ corresponding to the solution $\lambda(t)$ of the new dual problem remain zero when $j \in I$, while some of the $c_j(t)$ with $j \notin I$ are nonconstant linear functions of t. Let J be the set of those j for which $c_j(t)$ are nonconstant functions of t. When $t = 0$, all $c_j(t)$ are nonpositive; it follows that we can easily find a value $\bar{t} \in \mathbf{R}$ of t such that all $c_j(\bar{t})$ are nonpositive, and at least one of $c_j(\bar{t})$ with $j \in J$ is zero — let this value of j be denoted by j_*. Setting $I_+ = I \cup \{j_*\}$, it is immediately seen that the columns $[A_+]_j$, $j \in I_+$, of A_+ are linearly independent, so that I_+ is a basis of the new problem.

> Indeed, since the columns a_j, $j \in I$, of A are linearly independent, so are the columns $[A_+]_j$, $j \in I$. It follows that the only possibility for the columns $[A_+]_j$, $j \in I_+$, to be linearly dependent is that $[A_+]_{j_*} = \sum_{j\in I} y_j[A_+]_j$ for some y_i. But then $\lambda^T(t)[A_+]_{j_*} = \sum_{j\in I} y_j\lambda^T(t)[A_+]_j = \sum_{j\in I} y_j[A_j^T\bar{\lambda} + tp_j] = \sum_{j\in I} y_j[A_j^T\bar{\lambda}]$, since $p_j = 0$ for $j \in I$. We see that $\lambda^T(t)[A_+]_{j_*}$, and thus $c_{j_*}(t)$, is independent of t, which is not the case.

By construction, setting $\tilde{\lambda} = \lambda(\bar{t})$, we have that $c - A_+^T\tilde{\lambda} \leq 0$ and $(c - A_+^T\tilde{\lambda})_{I_+} = 0$, i.e., we have at our disposal a dual basic feasible solution associated with the basis I_+ of the new problem. We can now use I_+ and $\tilde{\lambda}$ to run the DSM on the new problem.

4.4 Exercises

Exercise 4.1.* Solve the LO program

$$\begin{array}{l} \max \ -6x_1 - 5x_2 - 2x_3 - x_4 - 2x_5 - x_6 \\ x_1 + x_2 + x_3 = 2 \\ x_4 + x_5 + x_6 = 3 \\ x_1 + x_4 = 1 \\ x_2 + x_5 = 2 \\ x \geq 0 \end{array}$$

by the Primal Simplex Method, the initial basis being $\{1, 2, 5, 6\}$.

Exercise 4.2.* Four students were solving a maximization LO program by the Primal Simplex method and arrived at intermediate tableaus as follows:

A.

	x_1	x_2	x_3	x_4	x_5	x_6
16	0	−2	0	0	4	0
$x_3 = 5$	0	−2	1	0	5	0
$x_1 = 2$	1	3	0	0	−6	0
$x_4 = 3$	0	1	0	1	4	0
$x_6 = -1$	0	1	0	0	2	1

B.

	x_1	x_2	x_3	x_4	x_5	x_6
16	0	−2	0	0	4	−1
$x_3 = 5$	0	−2	1	0	5	0
$x_1 = 2$	1	3	0	0	−6	0
$x_4 = 3$	0	1	0	1	4	0
$x_6 = 1$	0	1	0	0	2	1

C.

	x_1	x_2	x_3	x_4	x_5	x_6
16	0	−2	0	0	4	0
$x_3 = 5$	0	−2	1	0	5	0
$x_1 = 2$	1	3	0	0	−6	0
$x_4 = 3$	0	1	0	1	4	0
$x_6 = 1$	0	1	0	0	2	1

D.

	x_1	x_2	x_3	x_4	x_5	x_6
16	0	−2	0	0	4	0
$x_3 = 5$	0	−2	1	0	5	0
$x_1 = 2$	1	3	0	0	−6	−1
$x_4 = 3$	0	1	0	1	4	0
$x_6 = 1$	0	1	0	0	2	1

It is known that exactly one of the students did not make a mistake. Identify the correct tableau and complete the solution process.

Exercise 4.3.* Four students were solving a maximization LO program by the Dual Simplex method and arrived at intermediate tableaus as follows:

A.

	x_1	x_2	x_3	x_4	x_5	x_6
16	0	−2	0	0	−4	0
$x_3 = 5$	0	−2	1	0	5	0
$x_1 = 2$	1	3	0	0	−6	0
$x_4 = 3$	0	1	0	1	4	0
$x_6 = -1$	0	1	0	0	−2	1

B.

	x_1	x_2	x_3	x_4	x_5	x_6
16	0	−2	0	0	−4	−1
$x_3 = 5$	0	−2	1	0	5	0
$x_1 = 2$	1	3	0	0	−6	0
$x_4 = 3$	0	1	0	1	4	0
$x_6 = -1$	0	1	0	0	−2	1

C.

	x_1	x_2	x_3	x_4	x_5	x_6
16	0	−2	0	0	4	0
$x_3 = 5$	0	−2	1	0	5	0
$x_1 = 2$	1	3	0	0	−6	0
$x_4 = 3$	0	1	0	1	4	0
$x_6 = -1$	0	1	0	0	−2	1

D.

	x_1	x_2	x_3	x_4	x_5	x_6
16	0	−2	0	0	−4	0
$x_3 = 5$	0	−2	1	0	5	0
$x_1 = 2$	1	3	0	0	−6	−1
$x_4 = 3$	0	1	0	1	4	0
$x_6 = -1$	0	1	0	0	−2	1

It is known that exactly one of the students did not make a mistake. Identify the correct tableau and complete the solution process.

Exercise 4.4.[†] (1) Consider the following parametric LO program:

$$\max_{x,s}\left\{c^T x - M\sum_{i=1}^{m} s_i : Ax + s = b, x \geq 0, s \geq 0\right\} \qquad (P_M)$$

with $b \geq 0$, along with the LO program

$$\max_{x}\left\{c^T x : Ax = b, x \geq 0\right\} \qquad (P)$$

with $m \times n$ matrix A of rank m. Prove that if (P) is solvable, then there exists $M_* > 0$ such that whenever $M > M_*$, problem (P_M) is solvable, the optimal values of (P_M) and (P) are the same, and all optimal solutions (x, s) to (P_M) are of the form $(x, 0)$, where x is an optimal solution to (P).

(2) The result of (1) suggests the following single-phase "Big M"-implementation of the PSM: Given a standard form LO program (P) with $b \geq 0$ (the latter can always be achieved by multiplying appropriate equality constraints by -1), we associate with it the program (P_M) for which we can easily point out an initial basis comprising (indexes of) the slack variables s along with the associated basic feasible solution $(x = 0, s = b)$. We then solve the resulting problem (P_M) thinking of M as about a large constant. From the description of the PSM it follows that M will affect only the subsequent vectors of reduced costs (and the decisions we make based on these costs) and will appear linearly in the reduced costs. Now, making decisions based on reduced costs requires to compare them with zero, and here we think about M as a large constant, meaning that reduced cost of the

form $a + pM$ (a and p are known reals) should be treated as positive when $p > 0$ and as negative when $p < 0$; when $p = 0$, the decision is made based on what a is. Assuming that (P) is solvable, by (1) the problem (P_M) for all large enough values of M is solvable, and every optimal solution to this problem is $(x, 0)$, where x is an optimal solution to (P). In other words, when (P) is solvable, running the big M method, modulo highly unlikely cycling, will result in arriving at a basic feasible solution with the s-components equal to 0 and the reduced costs of the form $a_i + p_i M$ where for every i either $a_i \leq 0$ and $p_i \leq 0$, or $p_i < 0$, meaning that the current basic solution is feasible for (P) and optimal for all problems (P_M) with large enough values of M and therefore is optimal for (P).

Use the big M version of the PSM to solve the following problem:

$$\begin{aligned}
&\max \; -2x_1 + x_2 + x_3 - 6x_4 \\
&x_1 + x_2 = 2 \\
&x_3 + x_4 = 2 \\
&x_1 + x_3 = 2 \\
&x \geq 0.
\end{aligned}$$

Chapter 5

The Network Simplex Algorithm

In this chapter, we present a version of the Simplex method for solving the *Network Flow* problem — the single-commodity version of the multicommodity network flow problem (p. 16 in Section 1.2). The presence of add an additional structure — the graph underlying the problem — allows for significant modifications and simplifications in the prototype Simplex algorithm and thus deserves a dedicated description.

Our presentation follows Bertsimas and Tsitsiklis (1997, Chapter 7), where an interested reader can find a much broader picture of network flow algorithms.

5.1 Preliminaries on graphs

5.1.1 *Undirected graphs*

An *undirected graph* $G = (\mathcal{N}, \mathcal{E})$ is a pair of two finite sets. The first, $\mathcal{N}$, is called the set of *nodes* and can be an arbitrary *nonempty* finite set; without loss of generality, we can identify the nodes with the integers $1, \ldots, m$, where $m = \mathrm{Card}\,\mathcal{N}$ is the number of elements in $\mathcal{N}$. The elements of the second set $\mathcal{E}$, called *arcs*, are *two-element subsets* of the set of nodes; thus, an arc is an *un*ordered pair $\{i, j\}$ of two *distinct* nodes i, j. Note that $\mathcal{E}$, in contrast to $\mathcal{N}$, is allowed to be empty. We say that the nodes i, j are *incident* to the arc $\{i, j\}$, and the arc *links* the nodes i, j.

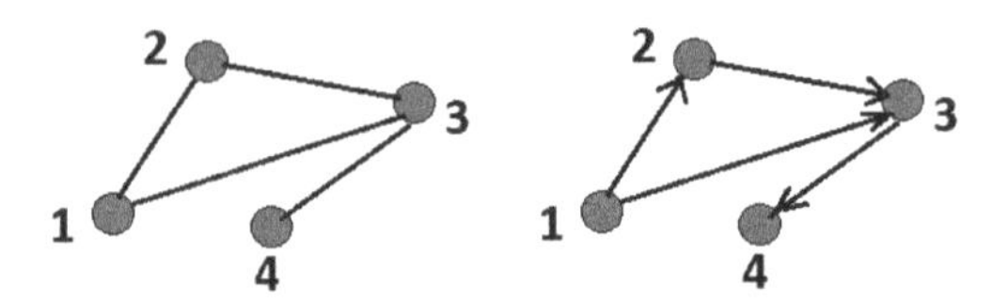

Fig. 5.1. Graphs: left — undirected graph; right — directed graph.

Walks, paths, cycles: A *walk* in an undirected graph G is an ordered collection $i_1, \ldots, i_k$ of nodes such that every two consecutive nodes in this collection are linked by an arc in G, that is, $\{i_s, i_{s+1}\} \in \mathcal{E}$, $1 \leq s \leq k-1$. A walk $i_1, \ldots, i_t$ is called a *path* if all the nodes $i_1, \ldots, i_t$ are distinct from each other. A walk $i_1, \ldots, t_t$ is called a *cycle* if the nodes $i_1, \ldots, i_{t-1}$ are distinct from each other, $i_t = i_1$ and, in addition, $t \geq 3$. For example, in the undirected graph depicted in Fig. 5.1(left):

- $\mathcal{N} = \{1,2,3,4\}$, $\mathcal{E} = \{\{1,2\},\{2,3\},\{1,3\},\{3,4\}\}$;
- $1,3,2,3,4$ is a walk, but not a path;
- $1,2,3,4$ is a path;
- $1,2,3,1$ is a cycle, while $1,2,3$ and $1,2,1$ are not cycles.

Connectedness: An undirected graph is called *connected* if there exists a walk passing through all nodes. For example, the graph in Fig. 5.1(left) is connected.

Leaves: A node in an undirected graph is called *isolated* if it is not incident to any arc (for a connected graph, such a node exists only in the trivial case of a single-node graph). A node is called *a leaf* if it is incident to exactly one arc. For example, node 4 of the graph in Fig. 5.1(left) is a leaf, and all other nodes are not leaves.

Trees: An undirected graph $G = (\mathcal{N}, \mathcal{E})$ is called a *tree* if it is connected and does not have cycles. The graph in Fig. 5.1(left) is not a tree; we can make it a tree by removing any one of the arcs $\{1,2\}$, $\{1,3\}$, $\{2,3\}$. Trees play an important role in the simplex-type network flow algorithms, and we summarize here the properties of trees needed in the sequel.

Theorem 5.1.

(i) *Every tree with more than one node has a leaf;*
(ii) *an undirected graph $(\mathcal{N}, \mathcal{E})$ is a tree with m nodes if and only if it is connected and has exactly $m-1$ arcs;*

(iii) *for every two distinct nodes i, j in a tree, there exists exactly one path which starts at i and ends at j;*
(iv) *when extending the set of arcs of a tree by adding a new arc, the resulting graph gets exactly one cycle.*[1]

Proof. (i) Assume that we are given a tree with more than one node and no leaves, and let us lead this assumption to a contradiction. Since our tree is connected and does not have leaves, every node is incident to at least two arcs. Now, let us walk along the tree as follows: we start from an arbitrary node; after arriving at a node along a certain arc, we leave it along another arc incident to this node (since every node is incident to at least two arcs, an "exiting arc" always can be found). Since the number of nodes is finite, eventually we will visit for the second time a node i which we have already visited; when it happens for the first time, the segment of our walk from leaving i on the first visit to the node to entering i on the second visit to it is a cycle (recall that we exit a node by an arc distinct from the one used to enter the node, so that the above segment cannot be of the form i, j, i and thus indeed is a cycle). Since a tree does not have cycles, we get a desired contradiction.

(ii) Let us prove by induction in m that every tree with m nodes has $m-1$ arcs. When $m = 1$, the statement clearly is true. Assuming that the statement is true for m node trees, let us prove that it is true for every tree with $m+1$ nodes. Thus, let $G = (\mathcal{N}, \mathcal{E})$ be a tree with $m+1$ nodes; we should prove that G has exactly m arcs. By (i), G has a leaf i; removing from the nodal set this node, and from the set of arcs the (unique) arc incident to the leaf i, we get an m-nodal graph G' which clearly is connected along with G; since every cycle in G' clearly is a cycle in G, G', along with G, has no cycles. Thus, G' is a tree, and by inductive hypothesis, G' has $m-1$ arcs, whence G has m arcs, as claimed. The induction is complete.

It remains to prove that every connected m-nodal graph $G = (\mathcal{N}, \mathcal{E})$ with $m-1$ arcs is a tree. If G has a cycle and we eliminate from $\mathcal{E}$ an arc from this cycle, we clearly keep the resulting graph connected; if it still has cycles, we can remove in a similar fashion

[1]It is assumed here that when counting cycles, we do not distinguish between the cycles like a, b, c, d, a and b, c, d, a, b (the same "loop" along which we move starting from different nodes), same as we do not distinguish between the cycles like a, b, c, d, a and a, d, c, b, a (the same "loop" along which we move in two opposite directions).

another arc, keeping the graph connected, and so on. As a result, we will end up with a *connected* m-node graph which has no cycles and thus is a tree; by the statement we have already proved, the resulting graph has $m-1$ arcs — as many as G itself. This is possible only when no arcs were actually removed, that is, G itself does not have cycles and thus is a tree.

(iii) Let G be a tree, and i, j be two distinct nodes of G. Since G is connected, there exists a walk which starts at i and ends at j. The smallest, in the number of nodes incident to it, walk of this type clearly is a path. Now, if there are two paths of this type distinct from each other, then we can build a walk from i to i, which first goes from i to j along the first path and then goes back from j to i along the "inverted" second path. Unless the paths are identical, the resulting "loop" clearly contains one or more cycles, which is impossible, since G is a tree.

(iv) Let G be a tree with m nodes (and thus, by (ii), $m-1$ arcs). When extending the set $\mathcal{E}$ of arcs of G with an arc $\{i, j\}$ which is not in the set, we get a connected graph G' which is not a tree (indeed, this graph has m nodes and m arcs, which is forbidden for a tree by (ii)). Since G' is connected and is not a tree, it has a cycle; this cycle clearly contains the arc $\{i, j\}$ (otherwise it would be a cycle in G, and G does not have cycles). Since all arcs $\{i_s, i_{s+1}\}$ in a cycle $i_1, \ldots, i_{t-1}, i_t = i_1$ are distinct from each other, the part of the cycle obtained when eliminating the arc $\{i, j\}$ is a path *in* G which links j and i. By (iii), such a path is unique, whence the cycle in question also is unique, since, by our analysis, it must be obtained by "closing" the added arc $\{i, j\}$ by the unique path from j to i in G. $\square$

Subgraphs of a graph: Let $G = (\mathcal{N}, \mathcal{E})$ be an undirected graph. An undirected graph $G' = (\mathcal{N}', \mathcal{E}')$ is called a *subgraph* of G if $\mathcal{N}' \subset \mathcal{N}$ and $\mathcal{E}' \subset \mathcal{E}'$. In other words, a subgraph is what we can get from G by eliminating from $\mathcal{N}$ part of the nodes to get $\mathcal{N}'$ and eliminating from $\mathcal{E}$ part of the arcs, *including all arcs which do not link pairs of nodes from* $\mathcal{N}'$, to get $\mathcal{E}'$. For example, the graph with the nodes 1,3,4 and the arcs $\{1, 3\}, \{1, 4\}$ is a subgraph of the graph presented in Fig. 5.1(left).

Spanning trees: A subgraph of an undirected graph $G = (\mathcal{N}, \mathcal{E})$ is called a *spanning tree* if it is a tree with the same set of nodes as

G; in other words, a spanning tree is a tree of the form $T = (\mathcal{N}, \mathcal{E}')$, where $\mathcal{E}' \subset \mathcal{E}$.

Theorem 5.2. *Let $G = (\mathcal{N}, \mathcal{E})$ be a connected undirected graph and $\mathcal{E}_0$ be a subset of arcs of G such that there is no cycle in G with all arcs belonging to $\mathcal{E}_0$. Then there exists a spanning tree $T = (\mathcal{N}, \mathcal{E}_1)$ of G such that $\mathcal{E}_0 \subset \mathcal{E}_1$.*

Proof. Let $G = (\mathcal{N}, \mathcal{E})$ and $\mathcal{E}_0$ be as in the premise of the theorem. If G is a tree, we can set $\mathcal{E}_1 = \mathcal{E}$, and we are done. Otherwise, G contains a cycle. The arcs in this cycle cannot all belong to $\mathcal{E}_0$, and thus the cycle contains an arc which does not belong to $\mathcal{E}_0$. Let us remove this arc from $\mathcal{E}$. The resulting graph G' can be a tree, and we can take as $\mathcal{E}_1$ the set of its arcs; otherwise we can repeat the above transformation with G' in the role of G. This process clearly cannot last forever, and when it stops, we have at our disposal a spanning tree $(\mathcal{N}, \mathcal{E}_1)$ with $\mathcal{E}_0 \subset \mathcal{E}_1$, as claimed. □

Corollary 5.1. *Let $G = (\mathcal{N}, \mathcal{E})$ be an undirected graph with m nodes, $m - 1$ arcs and no cycles. Then G is a tree.*

Proof. All we should prove is that G is connected. Assume that this is not the case. Then G can be split into $k \geq 2$ connected components — *connected* subgraphs $G_1, \ldots, G_k$ — in such a way that

- the nodal sets $\mathcal{N}_1, \ldots, \mathcal{N}_k$ of the subgraphs form a partition of the nodal set $\mathcal{N}$ of G into nonoverlapping components;
- the set $\mathcal{E}_\ell$ of arcs of G_ℓ comprises *all* arcs of G linking the nodes from $\mathcal{N}_\ell$, $1 \leq \ell \leq k$;
- we have $\mathcal{E} = \mathcal{E}_1 \cup \cdots \cup \mathcal{E}_k$, that is, there are no arcs in G which start in one of the sets $\mathcal{N}_\ell$ and end in another of these sets.

To get this partition, let us take a node and define $\mathcal{N}_1$ as the set of all nodes which can be reached from this node by a walk. Defining $\mathcal{E}_1$ as the set of all arcs of G which link the nodes from $\mathcal{N}_1$, note that by construction of $\mathcal{N}_1$, no other arc is incident to a node from $\mathcal{N}_1$. If the resulting graph $(\mathcal{N}_1, \mathcal{E}_1)$ differs from G, the remaining nodes and arcs of G form a graph, and we can repeat the same construction for this graph, and so on.

Let the cardinalities of $\mathcal{N}_\ell$ be m_ℓ. Since G does not have cycles, nor do the components of G, and since they are connected, they are trees.

By Theorem 5.1, ℓth component has $m_\ell - 1$ arcs, and thus the total number of arcs in G is $\sum_{\ell=1}^{k} m_\ell - k = m - k$. Since $k > 1$ and G has $m - 1$ arcs, we get a contradiction. □

5.1.2 *Directed graphs*

A *directed graph* is a pair of two finite sets, the (nonempty) *set of nodes* $\mathcal{N}$ and (possibly empty) *set of arcs* $\mathcal{E}$, where an arc is an *ordered pair* (i, j) of two nodes *distinct* from each other, $i, j \in \mathcal{N}$. We say that an arc $(i, j) \in \mathcal{E}$ *starts* at node i, *ends* at node j and *links* nodes i and j. For example, the oriented graph depicted in Fig. 5.1(right) has the nodal set $\mathcal{N} = \{1, 2, 3, 4\}$, and the set of arcs $\mathcal{E} = \{(1, 2), (1, 3), (2, 3), (3, 4)\}$. For a oriented graph, it is legitimate to include "inverse to each other" pairs of arcs (i, j) and (j, i) — an option impossible for an undirected graph. However, in both directed and indirected cases we forbid arcs with endpoints identical to each other.

A *walk* in an oriented graph is a sequence $i_1, \ldots, i_t$ of nodes *augmented with a sequence* $\gamma_1, \ldots, \gamma_{t-1}$ *of arcs* such that for every $s < t$ either $\gamma_s = (i_s, i_{s+1})$, or $\gamma_s = (i_{s+1}, i_s)$. Informally speaking, when walking along a directed graph, we are allowed to move through arcs, but not necessarily in the directions of these arcs. The arcs γ_s along which we move in the direction of the arc (that is, $\gamma_s = (i_s, i_s + 1) \in \mathcal{E}$) are called *forward* arcs of the walk, and the arcs γ_s along which we move in the direction opposite to the one of the arcs, (i.e., $\gamma_s = (i_{s+1}, i_s) \in \mathcal{E}$) are called *backward arcs*. A walk $i_1, \gamma_1, i_2, \gamma_2, \ldots, i_{t-1}, \gamma_{t-1}, i_t$ is called a *path* if the nodes $i_1, \ldots, i_t$ are distinct from each other. For example, the sequence $i_1 = 1, \gamma_1 = (1, 2), i_2 = 2, \gamma_2 = (2, 3), i_3 = 3, \gamma_3 = (1, 3), i_4 = 1$ is a walk in the oriented graph depicted in Fig. 5.1(right), the forward arcs being γ_1, γ_2 and the backward arc being γ_3. This walk is not a path; to get a path, we could, e.g., eliminate from the walk the arc γ_3 and the last node $i_4 = 1$.

Given a directed graph, we can convert it into an indirected one, keeping the set of nodes intact and passing from oriented arcs (i.e., ordered pairs of nodes) to their unordered counterparts; needless to say, the arcs opposite to each other (i, j), (j, i) in the oriented graph, if any are present, lead to the same arc $\{i, j\} = \{j, i\}$ in the undirected graph. Note that the indirected graph in Fig. 5.1 (left) is obtained in the outlined way from the directed graph depicted in the right half of the same figure.

A directed graph is called *connected* if there is a walk passing through all the nodes, or, which is the same, if the indirected counterpart of the graph is connected.

5.2 The network flow problem

Recall the single-commodity version of the Multicommodity Network Flow problem (p. 16 in Section 1.2):

> *Given*
>
> - *a directed graph* $G = (\mathcal{N} = \{1, \ldots, m\}, \mathcal{E})$,
> - *a vector* s *of external supplies at the nodes of* G,
> - *a vector* u *of arc capacities, and*
> - *a vector* c *of transportation costs of the arcs,*
>
> *find a feasible flow* f *with the smallest possible transportation cost.*

Recall that

- a *supply* s is a vector with entries indexed by the nodes of G;
- a *flow* f is *any* vector f_{ij} with entries indexed by the arcs (i, j) of G;
- a flow is *feasible* if it
  - is nonnegative and respects the capacities of the arcs:

$$0 \leq f_{ij} \leq u_{ij} \quad \forall (i, j) \in \mathcal{E}$$

  - respects the *flow conservation law*

$$\forall i \leq n : s_i + \sum_{j:(j,i)\in\mathcal{E}} f_{ji} = \sum_{j:(i,j)\in\mathcal{E}} f_{ij}$$

  or, in words, *for every node, the sum of the external supply at the node plus the total incoming flow of the node is equal to the total outgoing flow of the node.*

Recall that the *incidence matrix* of a directed graph $G = (\mathcal{N}, \mathcal{E})$ with $n = \operatorname{Card} \mathcal{E}$ arcs and $m = \operatorname{Card} \mathcal{N}$ nodes is the $m \times n$ matrix

$P = [P_{i\gamma}]$ with the rows indexed by the nodes $i = 1, \dots, m$ of the graph, and the columns indexed by the arcs $\gamma \in \mathcal{E}$, defined by the relation

$$P_{i\gamma} = \begin{cases} 1, \text{ node } i \text{ starts arc } \gamma, \\ -1, \text{ node } i \text{ ends arc } \gamma, \\ 0, \text{ in all other cases,} \end{cases}$$

and in terms of this matrix, the flow conservation law reads

$$Pf = s.$$

Thus, the Network Flow problem reads

$$\min_f \left\{ c^T f := \sum_{\gamma \in \mathcal{E}} c_\gamma f_\gamma : \begin{array}{c} Pf = s \\ 0 \le f \le u \end{array} \right\}. \tag{5.1}$$

5.3 The network simplex algorithm

Here, we present a variant of the Primal Simplex method aimed at solving the *uncapacitated* version of the network flow problem (5.2), that is, the problem

$$\min_f \left\{ c^T f := \sum_{\gamma \in \mathcal{E}} c_\gamma f_\gamma : Pf = s, f \ge 0 \right\}, \tag{5.2}$$

obtained from (5.2) by letting all arc capacities be $+\infty$.

5.3.1 *Preliminaries*

From now on, we make the following

Assumption A: *The graph G is connected, and* $\sum_{i \in \mathcal{N}} s_i = 0$.

Note that when G is not connected, it can be split into connected components. It is clear that the network flow problems (5.2), (5.1) reduce to series of uncoupled similar problems for every one of the components, so that the assumption that G is connected in fact does not reduce generality. Further, the sum of rows of an incidence matrix clearly is 0, meaning that Pf is a vector with zero sum of entries for every flow f. In other words, the assumption $\sum_{i \in \mathcal{N}} s_i = 0$ is *necessary* for a network flow problem, capacitated or uncapacitated alike, to be feasible. The bottom line is that Assumption A in fact does not restrict generality.

5.3.2 *Bases and basic feasible solutions*

We intend to solve problem (5.1) by the Primal Simplex method somehow adjusted to the specific structure of the problem. Note that this problem is in the standard form, as is needed by the method. It remains to take care of the rank of the constraint matrix (the method "wants" it to be equal to the number of rows in the matrix) and to understand what are its bases. Our immediate observation is that the constraint matrix P arising in (5.1) has linearly dependent rows — their sum is the zero vector, in contrast to what we need to run the Simplex method. The remedy is simple: linear dependence of the vectors of coefficients in equality constraints of an LO problem, depending on the right-hand side, either makes the problem infeasible, or allows to eliminate the equality constraints which are linear combinations of the remaining equality constraints. Under Assumption A, every one of the equality constraints in (5.1) is just the minus sum of the remaining equality constraints (recall that $\sum_i s_i = 0$), and therefore when eliminating it, we get an equivalent problem. For the sake of definiteness, let us eliminate the last, m-th equality constraint, thus passing to the equivalent problem

$$\min_f \left\{c^T f : Af = b, f \geq 0\right\}, \tag{5.3}$$

where A is the $(m-1) \times n$ matrix comprising the first $m-1$ rows of P, and $b = [s_1; \ldots; s_{m-1}]$. We shall see in a while that the resulting constraint matrix A has linearly independent rows, as required by the Simplex method. This result will be obtained from a very instructive description of the basic feasible solutions we are about to derive.

5.3.2.1 *Basic solutions are tree solutions*

Recall that the Primal Simplex method works with *bases* — sets I of (indexes of) the columns of A such that the corresponding submatrix of A is square and nonsingular — and with the corresponding basic solutions f^I uniquely defined by the requirements that $Af^I = b$ and the entries of f^I with indexes outside of I are zeros. In the case of problem (5.3), the bases should be sets of $m-1$ arcs of G (since these are the arcs which index the columns in A) such that the corresponding $m-1$ columns of A are linearly independent. We are about to describe these bases and to present a simple specialized algorithm for finding the associated basic solutions.

Let $I \subset \mathcal{E}$ be a set of $m - 1$ arcs of G such that the undirected counterpart G_I associated with the subgraph $G_I = (\{1, \ldots, m\}, I)$ is a tree; in the sequel, we shall express this fact by the words "the arcs from I form a tree when their directions are ignored." Note that by Theorem 5.1, this tree has $m-1$ arcs, meaning that the set I (which contains $m-1$ oriented arcs) does not contain pairs of arcs inverse to each other, (i,j), (j,i), and thus different arcs in I induce different nonoriented arcs in G_I.

Lemma 5.1. *Let I be the set of $m-1$ arcs in G which form a tree when their directions are ignored. Whenever a vector of external supplies s satisfies $\sum_i s_i = 0$, there exists an associated flow f (that is, a flow satisfying $Pf = s$) such that $f_\gamma = 0$ when $\gamma \notin I$, and such a flow is unique.*

Proof. Let B be the $(m-1) \times (m-1)$ submatrix of A comprising columns which are indexed by arcs from I. We should prove that whenever $b \in \mathbf{R}^{m-1}$, the system of equations $Bx = b$ has exactly one solution, or, equivalently, that B is nonsingular. Let us call the last node of G the *root node*; this is the only node which does not correspond to a row in A. By Theorem 5.1, for every other node there exists a unique path in G_I which links the node with the root node. We can clearly renumber the nodes of G in such a way that the new indexes i' of the nodes i would increase along every such path.[2] We then associate with every arc $\gamma = (i,j) \in I$ the serial number

[2]This is how it can be done: let $d(i)$ be the "distance" from a node i to the root node in G_I, that is, the number of arcs in the unique path of G_I which links i and the root node. We give to the root node index m (as it used to be). We then order all the nodes according to their distances to the root, ensuring that these distances form a nonascending sequence. The new index i' of a node i is the serial number of this node in the above sequence.

$\min[i', j']$. Clearly, different arcs in I get different numbers (otherwise G_I would have cycles). The incidence matrix P' of the graph with renumbered nodes is obtained from P by permuting rows, and this permutation does not move the last row (since by construction $m' = m$), so that the matrix A' which we get when eliminating the last row from P' is obtained from A by permuting rows. Consequently, the submatrix of A' comprising columns indexed by $\gamma \in I$ is obtained from B by permuting rows. Numbering the columns in this submatrix of A' according to the serial numbers we gave to the arcs from I, we get a $(m-1) \times (m-1)$ matrix B' which is obtained from B by permuting rows and then permuting columns; clearly, B is nonsingular if and only if B' is so. Now, ℓth column of B' corresponds to an arc (i, j) from I such that $\min(i', j') = \ell$; the nonzero entries in the corresponding column of P have row indexes i and j, and in P' — row indexes i' and j'; since $\ell = \min[i', j']$, one of these row indexes is ℓ, and the other one is $> \ell$, meaning that B' is lower-triangular with nonzero diagonal entries. Thus, B', and then B, is nonsingular. □

Corollary 5.2. *The rank of the matrix A in* (5.3) *is $m-1$, that is, the rows of A are linearly independent.*

Proof. Since G is connected, there is a set I of $m-1$ arcs of G which form a tree after their directions are ignored (apply Theorem 5.2 to an empty set $\mathcal{E}_0$ of arcs). By Lemma 5.1, the columns of A indexed by the $m-1$ arcs of I are linearly independent. Thus, the $(m-1) \times n$ matrix A has a $(m-1) \times (m-1)$ nonsingular submatrix. □

Now, we are ready to describe the bases of A. Observe that by definition of a basis, this is an $(m-1)$-element subset of indices of the columns in A (i.e., an $(m-1)$-element set of arcs in G) such that the corresponding columns of A are linearly independent.

Theorem 5.3. *An $(m-1)$-element set I of arcs in G is a basis of A if and only if the arcs from I form a tree when their directions are ignored.*

Proof. In one direction — *if* $m-1$ arcs of G form a tree after their directions are ignored, *then* the set I of these arcs is a basis of A — the fact was already proved. To prove it in the opposite direction, let I

be a set of $m-1$ arcs which is a basis; we should prove that after the directions of these arcs are ignored, they form a tree. First, we claim that I does not contain any pair of "inverse to each other" arcs (i,j), (j,i). Indeed, for such a pair, the sum of the corresponding columns of P (and thus of A) is zero, while the columns of A indexed by arcs from I are linearly independent. Since I does not contain pairs of arcs inverse to each other, different arcs from I induce different arcs in G_I, meaning that the undirected m-node graph G_I has $m-1$ arcs. We want to prove that the latter graph is a tree; invoking Corollary 5.1, all we need to prove is that G_I has no cycles. Assuming that this is not the case, let $i_1, i_2, \ldots, i_{t-1}, i_t = i_1$, $t \geq 4$, be a cycle in G_I, meaning that I contains arcs $\gamma_1,\ldots,\gamma_{t-1}$ where γ_ℓ is either $(i_\ell, i_{\ell+1})$, or $(i_{\ell+1}, i_\ell)$. Observe that all $t-1$ arcs γ_ℓ are distinct from each other. Now, let ϵ_ℓ be equal to 1 when $\gamma_\ell = (i_\ell, i_{\ell+1})$ and $\epsilon_\ell = -1$ otherwise. It is immediately seen that the flow f with $f_\gamma = \epsilon_\ell$ for $\gamma = \gamma_\ell$ and $f_\gamma = 0$ when γ is distinct from $\gamma_1, \ldots, \gamma_{t-1}$ satisfies $Pf = 0$, meaning that the columns of P indexed by $\gamma_1, \ldots, \gamma_{t-1}$ are linearly dependent. Since $\gamma_\ell \in I$, the columns of P indexed by $\gamma \in I$ are also linearly dependent, which is a desired contradiction. □

Integrality of basic feasible solutions: As a byproduct of the above considerations, we get the following important result:

Proposition 5.1. *Let the right-hand side vector b in* (5.3) *be integral. Then every basic solution to the problem, feasible of not, is an integral vector.*

Proof. By Theorem 5.3, the bases of A are exactly the $(m-1)$-element sets of arcs in G which form trees when the directions of the arcs are ignored. From the proof of Lemma 5.1 it follows that the associated $(m-1)\times(m-1)$ submatrices B of A become lower triangular with nonzero diagonal entries after reordering of rows and columns. Since every entry in A is either 1, or -1, or 0, every lower triangular matrix B' in question has integral entries and diagonal entries ± 1, whence the entries in the inverse of B' are integral as well. Therefore, if B is a basic submatrix of A and b is an integral vector, the vector $B^{-1}b$ is integral, meaning that the nonzero entries in every basic solution to $Ax = b$ are integral. □

Algorithm for building basic solutions: Theorem 5.3 describes the basic solutions to the uncapacitated network flow problem (5.3) — these are exactly the *tree solutions*, which can be obtained as follows:

- we choose in the set $\mathcal{E}$ of arcs of G a subset I of $m-1$ arcs which form a tree when their directions are ignored;
- the tree solution associated with I is a flow f such that $f_\gamma = 0$ for $\gamma \notin I$ and $Af = b := [s_1; \ldots; s_{m-1}]$. These conditions uniquely define f (Lemma 5.1), and in fact f is given by the following simple algorithm. Let G_I be the m-node tree associated with I. To get f, we process this tree in $m-1$ steps as follows:
  - At the beginning of step $t = 1, 2, \ldots, m-1$ we have at our disposal a $m-t+1$-node subgraph $G^t = (\mathcal{N}^t, \mathcal{E}^t)$ of G_I which is a tree, and a vector s^t of supplies at the nodes of G^t satisfying the relation $\sum_{i\in\mathcal{N}^t} s_i^t = 0$. At the first step, G^1 is G_I, and $s^1 = s$.
  - At step t, we act as follows:
    1. We identify a leaf of G^t, let it be denoted by $\bar{i}$; such a leaf exists by Theorem 5.1. By the definition of a leaf, in G^t there exists a unique arc $\{j, \bar{i}\}$ incident to the leaf. Since G^t is a subgraph of G_I, this arc corresponds to an arc $\gamma \in I$, and this arc is either $(j, \bar{i})$, or $(\bar{i}, j)$. In the first case, we set $f_\gamma = -s^t_{\bar{i}}$, in the second, we set $f_\gamma = s^t_{\bar{i}}$.
    2. We eliminate from G^t the node $\bar{i}$ and the incident arc $\{j, \bar{i}\}$, thus getting a graph G^{t+1} (which is clearly a tree).
    3. We further convert s^t into s^{t+1} as follows: for every node k of G^{t+1} which is distinct from j, we set $s_k^{t+1} = s_k^t$, and for the node j we set $s_j^{t+1} = s_j^t + s^t_{\bar{i}}$; note that the sum of entries in the resulting vector s^{t+1} is the same as the similar sum for s^t, that is, it is 0.
    4. Step t is completed; when $t < m-1$, we pass to step $t+1$, otherwise we terminate.

  Note that upon termination, we have at our disposal the entries f_γ for all $\gamma \in I$. Setting $f_\gamma = 0$ for $\gamma \notin I$, we get a flow f; it is immediately seen that this flow satisfies $Af = b$, that is, f is the tree solution we are looking for.

Illustration: Let us illustrate the above algorithm for finding a tree solution. Our data are as follows:

- G is the 4-node-oriented graph shown in Fig. 5.1(right):

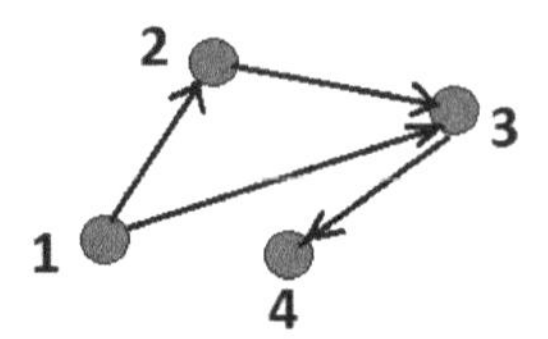

- $I = \{(1,3),(2,3),(3,4)\}$ (these arcs clearly form a tree when their directions are ignored).
- $s = [1; 2; 3; -6]$.

The algorithm works as follows:

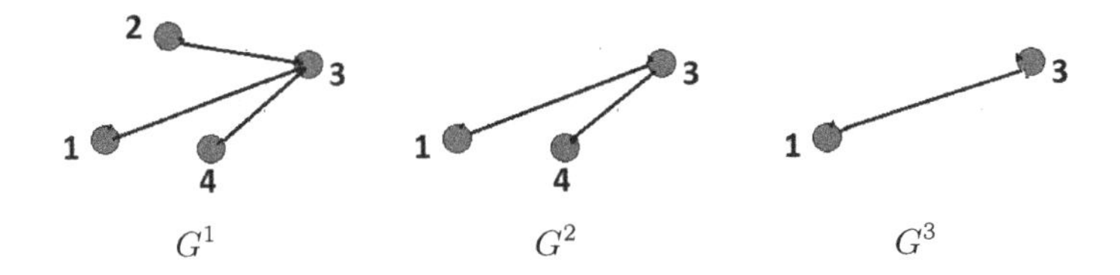

- $G^1 = G_I$ is the undirected graph with the nodes 1,2,3,4 and the arcs $\{1,3\},\{2,3\},\{3,4\}$, and $s^1 = [1; 2; 3; -6]$.
- At the first step, we choose a leaf in G^1, let it be the node 2, and the unique arc $\{2,3\}$ incident in G^1 to this node. The corresponding arc in G is $\gamma = (2,3)$. We set $f_{2,3} = s_2^1 = 2$, eliminate from G^1 the node 2 and the arc $\{2,3\}$, thus obtaining $G^2 = (\{1,3,4\},\{\{1,3\},\{3,4\}\})$, and set $s_1^2 = 1$, $s_3^2 = 2+3 = 5$, $s_4^2 = -6$. The first step is completed.
- At the second step, we choose in G^2 a leaf, let it be the node 4, and the unique arc $\{3,4\}$ incident to this node; the corresponding arc in G is $(3,4)$. We set $f_{3,4} = -s_4^2 = 6$, eliminate from G^2 the node 4 and the arc $\{3,4\}$, thus obtaining $G^3 = (\{1,3\},\{\{1,3\}\})$, and set $s_1^3 = s_1^2 = 1$, $s_3^3 = s_3^2 + s_4^2 = -1$. The second step is completed.
- At the third step, we take a leaf in G^3, let it be the node 3, and the arc $\{1,3\}$ incident to this node. The associated arc in G is $(1,3)$ and we set $f_{1,3} = -s_3^3 = 1$.

  Finally, we augment the entries $f_{2,3} = 2$, $f_{3,4} = 6$, $f_{1,3} = 1$ we have built with $f_{1,2} = 0$, thus getting a flow satisfying the conservation law, the vector of external supplies being s, and vanishing outside of the arcs from I.

5.3.3 *Reduced costs*

As we remember, at every step of the Primal Simplex method we have at our disposal a basis I of A along with the associated basic solution f^I, and the vector c^I of *reduced costs* of the form $c - A^T\lambda_I$, where λ_I is uniquely defined by the requirement that the reduced costs with indexes from I are equal to 0. In the case of problem (5.3), it is convenient to write the reduced costs in the equivalent form $c - P^T\lambda$, where $\lambda \in \mathbf{R}^m$ is normalized by the requirement $\lambda_m = 0$. Recalling the structure of P, the requirements specifying the reduced cost c^I become

$$\begin{aligned} c^I_{ij} &:= c_{ij} + \lambda_j - \lambda_i, \ (i,j) \in \mathcal{E} & (a) \\ c_{ij} &= \lambda_i - \lambda_j, \ (i,j) \in I & (b) \end{aligned} \qquad (5.4)$$

We are about to point out a simple algorithm for building the required vector λ. Recall that the bases I are exactly the collections of $m-1$ arcs of G which form a tree after their directions are ignored. Given such a collection, a byproduct of the algorithm for building the associated feasible solution is the collection of subgraphs $G^t = (\mathcal{N}^t, \mathcal{E}^t)$, $t = 1, \ldots, m-1$, of G_I, every one of the subgraphs being a tree, such that $G^1 = G_I$ and for every $t < m-1$, G^{t+1} is obtained from G^t by eliminating a leaf node $\bar{\imath}_t$ of G^t along with the unique arc $\{\bar{\jmath}_t, \bar{\imath}_t\}$ of G^t which is incident to this node. It is immediately seen that *every* arc in G_I is of the form $\{\bar{\jmath}_t, \bar{\imath}_t\}$ for certain t. Given G^t, $t = 1, \ldots, m-1$, we can build the vector λ_I of Lagrange multipliers as follows:

- [first step] G^{m-1} is a 2-node tree, let these nodes be α and β. Without loss of generality we can assume that (α, β) is an arc in I. Let us set $\bar{\lambda}_\alpha = c_{\alpha\beta}$, $\bar{\lambda}_\beta = 0$, thus ensuring $c_{\alpha\beta} = \bar{\lambda}_\alpha - \bar{\lambda}_\beta$.
- [ℓth step, $m-1 \geq \ell \geq 2$] At the beginning of step ℓ we already have at our disposal the Lagrange multipliers $\bar{\lambda}_i$ associated with the nodes i of $G^{m-\ell+1}$, and these multipliers satisfy the relations $c_{ij} = \bar{\lambda}_i - \bar{\lambda}_j$ for all arcs $\gamma = (i,j) \in I$ which link the nodes of $G^{m-\ell+1}$. At the ℓth step, we take the only node, $\bar{\imath} = \bar{\imath}_{m-\ell}$, of the graph $G^{m-\ell}$ which is not a node of $G^{m-\ell+1}$, and identify the (unique!) incident to this node arc $\{\bar{\jmath}, \bar{\imath}\}$ in $G^{m-\ell}$; note that by construction of the graphs G^t, $\bar{\jmath}$ is a node of $G^{m-\ell+1}$, so that the Lagrange multiplier $\bar{\lambda}_{\bar{\jmath}}$ is already defined. Now, the arc γ in I

associated with the arc $\{\bar{j},\bar{i}\}$ of $G^{m-\ell}$ is either $(\bar{j},\bar{i})$ or $(\bar{i},\bar{j})$. In the first case, we set $\bar{\lambda}_{\bar{i}} = \bar{\lambda}_{\bar{j}} - c_{\bar{j},\bar{i}}$, and in the second case, we set $\bar{\lambda}_{\bar{i}} = \bar{\lambda}_{\bar{j}} + c_{\bar{i},\bar{j}}$, thus ensuring in both cases that c_γ is the difference of the Lagrange multipliers associated with the start- and the end-nodes of γ. As a result of step ℓ, we have defined the multipliers $\bar{\lambda}_i$ for all nodes of $G^{m-\ell}$, and have ensured, for every arc $\gamma \in I$ linking the nodes from $G^{m-\ell}$, that the transportation cost c_γ of this arc is the difference between the multipliers associated with the start- and the end-node of the arc. When $\ell < m-1$, we pass to the next step, otherwise we terminate.

As a result of the outlined algorithm, we get a vector $\bar{\lambda}$ of multipliers associated with the nodes of G and satisfying the requirements (5.4(b)). Subtracting from all entries of $\bar{\lambda}$ the quantity $\bar{\lambda}_m$, we keep (5.4(b)) intact and meet the normalization requirement $\lambda_m = 0$ on the multipliers, thus obtaining λ_I.[3]

Illustration: Let us equip the graph G depicted in Fig. 5.1(right)

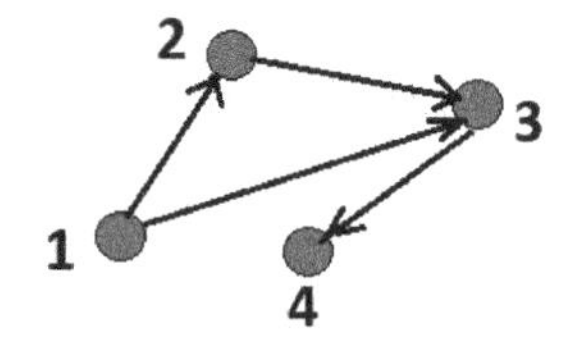

with transportation costs $c_{1,2} = 1$, $c_{2,3} = 4$, $c_{1,3} = 6$, $c_{3,4} = 8$ and compute the vector of reduced costs, the base I being the same set of arcs $\{(1,3),(2,3),(3,4)\}$ we used when illustrating the algorithm for building basic solutions, see p. 333. The corresponding subgraphs of G_I are $G^1 = (\{1,2,3,4\},\{\{1,3\},\{2,3\},\{3,4\}\})$, $G^2 = (\{1,3,4\},\{\{1,3\},\{3,4\}\})$, $G^3 = (\{1,3\},\{\{1,3\}\})$:

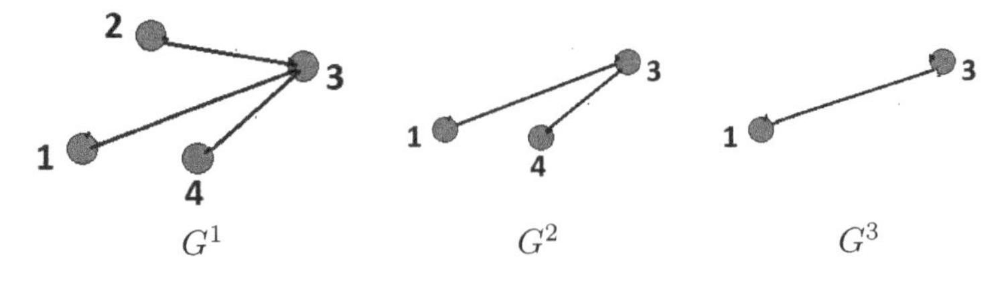

G^1 G^2 G^3

[3]In actual computations, there is no reason to care for normalization of $\lambda_m = 0$ of a vector of Lagrange multipliers, since all that matters — the associated reduced costs (5.4(b)) — depend only on the differences between the entries in λ.

We start with looking at G^3 and setting $\bar{\lambda}_1 = c_{1,3} = 6$, $\bar{\lambda}_3 = 0$. At the next step we look at G^2; the node of G^2 which is not in G^3 is 4, and we set $\bar{\lambda}_4 = \bar{\lambda}_3 - c_{3,4} = 0 - 8 = -8$. At the last step, we look at the graph $G^1 = G_I$; the node of this graph which is not in G^2 is 2, and we set $\bar{\lambda}_2 = \bar{\lambda}_3 + c_{2,3} = 0 + 4 = 4$. The Lagrange multipliers (before normalization) form the vector $\bar{\lambda} = [6; 4; 0; -8]$, and after normalization — $\lambda = \lambda_I = [14; 12; 8; 0]$. The reduced costs are

$$
\begin{aligned}
(c^I)_{1,2} &= c_{1,2} + \lambda_2 - \lambda_1 = 1 + 12 - 14 = -1, \\
(c^I)_{1,3} &= c_{1,3} + \lambda_3 - \lambda_1 = 6 + 8 - 14 = 0, \\
(c^I)_{2,3} &= c_{2,3} + \lambda_3 - \lambda_2 = 4 + 8 - 12 = 0, \\
(c^I)_{3,4} &= c_{3,4} + \lambda_4 - \lambda_3 = 8 + 0 - 8 = 0.
\end{aligned}
$$

Note that we could skip computation of the reduced costs associated with the arcs from I — we know in advance that they are zeros.

5.3.4 *Updating basic feasible solution*

We are now ready to describe a step of the Network Simplex algorithm. As we remember from the general description of the Primal Simplex algorithm, this step is as follows (in the following description, the general-case rules (in Italics) are accompanied by their "network flow translation" (in Roman)):

(1) *At the beginning of the step, we have at our disposal the current basis I along with the associated basic feasible solution.*
At the beginning of the step, we have at our disposal a set I of $m - 1$ arcs of G which form a tree after their directions are ignored, along with the flow f^I which vanishes outside of the arcs from I, satisfies $Af^I = b$, and is nonnegative.

(2) *At the step, we start with computing the vector of reduced costs.*
We apply the algorithm presented on p. 335 to get the vector of Lagrange multipliers $\lambda \in \mathbf{R}^n$ such that

$$c_{ij} = \lambda_i - \lambda_j \quad \forall (i, j) \in I,$$

and form the reduced costs

$$c^I_{ij} = c_{ij} + \lambda_j - \lambda_i, \quad \forall (i, j) \in \mathcal{E}. \tag{5.5}$$

If all the reduced costs are nonnegative, we terminate — f^I is an optimal solution to (5.3), *and λ is the corresponding "optimality certificate" — an optimal solution to the dual problem.* (Note that we are solving a minimization problem, so that optimality corresponds to nonnegativity of all reduced costs). *Otherwise, we identify a variable with negative reduced cost; this is the nonbasic variable to enter the basis.*

We terminate when $c^I \geq 0$, otherwise find an arc $\bar{\gamma} = (\bar{i}, \bar{j})$ such that $(c^I)_{\bar{i}\bar{j}} < 0$.

(3) *We start to increase the value of the nonbasic variable entering the basis, updating the values of the basic variables in order to maintain the equality constraints, until*

- *either one of the basic variables is about to become negative; this variable leaves the basis. In this case, we update the basis by including the variable which enters the basis and eliminating the variable which leaves it, updating accordingly the basic feasible solution, and then pass to the next step of the method;*
- *or it turns out that when the variable entering the basis increases, no basic variables are about to become negative. In this case, we have found a recessive direction of the feasible domain X along which the objective decreases; this direction certifies that the problem is unbounded, and we terminate. Recall that according to the general theory of the Primal Simplex method, the recessive direction in question generates an extreme ray in* Rec(X).

The "network translation" of the latter rules is as follows. There are two possibilities:

A. The "index" — the arc $\bar{\gamma} = (\bar{i}, \bar{j})$ — of the variable $f_{\bar{\gamma}}$ which enters the basis is *not* the arc inverse to one of the arcs in I;
B. The arc $\bar{\gamma}$ *is* inverse to one of the arcs from I.

Case A: In this case, the arc $\bar{\gamma}$ induces a new arc, $\{\bar{i}, \bar{j}\}$, in the undirected graph G_I, thus converting this tree into a new graph G_I^+ which is not a tree anymore. By Theorem 5.1, G_I^+ has exactly one cycle, which includes this new arc $\{\bar{i}, \bar{j}\}$. "Lifting" this cycle to G, we get a cycle C in G — a sequence $i_1, \ldots, i_t = i_1$ of nodes of G such that

- the nodes $i_1, \dots, i_{t-1}$ are distinct from each other,
- $i_1 = \bar{i}, i_2 = \bar{j}$, so that i_1 and i_2 are linked by the arc $\bar{\gamma}$, and
- every other pair of consecutive nodes $i_2, i_3, i_3, i_4, \dots, i_{t-1}, i_t$ in our sequence $i_1, \dots, i_t = i_1$ is linked by an arc from I. For a pair of consecutive nodes i_s, i_{s+1}, $2 \le s < t$, the corresponding arc is either (i_s, i_{s+1}) ("forward arc"), or (i_{s+1}, i_s) ("backward arc").

Let Γ be the collection of arcs forming the above cycle C; one of them is $\bar{\gamma}$, and it is a forward arc of the cycle, and the remaining arcs belong to I, some of them being forward, and some of them being backward. Consider the flow h given by

$$h_\gamma = \begin{cases} +1, \ \gamma \in \Gamma \text{ is a forward arc of } C, \\ -1, \ \gamma \in \Gamma \text{ is a backward arc of } C, \\ \ \ 0, \text{ all other cases.} \end{cases}$$

It is immediately seen that $Ph = 0$, whence $Ah = 0$ as well, whence, in turn, $A(f^I + th) = b := [s_1; \dots; s_{m-1}]$ for all t, that is, $f^I(t) := f^I + th$ is a flow which satisfies the equality constraints in (5.3). When $t = 0$, this flow is feasible, and when t grows, starting with 0, the cost of this flow decreases. Indeed, the change in the cost $c^T f^I(t) - c^T f^I$ is the same as when the original costs c are replaced with the reduced costs c^I.[4] Since the reduced costs corresponding to all arcs in C except for the arc $\bar{\gamma}$ are zero, we conclude that

$$c^T f^I(t) - c^T f^I = [c^I]^T f^I(t) - [c^I]^T f^I = [c^I]^T [th] = t[c^I]^T h \underbrace{=}_{(a)} tc^I_{\bar{\gamma}} h_{\bar{\gamma}} \underbrace{=}_{(b)} tc^I_{\bar{\gamma}}$$

((a) is due to the fact that $c^I_\gamma = 0$ for all arcs γ in C distinct from $\bar{\gamma}$, since these arcs are in I; (b) is due to $h_{\bar{\gamma}} = 1$ by construction of h; note that $\bar{\gamma}$ is a forward arc in C). Since $c^I_{\bar{\gamma}} < 0$, we see that the direction h is a direction of improvement of the objective: when t is positive, $c^T f^I(t) < c^T f^I$.

---

[4]This fact is given by the theory of the Primal Simplex method: on the set of solutions to the system of equality constraints of a problem in the standard form, the objective given by the original cost vector differs by a constant from the objective given by the vectors of reduced costs, so that the changes in both objectives when passing from one solution satisfying the equality constraints to another solution with the same property are equal to each other.

Now, tho cases are possible:

A.1. $h \geq 0$ (this is so if and only if all arcs of G comprising the cycle C are forward). In this case, $f^I(t)$ is a feasible solution of (5.3) for all $t > 0$, and as $t \to \infty$, the cost of this feasible solution goes to $-\infty$, meaning that the problem is unbounded below. In this case, we terminate with a certificate h of unboundedness in our hands.
A.2. Some entries in h are negative (this happens when C includes backward arcs). In this case, when increasing t starting from $t = 0$, the flow $f^I(t)$ eventually looses nonnegativity. Since the $\bar{\gamma}$-component of the flow $f^I(t)$ is equal to t, the entries in the flow which can eventually become negative are among those with indexes in I. We can easily compute the largest $t = \bar{t}$ for which $f^I(t)$ is still nonnegative:

$$\bar{t} = \min\{f^I_\gamma : \gamma \in I \text{ is a backward arc of } C\}.$$

In the flow $f^I(\bar{t})$, one (or more) of the components of $f^I_\gamma(\bar{t})$ with indexes $\gamma \in I$ become zero. We take one of the corresponding arcs, let it be denoted $\widetilde{\gamma}$, and claim that it leaves the basis, while $\bar{\gamma}$ enters it, so that the updated basis is $I^+ = (I\backslash\{\widetilde{\gamma}\}) \cup *\{\bar{\gamma}\}$, and the new basic feasible solution is $f^I(\bar{t})$. Observe that I^+ is indeed a basis.

This observation (which is readily given by the general theory of the Primal Simplex method; recall that we are just "translating" this method to the particular case of network problem (5.3)) can be justified "from scratch," namely, as follows: the undirected counterpart G_{I^+} of the graph obtained from G when eliminating all arcs except for those from I^+ is obtained from the tree G_I as follows:

- we first add to G_I a new arc $\{\bar{i}, \bar{j}\}$ (it is induced by $\bar{\gamma}$); the resulting graph G_I^+ has m arcs and a unique cycle induced by the cycle C in G;
- to get G_{I^+}, we eliminate from the graph G_I^+ an arc which belongs to the unique cycle of G_I^+, thus getting a graph with $m - 1$ nodes. The arc we eliminate from G_I^+ is induced by the arc $\widetilde{\gamma}$ in C and is different from the arc $\{\bar{i}, \bar{j}\}$.

Since G_I^+ clearly is connected along with G_I, and eliminating from a connected graph an arc belonging to a cycle, we preserve connectivity, G_{I^+} is a connected graph with m nodes and $m - 1$ arcs and thus is a tree (Theorem 5.1), as claimed.

Thus, in the case of A.2 we convert the current basis I and the associated basic feasible solution f^I into a new basis I^+ and the corresponding new basic feasible solution $f^{I^+} = f^I_{\bar{t}}$, and can pass to the next step of the algorithm.

Remark. Note that in the case of $\bar{t} > 0$, the cost of the new basic feasible solution is strictly smaller than the cost of the old solution, in full accordance with the general theory of the Primal Simplex method. It may happen, however, that $\bar{t} = 0$, in which case the basic feasible solution remains intact, and only the basis changes; this, in full accordance with the general theory, can happen only when the basic feasible solution f^I we are updating is degenerate — has less than $m - 1$ positive entries.

Case B: Recall that this is the case when the arc $\bar{\gamma} = (\bar{i}, \bar{j})$ with negative reduced cost $c^I_{\bar{\gamma}}$ — the index of the variable which enters the basis, in the general Simplex terminology — is inverse to one of the arcs from I. In this case, we act in the same fashion as in the case A, although our life becomes simpler. The analogy of the cycle C is now the loop $\bar{\gamma} = (\bar{i}, \bar{j})$, $\widehat{\gamma} = (\bar{j}, \bar{i})$ (recall that we are in the situation when the second arc in this loop belongs to I). The flow h satisfying $Ph = 0$ is now the flow with exactly two nonzero components, $h_{\bar{\gamma}}$ and $h_{\widehat{\gamma}}$, both equal to 1; and since $c^I_{\bar{\gamma}} < 0$, $c^I_{\widehat{\gamma}} = 0$, adding to f^I a positive multiple th of the "circulation" h, we, same as in the case of A, reduce the cost of the flow, keeping it feasible; thus, we have discovered a recessive direction of the feasible set of the problem. Along this direction, the objective decreases, that is, we can terminate with a certificate of unboundedness in out hands.

The description of a step in the Network Simplex algorithm as applied to the uncapacitated network flow problem is completed.

Illustration: Let us carry out a step of the Network Simplex algorithm as applied to the directed graph depicted in Fig. 5.1(right)

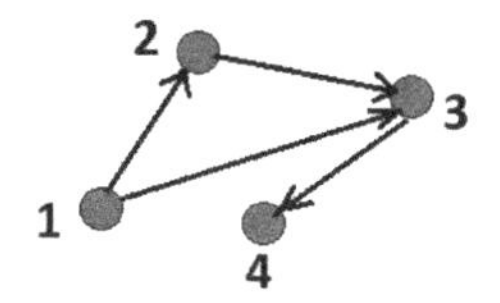

the external supplies and the transportation costs being as in the illustrations on pp. 334 and 336, that is:

$$s = [1; 2; 3; -6]; \quad c_{1,2} = 1, \quad c_{2,3} = 4, \quad c_{1,3} = 6, \quad c_{3,4} = 8.$$

Let the current basis be $I = \{(1,3),(2,3),(3,4)\}$; the associated graph G_I is the undirected graph with nodes $1,2,3,4$ and arcs $\{1,3\},\{2,3\},\{3,4\}$. The corresponding basic solution was computed in the illustration on p. 334, it is

$$f^I_{1,2} = 0, \quad f^I_{1,3} = 1, \quad f^I_{2,3} = 2, \quad f^I_{3,4} = 6;$$

as we see, this solution is feasible. The reduced costs associated with the basis I were computed in the illustration on p. 336; they are

$$c^I_{1,2} = -1, \quad c^I_{1,3} = 0, \quad c^I_{2,3} = 0, \quad c^I_{3,4} = 0.$$

The reduced cost associated with the arc $(1,2)$ is negative, meaning that the solution f^I is perhaps nonoptimal, and more steps of the algorithm are needed. We enter the arc $\bar{\gamma} = (1,2)$ into the basis (that is, the variable $f_{1,2}$ should become a basic variable). When adding to G_I the nonoriented arc $\{1,2\}$ associated with the oriented arc $\bar{\gamma}$ in G, we get a graph G_I^+ which is a nonoriented counterpart of G; in full accordance with our theory, it has a single cycle $1,2,3,1$:

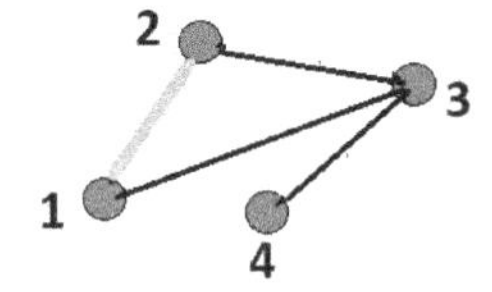

The corresponding cycle C in G is $1,(1,2),2,(2,3),3,(1,3),1$; the arcs $(1,2)$ and $(2,3)$ are forward, and the arc $(1,3)$ is backward. We are in Case A; the improving direction h is $h_{1,2} = 1$, $h_{1,3} = -1$, $h_{2,3} = 1$, $h_{3,4} = 0$; as it should be, this direction satisfies $Ph = 0$ (that is, it is a flow meeting the conservation law, the external supplies being zero). We now add a nonnegative multiple th of the flow h to our "old" basic feasible flow f^I, choosing the largest possible t for which the updated flow is nonnegative. Looking at f^I and h, we get $t = 1$; the variable which leaves the basis is $f_{1,3}$, the variable which enters the basis is $f_{1,2}$ (in terms of the indexes: the arc which should be added to I is $\bar{\gamma} = (1,2)$, the arc which leaves I is $\widetilde{\gamma} = (1,3)$). The new basis and basic feasible solution are

$$I^+ = \{(1,2),(2,3),(3,4)\}, \quad f^{I^+}_{1,2} = 1, \quad f^{I^+}_{1,3} = 0, \quad f^{I^+}_{2,3} = 3, \quad f^{I^+}_{3,4} = 6.$$

The step is completed.

Let us carry out the next step of the algorithm. To this end, let us compute the reduced costs associated with our new basis I^+. Applying the algorithm presented on p. 336, we get

$$\lambda_1 = 13, \quad \lambda_2 = 12, \quad \lambda_3 = 8, \quad \lambda_4 = 0,$$
$$c_{1,2}^{I+} = c_{2,3}^{I+} = c_{3,4}^{I+} = 0, \quad c_{1,3}^{I+} = c_{1,3} + \lambda_3 - \lambda_1 = 6 + 8 - 13 = 1.$$

The reduced costs are nonnegative, meaning that the flow f^{I^+} is optimal.

5.3.5 *Network simplex algorithm: Summary and remarks*

Summary of the algorithm: We have described in detail the "building blocks" of the network-oriented implementation of the Primal Simplex method. When put together, we get an algorithm which can be summarized as follows:

(1) At the beginning of a step, we have at our disposal current basis I, which is the set of $m-1$ arcs in G which form a tree after their directions are ignored, along with the corresponding basic feasible solution f^I; this is a feasible flow such that $f_\gamma^I = 0$ for all arcs $\gamma \notin I$.

(2) At the step, we

(a) use the algorithm presented on p. 336 to compute the Lagrange multipliers $\lambda_1, \ldots, \lambda_m$ such that

$$c_{ij} = \lambda_i - \lambda_j \quad \forall (i,j) \in I,$$

and then compute the reduced costs

$$c_{ij}^I = c_{ij} + \lambda_j - \lambda_i, \; (i,j) \in \mathcal{E}.$$

The reduced costs associated with the arcs $\gamma \in I$ are zero. If all remaining reduced costs are nonnegative, we terminate — f^I is an optimal solution to (5.3), and the vector $\lambda^I = [\lambda_1 - \lambda_m; \ldots; \lambda_{m-1} - \lambda_m]$ is the optimal solution to the dual problem.

(b) If there are negative reduced costs, we pick one of them, let it be $c_{\bar{i}\bar{j}}^I$, and add to I the arc $\bar{\gamma} = (\bar{i}, \bar{j})$ ("variable $f_{\bar{\gamma}}$ enters the basis"). We extend (it is always possible) the arc $\bar{\gamma}$ by arcs from I to get a loop $i_1 := \bar{i}, \gamma_1 := \bar{\gamma}, i_2 := \bar{j}, \gamma_2, i_3, \gamma_3, \ldots, \gamma_{t-1}, i_t = \bar{i}$; here $\gamma_2, \ldots, \gamma_{t-1}$ are distinct arcs from I, the nodes $i_1 = \bar{i}, i_2 = \bar{j}, i_3, \ldots, i_{t-1}$ are distinct from each other, and for every $s < t$ γ_s is either the arc (i_s, i_{s+1}) ("forward arc"), or the arc (i_{s+1}, i_s) ("backward arc"). This loop gives rise to a flow h which is equal to 1 at every forward arc and equal to -1 at every backward arc of the loop, and vanishes in all arcs not belonging to the loop; this flow satisfies the relation $Ph = 0$ and $[c^I]^T h = c^I_{\bar{\gamma}} < 0$. It follows that a flow of the form $f^I(t) := f^I + th$ with $t \geq 0$ satisfies the flow conservation law: $Pf^I(t) = s$, and its cost strictly decreases as t grows.

(c) It may happen that the flow $f^I(t)$ remains nonnegative for all $t \geq 0$. In this case, we have found a ray in the feasible set along which the objective tends to $-\infty$, meaning that the problem is unbounded; if it happens, we terminate, h being the unboundedness certificate.

An alternative is that some of the entries in $f^I(t)$ decrease as t increases; this clearly can happen only for the entries corresponding to arcs from I (since the only other entry in $f^I(t)$ which is not identically zero is the entry indexed by $\bar{\gamma}$, and this entry is equal to t (recall that $\bar{\gamma}$, by construction, is a forward arc in the above loop, and thus $h_{\bar{\gamma}} = 1$) and thus increases with t). We find the largest value $\bar{t}$ of t such that the flow $f^I(\bar{t})$ is nonnegative, and eliminate from I an arc $\widehat{\gamma}$ where the flow $f^I(t)$ is "about to become negative" at $t = \bar{t}$ (i.e., $\widehat{\gamma}$ is such that the corresponding component of $f^I(t)$ decreases with t and becomes 0 when $t = \bar{t}$). We set $I^+ = [I \cup \{\bar{\gamma}\}] \backslash \{\widehat{\gamma}\}$; this is our new basis (and it is indeed a basis), the corresponding basic feasible solution being the flow $f^{I^+} = f^I(\bar{t})$, and then loop to the next step of the algorithm.

Remarks. A. **Correctness and finite termination:** The Network PSM is just an adaptation of the general-purpose PSM to the special case of the uncapacitated Network Flow problem (5.3); as such, it inherits the fundamental properties of the PSM:

- the only possibilities to terminate are
  - either to produce an optimal basic feasible solution to the problem along with the optimality certificate a feasible solution to

the dual problem which, taken along with the primal solution, satisfies the complementary slackness condition and thus is an optimal solution to the dual problem,
- ∘ or to detect that the problem is unbounded and to produce the corresponding certificate.

- the method is monotone — as a result of a step, the value of the objective either decreases or remains the same. The latter option takes place only if the basic feasible solution the step starts with is degenerate — has less than $m-1$ positive entries. If there are no degenerate feasible solutions at all, the method terminates in finite time. There are also examples which show that, same as in the case of a general-purpose PSM, in the presence of degenerate solutions the Network PSM can loop forever, unless appropriate care is taken of the "ties" in the pivoting rules.

B. **Oriented cycles of negative cost:** The unboundedness certificate, if any, is an oriented cycle in G with negative cost, that is, a collection of t arcs $(i_1, i_2), (i_2, i_3), \ldots, (i_t, i_1)$ such that the sum of the associated reduced costs is negative. Now, observe that from (5.5) it follows that *for every oriented cycle and for every basis I of (5.3), the sum of reduced costs of all arcs of the cycle is equal to the sum of the original costs of the arcs.* It follows that the certificate of unboundedness produced by the Network PSM is an oriented cycle of negative total costs. It is also clear that whenever a feasible problem (5.3) admits an oriented cycle of negative total costs, the problem is unbounded.

C. **Integrality:** From the description of the Network PSM it follows that *all arithmetic operations used in the computations are additions, subtractions and multiplications; no divisions are involved.* Therefore, *when the vector of external supplies is integral, so is the optimal solution (if any) reported by the Network PSM* — the fact already known to us from Proposition 5.1. We can augment this statement by noting that *when all costs c_{ij} are integral, so are all reduced costs and the associated λ_i's* (see (5.5)) *and, in particular, the optimal dual solution (if any) returned by the algorithm is integral* — this observation is readily given by the algorithm for computing reduced costs presented in Section 5.3.3, see p. 335.

Needless to say, absence of divisions implies that in the case of integral data all computations in the Network PSM are computations with integral operands and results, which eliminates a lot of numerical issues related to rounding errors.[5]

D. **Getting started:** We have not explained yet how to initialize the Network PSM — from where to take the initial basic feasible solution. The simplest way to do it is as follows. Let us augment the m-vertex graph G by an additional node, let its index be $m+1$, and m additional artificial arcs, namely, as follows. For every original node i with nonnegative external supply, we add to the graph the arc $(i, m+1)$, and for every original node i with negative external supply — the arc $(m+1, i)$. Finally, we assign the new node $m+1$ with zero external supply. For the extended graph, finding the initial basic feasible solution is trivial: the corresponding basis I is just the basis comprising the m artificial arcs,[6] and the solution is to send s_i units of flow from node i to node $m+1$ along the arc $(i, m+1)$, when s_i is ≥ 0, and $-s_i$ units of flow through the arc $(m+1, i)$, when s_i is negative. Now, it is easily seen that if the original problem is feasible and the transportation costs for the added m arcs are large enough, every optimal solution to the augmented problem will induce zero flows in the artificial — "expensive" — arcs and thus will be an optimal solution to the problem of actual interest.

5.4 Capacitated network flow problem

In this section, we present a modification of the Network PSM aimed at solving the *capacitated Network Flow problem* which is obtained from (5.3) by assuming that some of the arcs have finite capacities, that is, the flows in these arcs are subject to upper bounds. In fact, it makes sense to consider a slightly more general situation where the flows in the arcs are subject to both lower and upper bounds, so that

[5]It should be mentioned, however, that without additional precautions the method may suffer from "integral overflows" — some intermediate results can become too large to be stored in a computer as integers.

[6]Note that our new graph has $m+1$ nodes, so that the basis should contain m arcs which form a tree after their directions are ignored; this is exactly the case with the proposed set of arcs.

the problem reads

$$\min_f \left\{c^T f : Af = b, u_\gamma \geq f_\gamma \geq \ell_\gamma, \gamma \in \mathcal{E}\right\}, \tag{5.6}$$

where, same as in (5.3), A is the $(m-1) \times n$ matrix obtained by eliminating the last row in the incidence matrix P of an m-node graph G with the set of arcs $\mathcal{E}$, and b is the "truncation" $[s_1; \ldots; s_{m-1}]$ of the vector of external supplies s, $\sum_{i=1}^m s_i = 0$. As for, the bounds u_γ, ℓ_γ, we assume that

B.1: some of ℓ_γ are reals, and some can be equal to $-\infty$, similarly, some of u_γ are reals, and some are $+\infty$,
B.2: $\ell_\gamma < u_\gamma$ for all γ ("no arc flows are fixed in advance"), and
B.3: for every γ, either ℓ_γ, or u_γ, or both these quantities are reals ("no free arc flows").

We also keep our initial assumption that G is connected; this assumption, as we remember, implies that the rows of A are linearly independent.

In connection with assumptions B.2–3 note that if in the original formulation of the problem there are arcs with fixed flow, we can eliminate these arcs at the price of updating accordingly the external supplies at the nodes linked by these arcs. Similarly, arcs (i, j) with no bounds on flow can be augmented by their inverse arcs (j, i); setting for these arcs $c_{ji} = -c_{ij}$ and requiring the nonnegativity of flows in both arcs (i, j) and (j, i), the pair of these arcs clearly mimicks the original arcs (i, j) with free flows. replaced with pairs of opposite arcs with nonnegative flows and inherited from the original arcs transportation costs. The outlined preprocessing of the original problem converts it into an equivalent problem satisfying assumptions B.2–3. Recall also that when the connectedness assumption is violated, the network flow problem in question decomposes into a bunch of uncoupled problems satisfying this assumption.

5.4.1 *Preliminaries: Primal Simplex method with bounds on variables*

When presenting the Network PSM for the capacitated Network Flow problem, it makes sense to start with the version of the general-purpose PSM aimed at handling bounds on the variables. Thus, consider an LO program

$$\mathrm{Opt} = \min_x \left\{c^T x : Ax = b,\, \ell_j \leq x \leq u_j, 1 \leq j \leq n\right\}, \tag{5.7}$$

where for every j either ℓ_j or u_j or both these quantities are finite, and $\ell_j < u_j$ for all j.[7] As for A, we assume that this is an $m \times n$ matrix with linearly independent rows (as we remember, the latter assumption is "for free").

Problem (5.7) is *not* in the standard form of maximizing a linear objective over nonnegative variables satisfying a system of linear equality constraints. The minor difference is that now we are minimizing, as required in a flow problem, rather than maximizing; the major difference is that we allow for two-sided bounds on the variables rather than restricting all of them to be nonnegative. One way to handle the problem via the PSM is to convert it into the standard form and to apply the method to the resulting problem. However, as far as the efficiency of the computational process is concerned, this option is not the best one; it is better to adjust the PSM to (5.7), which, as we shall see, requires just minor and mostly terminological modifications of the method.

5.4.1.1 *Bases and basic solutions to* (5.7)

The feasible set of (5.7) clearly does not contain lines; thus, if the problem is solvable, it admits a solution which is a vertex of the feasible set $X = \{x : Ax = b, \ell \leq x \leq u\}$. As in the basic version of the method, the PSM as applied to (5.7) travels along the vertices, and our first step is to describe these vertices. By algebraic characterization of extreme points, a point $v \in X$ is a vertex of X iff among the constraints defining X and active at v (i.e., satisfied at v as equalities) there are $n = \dim x$ constraints with linearly independent vectors of coefficients. Part of these n active constraints with independent vectors of coefficients, say, $m' \leq m$ of them, are the

[7]In connection with our restrictions on the bounds, note that when the original problem has free variables (those with $\ell_i = -\infty$, $u_i = +\infty$, we can replace these variables with the differences between pairs of new variables restricted to be nonnegative; and as for fixed variables (those with $\ell_i = u_i \in \mathbf{R}$) in the original problem, they can be eliminated at the price of appropriate modification of the right-hand side vector.

equality constraints $a_p^T x = b_p$, $p \in P$, and $n - m'$ remaining constraints in this group come from the bounds. Since $\ell_j < u_j$, the lower and the upper bounds on the same variable cannot become active simultaneously; thus, $n - m'$ entries in v should be on their upper or lower bounds, let the indexes of these constraints form a set J. Let $J' = \{1, \ldots, n\} \backslash J$. We claim that either $J' = \emptyset$ or the columns of A with indexes from J' are linearly independent. Indeed, assuming the opposite, we can find a nonzero vector h such that $h_j = 0$, $j \notin J'$, and $Ah = 0$, implying that when $t > 0$, the vectors $v \pm th$ satisfy the constraints $Ax = b$. The entries in both these vectors with indexes from J are the same as in v, and the entries with indexes j from J' satisfy, for small positive t, the bounds $\ell_j \leq v_j \pm th_j \leq u_j$, since for these j we have $\ell_j < v_j < u_j$. We see that the vectors $v \pm th$ for small positive t belong to X, which is impossible, since v is a vertex of X, and $th \neq 0$ when $t > 0$. Since J' is either empty, or the columns A_j of A with indexes $j \in J'$ are linearly independent, on the one hand, and the m rows of A are linearly independent, on the other hand, J' can be extended to a basis I of A. Thus,

> Every vertex v of the feasible set X of (5.7) can be associated with a basis I of A (a set of m distinct indexes from $\{1, \ldots, n\}$ such that the columns of A with these indexes are linearly independent) in such a way that every entry v_i in v with index outside of I is either ℓ_i or u_i.

This observation can be reversed:

> Let I be a basis of A and v be a solution to the system of equations $Ax = b$ such that $\ell_i \leq v_i \leq u_i$ for every i, with one of these two inequalities being equality when $i \notin I$. Then v is a vertex of X.

Indeed, for v of the above type, assuming $v \pm h \in X$, we should have $h_i = 0$, $i \notin I$, since for these i v_i is either on its upper or its lower bound, and $v_i \pm h_i$ should obey this bound. Taking into account that $A[v \pm h] = b$, that is, $Ah = 0$, we conclude that $\sum_{i \in I} h_i A_i = 0$, implying $h_i = 0$, $i \in I$, since I is a basis. The bottom line is that $v \pm h \in X$ implies $h = 0$, so that v is a vertex of X.

Given a basis I and setting every entry in x with a nonbasic index to ℓ_i or to u_i, there is exactly one option to define the "basic" entries in x — those with indexes from I — which results in a vector satisfying the equality constraints in (5.7); specifically, the basic part of x should be

$$x_I = A_I^{-1}\left[b - \sum_{i\notin I} x_i A_i\right].$$

The resulting solution x^I is called the *basic solution* associated with basis I *and* the way in which we set the nonbasic entries in x^I — those with indexes outside of I — to the values ℓ_i and u_i. Note that feasibility of a basic solution x^I depends solely on what are the basic (those with indexes from I) entries in this solution: the solution is feasible iff these entries are between the corresponding bounds ℓ_i, u_i.

The bottom line of our discussion is as follows:

(!) Extreme points of the feasible set of (5.7) are exactly the *basic feasible solutions*, that is, the just-defined basic solutions which happen to be feasible.

This conclusion resembles pretty much the similar conclusion for the standard form LO's. The only difference is that in the standard case a basic solution is uniquely defined by the basis, while now it is uniquely defined by the basis *and* the way in which we set the nonbasic entries to their upper and lower bounds.

5.4.1.2 *Reduced costs*

Given a basis I, we define the associated vector c^I of reduced costs exactly as in the case of a standard form problem, that is, as

$$c^I = c - A^T A_I^{-T} c_I, \tag{5.8}$$

where c_I is the m-dimensional vector comprising the entries in c with indexes from I. In other words, c^I is obtained from c by adding a linear combination of rows of A, and this combination is chosen in such a way that the basic entries in c^I are zeros. We clearly have $c^T x = [c^I]^T x$ for all x with $Ax = 0$, or, equivalently,

$$Ax' = Ax'' = b \quad \Rightarrow \quad c^T[x' - x''] = [c^I]^T[x' - x'']. \tag{5.9}$$

Let us make the following observation:

Lemma 5.2. *Let I be a basis, x^I be an associated basic feasible solution (that is, x^I is feasible and nonbasic entries in x^I sit on the* *bounds), and c^I be the vectors of reduced costs associated with I. Then the condition*

$$\forall i \notin I: \text{ either } (x_i^I = \ell_i \text{ and } c_i^I \geq 0), \text{ or } (x_i^I = u_i \text{ and } c_i^I \leq 0), \tag{5.10}$$

is sufficient for x^I to be optimal.

Proof. The fact can be easily derived from considering the dual problem, but we prefer a self-contained reasoning as follows: since x^I is feasible, all we need is to verify that if y is a feasible solution, then $c^T x^I \leq c^T y$. By (5.9), we have $c^T[y - x^I] = [c^I]^T[y - x^I] = \sum_{i \notin I} c_i^I [y_i - x_i^I]$ (we have taken into account that $c_i^I = 0$ when $i \in I$). For $i \notin I$, we have

- either $x_i^I = \ell_i$, $y_i \geq \ell_i$ and $c_i^I \geq 0$,
- or $x_i^I = u_i$, $y_i \leq u_i$ and $c_i^I \leq 0$,

and in both cases $c_i^I[y_i - x_i^I] \geq 0$, whence $\sum_{i \notin I} c_i^I [y_i - x_i^I] \geq 0$. □

5.4.1.3 *A step of the algorithm*

We are ready to describe a step of the PSM as applied to (5.7). At the beginning of the step, we have at our disposal a basis I along with an associated *feasible* basic solution x^I and the vector of reduced costs c^I. At the step, we act as follows:

(1) We check whether x^I, c^I satisfy the sufficient condition for optimality (5.10). If it is the case, we terminate, x^I being the resulting optimal solution and c^I being the optimality certificate.
(2) If (5.10) is not satisfied, we identify a nonbasic index j such that either

- Case A: $x_j^I = \ell_j$ and $c_j^I < 0$, or
- Case B: $x_j^I = u_j$ and $c_j^I > 0$.

In the case of A, we proceed as follows: we define a parametric family $x^I(t)$ of solutions by the requirements that $x_j^I(t) = \ell_j + t$, $Ax^I(t) = b$ and $x_k^I(t) = x_k^I(t)$ for all nonbasic entries with indexes k different from j. We clearly have $x^I(0) = x^I$; as $t \geq 0$ grows, the entries $x_k^I(t)$ with indexes $k \in I \cup \{j\}$ somehow vary, while the remaining entries stay intact, and we have $Ax^I(t) = b$ for all t.

Note that these requirements specify $x^I(t)$ in a unique fashion, completely similar to what we have in the case of a standard form problem:

$$x^I(t) = x^I - tA_I^{-1}A_j.$$

Observe that the objective, as evaluated at $x^I(t)$, strictly improves as t grows:

$$\begin{aligned}
c^T[x^I(t) - x^I] &= [c^I]^T[x^I(t) - x^I] \\
&= \sum_{i\in I} \underbrace{c_i^I}_{=0}[x_i^I(t) - x_i^I] + \underbrace{c_j^I}_{<0} t \\
&\quad + \sum_{i\notin I\cup\{j\}} c_i^I \underbrace{[x_i^I(t) - x_i^I]}_{=0}.
\end{aligned}$$

Now, when $t = 0$, $x^I(t) = x^I$ is feasible. It may happen that $x^I(t)$ remains feasible for all $t > 0$; in this case, we terminate with the claim that the problem is unbounded. An alternative is that eventually some entries $x_i^I(t)$ leave the feasible ranges $[\ell_i, u_i]$; this, of course, can happen only with the entries which do depend on t (these include the entry with index j and those of the basic entries for which the corresponding coordinate in $A_I^{-1}A_j$ is nonzero). We can easily find the largest $t = \bar{t}$ for which $x^I(t)$ is still feasible, but one or more of the coordinates in $x^I(\bar{t})$ are about to become infeasible. Denoting by i the index of such a coordinate, we set $I^+ = [I \cup \{j\}]\backslash\{i\}$, $x^{I^+} = x^I(\bar{t})$. It is immediately seen that I^+ is a basis, and x^{I^+} is the associated basic feasible solution, and we pass to the next step.

In the case of B, our actions are completely similar, with the only difference that now, instead of requiring $x_j^I(t) = \ell_j + t$, we require $x_j^I(t) = u_j - t$.

Note that the construction we have presented is completely similar to the one in the case of a standard form program. The only new possibility we meet is that as a result of a step, the basic solution is changed, but the basis remains the same. This possibility occurs when the entry $x_i^I(t)$ which is about to become infeasible when $t = \bar{t}$ is nothing but the jth entry (which indeed is possible when both ℓ_j and u_j are finite). In this case, as a result of a step, the basis remains intact, but one of the nonbasic variables jumps from one of its bounds to another bound, and the objective is strictly improved.

Finally, observe that the presented method is monotone. Moreover, the objective strictly improves at every step which indeed changes the current basic feasible solution, which definitely is the case when this solution is nondegenerate, i.e., all basic entries in the solution are strictly within their bounds. What does change at a step, which does not update the basic feasible solution, is the basis. These remarks clearly imply our standard conclusions on the correctness of the results reported by the method upon termination (if any) and on finiteness of the method in the nondegenerate case.

5.4.2 *Network PSM for capacitated network flow problem*

The network PSM for the capacitated Network Flow problem is nothing but the specialization of the just-outlined PSM for solving (5.7) for the case of problem (5.6). We already have at our disposal all "network-specific" building blocks (they are inherited from the uncapacitated case), same as the blueprint of how these blocks should be assembled (see the previous Section), so that all we need is just a brief summary of the algorithm. This summary is as follows:

(1) At the beginning of an iteration, we have at our disposal a basis I — a collection of $m-1$ arcs in G which form a tree when their directions are ignored — along with the corresponding basic feasible flow f^I — a feasible solution f^I to (5.6) such that the flows f^I_γ in the arcs distinct from those from I are at their upper or lower bounds.

(2) At a step we act as follows:

(a) We compute the vector of reduced costs $c^I_{ij} = c^I_{ij} + \lambda_j - \lambda_i$ according to the algorithm presented in Section 5.3.3, so that $c^I_\gamma = 0$ when $\gamma \in I$.

(b) If for every arc $\gamma \notin I$ we have either $c^I_\gamma \geq 0$ and $f^I_\gamma = \ell_\gamma$ or $c^I_\gamma \leq 0$ and $f^I_\gamma = u_\gamma$, we terminate — f^I is an optimal solution, and c^I is the associated optimality certificate.

(c) If we do not terminate according to the previous rule, we specify an arc $\bar{\gamma} := (\bar{i}, \bar{j}) \notin I$ such that either

- Case A: $c^I_{\bar{\gamma}} < 0$ and $f^I_{\bar{\gamma}} = \ell_{\bar{\gamma}}$, or
- Case B: $c^I_{\bar{\gamma}} > 0$ and $f^I_{\bar{\gamma}} = u_{\gamma}$.

In what follows we assume that we are in the case of A (our actions in the case of B are completely "symmetric").

(d) Same as in the uncapacitated case, we find a loop $i_1 := \bar{i}, \gamma_1 := \bar{\gamma}, i_2 := \bar{j}, \gamma_2, i_3, \ldots, i_{t-1}, \gamma_{t-1}, i_t = i_1$, where $i_1, \ldots, i_{t-1}$ are distinct nodes of G and $\gamma_2, \ldots, \gamma_{t-1}$ are distinct from each other arcs from I such that either $\gamma_s = (i_s, i_{s+1})$ (forward arc) or $\gamma_s = (i_{s+1}, i_s)$ (backward arc) for every s, $1 \leq s \leq t-1$. We define flow h which vanishes outside of the arcs from the loop, and in the arc γ from the loop is equal to 1 or -1 depending on whether the arc is or is not a forward one. We further set $f^I(t) = f^I + th$, thus getting a flow such that

$$f^I(0) = f^I \quad \text{and} \quad Af^I(t) = b \, \forall t, \, c^T[f^I(t) - f^I] = \underbrace{c^I_{\bar{\gamma}}}_{<0} t.$$

(e) It may happen that the flow $f^I(t)$ remains feasible for all $t \geq 0$. In this case, we have found a feasible ray of solutions to (5.6) along which the objective goes to $-\infty$, and we terminate with the conclusion that the problem is unbounded. An alternative is that as $t \geq 0$ grows, some of the entries $f^I_\gamma(t)$ eventually become infeasible. In this case, it is easy to identify the smallest $t = \bar{t} \geq 0$ for which $f^I(\bar{t})$ is still feasible, same as it is easy to identify an arc $\widehat{\gamma}$ from the above loop such that the flow $f^I_{\widehat{\gamma}}(t)$ at $t = \bar{t}$ is about to become infeasible. We add to I the arc $\bar{\gamma}$ and delete from the resulting set the arc $\widehat{\gamma}$, thus getting a new basis, the corresponding basic feasible solution being $f^{I^+} = f^I(\bar{t})$, and pass to the next step.

The resulting algorithm shares with its uncapacitated predecessor all the properties mentioned on p. 344, in particular, it produces upon termination (if any) an *integral* optimal flow, provided that all entries in b and all finite bounds ℓ_i, u_i are integral. To initialize the algorithm, one can use the same construction as in the uncapacitated case, with zero lower and infinite upper bounds on the flows in the artificial arcs.

5.5 Exercises

Exercise 5.1.* Recall the Transportation problem:

> There is a single product, p suppliers with positive supplies $a_1, \ldots, a_p$ of the product and q customers with positive demands $b_1, \ldots, b_q$ for the product, with the total supply equal to the total demand:
>
> $$\sum_{i=1}^{p} a_i = \sum_{j=1}^{q} b_j.$$
>
> The cost of shipping a unit of product from supplier i to customer j is a real c_{ij}. Find the cheapest shipment from the suppliers to the customers, i.e., find the amounts $x_{ij} \geq 0$ of product to be shipped from supplier i to customer j in such a way that every supplier i ships to the customers totally a_i units of product, every customer j gets totally b_j units of product, and the total transportation cost $\sum_{i,j} c_{ij}x_{ij}$ is as small as possible.

(1) Pose the problem as an uncapacitated Network Flow problem on an appropriate graph.
(2) Show that the problem is solvable.
(3) Prove that there exists an optimal solution x^* to the problem with the following properties:
   (a) the total number of nonzero shipments is $\leq p + q - 1$,
   (b) if i_1, i_2 are two distinct suppliers and j_1, j_2 are two distinct customers, then among the four shipments $x^*_{i_\mu, j_\nu}$, $1 \leq \mu$, $\nu \leq 2$, at least one is equal to zero.

Exercise 5.2.* Prove that every solvable capacitated Network Flow problem with nonzero vector of external supplies is equivalent to a capacitated Network Flow problem where the vector of external supplies has at most 2 nonzero entries summing up to 0.

Exercise 5.3.* Consider a transportation problem

$$\mathrm{Opt} = \min_{x_{ij}} \left\{ \sum_{i=1}^{p}\sum_{j=1}^{q} c_{ij}x_{ij} : \begin{array}{ll} \sum_{j=1}^{q} x_{ij} = a_i, & 1 \leq i \leq p \\ \sum_{i=1}^{p} x_{ij} = b_j, & 1 \leq j \leq q \\ x_{ij} \geq 0 & \end{array} \right\},$$

where $a_i \geq 0$, $b_i \geq 0$ and $\sum_i a_i = \sum_j b_j$.

Assume that supplies a_i and the demands b_j are somehow decreased (but remain nonnegative), while the new total supply is equal to the new total demand. Is it true that the optimal value in the problem cannot increase?

Exercise 5.4.[†] The Maximal Flow problem is as follows:

> *Given an oriented graph $G = (\mathcal{N}, \mathcal{E})$ with arc capacities $\{u_\gamma \in (0, \infty) : \gamma \in \mathcal{E}\}$ and two selected nodes distinct from each other: source $\underline{i}$ and sink $\overline{i}$, find the largest flow from source to sink, that is, find the largest s such that the vector of external supplies "s at the source, $-s$ at the sink, zero at all remaining nodes" fits a flow f satisfying the conservation law and obeying the bounds $0 \leq f_\gamma \leq u_\gamma$ for all $\gamma \in \mathcal{E}$.*

(1) Write down the Maximal Flow problem as an LO program in variables s (external supply at the source) and f (flow in the network) and write down the dual problem.

(2) Let us define a *cut* as a partition of the set $\mathcal{N}$ of nodes of G into two nonoverlapping subsets $\underline{I}$ and $\overline{I} = \mathcal{N} \backslash \underline{I}$ such that the source is in $\underline{I}$, and the sink is in $\overline{I}$. For a cut $(\underline{I}, \overline{I})$, let the capacity $U(\underline{I}, \overline{I})$ of the cut be defined as

$$U(\underline{I}, \overline{I}) = \sum_{\gamma=(i,j)\in\mathcal{E}: i\in\underline{I}, j\in\overline{I}} u_\gamma.$$

Prove that if (s, f) is a feasible solution to the LO reformulation of the Maximal Flow problem, then for every cut $(\underline{I}, \overline{I})$ one has

$$s \leq U(\underline{I}, \overline{I}).$$

Prove that the inequality here is an equality if and only if the flow f_γ in every arc γ which starts in $\underline{I}$ and ends in $\overline{I}$ ("forward arc of the cut") is equal to u_γ, and the flow f_γ in every arc γ which starts in $\overline{I}$ and ends in $\underline{I}$ ("backward arc of the cut") is zero; if it is the case, s is the optimal value in the Maximal Flow problem.

(3) *Invoking the LP Duality Theorem, prove the famous *Max Flow–Min Cut Theorem: If the Max Flow problem is solvable, then its optimal value is equal to the minimum, over all cuts, capacity of a cut.*

Exercise 5.5.† Consider the *Assignment problem* as follows:

> *There are p jobs and p workers. When job j is carried out by worker i, we get profit c_{ij}. We want to associate jobs with workers in such a way that every work is assigned to exactly one worker, and no worker is assigned two or more jobs. Under these restrictions, we seek to maximize the total profit.*

(1) Let G be a graph with $2p$ nodes, p of them representing the workers, and remaining p representing the jobs. Every worker node i, $1 \leq i \leq p$, is linked by an arc $(i, p+j)$ with every job node $p+j$, $1 \leq j \leq p$, the capacity of the arc is $+\infty$, and the transportation cost is $-c_{i,p+j}$. The vector of external supplies has entries indexed by the worker nodes equal to 1 and entries indexed by the job nodes equal to -1.

Prove that the basic feasible solutions $f = \{f_{i,p+j}\}_{1 \leq i,j \leq p}$ of the resulting Network Flow problem are exactly the assignments, that is, flows given by permutations $i \mapsto \sigma(i)$ of the index set $\{1, 2, \ldots, p\}$ according to $f_{i,p+\sigma(i)} = 1$ and $f_{i,p+j} = 0$ when $j \neq \sigma(i)$.

(2) Extract from the previous item that the Assignment problem can be reduced to a Network Flow problem (more exactly, to finding an optimal basic feasible solution in a Network Flow problem).

Pay attention to the following two facts:

- every spanning tree of the graph in question has $2p-1$ arcs, while every basic feasible solution has just p nonzero entries, meaning that when $p > 1$, all basic feasible solutions are degenerate;
- the flows $\{f_{i,p+j}\}_{i,j=1}^{p}$ on G can be associated with $p \times p$ matrices $F_{ij} = f_{i,p+j}$, and with this interpretation, the feasible set of the above Network Flow problem is nothing but the set of all double stochastic $p \times p$ matrices. Thus, the fact established in the first item: "the extreme points of the feasible set of the problem (i.e., the vertices of its feasible set) are exactly the assignments" recovers anew the Birkhoff Theorem: "The extreme points of the set of double stochastic matrices are exactly the permutation matrices."

Exercise 5.6.† Consider the following modification of the Assignment problem:

(!) *There are q jobs and $p \geq q$ workers; assigning worker i with job j, we get profit c_{ij}. We want to assign every job with exactly one worker in such a way that no worker is assigned to two or more jobs (although some of the workers can be assigned to no jobs at all) and want to maximize the total profit under this assignment.*

(1) Reformulate the problem as an Assignment problem.
(2) Assume that all c_{ij} are either 0 or 1; let us say that $c_{ij} = 1$ means that worker i knows how to do job j, and $c_{ij} = 0$ means that worker i does not know how to do job j, and that an assignment ("exactly one worker for every job, no worker with more than one job assigned") is good if every job is assigned to a worker who knows how to do this job. Assume that at least one worker knows how to do at least one job, that is, not all c_{ij} are zeros.

Consider the network with $p+q+2$ nodes: the source (node 0), p worker nodes $(1, 2, \ldots, p)$, q job nodes $(p + 1, \ldots, p + q)$ and the sink (node $p + q + 1$). The arcs, every one of capacity 1, are as follows:

- p arcs "source $\mapsto$ worker node (i.e., arcs $(0, i)$, $1 \leq i \leq p$);
- the arcs "worker $i \mapsto$ "job j which the worker i knows how to do" (i.e., arcs $(i, p+j)$, $1 \leq i \leq p$, $1 \leq j \leq q$, corresponding to the pairs (i, j) with $c_{ij} = 1$);
- q arcs "job node $\mapsto$ sink" (i.e., arcs $(p+j, p+q+1)$, $1 \leq j \leq q$)

along with the Maximal Flow problem on this network.

(a) Prove that the existence of a good assignment is equivalent to the fact that the magnitude of the maximal flow in the above Max Flow problem is equal to q.
(b) Use the Max Flow–Min Cut Theorem to prove the following fact:

(!!) *In (!) with zero/one c_{ij}, an assignment with profit q (i.e., an assignment where every job j is assigned to a worker who knows how to do this job) exists if and only if for every subset S of the set $\{1, \ldots, q\}$ of jobs the total number of workers who know how to do a job from S is at least the cardinality of S.*

Note: (!!) is called the *Marriage Lemma*, according to the following interpretation: there are q young ladies and p young men, some of the ladies being acquainted with some of the men. When is it possible to select for every one of the ladies a bridegroom from the above group of men in such a way that different ladies get different bridegrooms, and every lady is acquainted with her bridegroom? The answer is: it is possible if and only if for every set S of the ladies, the total number of men acquainted each with at least one of the ladies from S is at least the cardinality of S.

Remark 5.1. Pay attention to the dramatic difference in descriptive and operational "powers" of elegant statements like the Marriage Lemma — the difference you could observe in many other results of our Course. The explanatory power of Marriage Lemma is really huge — it points out an evident obstacle to the possibility of a "successful marriage" and demonstrates that this evident obstacle is the only one preventing such a marriage. This being said, the algorithmic, operational power of this result is nearly nonexisting: in order to use it to find an obstacle (or to see that none exists), we were supposed to look through astronomically many subsets of ladies!

Part III

Complexity of Linear Optimization and the Ellipsoid Method

Chapter 6

Polynomial Time Solvability of Linear Optimization

6.1 Complexity of LO: Posing the question

We know that the Simplex method equipped with an anti-cycling strategy is a *finite* algorithm: assuming precise arithmetics, it solves *exactly* every LO program in *finitely many arithmetic operations.* Once upon a time this property was considered the most desirable property a computational routine could possess. However, actual computations require from a good computational routine that it not only be able to solve problems, but also to do so "in reasonable time." From a practical perspective, the promise that a particular algorithm, as applied to a particular problem, will eventually solve it, but that the computation will take 10^{10} years, is not worth much. At the theoretical level, the purely practical by its origin question of "how long a computation will take" is posed as investigating the *complexity* of problem classes and solution algorithms. Speaking informally, the complexity is defined as follows:

(1) We are interested in solving problems from certain family $\mathcal{P}$ (e.g., the family of all LO programs in the standard form); let us call this family a *generic problem*, and a particular member $p \in \mathcal{P}$ of this family (in our example, a particular LO program in the standard form) — an *instance* of $\mathcal{P}$. We assume that within the family an instance is identified by its *data*, which form an array of numbers. For example, the data of an LO in the standard

form comprise two integers m, n — the sizes of the constraint matrix A, two linear arrays storing the entries of the objective c and the right-hand side vector b, and one two-dimensional array storing the constraint matrix A. In fact, we lose nothing by assuming that the data data(p) of an instance is a single one-dimensional array, i.e., a vector. For example, in the LO case, we could start this vector with m and n, then write one by one the entries of c, then the entries of b, and then the entries of A, say, column by column. With this convention, given the resulting data vector, one can easily convert it into the usual representation of an LO program by two vectors and a matrix.

Similarly, we can assume that candidate solutions to instances are linear arrays — vectors (as it indeed is the case with LO).

(2) Now, a solution method $\mathcal{B}$ for $\mathcal{P}$ can be thought of as a program for a computer. Given on input the data vector data(p) of an instance $p \in \mathcal{P}$ and executing this program on this input, the computer should eventually terminate and output the vector representing the solution of p we are looking for, or, perhaps, a valid claim "no solution exists." For example, in the LO situation the result should be either the optimal solution to the LO instance the algorithm is processing, if any, or the correct claim "the instance is unsolvable," perhaps, with explanation why (infeasibility or unboundedness).

(3) We can now measure the complexity $\mathcal{C}(\mathcal{B}, p)$ of $\mathcal{B}$ as applied to an instance $p \in \mathcal{P}$ as the *running time* of $\mathcal{B}$ as applied to the instance, that is, as the total time taken by elementary operations performed by the computer when processing the instance. Given $\mathcal{B}$, this complexity depends on the instance p in question, and it makes sense to "aggregate" it into something which depends only on the pair $(\mathcal{B}, \mathcal{P})$. A natural way to do it is to associate with an instance p its "size" Size(p), which is a positive integer somehow quantifying the "volume of the data" in p. For example, we can use, as the size, the number of data entries specifying the instance within $\mathcal{P}$, that is, the dimension of the data vector data(p). Another, more realistic, as far as digital computations are concerned, way to define the size is to assume that the data entries admit finite binary representation, e.g., are integers or

rationals (i.e., pairs of integers), and to define the size of an instance p as the total number of bits in the vector data(p); this is called the *bit size* of p.

After the instances of $\mathcal{P}$ are equipped with sizes, we can define the *complexity of a solution algorithm* $\mathcal{B}$ as the function

$$\mathcal{C}_{\mathcal{B}}(L) = \sup_{p} \{\mathcal{C}(\mathcal{B}, p) : p \in \mathcal{P}, \mathrm{Size}(p) \leq L\},$$

which shows how the worst-case, over instances of sizes not exceeding a given bound L, running time of $\mathcal{B}$, as applied to an instance, depends on L.

We can then decide which types of complexity functions are, and which are not, appropriate for us, thus splitting all potential solution algorithms for $\mathcal{P}$ into *efficient* and *inefficient*, and investigate questions like

- *Whether such and such a solution algorithm $\mathcal{B}$, as applied to such and such a generic problem $\mathcal{P}$, is efficient? What is the corresponding complexity?*
  or
- *Whether such and such a generic problem $\mathcal{P}$ is efficiently solvable, that is, admits an efficient solution algorithm?*

A reader could argue why should we quantify complexity in the worst-case fashion, and not by the "average" or the "with high probability" behavior of the algorithm. Practically speaking, a claim like "such and such an algorithm solves instances of size L in time which, *with probability 99.9%*, does not exceed $0.1L^2$" looks much more attractive than the claim "the algorithm solves *every* instance of size L in time not exceeding $100L^3$." The difficulty with the probabilistic approach is that most often than not we cannot assign the data of instances with a meaningful probability distribution. Think of LO: LO programs arise in an extremely wide spectrum of applications, and what is "typical" for the Diet problem modeling nutrition of chickens can be quite atypical for "the same" problem as applied to nutrition of cows, not speaking about problems coming from production planning. As a result, a meaningful probabilistic approach can make sense (and indeed makes sense) only when speaking about relatively narrow problems coming from an "isolated" and well-understood source. A typical alternative — to choose a data distribution on the basis of mathematical convenience — usually leads to "efficient" algorithms poorly performing in actual applications.

Polynomial time algorithms and polynomially solvable generic problems: As a matter of fact, there exists a complete consensus on what should be treated as an "appropriate complexity" — this is *polynomially* of the complexity in L, that is, the existence of a polynomial in L upper bound on this function. Thus, a solution algorithm $\mathcal{B}$ for a generic problem $\mathcal{P}$ is called *polynomial*, or *polynomial time*, if

$$\mathcal{C}_{\mathcal{B}}(L) \leq cL^d$$

for fixed constants c, d and all $L = 1, 2, \ldots$. A generic problem $\mathcal{P}$ is called *polynomially solvable*, or *computationally tractable*, if it admits a polynomial time solution algorithm.

Informally, the advantage of polynomial time algorithms as compared to their most typical (and usually easy-to-build) alternatives, the exponential time algorithms with the complexity $\exp\{O(1)L\}$, can be illustrated as follows. Imagine that you are solving instances of a certain generic problem on a computer, and then get a 10 times faster computer. If you use a polynomial time algorithm, this constant factor improvement in the hardware will result in a constant *factor* increase in the size of an instance you can process in a given time, say, in one hour. In contrast to this, with an exponential algorithm, 10-fold improvement in the performance of the hardware increases the size L of an instance you can solve in one hour *only additively*: $L \mapsto L + \text{const}$.

There are deep reasons for why the theoretical translation of the informal notion "computationally tractable generic problem" is polynomial time solvability. The most important argument in favor of formalizing tractability as polynomial solvability is that with this formalization the property to be tractable becomes independent of how exactly we encode candidate solutions and data of instances by linear arrays; as a matter of fact, all natural encodings of this type can be converted one into another by polynomial time algorithms, so that the property of a problem to be tractable turns out to be independent of how exactly the data/candidate solutions are organized. Similarly, the property of an algorithm to be polynomial time is insensitive to minor details of what are the elementary operations of the computer we use. As a result, theoretical investigation of complexity can ignore "minor technical details" on how exactly we encode the data and what exactly our computer can do, thus making the investigation humanly possible.

For example, a natural ("compact") way to store the data of a multicommodity flow problem is essentially different from a natural way to store the data of the LO-standard form representation of the same problem. Indeed, to specify a problem with N commodities on a graph with n nodes and m arcs, we need to point out the incidence matrix of the graph (mn data entries when we store the matrix as a full one), the supply of each commodity at each node (totally Nn data entries) and Nm cost coefficients; all this amounts to $N(m+n)+mn$ data entries. In the standard form representation of the same problem, there are more than Nm variables and more than n equality constraints, so that the number of data entries in the constraint matrix is more than Nmn; when N, m, n are large, the number of data entries in the standard form representation is by orders of magnitude larger than in the former, natural and compact, representation of the problem's data. Nevertheless, the standard form data clearly can be produced in a polynomial time fashion from the data of the compact representation of the problem. As a result, if we know a polynomial time algorithm $\mathcal{B}$ for solving multicommodity flow problems represented in the standard form, we automatically know a similar algorithm for these problems represented in the compact form; to build such an algorithm, we first convert in polynomial time the "compact form data" into the standard form one, and then apply to the resulting standard form problem the algorithm $\mathcal{B}$. The overall computation clearly is polynomial time. Note that conclusions of this type are "automatically true" due to our somehow "loose" definition of efficiency. Were we defining an efficient algorithm as one with the complexity bounded by a *linear* function of L, similar conclusions would be impossible; say, it would be unclear why efficient solvability of the problem in the standard form implies efficient solvability in the compact form, since the conversion of the data from the latter form to the former one takes nonlinear, although polynomial, time.

From the viewpoint of actual computations, the simple dichotomy "polynomial–nonpolynomial" is indeed loose: we would not be happy when running an algorithm with complexity of order of L^{100}. However, the loose definition of efficiency of an algorithm as polynomiality of its running time captures perhaps not all, but at least the most crucial components of the diffuse real-life notion of efficiency. While a polynomial complexity bound *may happen* to be too big from a practical viewpoint, a typical nonpolynomial bound — the exponential bound $\exp\{O(1)L\}$ — definitely *is* too big from this viewpoint. It should be added that, as a matter of fact, the complexity bounds for the vast majority of known polynomial time algorithms are polynomials of quite moderate order, like 2 and 3, so that discussions on what is better in applications — a polynomial time algorithm with complexity L^{100} or a nonpolynomial time algorithm with complexity $\exp\{0.1L\}$ — have more or less empty scope. What indeed is true is that *after* we know that a generic problem is polynomially solvable, we can start to bother about the degrees of the related polynomials, about what is the best, from the practical viewpoint, way to organize the data, etc.

6.1.1 *Models of computations*

The informal "definition" of complexity we have presented is incomplete, and the crucial yet missing element is *what is our model of computations*, what is the "computer" and its "elementary operations" we were speaking about. In our context, one could choose between two major models of computation,[1] which we will refer to as the *Rational Arithmetic* and *Real Arithmetic* models.

Real Arithmetic model of computations: You can think about this model of a computation on an idealized *Real Arithmetic computer* capable of storing indefinitely many real numbers and carrying out *precisely* elementary operations with these numbers — the four arithmetic operations, comparisons, and computing the values of elementary functions, like $\surd$, exp, sin, etc. Execution of such an operation takes unit time.

Rational Arithmetic model of computations: Here, we are computing on a *Rational Arithmetic computer* capable of storing indefinitely many finite binary words. We interpret these words as encoding pairs of binary-represented integers (say, "00" encodes bit 0, "11" encodes bit 1, "10" encodes sign $-$, and "01" separates bits and signs of the first member of the pair from those of the second member. For example, the pair of integers $(3, -2)$ will be encoded as

$$\overbrace{1111}^{a}\,\overbrace{01}^{b}\,\overbrace{10}^{c}\,\overbrace{1100}^{d},$$

where a encodes the binary representation '11' of 3, b plays the role of comma, c encodes the sign $-$, and d encodes the binary representation '10' of 2. We shall interpret the integers encoded by a binary word as the

[1] In the presentation to follow, we intend to explain the essence of the matter and skip some of the technicalities.

numerator and denominator of a fraction representing a rational number. The operations of the Rational Arithmetic computer are the four arithmetic operations with rational numbers encoded by the operands and comparisons ($>$, $=$, $<$) of the rational numbers encoded by the operand with 0. The time taken by an elementary operation is now not a constant, but depends on the total binary size ℓ of the operands; all that matters in the sequel is that this time is bounded from above by a fixed polynomial of ℓ.

Note that our assumption on how long an operation takes corresponds to the situation when the true elementary operations, those taking unit time, are "bit-wise" — operate with operands of once-forever-fixed binary length, say, with bytes. With a rich enough family of these bit-wise operations, it is still possible to carry our arithmetic operations with rational numbers, but such operations, rather than being "built into the hardware," are implemented as execution of micro-programs which use the bit-wise operations only. As a result, time taken by, say, addition of two rational numbers is not a constant, it grows with the total binary size of the operands, as it is the case with computations by hand.

Needless to say, the Rational Arithmetic computer is much closer to a real-life computer than the Real Arithmetic computer is. Nevertheless, both models are meaningful idealizations of the process of actual computations; which one of the models to use depends on the subject area we intend to analyze. For example, the Rational Arithmetic model is *the* model used in Combinatorial Optimization which operates with fully finite entities and hardly needs either rational fractions, or reals. In contrast to this, in Continuous Optimization, especially, beyond LO, it would be a complete disaster — in this area, to bother all the time about keeping the results of all computations in the field of rational numbers would be a severe obstacle to inventing and analyzing algorithms, and here a Real Arithmetic model of computations is really indispensable.

The existence of several models of computation makes the question "Is a given generic problem $\mathcal{P}$ polynomially solvable or not" somehow ill-posed. Assume that we use the Rational Arithmetic model of computations and want to know what is the solvability status of a generic problem like "Solving square nonsingular systems of linear equations," or "Solving LO programs." First of all, we should restrict ourselves to instances where the data vector is rational, since otherwise we cannot even load the data in our computer. Second, we should measure the size of an instance, taking into account the total binary lengths of the entries in the data vector, not only their number. Otherwise, solving the linear equation as simple as $ax = b$, a and b being positive rationals, would be an intractable computational task: to solve the equation, we should compute the numerator and the denominator of the rational number b/a, and the larger the running time of such a division on a Rational Arithmetic computer, the larger the total bit size of the four integers hidden in a, b. Thus, our equation cannot be solved in time polynomial in the *number* of data entries. To overcome this difficulty, the standard definition of the size of an instance p is the *bit length of the corresponding data*, that is, the total number of bits in the binary representation of the (rational!) entries of the data vector data(p). We will refer to the resulting complexity model as to the *Combinatorial one*. Thus,

> *In the Combinatorial Complexity model:*
>
> - *all instances of a generic problem have rational data vectors,*
> - *the size of an instance is the total bit length of its data vector, and*
> - *computations are carried out on a Rational Arithmetic computer, so that the running time is quantified by the total number of bit-wise operations performed in the course of the solution process. In particular, the running time of an arithmetic operation with rational operands is a fixed polynomial of the total binary length of the operands.*

Now, assume that we use the Real Arithmetic model of computations and address "the same" generic problems "Solving square nonsingular system of linear equations" and "Solving LO programs." Now, there is no necessity to require from the data of the instances to be rational. Likewise, it is natural to define the size of an instance as the number of entries (i.e., the dimension) of the corresponding data vector. We will refer to this complexity model as the *Real Arithmetic* one. Thus,

> *In the Real Arithmetic Complexity model:*
>
> - *instances of a generic problem may have real data vectors,*
> - *the size of an instance is the dimension of its data vector, and*
> - *computations are carried out on a Real Arithmetic computer, so that the running time is quantified by the total number of real arithmetic operations performed in the course of the solution process. In particular, it takes unit time to perform an arithmetic operation with real operands.*

For more details on the Real Arithmetic Complexity model (including, in particular, the case where we seek for an *approximate* solution of a prescribed accuracy rather than for a precise solution — the situation typical for nonlinear optimization programs which cannot be solved exactly in finite number of arithmetic operations), see (Ben-Tal and Nemirovski, 2001a).

Note that when passing from investigating tractability of a generic problem, say, "Solving square nonsingular systems of linear equations" in the framework of CCM (Combinatorial Complexity model) to investigating "the same" problem in the framework of RACM (Real Arithmetic Complexity model), the task — to build a polynomial time solution algorithm — simplifies in some aspects and becomes more complicated in other aspects. Simplification comes from the fact that now we have at our disposal much more powerful computers. Complication comes from two sources. First, we extend the family of instances by allowing for real data (this explains the quotation marks in the above "the same"); usually, this is a minor complication. Second, and much more important, complication comes from the fact that even for instances with rational data we now want much more than before: we want the running time of our new powerful computer to be bounded by a *much smaller* quantity than before — by polynomial in the dimension of the data vector instead of polynomial in the bit size of this vector. As a result, polynomial solvability of the problem in one of the models does not say much about polynomial solvability of "the same" problem in the second model. For example, the problem of solving a square nonsingular system of linear equations is tractable — polynomially solvable — in both CCM and RACM, and the efficient solution algorithm can be chosen to be "the same," say, Gaussian elimination. However, the meaning of the claim "Gaussian elimination allows to solve square nonsingular systems of linear equations in polynomial time" heavily depends on what is the model of complexity in question. In CCM, it means that the *"bit-wise"* effort of solving a system is bounded by a polynomial in the *total bit size* of the data; from this fact we cannot make any conclusion on how long it takes to solve a single linear equation with one variable — the answer depends on how large is the binary length of the data. The same claim in RACM means that an $n \times n$ system of linear equations can be solved in a polynomial in n time of operations of the precise Real Arithmetics (just one division in the case of a 1×1 system). At the same time, the claim does not say how long it will take to solve a system with rational data on a "real life" computer, which in fact is a Rational Arithmetics one. That the latter shortcoming of the "RACM tractability" indeed is an important "real life" phenomenon, can be witnessed by everybody with even minor experience in applications of Numerical Linear Algebra — just ask him/her about rounding errors, condition numbers, ill-posed systems and the like.

Now, what about tractability status of the main hero of our book — LO? The situation is as follows.

- The question whether LO admits a polynomial time solution algorithm in the Real Arithmetic Complexity model has remained open for over 5 decades; in somehow refined form, it is included in the "short list," compiled by the famous mathematician, Fields Laureate Stephen Smale, of the major open problems with which Mathematics entered the XXI century. The strongest result in this direction known so far is that *when restricting the entries of the constraint matrix in a standard form LO to be integers varying in a once-forever-fixed finite range, say,* $0/1$ *or* $0, 1, \ldots, 10$ *and allowing the right-hand side and the objective vectors to be real, we get an RACM generic problem that is polynomially solvable* (Tardos, 1985; Vavasis and Ye, 1996).
- The question whether LO admits a polynomial time solution algorithm in the Combinatorial Complexity model of computations remained open for about three decades; it was answered positively by Khachiyan (1979); his construction, to be reproduced later in this chapter, heavily exploits the Real Arithmetic Ellipsoid Algorithm for convex optimization proposed in Nemirovski and Yudin (1976) and independently, slightly later, in Shor (1977).

6.1.2 *Complexity status of the Simplex method*

In spite of the exemplary performance exhibited by the Simplex method in practical computations, this method is not polynomial either in RACM, or in CCM. Specifically, as early as in the mid-1960s, Klee and Minty presented examples of simple-looking LO programs ("Klee–Minty cubes," see Bertsimas and Tsitsiklis (1997, Section 3.7)) where the method exhibits exponential time behavior. More precisely, Klee and Minty discovered a series of concrete LO instances p_1, $p_2, \ldots$, n being the design dimension of p_n, with the properties as follows:

- both the bit length and the dimension of the data vector of p_n are polynomial (just quadratic) in n;
- the feasible set of p_n is a polytope given by $2n$ linear inequality constraints and possessing 2^n vertices, all of then nondegenerate;
- as applied to p_n, the Simplex method, started at an appropriately chosen vertex and equipped with appropriate pivoting rules compatible with the method's description, arrives at the optimal vertex after 2^n iterations, that is, it visits all 2^n vertices one by one.

Since Klee and Minty presented their examples, similar "bad examples" were built for all standard pivoting rules. Strictly speaking, these examples do not prove that the Simplex method cannot be "cured" to become a polynomial. Recall that the method is not a single fully determined algorithm; it is a family of algorithms, a particular member of this family being specified by the pivoting rules in use. These rules should be compatible with a certain "conceptual scheme" we have presented when describing the method, but they are not uniquely defined by this scheme. And nobody has proved yet that this scheme does not allow for implementation that does lead to a polynomial time algorithm. What could be fatal in this respect, at least in the RACM framework, is the "heavy failure to be true" of the famous *Hirsch Conjecture* as follows:

> *Let X be a polyhedral set given by m linear inequality constraints on n variables and not containing lines. Every two vertices of X can be linked to each other by a path of at most $m - n$ consecutive edges (one-dimensional faces such that the end vertex of an edge from the path is the start vertex in the next edge from it).*

If this long-standing conjecture is in fact "severely wrong," and the maximin number

$$N(m, n) = \max_{X,v,v'} \min_{\text{paths}} \{\# \text{ of path edges}\}$$

of edges in an edge path (the maximum is taken over all "line-free" polyhedra X given by m linear inequalities on n variables and over all pairs of vertices v, v' of X, the minimum is taken over all edge paths on X linking v and v') grows with m, n faster than every polynomial of $m + n$, then the Simplex method cannot be cured to become a polynomial time one. Indeed, given the polyhedral set X along with its vertices v, v' corresponding to the above maximin, we can make v' the unique optimal solution to $\min_X c^T x$ (Proposition 3.3), and v — the vertex the Simplex method starts with when solving this program. Whatever be the pivoting rules, the number of vertices visited by the Simplex method when solving the program clearly will be bounded from below by $N(m, n)$, and the latter quantity, if Hirsch was severely mistaken in his conjecture, grows with m, n in a super-polynomial fashion.

The current status of the Hirsch Conjecture is as follows. For polytopes (i.e., nonempty and bounded polyhedral sets) its validity remained open till 2011, when Francisco Santos built a polytope with $m = 86$ and $n = 43$ serving as a counterexample. For unbounded polytopes, it is disproved by Klee and Walkup (1967) who demonstrated that

$$H(m, n) \geq m - n + \text{floor}(n/5).$$

These counterexamples to the original conjecture, however, do not lead to any fatal consequences regarding potential polynomiality of the Simplex method. As far as an upper bound on $H(m, n)$ is concerned, the best-known bound, due to Kalai and Kleitman (1963), is

$$N(m, n) \leq n^{\log_2 m};$$

while being sub-exponential, it is not polynomial. The bottom line is that making the Simplex method polynomial is not easier than proving a somehow relaxed (a polynomial in m, n instead of $m - n$) version of the long-standing Hirsch Conjecture.

6.1.3 **Classes* **P** *and* **NP**

Here, we briefly explain the notions of complexity classes P and NP which are of paramount importance in CCM; while we do not directly use them in what follows, it would be unwise to omit this topic. In what follows, we deal with the Combinatorial Complexity model only.

Consider a generic problem $\mathcal{P}$ with rational data, and let it be a *decision problem*, meaning that a candidate solution to an instance is just the answer "yes, the instance possesses certain property P," or "no, the instance does not possess the property P;" the property in question is part of the description of the generic problem $\mathcal{P}$. Examples are:

- [Shortest path] "Given a graph with arcs assigned nonnegative integral weights, two nodes in the graph and an integer M, check whether there exists a path of length[2] $\leq M$ linking these two nodes."
- [Traveling salesman] "Given a graph with arcs assigned nonnegative integral weights and an integer M, check whether there exists a tour (cyclic path which visits every node exactly once) of length $\leq M$."
- [Stones] "Given positive integral weights $a_1, \ldots, a_n$ of n stones, check whether the stones can be split into two groups of equal total weights."
- [Decision version of LO] "Given an LO program with rational data, check its feasibility."

A decision problem is in class NP if the positive answer is "easy to certify." Specifically, there exists a predicate $\mathcal{A}(x, y)$ (a function of two finite binary words x, y taking values 0 and 1) and a polynomial $\pi(\ell)$ such that

(1) The predicate is polynomially computable — there exists a code for the Rational Arithmetic computer which, given on input a pair of binary words x, y, computes $A(x, y)$ in the polynomial in $\ell(x) + \ell(y)$ number of bit-wise operations; here and in what follows $\ell(\cdot)$ is the bit length of a binary word;
(2) For every x, if there exists y such that $\mathcal{A}(x, y) = 1$, then there exists a "not too long" y' with the same property $\mathcal{A}(x, y') = 1$, specifically, such that $\ell(y) \leq \pi(\ell(x))$;
(3) An instance p of $\mathcal{P}$ possesses property P iff the data data(p) of the instance can be augmented by certain y in such a way that $\mathcal{A}(\text{data}(p), y) = 1$.

We assume that the predicate $\mathcal{A}$, the polynomial time algorithm computing this predicate and the polynomial π are part of the description of the generic problem $\mathcal{P}$.

Comments: A. $\mathcal{P}$ is a generic program with rational data, so that data(p) is a finite sequence of integers; we can always encode such

[2]Defined as the sum of lengths of edges constituting the path.

a sequence by a finite binary word (cf. encoding of $(3,-2)$ above). After the encoding scheme is fixed, we can think of data(p) as of a binary word, and (3) above makes sense.

B. The structure of the above definitions is well known to us. They say that presence of the property P in an instance p is easy to certify: a certificate is a binary word y which, taken along with the data of p, satisfies the condition $\mathcal{A}(\text{data}(p), y) = 1$, and this certification scheme is complete. Moreover, given data(p) and candidate certificate y, it is easy to verify whether y is a valid certificate, since $\mathcal{A}(\cdot,\cdot)$ is polynomially computable. Finally, given p, we can point out a *finite* set $C(p)$ of candidate certificates such that if p possesses property P, then the corresponding certificate can be found already in $C(p)$. Indeed, it suffices to take, as $C(p)$ the set of all binary words y with $\ell(y) \leq \pi(\ell(x))$, see (2).

It is easily seen that all decision problems we have listed above as examples (except, perhaps, the decision version of LO) are in the class NP. For example, in the Shortest path problem, a candidate certificate y encodes a loop-free path in a graph.[3] In order to compute $\mathcal{A}(\text{data}(p), y)$, we first check whether y is a loop-free path in the graph represented by p, and then compute the length of this path and compare it with the threshold M (which is part of the data of p); you can easily verify that this construction meets all the requirements in (1)–(3).

The only example in our list for which its membership in NP is not completely evident is the decision version of LO with rational data. Of course, we understand how to certify the feasibility of an LO program — just by pointing out its feasible solution. The question, however, is why can such a certificate, if it exists, be chosen to be rational with a "not too long" binary representation? What we need here is the following statement:

> (!) *If a system of linear inequalities with rational data of total bit length L is feasible, then it admits a rational solution with total bit length of the entries bounded by a fixed polynomial of L.*

The validity of this statement can be easily derived from what we already know, specifically, from algebraic characterization of extreme

[3] That is, a finite sequence of positive integers distinct from each other, like (1,17,2), meaning that the consecutive arcs in the path are (1,17) and (17,2); we assume that the nodes in a graph are identified by their serial numbers.

points of polyhedral sets, see Step 1 in the proof of the following Theorem 6.2. With (!) at our disposal, the fact that the decision version of LO with rational data is in NP becomes clear.

The class P is, by definition, a subclass of NP comprising all *polynomially solvable* problems in NP.

A decision problem $\mathcal{P} \in$ P clearly admits a finite solution algorithm. Indeed, given an instance p of $\mathcal{P}$, we can look, one by one, at all candidate certificates y with $\ell(y) \leq \pi(L)$, where $L = \ell(\text{data}(p))$ is the bit size of p. Each time we check whether y is a valid certificate (this reduces to computing $\mathcal{A}(\text{data}(p), y)$, which is easy). When a valid certificate is found, we terminate and claim that p possesses property P, and if no valid certificate is found when scanning all $2^{\pi(L)}$ of our candidates, we can safely claim that p does not possess the property. While being finite, this algorithm is *not* polynomial, since the number $2^{\pi(L)}$ of candidates we should scan is not bounded by a polynomial of the bit size L of p. Thus, while every problem belonging to NP admits a finite solution algorithm, it is unclear whether it is tractable — that is, belongs to P.

The question whether P = NP is *the major open question* in Complexity Theory and in Theoretical Computer Science; it also belongs to Smale's list of major mathematical challenges. The extreme importance of this question stems from the fact that *essentially every problem in Combinatorics, Boolean and Integer Programming, etc., can be reduced to solving a "small series" of instances of an appropriate decision problem from* NP, specifically, in such a way that a polynomial time solution algorithm for the decision problem can be straightforwardly converted into a polynomial time solution algorithm for the original problem.

For example, the optimization version of the Traveling salesman problem "given a graph with arcs assigned nonnegative integral lengths, find the shortest tour" can be reduced to small series of instances of the decision version of the same problem, where instead of minimizing the length of a tour one should check whether there exists a tour of the length not exceeding a given bound (think what this reduction is). Similarly, *solving* LO programs with rational data reduces to a small series of decision problems "check whether a given system of linear inequalities with rational data is feasible," see Step 2 and Step 3 in the proof of the following Theorem 6.2.

Thus, if NP were equal to P, our life would become a paradise — there would be no difficult, at least from the theoretical viewpoint, computational problems at all! Unfortunately, even after over several decades of intensive studies, we do not know whether P=NP. The common belief is that this is not the case, and the reason stems from the fundamental discovery of NP-*complete* problems.

Polynomial reducibility and NP-completeness: Consider two generic decision problems with rational data, $\mathcal{P}$ and $\mathcal{P}'$. We say that $\mathcal{P}'$ can be polynomially reduced to $\mathcal{P}$ if the data data(p') of every instance $p' \in \mathcal{P}'$ can be converted *in polynomial time* into the data of an instance $p[p']$ of $\mathcal{P}$ such that the answers in p' and in $p[p']$ are the same — or both are "yes," or both are "no." In this situation, tractability of $\mathcal{P}$ automatically implies tractability of $\mathcal{P}'$; indeed, to solve an instance p' of $\mathcal{P}'$, we first convert its data in polynomial time in the data of $p[p']$ and then solve $p[p']$ by a polynomial time solution algorithm for $\mathcal{P}$ ($\mathcal{P}$ is tractable!); the answer we get is the desired answer for p', and the overall computation is polynomial.[4] In particular, if $\mathcal{P}$ and $\mathcal{P}'$ are polynomially reducible to each other, their tractability status is the same.

Now, a problem $\mathcal{P}$ from the class NP is called NP-*complete* if *every* problem from NP is polynomially reducible to $\mathcal{P}$. Thus, all NP-complete problems have the same tractability status; moreover, if one of them is polynomially solvable, then so are *all* problems from NP, and thus P=NP.

The major discovery in the Complexity Theory we have mentioned is that NP-*complete problems do exist.* Moreover, it turned out that nearly all difficult problems in Combinatorics — those for which no polynomial time algorithms were known — are NP-complete — "as difficult as a problem from NP can be." For example, in our list of examples, the Shortest path problem is polynomially solvable, while the Traveling salesman and the Stones problems are NP-complete. Eventually, nearly all interesting problems from NP fall in one of just two groups: NP-complete and polynomially solvable. There is just a handful of exceptions — problems from NP for which still it is not known whether they are NP-complete or polynomially solvable.

Now, we can explain why the common expectation is that NP$\neq$P (or, which is the same, that NP-complete problems are

[4]To see this, note that the bit size of $p[p']$ is bounded by a polynomial of the bit size of p'; indeed, in the CCM, the bit size of the output of a polynomial time computation is bounded by a polynomial of the bit size of the input (it takes unit time just to write down a single bit in the output).

intractable), so that in our current practice, the verdict "problem $\mathcal{P}$ is NP-complete" is interpreted as lack of hope to solve the problem efficiently.[5] As we have already mentioned, the vast majority of NP-problems are known to be NP-complete and thus, in a sense, form a single problem. This "single problem" in various forms arises in, and often is vitally important for, a huge spectrum of applications. Due to their vital importance, these problems were subject of intensive research of many thousands of excellent scholars over decades. Since all these scholars were in fact solving the same problem, it is highly unlikely that all of them overlooked the possibility of solving this problem efficiently.

6.2 The Ellipsoid Algorithm

In this section, important by its own right, we take a crucial step toward demonstrating that LO is polynomially solvable in the Combinatorial Complexity model. Surprisingly, this step seems to have nothing to do with LO; the story to be told is about a particular algorithm for solving black-box-oriented convex problems.

6.2.1 *Problem and assumptions*

The Ellipsoid Algorithm (EA) is a "universal" method for solving convex optimization problems in the form

$$\mathrm{Opt} = \min_{x \in X} f(x), \tag{6.1}$$

where $X \subset \mathbf{R}^n$ is a *solid* — a closed convex set with a nonempty interior, and $f : \mathbf{R}^n \to \mathbf{R}$ is a convex function. To stick to the policy "minimum of Calculus" we follow in our book, we assume from now on that $f(x)$ is the maximum of finitely many differentiable convex functions $f_i(\cdot) : \mathbf{R}^n \to \mathbf{R}$, $i \leq M$:

$$f(x) = \max_{1 \leq i \leq M} f_i(x), \tag{6.2}$$

[5] Meaning absence of polynomial time complexity *guarantees*; there are algorithms in Combinatorics which succeed in solving in reasonable time typical, from the practical viewpoint, instances of various NP-complete problems.

while X is cut off $\mathbf{R}^n$ by finitely many constraints $g_i(x) \le 0$, where $g_i(\cdot) : \mathbf{R}^n \to \mathbf{R}$ are differentiable convex functions:

$$X = \{x \in \mathbf{R}^n : g_i(x) \le 0, \, 1 \le i \le N\}. \tag{6.3}$$

Black box representation of (6.1): In what follows, we assume that the descriptions of f as the maximum of finitely many differentiable convex functions and of X as the set given by finitely many convex inequality constraints with differentiable right-hand sides "exist in nature," but are not necessarily accessible for a solution algorithm. For the latter, the problem is given by two *oracles*, or *black boxes* — routines the algorithm can call to get their outputs. The oracles are as follows:

- *Separation Oracle*, which, given on input a query point $x \in \mathbf{R}^n$, "says" on the output whether or not $x \in X$, and if it is not the case, returns a *separator* — a nonzero vector $e \in \mathbf{R}^n$ such that

  $$e^T(x - x') > 0 \quad \forall x' \in X; \tag{6.4}$$

  Note that $e \neq 0$, since X is nonempty.
- *First-Order Oracle*, which, given on input a query point $x \in \mathbf{R}^n$, returns the value $f(x)$ and a *subgradient* $f'(x)$ of f at x.

  A *subgradient* of a convex function $h : \mathbf{R}^n \to \mathbf{R} \cup \{+\infty\}$ at a point $x \in \operatorname{Dom} h$ is, by definition, a vector $e \in \mathbf{R}^n$ such that $h(y) \ge h(x) + e^T(y - x)$ for all y. If a convex function h is differentiable at a point x from its domain, one can take, as a subgradient of h at x, the gradient $\nabla h(x)$ of h at x; the easy-to-prove fact is that $\nabla h(x)$ is indeed a subgradient:

  $$h(y) \ge h(x) + [\nabla h(x)]^T(y - x) \, \forall y.$$

  This is called *gradient inequality for convex functions*; it expresses the quite intuitive fact that the graph of a convex function is above its tangent hyperplane. In fact, subgradients do exist at every *interior* point of the domain of a convex function. A subgradient of a given function at a given point is not necessarily unique (consider what are the subgradients of the function $h(x) = |x|$ at $x = 0$).

One of the ways to build a Separation and a First-Order oracle is as follows:

- *Separation Oracle*: given on input a point x, we compute one by one the quantities $g_i(x)$ and check whether they are nonpositive. If that is the case for all i, then $x_i \in X$, and we report this outcome to the algorithm. If $g_{i_*}(x) > 0$ for certain i_*, we report to the algorithm that $x \notin X$ and return, as a separator, the vector $e = \nabla g_{i_*}(x)$. This indeed is a separator, since by gradient inequality

$$g_{i_*}(y) \geq g_{i_*}(x) + e^T(y - x),$$

meaning that if $y \in X$, so that $g_{i_*}(y) \leq 0$, we have $e^T(x - y) \geq g_{i_*}(x) - g_{i_*}(y) \geq g_{i_*}(x) > 0$.

- *First-Order Oracle*: given on input a point x, we compute one by one all the quantities $f_i(x)$ and pick the largest of them, let it be $f_{i_*}(x)$; thus, $f(x) = \max_i f_i(x) = f_{i_*}(x)$. We return to the algorithm the real $f_{i_*}(x)$ as the value, and the vector $\nabla f_{i_*}(x)$ — as a subgradient of f at x. To justify correctness of this procedure, note that by the gradient inequality as applied to the convex and differentiable function $f_{i_*}(\cdot)$, we have $f_{i_*}(y) \geq f_{i_*}(x) + [\nabla f_{i_*}(x)]^T(y-x)$ for all y; the left-hand side in this inequality is $\leq f(y)$, and the right-hand side is equal to $f(x) + [\nabla f_{i_*}(x)]^T(y - x)$, so that $f(y) \geq f(x) + [\nabla f_{i_*}(x)]^T(y - x)$ for all y, as claimed.

***A priori* information on (6.1)** comprises two positive reals $R \geq r$ such that X is contained in the Euclidean ball $B^R = \{x \in \mathbf{R}^n : \|x\|_2 \leq R\}$, and there exists $\bar{x}$ such that the Euclidean ball $\{x : \|x - \bar{x}\|_2 \leq r\}$ is contained in X. Recall that X was assumed to be bounded and with a nonempty interior, so that the required R and r "exist in nature." We assume that these "existing in nature" quantities are known to the solution algorithm (note: we do *not* assume that the center $\bar{x}$ of the second ball is known to the algorithm). It should be mentioned that the *a priori* knowledge of R is crucial, while the knowledge of r does not affect the iterations and is used in the termination criterion only. In fact, with appropriate modification of the method (see Nemirovski *et al.* (2010)), we can get rid of the necessity to know r.

In addition, we assume that we are given an *a priori* bound $V < \infty$ on the *variation* $V_X(f) := \max_X f - \min_X f$ *of the objective on the feasible set.*

Here is a simple way to build V. Calling the First-Order Oracle at $x = 0$, we get $f(0)$ and $f'(0)$, thus getting an affine lower bound $\ell(x) = f(0) + [f'(0)]^T x$ on $f(x)$. Minimizing this bound over the ball B^R which contains X, we get a lower bound $\underline{\mathrm{Opt}} = f(0) - R\|f'(0)\|_2$

on Opt. Further, we build $n+1$ points $x_0, \ldots, x_n$ in such a way that their convex hull Δ contains the ball B^R and thus contains X. Calling the First-Order Oracle at these $n+1$ points, we build the quantity $F = \max_{0 \leq i \leq n} f(x_i)$. By Jensen's inequality, $f(x) \leq F$ for all $x \in \Delta$ and thus for all $x \in X$ (since $X \subset B_R \subset \Delta$). It follows that the quantity $F - \underline{\text{Opt}} \geq \max_{x \in X} f(x) - \text{Opt}$ can be taken as V.

The goal: Our goal is, given access to the oracles, the outlined a priori information and an $\epsilon > 0$, to find an *ϵ-solution* to (6.1), that is, a feasible solution x_ϵ such that $f(x_\epsilon) \leq \text{Opt} + \epsilon$.

6.2.2 *The Ellipsoid Algorithm*

After we have specified our goal — solving (6.1) within a given accuracy $\epsilon > 0$, our oracle-based "computational environment," and our a priori information, we are ready to represent the Ellipsoid Algorithm. We start with recalling what an ellipsoid is.

Ellipsoids in $\mathbf{R}^n$: An ellipsoid E in $\mathbf{R}^n$ is, by definition, a set representable as the image of the unit Euclidean ball under an invertible affine transformation $x \mapsto Bx + c$:

$$E = \{x = Bu + c : u^T u \leq 1\} \qquad [B : n \times n \text{ nonsingular}].$$

Recall that under an invertible affine transformation $x \mapsto Bx + c$ the n-dimensional volumes of bodies in $\mathbf{R}^n$ are multiplied by $|\operatorname{Det}(B)|$. Taking, as the unit of n-dimensional volume, the volume of the unit n-dimensional ball, we therefore have

$$\text{Vol}(E) = |\operatorname{Det}(B)|.$$

We need the following.

Lemma 6.1. *Given an ellipsoid $E = \{Bu + c : u^T u \leq 1\}$ in $\mathbf{R}^n$, $n > 1$, and a nonzero e, let $\widehat{E} = \{x \in E : e^T x \leq e^T c\}$ be the half-ellipsoid cut off E by the linear inequality $e^T x \leq e^T c$ in variable x* (geometrically: we take the hyperplane $\{x : e^T x = e^T c\}$ passing through the center c of the ellipsoid E. This hyperplane splits E in two parts, and $\widehat{E}$ is one of these parts). *Then $\widehat{E}$ is contained in an explicitly given ellipsoid E^+ with the volume strictly less of the one* *of E, specifically, in the ellipsoid*

$$
\begin{aligned}
E^{+} &= \{x = B_{+}u + c_{+} : u^T u \leq 1\}, \text{ where} \\
c_{+} &= c - \frac{1}{n+1} Bp, \\
B_{+} &= B\left(\frac{n}{\sqrt{n^2-1}}(I_n - pp^T) + \frac{n}{n+1} pp^T\right), \\
&= \frac{n}{\sqrt{n^2-1}} B + \left(\frac{n}{n+1} - \frac{n}{\sqrt{n^2-1}}\right)(Bp)p^T, \\
p &= \frac{B^T e}{\sqrt{e^T B B^T e}}.
\end{aligned}
\tag{6.5}
$$

The volume of the ellipsoid E^{+} satisfies

$$
\begin{aligned}
\mathrm{Vol}(E^{+}) &= \left(\frac{n}{\sqrt{n^2-1}}\right)^{n-1} \frac{n}{n+1} \mathrm{Vol}(E) \\
&\leq \exp\{-1/(2(n+1))\} \mathrm{Vol}(E).
\end{aligned}
\tag{6.6}
$$

Proof. We only sketch the proof, leaving completely straightforward, although somehow tedious, computations to the reader. The key observation is that we can reduce our statement to the one where E is the unit Euclidean ball. Indeed, E is the image of the unit ball $U = \{u : \|u\|_2 \leq 1\}$ under the mapping $u \mapsto Bu + c$; $\widehat{E}$ is the image, under the same mapping, of the half-ball $\widehat{U} = \{u \in W : f^T u \leq 0\}$, where $f = B^T e$. Choosing appropriately the coordinates in the u-space, we can assume that $\widehat{U} = \{u \in U : u_n \leq 0\}$. Now, invertible affine mappings preserve the ratio of volumes and map ellipsoids onto ellipsoids. It follows that in order to cover $\widehat{E}$ by a "small" ellipsoid E^{+}, it suffices to cover the half-ball $\widehat{U}$ by a "small" ellipsoid U^{+} and to take, as E^{+}, the image of U^{+} under the mapping $u \mapsto Bu + c$ (see Fig. 6.1). Let us take, as U^{+}, the smallest volume ellipsoid containing $\widehat{U}$. The latter is easy to specify: by symmetry, its center should belong to the nth coordinate axis in the u-space, and U^{+} should be the result of rotating a 2D ellipse shown in Fig. 6.1 around this axis. "By minimality," this ellipse should look as shown in the picture: its boundary should pass through the "South pole" $[0; -1]$ and the "equatorial points" $[\pm 1; 0]$ of the unit circle, which specifies the

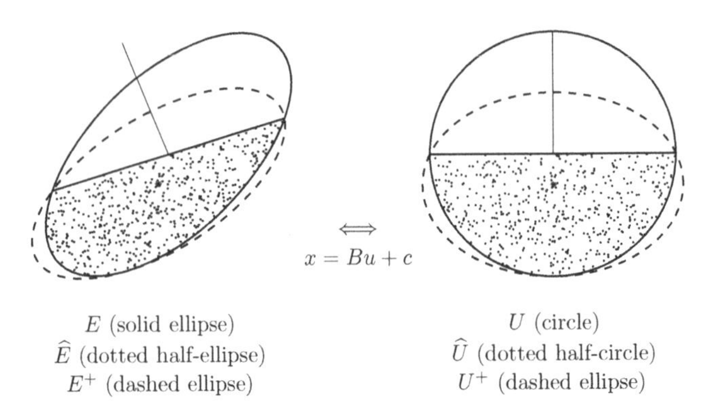

Fig. 6.1. Building a "small" ellipsoid containing a half-ellipsoid.

ellipse up to the "height" h of its center. The half-axes of the ellipse are easy-to-compute functions $p(h)$ and $q(h)$ of h; the ellipsoid U^+ has $n-1$ of its half-axes equal to the largest of the half-axes of the ellipse, let it be $p(h)$, and one half-axis equal to the smaller half-axis $q(h)$ of the ellipse. Since the volume of an ellipsoid is proportional to the product of its half-axes, and we want to minimize the volume, we need to minimize the univariate function $p^{n-1}(h)q(h)$. This problem has a closed-form solution, and this solution underlies the formulas in (6.5). □

The Ellipsoid Algorithm: Now, we are ready to present the EA. The idea of the method is very simple and looks as a natural multi-dimensional extension of the usual Bisection. Specifically, we build a sequence of ellipsoids of decreasing volume which "localize" the optimal set of the problem; it turns out that when the latter is localized in an ellipsoid of a small enough volume, we automatically have at our disposal a good approximate solution. In what follows, text in Italics constitutes the description of the algorithm, while text in Roman contains explanations, comments, etc. We describe the method in geometric terms; their algebraic (and thus algorithmic) "translation" is readily given by formulas in (6.5).

- ***Initialization:*** *We specify E_0 as the ball $B^R = \{x \in \mathbf{R}^n : \|x\|_2 \leq R\}$.*

Note that this ball contains X and thus contains the optimal set of (6.1).

- ***Step t* = 1,2,...:** *At the beginning of step t, we have at our disposal the current localizer — an ellipsoid E_{t-1}; let x_t be the center of this ellipsoid. At step t, we act as follows:*

(1) *We call the Separation Oracle, x_t being the input. If the oracle reports that $x_t \in X$, we call step t productive and go to rule* 2, *otherwise we call step t nonproductive, denote by e_t the separator reported by the oracle and go to rule* 3.

(2) *We call the First-Order Oracle, x_t being the input, and set $e_t = f'(x_t)$. If $e_t = 0$, we terminate and claim that x_t is an optimal solution to* (6.1), *otherwise we go to rule* 3.

Note that we arrive at rule 2 only when $x_t \in X$. If $e_t := f'(x_t) = 0$, then by the definition of a subgradient, $f(y) \geq f(x_t) + [f'(x_t)]^T(y - x) = f(x_t)$ for all y, meaning that x_t is a global minimizer of f; since we are in the situation when x_t is feasible for (6.1), the claim we make upon termination according to rule 2 is correct.

(3) *We set $\widehat{E}_t = \{x \in E_{t-1} : e_t^T x \leq e_t^T x_t\}$, build the ellipsoid E_t which covers $\widehat{E}_t$ according to the recipe from Lemma 6.1 and go to rule* 4. Since a separator is always nonzero and in view of the termination rule in 2, we arrive at rule 3 with $e_t \neq 0$, which makes $\widehat{E}$ a half-ellipsoid and thus makes Lemma 6.1 applicable. Besides this, we note — and this explains the idea of the construction — that $\widehat{E}_t$ (and thus $E_t \supset \widehat{E}_t$) inherits the property of E_{t-1} to localize the optimal set of (6.1). Indeed, if step t is nonproductive, then e_t separates x_t and X, whence $e_t^T(x_t - x) > 0$ for all $x \in X$. Looking at the formula defining $\widehat{E}_t$, we see that what we cut off E_{t-1} when passing from the ellipsoid E_{t-1} to its half $\widehat{E}_t$ is outside of X, i.e., not only optimal, but even feasible solutions to the problem which "sit" in E_{t-1} sit also in $\widehat{E}$. Now, let t be a productive step. In this case, x_t is a feasible solution to the problem, and $e_t = f'(x_t)$. By the definition of a subgradient, we have $f(x) \geq f(x_t) + [f'(x_t)]^T(x - x_t)$, meaning that for all points $x \in E_{t-1} \backslash \widehat{E}_t$, $f(x) > f(x_t)$ holds. Since x_t is feasible, it follows

that when passing from E_{t-1} to $\widehat{E}_t$, no optimal solutions are thrown away.

(4) *We check whether the volume of the ellipsoid E_t satisfies the inequality*

$$\mathrm{Vol}^{1/n}(E_t) < \frac{r\epsilon}{V}. \tag{6.7}$$

If it is the case, we terminate and output, as the approximate solution $\widehat{x}$ generated by the method, the best — with the smallest value of f — of the points x_τ corresponding to productive steps $\tau \leq t$. Otherwise, we pass to step $t+1$.

We shall check in a short while that the termination rule 4 is well defined, meaning that upon termination according to this rule, the set of productive steps performed so far is nonempty, so that the approximate solution is well defined.

Correctness and performance guarantees: We are about to prove the following.

Theorem 6.1. *Under the assumptions from Section* 6.2.1, *given a required accuracy ϵ, $0 < \epsilon < V$, the Ellipsoid Algorithm terminates after finitely many steps and returns a feasible approximate solution x_ϵ to* (6.1) *such that $f(x_\epsilon) \leq \mathrm{Opt} + \epsilon$. The number $N(\epsilon)$ of steps before termination can be upper bounded as*

$$N(\epsilon) \leq O(1)n^2 \ln\left(2 + \frac{V}{\epsilon} \cdot \frac{R}{r}\right), \tag{6.8}$$

where $O(1)$ is an absolute constant, and V is an a priori upper bound on $\max_{x \in X} f(x) - \mathrm{Opt}$, R is an a priori upper bound on the size of X (specifically, such that X is contained in $E_0 = \{x : \|x\|_2 \leq R\}$), and $r > 0$ is an a priori lower bound on the largest of radii of Euclidean balls contained in X.

A step of the method requires at most two calls to the Separation and the First-Order oracles, augmented by $O(1)n^2$ arithmetic operations to process the answers of the oracles.

Proof. The last statement, describing the effort per step, is readily given by the description of the method and formulas (6.5) explaining how to convert the description of E_{t-1} into the description of E_t.

To prove the upper bound on the number of steps, note that in view of (6.6) we have $\mathrm{Vol}(E_t) \leq \mathrm{Vol}(E_0)\exp\{-\frac{t}{2(n+1)}\} = R^n \exp\{-\frac{t}{2(n+1)}\}$; since at every step except for the last one the inequality opposite to (6.7) holds true, (6.8) follows.

It remains to verify that the algorithm does produce a result, and this result is a feasible ϵ-solution to the problem. There is nothing to prove if the algorithm terminates according to rule 2 (indeed, it was explained in the comment to this rule that in this case the algorithm returns an optimal solution to the problem). Thus, assume that the algorithm does not terminate according to rule 2. Let x_* be an optimal solution to (6.1) and let $\theta = \epsilon/V$, so that $0 < \theta < 1$. Let us set

$$X_* = (1-\theta)x_* + \theta X,$$

so that $\mathrm{Vol}(X_*) = \theta^n \mathrm{Vol}(X) \geq \theta^n r^n$. Let T be the termination step. We claim that E_T cannot contain X_*. Indeed, otherwise we would have $\mathrm{Vol}(X_*) \leq \mathrm{Vol}(E_T)$, whence $\mathrm{Vol}(E_T) \geq \theta^n r^n = (\epsilon/V)^n r^n$. But this contradicts the termination criterion (6.7) in rule 4, and this criterion should be satisfied at step T, since this is the termination step.

We have proved that $X_* \backslash E_T \neq \emptyset$, so that there exists $y \in X_*$ such that $y \notin E_T$. By construction of X_* we have $y = (1-\theta)x_* + \theta z$ for some $z \in X$, whence, in particular, $y \in X$ (since X is convex). Thus, $y \in X \subset E_0$ and $y \notin E_T$; thus, there is a step $\tau \leq T$ such that $y \in E_{\tau-1}$ and $y \notin E_\tau$. Since $E_\tau \supset \widehat{E}_\tau$, we conclude that $y \notin \widehat{E}_\tau$. The bottom line is that at certain step $\tau \leq T$ the point y satisfies the inequality $e_\tau^T y > e_\tau^T x_\tau$ (see the relation between $E_{\tau-1}$ and $\widehat{E}_\tau$). Now, τ is a productive step, since otherwise e_τ would separate x_τ and X, meaning that $e_\tau^T x' < e_\tau^T x_\tau$ for all $x' \in X$, and in particular for $x' = y$ (we have seen that $y \in X$). Since τ is a productive step, the result $\widehat{x}$ of the algorithm is well defined, is feasible and satisfies the relation $f(\widehat{x}) \leq f(x_\tau)$, see rule 4. Besides this, e_τ is a subgradient of f at x_τ, so that the relation $\mathrm{e}_\tau^T y > e_\tau^T x_\tau$ implies that $f(y) - f(x_\tau) \geq e_\tau^T[y - x_\tau] > 0$. It follows that

$$\begin{aligned} f(\widehat{x}) &\leq f(x_\tau) \leq f(y) = f((1-\theta)x_* + \theta z) \leq (1-\theta)f(x_*) + \theta f(z) \\ &= f(x_*) + \theta(f(z) - f(x_*)), \end{aligned}$$

where the third inequality is due to the fact that f is convex. It remains to note that the last quantity in the chain, due to the origin of V, is $\leq f(x_*) + \theta V = f(x_*) + \epsilon$. Thus, $f(\widehat{x}) \leq \mathrm{Opt} + \epsilon$. □

Discussion. Let us explain why the statement of Theorem 6.1 is truly remarkable. The point is that in the complexity bound (6.8) the desired accuracy ϵ and the "data-dependent" parameters V, R, r are under the logarithm; the only parameter which is not under the logarithm is the *structural* parameter n. As a result, the iteration count $N(\epsilon)$ is "nearly independent" of ϵ, V, R, r and is polynomial — just quadratic — in n. Another good news is that while we cannot say exactly how large the computational effort is per step — it depends on "what goes on inside the Separation and First-Order oracles," we do know that *modulo the computations carried out by the oracles*, the Real Arithmetic complexity of a step is quite moderate — just $O(1)n^2$. Now, assume that we use the Real Arithmetic Complexity model and want to solve a generic problem $\mathcal{P}$ with instances of the form (6.1)–(6.3). If the functions f_i and g_i associated with an instance $p \in \mathcal{P}$ of the problem are *efficiently computable* (that is, given x and the data of an instance $p \in \mathcal{P}$, we can compute the values taken at x and the gradients taken at x of all f_i and g_i in polynomial in $\text{Size}(p) = \dim \text{data}(p)$ time), then the RACM-complexity of an iteration becomes polynomial in the size of the instance. Under mild additional assumptions on $\mathcal{P}$ (see Ben-Tal and Nemirovski (2001a, Chapter 5)), polynomially of the per-iteration effort combines with low — just logarithmic — dependence of the iteration count $N(\epsilon)$ on ϵ to yield an *algorithm which, given on input the data of an instance $p \in \mathcal{P}$ and a desired accuracy $\epsilon > 0$, finds an ϵ-solution to p in a number of Real Arithmetic operations which is polynomial in the size of the instance and in the number* $\ln(1/\epsilon)$ *of accuracy digits we want to get.*[6]

[6]In the RACM framework, algorithms of this type usually are called polynomial time ones. The reason is that aside from a handful of special cases, generic computational problems with real data do not admit finite (i.e., with a finite on every instance running time) algorithms capable to solve the instances *exactly*. Therefore, it makes sense to relax the notion of efficient — polynomial time — algorithm, replacing the ability to solve instances *exactly* in RACM-time polynomial in their sizes, with the ability to solve instances within a prescribed accuracy $\epsilon > 0$ in RACM-time polynomial in the size of an instance *and* $\ln(1/\epsilon)$. Of course, such a relaxation requires to decide in advance how we quantify the accuracy of candidate solutions, see Ben-Tal and Nemirovski (2001a, Chapter 5) for details.

Historically, the Ellipsoid Algorithm was the first universal method for solving black-box-represented convex problems with iteration count polynomial in the problem's dimension n and in the number $\ln(1/\epsilon)$ of accuracy digits we want to get, and with polynomial in n "computational overhead" (computational effort modulo the one of the oracles) per iteration.[7] As we shall see in a while, these properties of the EA are instrumental when demonstrating CCM-tractability of LO with rational data.

6.3 Polynomial solvability of LO with rational data

Our goal here is to prove the following fundamental fact:

Theorem 6.2 (L.G. Khachiyan, 1979). *In the Combinatorial Complexity model, Linear Optimization with rational data is polynomially solvable.*

Proof. In what follows, "polynomial time" always means "polynomial time in the Combinatorial Complexity model of computations."

Step I: Checking feasibility of a system of linear inequalities: We start — and this is the essence of the proof — with demonstrating polynomial solvability of the following *feasibility problem*:

> *Given a system of linear inequalities $Ax \le b$ with rational data, check whether the system is or is not feasible.*

Note that we do not want to *certify* the answer: all we want is the answer itself.

Multiplying the constraints by appropriate positive integers, we can make all the entries in A and in b integral; it is easy to verify that this transformation can be carried out in polynomial time. Further, we can get rid of the kernel of A by eliminating, one by one, columns of A which are linear combinations of the remaining columns. This is a simple Linear Algebra routine which takes polynomial time. Thus, we lose nothing by assuming that the original system of constraints is with integer data, and A has trivial kernel. In the sequel, we denote

[7]Now, we know a handful of other methods with similar properties.

by m and n the sizes (numbers of rows and columns) of A, and by L, the bit size of $[A, b]$.

Our plan of attack is as follows: We observe that feasibility of the system $Ax - b$ is equivalent to the fact that the simple piecewise linear function

$$f(x) = \max_{1 \le i \le m} \left[a_i^T x - b_i\right],$$

is not everywhere positive. We intend to prove two facts:

A. If f is not everywhere positive, that is, attains nonpositive values, it attains nonpositive values already in the box $X = \{x : \|x\|_\infty := \max_i |x_i| \le M = 2^{O(1)L}\}$.

An immediate consequence is that *the system $Ax \le b$ is feasible iff the optimal value in the convex optimization problem*

$$\text{Opt} = \min_x \left\{ f(x) = \max_{1 \le i \le m} [a_i^T x - b_i] : \|x\|_\infty \le M = 2^{O(1)L} \right\} \tag{6.9}$$

is ≤ 0.

B. If the optimal value in (6.9) is not positive, it is not too close to 0:

$$\text{Opt} > 0 \Rightarrow \text{Opt} \ge \alpha := 2^{-O(1)L}. \tag{6.10}$$

Taking A and B for granted, a polynomial time algorithm for solving the feasibility problem can be built as follows: We apply to (6.9) the Ellipsoid Algorithm, the target accuracy being $\epsilon = \alpha/3$. As a result, we will get an approximate solution x_ϵ to the problem such that $\text{Opt} \le f(x_\epsilon) \le \text{Opt} + \epsilon = \text{Opt} + \alpha/3$. Therefore, computing $f(x_\epsilon)$, we get an approximation of Opt accurate within the error $\alpha/3$. Since, by B, either $\text{Opt} \le 0$, or $\text{Opt} \ge \alpha$, this approximation allows us to say *exactly* which one of the options takes place: if $f(x_\epsilon) \le \alpha/2$, then $\text{Opt} \le 0$, and thus, by A, the answer in the feasibility problem is "yes," otherwise $\text{Opt} \ge \alpha$, and the answer in the problem, by the same A, is "no."

It remains to understand how to apply the Ellipsoid Algorithm to (6.9). First, what we need are a Separation oracle for the box $\{x : \|x\|_\infty \le M\}$ and a First-Order oracle for f; these oracles can be built according to the recipe described in Section 6.2.1. Note that a single call to an oracle requires no more than mn operations. Next, what we need is the *a priori* information R, r, V. This information is immediate: we can take $R = M\sqrt{n} = \sqrt{n}2^{O(1)L}$, $r = M$ and $V = 2M(n+1)2^L$, where the latter bound comes from the evident observation that the magnitudes of all entries in A and b do not exceed 2^L, so that $|a_i^T x - b_i|$ is bounded from above by $2^L(Mn+1)$ provided $\|x\|_\infty \leq M$.

Invoking the complexity bound for the Ellipsoid method and taking into account that $M = 2^{O(1)L}$ the total number of iterations before termination is polynomial in n and L (since all exponents $2^{O(1)L}$ will appear under the logarithm), and thus is polynomial in L due to $mn \leq L$ (indeed, we have more than $m(n+1)$ entries in the data $[A, b]$, and it takes at least one bit to represent a single entry). Since the number of arithmetic operations at a step of the method (including those spent by the oracles) is $O(1)(mn + n^2)$, we conclude that *the total number of arithmetic operations when solving the feasibility problem is polynomial in* L.

The last difficulty we need to handle is that the Ellipsoid Algorithm is a *Real Arithmetic* algorithm — it operates with real numbers, and not with rational ones, and assumes precise real arithmetics (exact arithmetic operations with reals, precise comparisons of reals and even precise taking of square root, since this operation appears in (6.5)), and thus polynomial in L total number of arithmetic operations we spoke about is the number of operations of *real arithmetics.* And what we want to get is an algorithm operating with rational operands and the cost of an arithmetic operation taking into account the bit sizes of the operands. Well, a straightforward, although tedious, analysis shows that we can achieve our goal — to approximate the optimal value of (6.9) within accuracy $\epsilon = \alpha/3$ in polynomial in L number of operations — when replacing *precise* real arithmetic operations with "actual" reals with *approximate* operations with *rounded* reals; specifically, it turns out that it suffices to operate with rationals with at most $O(1)nL$ binary digits before and after the dot, and to carry out the arithmetic operations and taking the square root approximately, within accuracy $2^{-O(1)nL}$. We can implement all these approximate arithmetic operations, and thus the entire computation, on a digital computer, with the "bitwise cost" of every approximate operation polynomial in nL (and thus in L), so that the total number of bitwise operations in the course of the computation will be polynomial in L, as required.

All that remains is to justify A and B. Let us start with A. Since A is with trivial kernel, the polyhedral set $X = \{x : Ax \leq b\}$ does not contain lines (Proposition 2.12). Therefore, *if X is nonempty, X contains an extreme point v* (Theorem 2.12). By algebraic characterization of extreme points (Proposition 2.9), we can choose from $a_1, \ldots, a_m$ n linearly independent vectors $\{a_i : i \in I\}$ such that $a_i^T v = b_i$ for $i \in I$. In other words, v is the unique solution of the linear system $A^I x = b^I$ of n linear equations with n variables and nonsingular matrix A^I such that $[A^I, b^I]$ is a submatrix in $[A, b]$. Applying Cramer's rule, we conclude that

$$v_i = \frac{\Delta_i}{\Delta},$$

where $\Delta = \text{Det } A^I \neq 0$ and Δ_i is the determinant of the matrix A_i^I obtained from A^I by replacing the ith column with b^I. Now, Δ as the determinant of a matrix with integer entries is an integer; being nonzero (A^I is nondegenerate!), its magnitude is at least 1. It follows that

$$|v_i| \leq |\Delta_i|.$$

Now, the determinant of an $n \times n$ matrix P_{ij} does not exceed the product of the $\|\cdot\|_1$-lengths of its rows.[8] If now P is integral (as is the case with A_i^I) and ℓ_{ij} is the bit length of P_{ij}, then $|P_{ij}| \leq 2^{\ell_{ij}-1}$,[9] and we see that $|\text{Det } P| \leq 2^{\sum_{i,j} \ell_{ij}}$. When $P = A_i^I$, $\sum_{i,j} \ell_{ij} \leq L$, since A_j^I is a submatrix of $[A, b]$, and we end up with $|\Delta_i| \leq 2^L$. The bottom line is that $\|v\|_\infty \leq 2^L$, and since $f(v) \leq 0$, A holds true with $O(1) = 1$.

[8]Indeed, opening parentheses in the product $\prod_{i=1}^n (\sum_j |P_{ij}|)$, we get the sum of moduli of all "diagonal products" forming the determinant of P, and a lot of other nonnegative products. In fact, $|\text{Det } P|$ admits a better upper bound $\prod_{i=1}^n \sqrt{P_{i1}^2 + \cdots + P_{in}^2}$ (Hadamard inequality; consider how to prove it), but for our purposes, the previous, much worse, bound is also enough.

[9]Indeed, the number of bits in a representation of p_j is at least $\log_2(|P_{ij}| + 2)$ bits to store the binary digits of $|P_{ij}|$ plus one bit to store the sign of P_{ij}.

Now, let us prove B. Thus, assume that Opt > 0, and let us prove that Opt is not "too small." Clearly,

$$\text{Opt} \geq \text{Opt}^+ = \inf_x f(x) = \inf_{[x;t]} \{t : A^+[x;t] := Ax - t\mathbf{1} \leq b\} \quad [\mathbf{1} = [1;\dots;1]]. \tag{6.11}$$

The LO program

$$\text{Opt}^+ = \min_{[x;t]} \{t : A^+[x;t] := Ax - t\mathbf{1} \leq b\} \tag{!}$$

clearly is feasible; we are in the situation when it is below bounded (indeed, its objective at every feasible solution is nonnegative, since otherwise Opt^+ would be negative, meaning that f attains negative values somewhere; but then, by already proved A, Opt ≤ 0, which is not the case). Being feasible and bounded, (!) is solvable, meaning that f attains its minimum on $\mathbf{R}^n$.[10] This minimum cannot be ≤ 0, since then, by **A**, Opt would be nonpositive, which is not the case. The bottom line is that the *LO program* (!) *has the strictly positive optimal value* Opt^+.

Now, we claim that the $A^+ = [A, -\mathbf{1}]$ has trivial kernel along with A. Indeed, otherwise there exists $[\bar{x};\bar{t}] \neq 0$ such that $A^+[\bar{x};\bar{t}] = 0$, we have $\bar{t} \neq 0$, since otherwise $\bar{x} \neq 0$ and $\bar{x}$ is in the kernel of A, and this kernel is trivial. We see that both $[\bar{x};\bar{t}]$ and minus this vector are recessive directions of the nonempty feasible set X^+ of (!); along one of these directions, the objective strictly decreases (since $\bar{t} \neq 0$), meaning that (!) is below unbounded, which is not the case.

Since (!) is solvable with positive optimal value and the feasible set of this program does not contain lines, among the optimal solutions there exists one, let it be denoted $z = [x_*;t_*]$, which is a vertex of the feasible set of (!). Since $\text{Opt}^+ > 0$, we see that $t_* > 0$. Now, we can act exactly as in the proof of A: since z is a vertex of the feasible set of (!), there exists a nonsingular $(n+1) \times (n+1)$ submatrix $\bar{A}$ of the integral $m \times (n+1)$ matrix A^+ such that $\bar{A}[x_*;t_*] = \bar{b}$, where

[10]In fact, our reasoning shows that *every below-bounded piecewise linear function attains its minimum on* $\mathbf{R}^n$.

$\bar{b}$ is a subvector of b. By Cramer's rule and due to $t_* > 0$,

$$t_* = \frac{\Delta_*}{\Delta} = \frac{|\Delta_*|}{|\Delta|},$$

where Δ_* is the determinant of the matrix obtained from $\bar{A}$ by replacing the last column with $\bar{b}$, and Δ is Det $\bar{A}$. Since $t_* > 0$, $|\Delta_*| > 0$; being integral (A^+ is integral along with A, b), we therefore have $|\Delta_*| \geq 1$. Further, the bit size of $[A^+, b]$ clearly is $O(1)L$, whence, by the same reasons as in the proof of **A**, $|\Delta| \leq 2^{O(1)L}$. In view of these observations, $t_* = \mathrm{Opt}^+ \geq 2^{-O(1)L}$, whence, invoking (6.11), $\mathrm{Opt} \geq 2^{-O(1)L}$, as claimed in B. □

Step II: From checking feasibility to building a solution: The remaining steps are much simpler than the previous one. After we know how to check in polynomial time the feasibility of a system S of m linear inequalities and equations, let us understand how to find its solution in the case when the system is feasible. It makes sense to denote the initial system by S_0; w.l.o.g., we can assume that the vectors a_i of coefficients in all constraints $a_i^T x \overset{=}{\leq} b_i$ constituting S_0 are nonzero. We start with checking the feasibility of the system. If S_0 is infeasible, we terminate; if it is feasible, we take the first inequality $a^T x \leq b$ in the system, if any exists, convert it into the equality $a^T x = b$ and check the feasibility of the resulting system. If it is feasible, it will be our new system S_1, otherwise the hyperplane $a^T x = b$ does not intersect the feasible set of S_0, and since this set is nonempty and is contained in the half-space $a^T x \leq b$, the inequality $a^T x \leq b$ in S_0 is in fact redundant: when we remove it from the system, the solution set does not change (why?). In the case in question, we eliminate from S_0 the redundant inequality $a^T x \leq b$ and call the resulting system of inequalities and equalities S_1. Note that in all cases S_1 is solvable, and every solution to S_1 is a solution to S_0. Besides this, the number of inequalities in S_1 is less than in S_0 by 1.

We now repeat the outlined procedure with S_1 in the role of S_0, thus obtaining a feasible system S_2; its solution set is contained in the one of the system S_1, and thus — in the solution set of S_0, and the number of inequalities in S_2 is less than in S_0 by 2. Iterating this construction, we end up with a system S_ℓ in at most m steps which is feasible, contains only equality constraints, and every solution to this system solves S_0 as well. It remains to note that to solve a feasible system of linear *equality* constraints with integer or rational data is a simple Linear Algebra problem which is polynomially solvable. Solving S_ℓ, we get a solution to the original system S_0.

Now, note that the bit sizes of all systems we deal with in the above process are the same as, or even smaller than, the bit size of the original system. Since feasibility checks take time polynomial in L, $m \leq L$, and the last system is polynomially solvable, the complexity of the entire process is polynomial in L.

Step III: Solving an LO program: Now, we can solve in polynomial time an LO program

$$\max_x \{c^T x : Ax \leq b\} \tag{P}$$

with rational data. Let L be the bit size of the program. We start with checking in polynomial time whether the program is feasible; if it is not, we terminate with the correct claim "(P) is infeasible," otherwise we check in polynomial time the feasibility of the dual problem

$$\min_\lambda \{b^T \lambda : \lambda \geq 0,\ A^T \lambda = c\}, \tag{D}$$

which clearly has the bit size $O(1)L$. If the dual problem is infeasible, we terminate with a correct claim "(P) is unbounded."

It remains to consider the case when both (P) and (D) are feasible and thus both are solvable (LO Duality Theorem). In this case, we put together the constraints of both programs and augment them with the "zero duality gap" equation, thus arriving at the system of linear constraints

$$Ax \leq b; \lambda \geq 0; A^T \lambda = c; b^T \lambda = c^T x \tag{S}$$

in variables x, λ. By LO Optimality conditions, every feasible solution $[x; \lambda]$ to this system comprises optimal solutions to (P) and to (D), so that (S) is feasible; the bit size of (S) clearly is $O(1)L$. Applying the construction from Step II, we can find a feasible solution to (S), and thus optimal solutions to (P) and to (D), in time polynomial in L. From the above description it is clear that the overall computational effort of solving (P) is also polynomial in L. Khachiyan's Theorem is proved. □

Certifying insolvability in polynomial time: The proof of Khachiyan's Theorem presents a CCM-polynomial time algorithm which, as applied to an LO program (P) with rational data, detects correctly whether the problem is infeasible, feasible and unbounded, or solvable; in the latter case, the algorithm produces an optimal solution to (P) and to the dual problem, thus certifying optimality of both the solutions. A natural question is, whether it is possible to get in polynomial time certificates for infeasibility/unboundedness of (P), when the problem is indeed infeasible or unbounded. The answer is "yes." Indeed, as it was explained in Section 3.1.3 (see p. 176), in order to certify that (P) is infeasible, it suffices to point out a vector $\lambda \geq 0$ such that $A^T\lambda = 0$ and $b^T\lambda < 0$ (cf. (3.9)), or, which is the same (why?), a λ such that

$$\lambda \geq 0, \quad A^T\lambda = 0, \quad b^T\lambda \leq -1. \tag{S}$$

After infeasibility of (P) is detected, we can solve the latter system of linear inequalities and equations as explained in Step 2 above; note that we are in the situation when this system is solvable (since the certification scheme we are utilizing is complete). The data of (S) is rational, and its bit size is $O(1)L$, where L is the bit size of (P). Applying construction from Step 2 to (S), we get in polynomial time a solution to this system, thus certifying that (P) is infeasible.

After feasibility of (P) is detected and certified by presenting a feasible solution (which can be done by the algorithm from Step 2) and the problem is found unbounded, in order to certify the latter conclusion it suffices to point out a recessive direction of (P) along which the objective of (P) increases (under circumstances, this is the same as to certify that (D) is infeasible). Thus, we need to find a solution y to the system

$$Ay \leq 0, \quad c^Ty > 0,$$

(cf. (3.11)), or, which is the same, a solution to the system

$$Ay \leq 0, \quad c^Ty \geq 1, \tag{S'}$$

which we already know to be solvable. By the same reasons as in the case of (S), system (S') can be solved in time polynomial in L.

6.4 The Ellipsoid Algorithm and computations

As we have seen, the Ellipsoid Algorithm, which by itself is a "universal" algorithm for solving convex optimization problems, implies a CCM-polynomial time solution algorithm for LO with rational data; from the academic viewpoint, this algorithm outperforms dramatically the Simplex method which has exponential worst-case complexity. In actual computations the situation is completely opposite: when solving real-life LO programs, the Simplex method dramatically outperforms the Ellipsoid one. The reason is simple: as a matter of fact, the Simplex method never works according to its exponential complexity bound; the empirically observed upper bound on the number of pivoting steps is something like $3m$, where m is the number of equality constraints in a standard form LO program. In contrast to this, the EA works more or less according to its theoretical complexity bound (6.8) which says that in order to solve within an accuracy $\epsilon \ll 1$ a convex program with n variables, one should run $O(1)n^2 \ln(1/\epsilon)$ iterations of the method, with at least $O(n^2)$ arithmetic operations (a.o.) per iteration, which amounts to the total of at least $O(1)n^4 \ln(1/\epsilon)$ a.o. While being nearly independent of ϵ, this bound grows rapidly with n and becomes impractically large when $n = 1000$, meaning that problems with "just" 1000 variables (which is a small size in the LO scale) are beyond the "practical grasp" of the algorithm.

The fact that the EA "fails" when solving medium- and large-scale LO is quite natural — this algorithm is black-box-oriented and thus "does not know" how to utilize the rich and specific structure of an LO program. Essentially, all that matters for the EA, as applied to an LO program $\min_x\{c^T x : a_i^T x - b_i \leq 0,\ 1 \leq i \leq M\}$, is that the program have convex objective and constraints, and that the data of the problem allow to compute the values and the gradients of the objective and the constraints at a given point. Note that the Simplex method is of a completely different nature: it is not black-box-oriented, it works directly on a program's data, and it is "tailored" to LO: you just cannot apply this method to a problem with, say, convex quadratic objective and constraints.

The importance of the EA is primarily of academic nature — this is the algorithm which underlies the strongest known theoretical tractability results in Linear and Convex Optimization. Citing L. Lovasz, as far as the efficient solvability of generic convex problems is concerned, the EA plays the role of an "existence theorem" rather than a tool of choice for practical computations.

In this respect, we should mention an additional important feature which has not attracted our attention yet. Assume we are solving problem (6.1)–(6.3) with a simple — just linear — objective f (which, as we remember from Chapter 1, is not an actual restriction) and a "complicated" feasible set X given by a finite, but perhaps very large, number M of linear inequality constraints $g_i(x) \leq 0$. Note that M does not directly affect the complexity characteristics of the algorithm; the constraints are "inside the Separation oracle," and the EA simply does not "see" them. It follows that if the constraints depending on X are "well organized," so that given x, there is a way to identify in polynomial time a constraint, if any, which is violated at x, or to detect correctly that all the constraints are satisfied at x, the arithmetic complexity of a step of the EA will be polynomial. There are important optimization problems, some of them of combinatorial origin, which fit this framework; the EA, same as in the LO case, allows to demonstrate tractability of these problems (for examples, see Grotschel *et al.* (1987).

In should be added that while low-dimensional convex problems are of rather restricted applied interest, "restricted interest" is not the same as "no interest at all." Low-dimensional (few tens of variables) convex problems do arise in may applications; whenever it happens, the EA becomes a "method of choice" allowing to get a high-accuracy solution in a quite reasonable time. The EA is especially well suited for solving low-dimensional convex problems with large number N of constraints in (6.3), since the complexity of the algorithm, even with a straightforward implementation of the Separation oracle ("given a query point, look through all the constraints to find a violated one or to conclude that all of them are satisfied"), is just linear in N.

6.4.1 *Illustration: Cutting stock problem*

Consider the following problem: a factory should produce m types of rectangular sheets of steel, all of them of common width (let it be 1) and prescribed heights $h_1 < h_2 < \cdots < h_m$; the requested number of sheets of height h_j is b_j. In full accordance with reality, let us assume that with a properly selected unit of height, all h_j are positive integers. The sheets are cut off a band of given integer

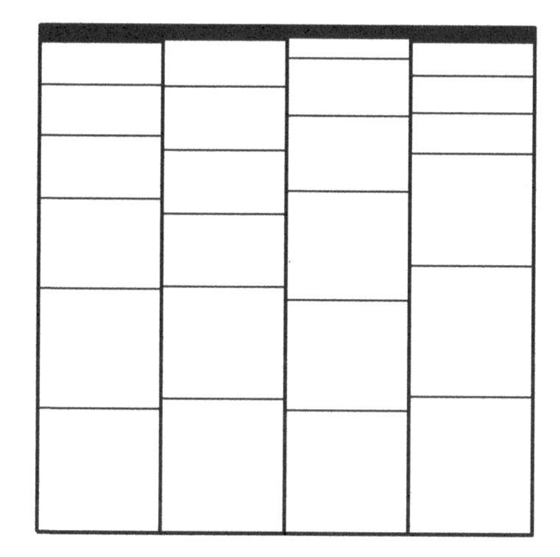

Fig. 6.2. Cutting stock. Top black rectangles: waste.

height $H \geq h_m$, and the manufacturing process goes as follows: we split the band into consecutive $H \times 1$ rectangles and cut off rectangle i $k_j^i \in \{0, 1, 2 \ldots\}$ sheets of height h_j, $j = 1, \ldots, m$ (see Fig. 6.2); this is possible if $\sum_j h_j k_j^i \leq H$. Let us call m-dimensional vector $k = [k_1; \ldots; k_m]$ *a pattern* when k is a nonzero nonnegative integral vector satisfying the constraint

$$\sum_{j=1}^{m} h_j k_j \leq H. \tag{6.12}$$

Patterns form a finite set $\mathcal{K}$, and we can cut off $H \times 1$ part of our band k_j sheets of height h_j, $j \leq m$, if and only if $k = [k_1; \ldots; k_m]$ is either the zero vector (of no interest for us) or is a pattern. Consequently, our production plan should be a vector x with integral entries $x_k \geq 0$, $k \in \mathcal{K}$, specifying the number of $H \times 1$ rectangles from which we intend to cut off k_j sheets of height h_j, $j \leq m$. Let us think about the patterns as the columns of $m \times N$, $N = \operatorname{card} \mathcal{K}$, matrix A; to meet the demand, x_i should be nonnegative integers satisfying the system of constraints $Ax \geq b$, where the entries in b are the required amounts of sheets of height h_j, $j \leq m$. Now, when utilizing pattern k on the $H \times 1$ part of our band, there will be *waste* $c_k := H - \sum_{j=1}^{m} h_j k_j^i$. A natural goal is to satisfy the demand when minimizing the objective $\sum_{k\in\mathcal{K}} C_k x_k$,

$$C_k = c_k + \beta,$$

where $\beta > 0$ represents expenses for processing the $H \times 1$ part of the band, and it is assumed w.l.o.g. that our loss per unit of waste is 1. Thus, our manufacturing problem can be modeled by the LO program

$$\min_x \left\{ \sum_{k\in\mathcal{K}} C_k x_k : Ax \geq b \right\}$$

to be solved in nonnegative integer vectors x. The standard cutting stock problem is obtained from this LO by relaxing the integrality constraint (which, as we shall eventually see, is of not much importance when b_j are large) and reads

$$\mathrm{Opt}(P) = \min_{x\in\mathbf{R}^N} \left\{ \sum_{k\in\mathcal{K}} C_k x_k : Ax \geq b, x \geq 0 \right\}. \tag{6.13}$$

This is just an LP in the standard form; however, normally this problem is out of reach of the standard LP algorithms due to the astronomically large width N of matrix A. For example, when $h_j = j$, $1 \leq j \leq m = 20$ and $H = 50$, $N = 1{,}186{,}802$; setting $h_j = j$, $1 \leq j \leq m = 30$ and $H = 100$, we get $N = 1{,}462{,}753{,}730$. Clearly, with this number of variables, even storing a candidate solution as a vector becomes impossible!

Utilizing duality: Note that the dual to (6.13) is the problem

$$\mathrm{Opt}(D) = \max_{\lambda} \left\{ b^T\lambda : \lambda \geq 0, A^T\lambda \leq C \right\}, \tag{D}$$

where C is the vector with entries C_k, $k \in \mathcal{K}$. This problem has just m variables, while the number of constraints $N + m$ can be astronomically large. The point, however, is that *the constraints of* (D) *are well organized*: given λ, it is easy to check whether λ is feasible for (D), and if this is not the case, to find a constraint which is violated at the point λ. Indeed, we start with checking whether $\lambda \geq 0$; if this is not the case, finding a violated constraint is trivial.

Assuming λ nonnegative, it is feasible if and only if $C_k - k^T\lambda \geq 0$ for all $k \in \mathcal{K}$, or, which is the same due to what C_k is, if and only if

$$\max_{k\in\mathcal{K}} \left\{ \sum_{j=1}^{m} [h_j + \lambda_j]k_j \right\} \leq H + \beta. \tag{6.14}$$

The maximum in the left-hand side, along with a maximizer, can be found by Dynamic Programming. Specifically, setting for $t = 0, 1, \ldots, m$

$$S_t(\eta) = \max_{k_1,\ldots,k_t} \left\{ \sum_{j=1}^{t} [h_j + \lambda_j]k_j : k_j \in \{0,1,\ldots\}, \sum_{j=1}^{t} k_j h_j \leq \eta \right\},$$
$$0 \leq \eta \leq H,$$

we have

$$\begin{aligned} S_0(\eta) &= 0,\ 0 \leq \eta \leq H, \\ S_t(\eta) &= \max_{0\leq j\leq \lfloor \eta/h_t \rfloor} [S_{t-1}(\eta - jh_t) + j[h_t + \lambda_t]], \\ &\quad 0 \leq \eta \leq H, 1 \leq t \leq m, \\ &\quad [\lfloor a \rfloor\text{: the largest integer} \leq a \in \mathbf{R}], \end{aligned}$$

implying that we can compute $S_t(\eta)$, $\eta \leq H$, recursively in $t \leq m$ at the "moderate" overall cost of $\mathcal{C} = O(1)m^2H^2$ arithmetic operations. Augmenting this recursive computation with simple backtracking (which increases the computational effort by at most an absolute constant factor), we will get at our disposal not only the real $f(\lambda) := S_m(H)$, but also a collection $k(\lambda) \in \mathcal{K}$ such that

$$f(\lambda) = \sum_{j=1}^{m} [h_j + \lambda_j]k_j(\lambda).$$

It remains to note that $\lambda \geq 0$ is feasible for (D) if and only if $f(\lambda) = S_m(H) \leq H + \beta$, see (6.14), and when this inequality is violated, the constraint $k^T(\lambda)\lambda \leq C_{k(\lambda)}$ from the system of constraints in (D) is violated as well.

The bottom line is that (D) *admits the Separation Oracle with "moderate" arithmetic cost*

$$\mathcal{C} = O(1)m^2H^2$$

of a call. This makes it natural to solve problem (D) by the Ellipsoid Algorithm.

Solving (D) by EA: With the Separation Oracle already built and the trivial First Order Oracle, it is easy to specify the remaining data — the parameters R, r, V (see p. 381) — required by EA as applied to (D):

(1) Columns of A contain the m basic orths of $\mathbf{R}^m$ (due to $h_j \leq h_m \leq H$) and $C_k \leq \overline{H} = H + \beta$, so that a feasible solution λ to (D) should satisfy $\lambda_j \geq 0$, $\lambda_j \leq \overline{H}$, $j \leq m$, implying that the feasible set Λ of (D) satisfies

$$\Lambda \subset \{\lambda \in \mathbf{R}^m : \|\lambda\|_2 \leq R\},\ R = \overline{H}\sqrt{m};$$

(2) When $0 \leq \lambda_j \leq \beta/H$, $j \leq m$, we have $k^T\lambda \leq \beta \leq C_k$, $k \in \mathcal{K}$, implying that Λ contains a Euclidean ball of radius

$$r = \frac{\beta}{2H};$$

(3) Due to item 1, the variation of the objective $b^T\lambda$ on Λ does not exceed

$$V = \left[\sum_j |b_j|\right]\overline{H}.$$

Given tolerance $\epsilon \in (0, \overline{H}\|b\|_1)$, we can solve (D) within accuracy ϵ in terms of the objective in

$$\mathcal{N} = O(1)m^2\ln(mH + \|b\|_1\overline{H}/\epsilon)$$

iterations with $O(1)m^2H^2$ arithmetic operations per iteration. Thus, we can find a feasible high-accuracy solution to (D) with "moderate" computational effort *completely independent of how large N is.*

Recovering a near-optimal primal solution: Finding a high-accuracy solution to (D) by itself is not what we want — we need a high-accuracy feasible solution to the *primal* problem (6.13). This goal can be achieved as follows. When processing (D) by EA, we, as a byproduct, identify certain columns of A, specifically, those yielded by the Separation Oracle we have built at the nonproductive steps of the algorithm. Augmenting these columns by the standard basic orths (which are the columns of A as well), we get at our disposal an $m \times n$, $n \leq \mathcal{N} + m$, submatrix $\overline{A} = [A_{k^1}, A_{k^2}, \ldots, A_{k^n}]$ of A. Replacing in the primal problem (6.13) A with $\overline{A}$ (that is, restricting in (6.13) the decision vector to have $x_k = 0$ for all k different from k^i, $i \leq n$), we get a "moderate size" standard form LP problem (P') with the dual

$$\max_{\lambda} \left\{ b^T \lambda : \lambda \geq 0, \overline{A}^T \lambda \leq \overline{C} \right\}, \quad \overline{C}_i = C_{k^i}, \quad i \leq n. \qquad (D')$$

(D') can be solved within accuracy ϵ by EA utilizing the straightforward Separation Oracle and the same parameters R, r, V as defined above. An immediate observation is that *the trajectory of EA as applied to (D) is the trajectory of EA as applied to (D') as well.* Verification is readily given by induction in the iteration number. Base is evident, and the inductive step is as follows:

> Assuming the iterates with indices $0, 1, \ldots, t-1$ and the reports of the respective First-Order and Separation Oracles at these iterates are the same for both trajectories — one generated by EA as working on (D) and the other one generated by EA as working on (D'), the iterates x_t in both trajectories will also be the same. If x_t was classified as feasible when solving (D), the same will happen when solving (D'). If the iterate was classified as infeasible when solving (D), so that when processing this iterate a violated at it constraint of (D) was discovered, this constraint is present in (D') as well, and we lose nothing by assuming that when processing this iterate by EA as applied to (D'), this is exactly the violated at x_t constraint reported by our "straightforward" Separation Oracle for (D'). Thus, we can assume w.l.o.g. that when processing iteration x_t when solving (D) and when solving (D'), reports of the oracles are the same, which justifies the inductive step.

We are nearly done. When applying EA to (D) and to (D'), both problems are solved within accuracy ϵ in terms of the (the same for both problems) objective by (the same for both problems) approximate solution generated by EA in course of $\mathcal{N}$ steps. As a result, the optimal values $\mathrm{Opt}(P) = \mathrm{Opt}(D)$ of (6.13) and $\mathrm{Opt}(P') = \mathrm{Opt}(D')$ of the $m \times n$ LP problem (P') are within ϵ from each other, implying that ϵ-optimal in terms of the objective feasible solution to (P') after augmenting by zero entries becomes 2ϵ-optimal in terms of the objective feasible solution to (P). But problem (P'), in contrast to (P'), has quite moderate sizes and can be solved within high accuracy by standard LO algorithms, including theoretically *and* practically efficient polynomial time interior point methods to be considered in Chapter 7. We end up with a computationally efficient scheme for solving the cutting stock problem to a whatever high accuracy, with computational effort completely independent of how large N is.

Numerical illustration: A couple of instructive numerical results illustrating the outlined scheme are displayed in Table 6.1. In this table, Obj is the value of the objective at the resulting feasible near-optimal solution to (6.13), δ is the certified (upper bound on the) nonoptimality of this solution in terms of the objective, CPU is the total CPU time of running EA and subsequent recovery of the near-optimal solution to (P') by commercial Interior Point solver `Mosek`. The remaining notation was described when presenting the algorithm. Computations were carried out on a standard desktop computer.

Concluding remarks: Strictly speaking, a natural model for the manufacturing problem we are considering is not problem (6.13) exactly, it is the refinement of this problem where the variables, in addition to meeting the constraints, should be integral. We are about to demonstrate that in the "mass production" case where b_i are large, ensuring integrality is not a big deal. Indeed, given a fractional near-optimal feasible solution to (P'), we can easily "refine" it — convert it into the feasible solution with at most $m+1$ nonzero

Table 6.1. Cutting stock problem via Ellipsoid Algorithm.

##	m	$h_j, j \leq m$	H	β	N	Obj	δ	$\mathcal{N}$	CPU, sec
1	20	$\equiv j$	50	0.1	1,186,802	6.0700	6.5e − 7	12,918	56"
2	30	$\equiv j$	100	0.1	1,462,753,730	10.0150	7.3e − 7	30,349	138"

entries, while preserving the value of the objective. Indeed, consider the system of linear equations $\overline{A}x = \overline{A}\bar{x}$, $\overline{C}^T x = \overline{C}^T \bar{x}$ in n variables x, where n is the design dimension of (P') and $\bar{x}$ is the near-optimal solution to (P') we have built when solving the problem within accuracy ϵ. Denoting by $\widehat{A}$ the matrix, and by $\widehat{b}$ the right-hand side of the resulting system of linear equations, observe that $\bar{x}$ is a nonnegative solution to this system. Let us look at the columns of $\widehat{A}$ corresponding to positive entries of $\bar{x}$. If the number of these columns is $> m+1$, so that the columns are linearly dependent, this is a simple Linear Algebra problem to find a nontrivial solution h to the system $\widehat{A}h = 0$ such that $h_i = 0$ whenever $\bar{x}_i = 0$. Now, we can add to $\bar{x}$ an easy-to-find multiple th of h in such a way that $\bar{x} + th$ (this vector automatically satisfies the constraints $\widehat{A}x = \widehat{b}$) is nonnegative and the number of its positive entries is less than the similar number for $\bar{x}$. If the number of nonzero entries in this new solution is still $> m+1$, we can repeat this procedure with the new solution in the role of $\bar{x}$, and proceed in this way until a nonnegative solution $\widehat{x}$ to the system $\widehat{A}x = \widehat{b}$ with at most $m+1$ positive entries is found. Recalling what $\widehat{A}$ and $\widehat{b}$ are, this solution is feasible for (P') and is "as near-optimal" as the solution $\bar{x}$ we started with. Now, let x^* be obtained from $\widehat{x}$ by rounding up to integers the fractional entries, if any. Passing from $\widehat{x}$ to integral solution x^*, we preserve feasibility and increase the objective by at most $(m+1)\overline{H}$, so that the resulting solution will definitely be feasible and optimal in terms of the objective, within accuracy $\bar{\epsilon} = \epsilon + (m+1)\overline{H}$, solution to the integral version of (P'). While by itself the nonoptimality $(m+1)\overline{H}$ can be large, its ratio to the optimal value of the problem of interest approaches zero as $\sum_i b_i$ grows. Thus, when b is large, nonoptimality of the resulting integral solution *measured in the relative scale* will be small, as claimed.

We should also add that the traditional "pivoting way" to handle standard LP's with a reasonable number m of equality constraints and huge design dimension N, called *column generation*, is applicable when columns of the constraint matrix are "well organized" in exactly the same way that in our illustration — there is a way, given a basis, to check with reasonable effort whether the associated vector of reduced costs certifies optimality of the current basic feasible solution (in a minimization problem, this means that all reduced costs are nonnegative), or that there exist positive reduced costs; in the latter case the index of such a reduced cost should also be provided by the check in question. When such a check is available, one can run PSM without operating with the entire A. Indeed, all that is needed to run a step of PSM associated with a given basis is access to the corresponding $m \times m$ submatrix of A and to the column of A, if any, corresponding to the reduced cost of the "wrong sign." Given such an access, the computational effort per iteration starts to depend solely on m and is not affected by N. A more detailed exposition of this and other large-scale versions of pivoting LO algorithms (it can be found in most of the existing textbooks on LO) goes beyond the scope of our book.

Part IV

Conic Programming and Interior Point Methods

Chapter 7

Conic Programming

Khachiyan's EA-based polynomial time algorithm for solving LO programs with rational data is of academic interest only; in the real-life LO, it is outperformed by far by "theoretically bad" pivoting algorithms. Nevertheless, the discovery of the polynomial time algorithm for LO triggered an intensive search for LO techniques which are both theoretically *and practically* efficient. The breakthroughs in this direction due to Karmarkar (1984), Renegar (1986) and Gonzaga (1986) led to the development of novel *Interior Point* polynomial time algorithms for LO, which were further extended onto nonlinear "well-structured" convex problems. The common name for the resulting Convex Programming techniques is *Interior Point Methods* (IPMs); this name is aimed at stressing the fact that even in the LO case, the methods, in contrast to the simplex-type ones, move along the interior of the feasible polyhedron rather than along its vertices.

As far as LO is concerned, state-of-the-art IPMs are quite competitive with the simplex-type algorithms. It suffices to say that they are the default options in modern commercial LO solvers.

While the very first IPMs were designed for LO, it turned out that their intrinsic nature has nothing to do with Linear Optimization, and the IPM approach can be extended, essentially, onto all convex programs; the corresponding theory was proposed in Nesterov and Nemirovskii (1994) and is now the standard way to treat IPMs. In this and subsequent IPM-oriented research it was discovered, in particular, that

- the most powerful IPMs are associated with representation of convex problems in a specific *conic form*, pretty similar to the form of a usual LO program. Aside of algorithmic issues, conic representation of convex programs allows for very instructive reformulation of convex programming theory, most notably its duality-related part;
- there exist three generic conic problems intrinsically very close to each other — Linear, Conic Quadratic and Semidefinite Optimization; taken together, these three problems allow to handle nearly all convex optimization models arising in applications. The specific intrinsic nature of these three problems, along with other important consequences, allows for a significant enrichment (primarily, due to Nesterov and Todd, 1997, 1998) of the general IPM theory, on the one hand, and for a unified and relatively nontechnical presentation of the basic part of this theory, on the other hand.

While our book is devoted to LO, it would be a pity to restrict ourselves to LO also when speaking about Interior Point Methods. Indeed, at a low "setup cost" of learning Conic Programming (a knowledge which is highly valuable by its own right) we can "kill two birds with one stone" — become acquainted with the basic IPM for Linear *and* Semidefinite Optimization.[1] To the best of our understanding, the only shortcoming of this course of actions is that while we will be able to understand *what* is going on, we will be unable to explain *why* we act as we do and where some "miracles" to be met come from. Answering these "why and where" questions would require lengthy excursions to the general theory of IPM as presented in Nesterov and Nemirovskii (1994), Renegar (2001), which would be too much for an introductory book on LO.

The course of our actions is as follows: we start with acquaintance with Conic Programming, with emphasis on Conic Programming Duality, which we see as the major part of the "descriptive component" of Conic Programming. As for the two other components,

[1]In our unified treatment, we are enforced to omit Conic Quadratic Optimization, where the IPM constructions, intrinsically the same as in LO and SDO, would require different notation.

(a) expressive abilities and applications,

and

(b) algorithmic toolbox,

of Conic Quadratic and Semidefinite Optimization (CQO and SDO, respectively), our presentation, while self-contained, is more sketchy. Specifically,

- we only touch the issue of expressive abilities of CQO/SDO and do not speak at all about their applications, referring the reader first and foremost to the truly exceptional, in our opinion, book Boyd and Vandenberghe (2004), and to Ben-Tal and Nemirovski (2001, 2022), Boyd *et al.* (1994);
- we restrict exposition of the algorithmic component of Conic Optimization to what is called "short step primal–dual path-following methods for LO and SDO." For a broader and more detailed presentation of IPMs, we refer the reader to books like Frenk *et al.* (2010), Nesterov and Nemirovskii (1994), Roos *et al.* (2005), Wright (1997), Ye (2011), and especially to the excellent book Nesterov (2018); the latter book covers, along with IPMs, efficient *first-order* algorithms of Convex Optimization, including the celebrated Nesterov Fast Gradient algorithms.

7.1 Conic programming: Preliminaries

7.1.1 *Euclidean spaces*

A *Euclidean space* is a finite-dimensional linear space over reals equipped with an *inner product* $\langle x, y\rangle_E$ — a real-valued function of $x, y \in E$ which is

- symmetric ($\langle x, y\rangle_E \equiv \langle y, x\rangle_E$),
- bilinear ($\langle \lambda u + \mu v, y\rangle_E = \lambda\langle u, y\rangle_E + \mu\langle v, y\rangle_E$, and similarly w.r.t. the second argument), and
- positive definite ($\langle x, x\rangle_E > 0$ whenever $x \neq 0$).

In the sequel, we usually shorten $\langle x, y\rangle_E$ to $\langle x, y\rangle$, provided that E is fixed by the context.

Example: The standard Euclidean space $\mathbf{R}^n$. This space comprises n-dimensional real column vectors with the standard coordinate-wise linear operations and the inner product $\langle x, y\rangle_{\mathbf{R}^n} = x^T y$. $\mathbf{R}^n$ is a universal example of a Euclidean space: for every Euclidean n-dimensional space $(E, \langle\cdot,\cdot\rangle_E)$ there exists a one-to-one linear mapping $x \mapsto Ax : \mathbf{R}^n \to E$ such that $x^T y \equiv \langle Ax, Ay\rangle_E$. All we need in order to build such a mapping is to find an *orthonormal basis* $e_1, ..., e_n$, $n = \dim E$, in E, that is, a basis such that $\langle e_i, e_j\rangle_E = \delta_{ij} \equiv \begin{cases} 1, & i = j \\ 0, & i \neq j \end{cases}$; such a basis always exists. Given an orthonormal basis $\{e_i\}_{i=1}^n$, a one-to-one mapping $A : \mathbf{R}^n \to E$ preserving the inner product is given by $Ax = \sum_{i=1}^n x_i e_i$.

Example: The space $\mathbf{R}^{m\times n}$ of $m \times n$ real matrices with the Frobenius inner product: The elements of this space are $m\times n$ real matrices with the standard linear operations and the inner product $\langle A, B\rangle_F = \mathrm{Tr}(AB^T) = \sum_{i,j} A_{ij}B_{ij}$.

Example: The space $\mathbf{S}^n$ of $n \times n$ real symmetric matrices with the Frobenius inner product: This is the subspace of $\mathbf{R}^{n\times n}$ comprising all symmetric $n \times n$ matrices; the inner product is inherited from the embedding space. Of course, for symmetric matrices, this product can be written down without transposition, as follows:

$$A, B \in \mathbf{S}^n \quad \Rightarrow \quad \langle A, B\rangle_F = \mathrm{Tr}(AB) = \sum_{i,j} A_{ij}B_{ij}.$$

The last example explains why we need Euclidean spaces instead of sticking to $\mathbf{R}^n$ with the standard inner product: we intend in the future to work also with the Euclidean space $(\mathbf{S}^n, \langle\cdot,\cdot\rangle_F)$; while it is possible to identify it with $\mathbf{R}^N$, $N = \frac{n(n+1)}{2}$, equipped with the standard inner product, it would be a complete disaster to work with "vector representations" of matrices from $\mathbf{S}^n$ instead of working with these matrices directly.

7.1.1.1 *Linear forms on Euclidean spaces*

Every homogeneous linear form $f(x)$ on a Euclidean space $(E, \langle\cdot,\cdot\rangle_E)$ can be represented in the form $f(x) = \langle e_f, x\rangle_E$ for certain vector $e_f \in E$ uniquely defined by $f(\cdot)$. The mapping $f \mapsto e_f$ is a one-to-one linear mapping of the space of linear forms on E onto E.

7.1.1.2 *Conjugate mapping*

Let $(E, \langle\cdot,\cdot\rangle_E)$ and $(F, \langle\cdot,\cdot\rangle_F)$ be Euclidean spaces. For a linear mapping $A : E \to F$ and every $f \in F$, the function $\langle Ae, f\rangle_F$ is a linear function of $e \in E$ and as such it is representable as $\langle e, A^* f\rangle_E$ for certain uniquely defined vector $A^* f \in E$. It is immediately seen that the mapping $f \mapsto A^* f$ is a linear mapping of F into E; the characteristic identity specifying this mapping is

$$\langle Ae, f\rangle_F = \langle e, A^* f\rangle_E \quad \forall (e \in E, f \in F).$$

The mapping A^* is called *conjugate* to A. It is immediately seen that the conjugation is a linear operation with the properties $(A^*)^* = A$, $(AB)^* = B^* A^*$. If $\{e_j\}_{j=1}^m$ and $\{f_i\}_{i=1}^n$ are orthonormal bases in E, F, then every linear mapping $A : E \to F$ can be associated with the matrix $[a_{ij}]$ ("matrix of the mapping in the pair of bases in question") according to the identity

$$A \sum_{j=1}^m x_j e_j = \sum_i \left[\sum_j a_{ij} x_j \right] f_i$$

(in other words, a_{ij} is the ith coordinate of the vector Ae_j in the basis $f_1, \dots, f_n$). With this representation of linear mappings by matrices, the matrix representing A^* in the pair of bases $\{f_i\}$ in the argument and $\{e_j\}$ in the image spaces of A^* is the transpose of the matrix representing A in the pair of bases $\{e_j\}$, $\{f_i\}$.

7.1.2 *Cones in Euclidean spaces*

A nonempty subset $\mathbf{K}$ of a Euclidean space $(E, \langle\cdot,\cdot\rangle_E)$ is called a cone if it is a convex set comprising rays emanating from the origin, or, equivalently, whenever $t_1, t_2 \geq 0$ and $x_1, x_2 \in \mathbf{K}$, we have $t_1 x_1 + t_2 x_2 \in \mathbf{K}$.

A cone $\mathbf{K}$ is called *regular* if it is closed, possesses a nonempty interior and is *pointed* — does not contain lines, or, which is the same, is such that $a \in \mathbf{K}$, $-a \in \mathbf{K}$ implies that $a = 0$.

Dual cone: If $\mathbf{K}$ is a cone in a Euclidean space $(E, \langle \cdot, \cdot \rangle_E)$, then the set

$$\mathbf{K}^* = \{e \in E : \langle e, h \rangle_E \geq 0 \, \forall h \in \mathbf{K}\}$$

is also a cone called the cone *dual* to $\mathbf{K}$. The dual cone is always closed. The cone dual to dual is the closure of the original cone: $(\mathbf{K}^*)^* = \mathrm{cl}\, \mathbf{K}$; in particular, $(\mathbf{K}^*)^* = \mathbf{K}$ for every closed cone $\mathbf{K}$. For a closed cone $\mathbf{K}$, the cone $\mathbf{K}^*$ possesses a nonempty interior iff $\mathbf{K}$ is pointed, and $\mathbf{K}^*$ is pointed iff $\mathbf{K}$ possesses a nonempty interior; in particular, $\mathbf{K}$ is regular iff $\mathbf{K}^*$ is so.

Example: Nonnegative ray and nonnegative orthants: The simplest one-dimensional cone is the nonnegative ray $\mathbf{R}_+ = \{t \geq 0\}$ on the real line $\mathbf{R}^1$. The simplest cone in $\mathbf{R}^n$ is the *nonnegative orthant* $\mathbf{R}^n_+ = \{x \in \mathbf{R}^n : x_i \geq 0, 1 \leq i \leq n\}$. This cone is regular and self-dual: $(\mathbf{R}^n_+)^* = \mathbf{R}^n_+$.

Example: Lorentz cone $\mathbf{L}^n$: The cone $\mathbf{L}^n$ "lives" in $\mathbf{R}^n$ and comprises all vectors $x = [x_1; ...; x_n] \in \mathbf{R}^n$ such that $x_n \geq \sqrt{\sum_{j=1}^{n-1} x_j^2}$; same as $\mathbf{R}^n_+$, the Lorentz cone is regular and self-dual.

By definition, $\mathbf{L}^1 = \mathbf{R}_+$ is the nonnegative orthant; this is in full accordance with the "general" definition of a Lorentz cone combined with the standard convention "a sum over an empty set of indices is 0."

Example: Semidefinite cone $\mathbf{S}^n_+$: The cone $\mathbf{S}^n_+$ "lives" in the Euclidean space $\mathbf{S}^n$ of $n \times n$ symmetric matrices equipped with the Frobenius inner product. The cone comprises all $n \times n$ symmetric *positive semidefinite* matrices A, i.e., matrices $A \in \mathbf{S}^n$ such that $x^T Ax \geq 0$ for all $x \in \mathbf{R}^n$, or, equivalently, such that all eigenvalues of A are nonnegative. Same as $\mathbf{R}^n_+$ and $\mathbf{L}^n$, the cone $\mathbf{S}^n_+$ is regular and self-dual.

Finally, we remark that the direct product of regular cones is regular, and the dual of this product is the direct product of the duals of the original cones.

When checking this absolutely evident statement, you should take into account how we take the direct product of Euclidean spaces, since without a Euclidean structure on the product of the Euclidean spaces embedding the cones we are multiplying, the claim about the dual of a direct product of cones becomes senseless. The Euclidean structure on the direct product $E = E_1 \times \cdots \times E_m$ of Euclidean spaces is defined as follows: vectors from E, by the definition of direct product, are ordered tuples $(x^1, \ldots, x^m)$ with $x^i \in E_i$, and we set

$$\langle (x^1, \ldots, x^m), (y^1, \ldots, y^m) \rangle_E = \sum_{i=1}^{m} \langle x^i, y^i \rangle_{E_i}.$$

with this definition, a direct product of the spaces $\mathbf{R}^{n_1}, \ldots, \mathbf{R}^{n_m}$ equipped with the standard inner products is $\mathbf{R}^{n_1+\cdots+n_m}$, also equipped with the standard inner product, and the direct product of the spaces $\mathbf{S}^{n_1}, \ldots, \mathbf{S}^{n_m}$ equipped with the Frobenius inner products can be viewed as the space $\mathbf{S}^{n_1,\ldots,n_m}$ of *block-diagonal* symmetric matrices with m diagonal blocks of sizes $n_1, \ldots, n_m$, equipped with the Frobenius inner product.

We have made several claims that are not self-evident, and here are their proofs (we slightly alter the order of claims and, aside of the latter item, assume w.l.o.g. that the Euclidean space in question is $\mathbf{R}^n$ with the standard inner product).

- *For every cone K, one has K^* is a closed cone, and $(K^*)^* = \operatorname{cl} K$.* The closedness of a dual cone is evident, same as the fact that $\operatorname{cl} K$ is a closed cone such that $(\operatorname{cl} K)^* = K^*$. Besides this, we clearly have $(K^*)^* \supset \operatorname{cl} K$. To prove that the latter $\supset$ is in fact $=$, assume that this is not the case, so that $(K^*)^*$ contains a vector $x \notin \operatorname{cl} K$. By Separation Theorem for Convex Sets (Theorem 2.14), there exists a linear form $e^T w$ such that $e^T x < \inf_{y \in \operatorname{cl} K} e^T y = \inf_{y z_1 K} e^T y$. But the infimum of a linear form $e^T y$ on a cone K is either $-\infty$ (this is the case when e has a negative inner product with certain vector from K, i.e., when $e \notin K^*$), or is 0 (this is the case when $e \in K^*$). We are in the case when the infimum $\inf_{y \in K} e^T y$ is $> e^T x$ and thus is finite, whence $e \in K^*$, the infimum is 0 and thus $e^T x < 0$, which is impossible due to $x \in (K^*)^*$. $\square$
- *For every cone $\mathbf{K}$, $\mathbf{K}^*$ is pointed iff* $\operatorname{int} \mathbf{K} \neq \emptyset$. Indeed, if $\operatorname{int} \mathbf{K}$ is nonempty and thus contains a ball B of radius $r > 0$, and $h, -h \in \mathbf{K}^*$, then the linear form $h^T x$ should be both nonnegative and nonpositive on B; but a vector h can be orthogonal to all vectors from a ball of positive radius iff $h = 0$. Thus, $\mathbf{K}^*$ is pointed. On the other hand, if $\operatorname{int} \mathbf{K} = \emptyset$, then $\operatorname{Aff}(\mathbf{K}) \neq \mathbf{R}^n$ due to Theorem 2.3. Since $0 \in \mathbf{K}$, $\operatorname{Aff}(\mathbf{K})$ is a linear subspace in $\mathbf{R}^n$, and since it differs from $\mathbf{R}^n$, its orthogonal complement does not reduce to $\{0\}$. In other words, there exists $h \neq 0$ which is orthogonal to $\operatorname{Aff}(\mathbf{K})$, whence $\pm h \in \mathbf{K}^*$, and the latter cone is not pointed. $\square$
- *For a closed cone $\mathbf{K}$, $\mathbf{K}^*$ has a nonempty interior iff $\mathbf{K}$ is pointed.* This is readily given by the previous item due to $\mathbf{K} = (\mathbf{K}^*)^*$.
- *The nonnegative orthant $\mathbf{R}^n_+$ is regular and self-dual.* This is evident.

- *The Lorentz cone* $\mathbf{L}^n$ *is regular and self-dual.* Regularity is evident. To prove self-duality, we should verify that given $[u;t]$ with $u \in \mathbf{R}^{n-1}$, the relation $[u;t]^T[v;\tau] \geq 0$ holds true for all $[v;\tau]$ with $\tau \geq \|v\|_2$ iff $t \geq \|u\|_2$, or, which is the same, to verify that for every vector $[u;t]$ one has $\inf_{[v;\tau]:\|v\|_2\leq\tau}[u;t]^T[v;\tau] \geq 0$ iff $t \geq \|u\|_2$. This is immediate, since

$$\inf_{[v;\tau]:\|v\|_2\leq\tau}[u;t]^T[v;\tau] = \inf_{\tau\geq 0}\left[t\tau + \inf_{v:\|v\|_2\leq\tau} u^T v\right] = \inf_{\tau\geq 0}\tau[t - \|u\|_2].$$

□

- *The cone* $\mathbf{S}^n_+$ *of positive semidefinite matrices in the space of* $\mathbf{S}^n$ *of symmetric* $n \times n$ *matrices equipped with the Frobenius inner product is regular and self-dual.* Regularity is evident. To prove self-duality, we should verify that if $B \in \mathbf{S}^n$, then $\mathrm{Tr}(BX) \equiv \langle B, X\rangle_{\mathbf{S}^n} \geq 0$ for all $X \in \mathbf{S}^n_+$ iff $B \in \mathbf{S}^n_+$. In one direction: setting $X = xx^T$ with $x \in \mathbf{R}^n$, we get $X \succeq 0$. Thus, if $B \in (\mathbf{S}^n_+)^*$, then $\mathrm{Tr}(xx^TB) = \mathrm{Tr}(x^TBx) = x^TBx \geq 0$ for all x, and thus $B \in \mathbf{S}^n_+$.[2] In the opposite direction: When $B \in \mathbf{S}^n_+$, then, by the Eigenvalue Decomposition Theorem, $B = \sum_{i=1}^n \lambda_i e_i e_i^T$ with orthonormal $e_1, \ldots, e_n$ and nonnegative λ_i (the latter in fact are eigenvalues of B). It follows that when $X \in \mathbf{S}^n_+$, then $\mathrm{Tr}(BX) = \sum_i \lambda_i \mathrm{Tr}(e_i e_i^T X) = \sum_i \lambda_i e_i^T X e_i$; when $X \in \mathbf{S}^n_+$, all terms in the resulting sum are nonnegative, and thus $\mathrm{Tr}(BX) \geq 0$ whenever $X \in \mathbf{S}^n_+$, that is, $B \in (\mathbf{S}^n_+)^*$.

We conclude our "executive summary" on cones in Euclidean spaces by the following

Proposition 7.1. *Let* $\mathbf{K}$ *be a cone in Euclidean space* E *and* $f \in \mathrm{int}\,\mathbf{K}^*$, *Then for some positive constant* c, *the following holds*:

$$h \in \mathbf{K} \Rightarrow \langle f, h\rangle_E \geq c\|h\|_E,\ \|h\|_E = \sqrt{\langle h, h\rangle_E}. \tag{7.1}$$

Indeed, since $f \in \mathrm{int}\,\mathbf{K}^*$, there exists $\rho > 0$ such that $\|\cdot\|_E$-ball of radius ρ centered at f belongs to $\mathbf{K}^*$. If now $h \in \mathbf{K}$, we have

$$\|e\|_E \leq \rho \Rightarrow f - e \in \mathbf{K}^* \Rightarrow 0 \leq \langle f, h\rangle_E - \langle e, h\rangle_E \Rightarrow \langle e, h\rangle_E \leq \langle f, h\rangle_E$$
$$\Rightarrow \rho\|h\|_E = \max_{e:\|e\|_E\leq\rho}\langle e, h\rangle_E \leq \langle f, h\rangle_E$$

so that (7.1) holds true for $c = \rho^{-1}$. □

[2]We have used a simple and useful identity: *when* P *and* Q *are matrices such that* PQ *makes sense and is a square matrix, so that* $\mathrm{Tr}(PQ)$ *makes sense, then* $\mathrm{Tr}(PQ) = \mathrm{Tr}(QP)$ (why?).

7.2 Conic problems

A *conic program* is an optimization program of the form

$$\mathrm{Opt}(P) = \min_x \left\{ \langle c, x\rangle_E : \begin{array}{l} A_i x - b_i \in \mathbf{K}_i,\ i = 1, \ldots, m, \\ Rx = r \end{array} \right\}, \qquad (P)$$

where

- $(E, \langle\cdot,\cdot\rangle_E)$ is a Euclidean space of *decision vectors* x and $c \in E$ is the *objective*;
- A_i, $1 \leq i \leq m$, are linear maps from E into Euclidean spaces $(F_i, \langle\cdot,\cdot\rangle_{F_i})$, $b_i \in F_i$ and $\mathbf{K}_i \subset F_i$ are regular cones;
- R is a linear mapping from E into a Euclidean space $(F, \langle\cdot,\cdot\rangle_F)$ and $r \in F$.

A relation $a - b \in \mathbf{K}$, where $\mathbf{K}$ is a regular cone, is often called *conic inequality* between a and b and is denoted $a \geq_{\mathbf{K}} b$; such a relation indeed preserves the major properties of the usual coordinate-wise vector inequality $\geq$. While in the sequel we do not use the notation $a \geq_{\mathbf{K}} b$, we do call a constraint of the form $Ax - b \in \mathbf{K}$ a *conic inequality constraint* or simply *conic constraint*.

Note that we can rewrite (P) equivalently as a conic program involving a *single* cone $\mathbf{K} = \mathbf{K}_1 \times \cdots \times \mathbf{K}_m$, specifically, as

$$\min_x \left\{ \langle c, x\rangle_E : \begin{array}{l} Ax - b \in \mathbf{K} = \mathbf{K}_1 \times \cdots \times \mathbf{K}_m, \\ Rx = r \end{array} \right\},$$

$$Ax - b = \begin{bmatrix} A_1 x - b_1 \\ \vdots \\ A_m x - b_m \end{bmatrix}; \qquad (P')$$

since the direct product of several regular cones clearly is regular as well, (P') is indeed a legitimate "single cone" conic program.

Note: Of course, we can express linear equality constraints $Rx = r$ by linear inequalities $[R; -R]x \leq [r; -r]$ and thus reduce the situation to the one where we want to minimize a linear objective under a single conic constraint. From time to time we shall use this possibility to save notation.

Examples: Linear, Conic Quadratic and Semidefinite Optimization: We will be especially interested in the following three generic conic problems:

- *Linear Optimization*, or *Linear Programming*: This is the family of all conic programs associated with nonnegative orthants $\mathbf{R}^m_+$, that is, the family of all usual LPs $\min_x\{c^Tx : Ax - b \geq 0, Rx = r\}$;
- *Conic Quadratic Optimization*, or *Conic Quadratic Programming*, or *Second-Order Cone Programming*: This is the family of all conic programs associated with the cones that are *finite direct products* of Lorentz cones, that is, the conic programs of the form

$$\min_x \left\{c^Tx : [A_1; ...; A_m]x - [b_1; ...; b_m] \in \mathbf{L}^{k_1} \times \cdots \times \mathbf{L}^{k_m}, Rx = r\right\},$$

where A_i are $k_i \times \dim x$ matrices and $b_i \in \mathbf{R}^{k_i}$. The "Mathematical Programming" form of such a program is

$$\min_x \left\{c^Tx : \|\bar{A}_i x - \bar{b}_i\|_2 \leq \alpha_i^T x - \beta_i,\ 1 \leq i \leq m\right\},$$

where $A_i = [\bar{A}_i; \alpha_i^T]$ and $b_i = [\bar{b}_i; \beta_i]$, so that α_i is the last row of A_i, and β_i is the last entry of b_i;
- *Semidefinite Optimization*, or *Semidefinite Programming*: This is the family of all conic programs associated with the cones that are *finite direct products* of Semidefinite cones, that is, the conic programs of the form

$$\min_x \left\{ c^Tx : A_i^0 + \sum_{j=1}^{\dim x} x_j A_i^j \succeq 0,\ 1 \leq i \leq m \right\}, \qquad (*)$$

where A_i^j are symmetric matrices of appropriate sizes.

Constraints of the form $A(x) \succeq B(x)$, where $A(\cdot)$ and $B(\cdot)$ are symmetric matrices affinely depending on decision vector $x \in \mathbf{R}^n$, are called *Linear Matrix Inequalities* (LMIs), giving rise to another name — "linear optimization under LMI constraints" — for programs $(*)$.

7.2.1 *Relations between LO, CQO and SDO*

Clearly, polyhedral representations of sets and functions are their $\mathcal{CQ}$- and $\mathcal{SD}$-representations as well — recall that the nonnegative ray is the same as the one-dimensional Lorentz and one-dimensional semidefinite cones; as a result nonnegative orthants "sit" in $\mathcal{CQ}$ and $\mathcal{SD}$,

- adding finite systems of scalar linear equality/inequality constraints to an $\mathcal{LO}/\mathcal{CQ}/\mathcal{SD}$ problem does not alter the problem's type.

For example, an LO program can be straightforwardly converted into a conic quadratic and into a semidefinite program. For example, the "single-cone" semidefinite reformulation of LO program

$$\min_x \left\{c^T x : Ax \geq b, Rx = r\right\} \tag{$*$}$$

is as follows: keep the objective and the linear equality constraints as they are, and put the entries of the m-dimensional vector $Ax - b$ on the diagonal of a diagonal $m \times m$ matrix $\mathcal{A}(x)$ which, of course, will depend affinely on x. Since a diagonal matrix is symmetric and is positive semidefinite iff its diagonal entries are nonnegative, $(*)$ is equivalent to the SDO program

$$\min_x \left\{c^T x : \mathcal{A}(x) \succeq 0, Rx = r\right\}.$$

A less trivial, but still simple, observation is that *conic quadratic representable sets/functions are semidefinite representable as well, with semidefinite representations readily given by conic quadratic ones.* The reason is that a Lorentz cone $\mathcal{L}^n$ is $\mathcal{SDO}$-representable — it is just the intersection of the semidefinite cone $\mathbf{S}^n_+$ and an appropriate linear subspace of $\mathbf{S}^n$ (this is completely similar to the fact that the nonnegative orthant $\mathbf{R}^n_+$ is the intersection of $\mathbf{S}^n_+$ and the subspace of $n \times n$ diagonal matrices). Specifically, given a vector $x \in \mathbf{R}^n$, let us build the $n \times n$ symmetric matrix

$$\text{Arrow}(x) = \left[\begin{array}{c|cccc} x_n & x_1 & x_2 & \cdots & x_{n-1} \\ \hline x_1 & x_n & & & \\ x_2 & & x_n & & \\ \vdots & & & \ddots & \\ x_{n-1} & & & & x_n \end{array}\right]$$

(blanks are filled with zeros).

Lemma 7.1. *Let $x \in \mathbf{R}^n$. The matrix* $\mathrm{Arrow}(x)$ *is positive semidefinite iff $x \in \mathbf{L}^n$. As a result, a conic quadratic representation of a set*

$$X = \{x : \exists w : A_i x + B_i w + b_i \in \mathcal{L}^{n_i},\ 1 \leq i \leq m\}$$

can be converted into a semidefinite representation of the same set, specifically, the representation

$$X = \{x : \exists w : \mathrm{Arrow}(A_i x + B_i w + b_i) \succeq 0,\ 1 \leq i \leq m\}.$$

The simplest way to prove the lemma is to extract it from the extremely important in its own right.

Lemma 7.2 (Schur Complement Lemma). *Consider the symmetric block matrix*

$$A = \left[\begin{array}{c|c} P & Q \\ \hline Q^T & R \end{array}\right]$$

with square diagonal blocks P, R, and assume that R is positive definite. Then A is positive semidefinite iff

$$P - QR^{-1}Q^T \succeq 0.$$

Proof. Observe, first, that a quadratic form

$$g(z) = z^T S z + 2b^T z : \mathbf{R}^k \to \mathbf{R}$$

with positive definite matrix S achieves its minimum at the point $\bar{z} = -S^{-1}b$ where the gradient of the form vanishes. Indeed, g is quadratic, so that

$$g(\bar{z} + h) = g(\bar{z}) + h^T \underbrace{\nabla g(\bar{z})}_{=0} + \frac{1}{2} h^T \nabla^2 g(\bar{z}) h = g(\bar{z}) + h^T S h \geq g(\bar{z}),$$

where the concluding inequality stems from $S \succeq 0$.

To prove SCL, let P be $p \times p$, and R be $r \times r$. Partitioning $(p+r)$-dimensional vector into consecutive blocks u and v of dimensions p and r, respectively, we have

$$\begin{aligned}
& A \succeq 0 \Leftrightarrow 0 \leq u^T P u + 2u^T Q v + v^T R v = [u; v]^T A [u; v]\ \forall (u, v) \\
& \Leftrightarrow 0 \leq \inf_v \left[u^T P u + 2u^T Q v + v^T R v\right] \geq 0\ \forall u \\
& \Leftrightarrow 0 \leq u^T P u + 2u^T Q[-R^{-1}Q^T u] + [-R^{-1}Q^T u]^T R [-R^{-1}Q^T u] \\
& \qquad\qquad\qquad\qquad\qquad\qquad\qquad\qquad [\text{observation above}] \\
& \Leftrightarrow 0 \leq u^T [P - QR^{-1}Q^T] u\ \forall u \Leftrightarrow P - QR^{-1}Q^T \succeq 0 \qquad \square
\end{aligned}$$

SCL ⇒ Lemma 7.1: The case of $n = 1$ is trivial, thus assume $n > 1$. In one direction: let $x \in \mathbf{L}^n$, and let us prove that $\mathrm{Arrow}(x) \succeq 0$. We have $x_n \geq 0$; when $x_n = 0$, from $x \in \mathbf{L}^n$ it follows that $x = 0$, so that $\mathrm{Arrow}(x) \succeq 0$. When $x_n > 0$, we can apply to $\mathrm{Arrow}(x)$ the SCL with 1×1 block $P = x_n$ and $(n-1) \times (n-1)$ block $R = x_n I_{n-1}$ to conclude that $\mathrm{Arrow}(x) \succeq 0$ if and only if $x_n \geq \sum_{i=1}^{n} x_i^2 / x_n$, or, which is the same, if and only if $x \in \mathbf{L}^n$. In the opposite direction: let $\mathrm{Arrow}(x) \succeq 0$, and let us prove that $x \in \mathbf{L}^n$. x_n is the diagonal entry of a positive semidefinite matrix and as such is nonnegative. As we have already seen, when $x_n > 0$, $\mathrm{Arrow}(x) \succeq 0$ iff $x \in \mathbf{L}^n$. It remains to consider the case of $x_n = 0$. In this case, from $\mathrm{Arrow}(x) \geq 0$ it follows that $x = 0$,[3] so that $x \in \mathbf{L}^n$. □

Remark. The possibility to convert straightforwardly LO and CQO to Semidefinite Optimization does not mean that this is the best way to solve LOs and CQO's in actual computations. The two former problems are somehow simpler than the latter one, and dedicated LO and CQO solvers available today in commercial packages can solve linear and conic quadratic programs much faster, and in a much wider range of sizes, than "universal" SDO solvers. This being said, when solving "moderate size" LOs and CQOs (what is "moderate," depends on the "fine structure" of a program being solved and may vary from few hundreds to few thousands of variables), it is very attractive to reduce everything to SDO and thus to use a single solver. This idea is implemented in the `cvx` package[4] which uses the calculus of semidefinite representable sets and functions to convert the input "high level" description of a problem into its "inner" SDO-reformulation which then is forwarded to an SDO solver. The input description of a problem utilizes full capabilities of MATLAB and thus is incredibly transparent and easy to use, making `cvx` an ideal tool for a classroom (and not only for it).

---

[3]We have used the following elementary fact: *Every diagonal entry in a positive semidefinite matrix $A = [a_{ij}]$ is nonnegative, and if it is zero, so are all entries in the row and column of this entry.* Indeed, for $A \succeq 0$ all principal minors should be nonnegative (Sylvester's criterion); nonnegativity of principal 1×1 minors means that $a_{ii} \geq 0$ for all i, and nonnegativity of principal 2×2 minors means that $a_{ij}^2 \leq a_{ii} a_{jj}$ for all $i \neq j$, so that $a_{ii} = 0$ implies that $a_{ij} = a_{ji} = 0$ for all j.

[4]"CVX: Matlab Software for Disciplined Convex Programming," Michael Grant and Stephen Boyd, http://cvxr.com/cvx/.

7.3 Conic duality

7.3.1 *Conic duality: Derivation*

The origin of conic duality is the desire to find a systematic way to bound from below the optimal value in a conic program

$$\begin{aligned}\mathrm{Opt}(P) = \quad & \min_x \left\{ \langle c, x\rangle_E : \begin{array}{l} A_i x - b_i \in \mathbf{K}_i,\ i = 1, \ldots, m, \\ Rx = r \end{array} \right\} \qquad (P) \\ \Leftrightarrow \quad & \min_x \left\{ \langle c, x\rangle_E : \begin{array}{l} Ax - b \in \mathbf{K} = \mathbf{K}_1 \times \cdots \times \mathbf{K}_m, \\ Rx = r \end{array} \right\} \qquad (P')\end{aligned}$$

$$Ax - b = \begin{bmatrix} A_1 x - b_1 \\ \vdots \\ A_m x - b_m \end{bmatrix}.$$

This way is based on the *linear aggregation* of the constraints of (P), as follows. Let $y_i \in \mathbf{K}_i^*$ and $z \in F$. By the definition of the dual cone, for every x feasible for (P), we have

$$\langle A_i^* y_i, x\rangle_E - \langle y_i, b_i\rangle_{F_i} \equiv \langle y_i, Ax_i - b_i\rangle_{F_i} \geq 0, \quad 1 \leq i \leq m,$$

and of course

$$\langle R^* z, x\rangle_E - \langle z, r\rangle_F = \langle z, Rx - r\rangle_F = 0.$$

Summing up the resulting inequalities, we get

$$\left\langle R^* z + \sum_i A_i^* y_i, x \right\rangle_E \geq \langle z, r\rangle_F + \sum_i \langle y_i, b_i\rangle_{F_i}. \qquad (C)$$

By its origin, this scalar linear inequality on x is a consequence of the constraints of (P), that is, it is valid for all feasible solutions x to (P). It may happen that the left-hand side in this inequality is, identically in $x \in E$, equal to the objective $\langle c, x\rangle_E$; this happens iff

$$R^* z + \sum_i A_i^* y_i = c.$$

Whenever this is the case, the right-hand side of (C) is a valid lower bound on the optimal value of (P). The dual program is nothing but the program

$$\mathrm{Opt}(D) = \max_{z,\{y_i\}} \left\{ \langle z, r\rangle_F + \sum_i \langle y_i, b_i\rangle_{F_i} : \begin{array}{r} y_i \in \mathbf{K}_i^*,\ 1 \le i \le m, \\ R^* z + \sum_i A_i^* y_i = c \end{array} \right\} \tag{D}$$

of maximizing this lower bound.

Remark. Note that the construction we have presented is completely similar to the one we used in Section 3.2.1 to derive the LO dual of a given LO program. The latter is the particular case of (P) where all $\mathbf{K}_i$ are nonnegative orthants of various dimensions, or, which is the same, the cone $\mathbf{K}$ in (P') is a nonnegative orthant. The only minor differences stem from the facts that now it is slightly more convenient to write the primal program as a minimization one, while in the LO we preferred to write down the primal program as a maximization one. Modulo this absolutely unessential difference, our derivation of the dual of an LO program is nothing but our present construction as applied to the case when all $\mathbf{K}_i$ are nonnegative rays. In fact, a reader will see that all Conic Duality constructions and results we are about to present mirror constructions and results of LO Duality already known to us.

Coming back to conic dual of a conic program, observe that by the origin of the dual we have

Weak Duality: *One has* $\mathrm{Opt}(D) \le \mathrm{Opt}(P)$.

Besides this, we see that (D) is a conic program. A nice and important fact is that *conic duality is symmetric.*

Symmetry of Duality: *The conic dual to (D) is (equivalent to) (P).*

Proof. In order to apply to (D) the outlined recipe for building the conic dual, we should rewrite (D) as a *minimization* program

$$-\mathrm{Opt}(D) = \min_{z,\{y_i\}} \left\{ \langle z, -r\rangle_F + \sum_i \langle y_i, -b_i\rangle_{F_i} : \begin{array}{r} y_i \in \mathbf{K}_i^*, 1 \le i \le m \\ R^* z + \sum_i A_i^* y_i = c \end{array} \right\}; \tag{D'}$$

the corresponding space of decision vectors is the direct product $F \times F_1 \times \cdots \times F_m$ of Euclidean spaces equipped with the inner product

$$\langle [z; y_1, \ldots, y_m], [z'; y'_1, \ldots, y'_m] \rangle = \langle z, z' \rangle_F + \sum_i \langle y_i, y'_i \rangle_{F_i}.$$

The above "duality recipe" as applied to (D') reads as follows: pick weights $\eta_i \in (\mathbf{K}_i^*)^* = \mathbf{K}_i$ and $\zeta \in E$, so that the scalar inequality

$$\underbrace{\langle \zeta, R^* z + \sum_i A_i^* y_i \rangle_E + \sum_i \langle \eta_i, y_i \rangle_{F_i}}_{=\langle R\zeta, z \rangle_F + \sum_i \langle A_i \zeta + \eta_i, y_i \rangle_{F_i}} \geq \langle \zeta, c \rangle_E \qquad (C')$$

in variables z, $\{y_i\}$ is a consequence of the constraints of (D'), and impose on the "aggregation weights" $\zeta, \{\eta_i \in \mathbf{K}_i\}$ an additional restriction that the left-hand side in this inequality is, identically in $z, \{y_i\}$, equal to the objective of (D'), that is, the restriction that

$$R\zeta = -r,\ A_i\zeta + \eta_i = -b_i,\ 1 \leq i \leq m,$$

and maximize under this restriction the right-hand side in (C'), thus arriving at the program

$$\max_{\zeta, \{\eta_i\}} \left\{ \langle c, \zeta \rangle_E : \begin{array}{l} \mathbf{K}_i \ni \eta_i = A_i[-\zeta] - b_i, 1 \leq i \leq m \\ R[-\zeta] = r \end{array} \right\}.$$

Substituting $x = -\zeta$, the resulting program, after eliminating η_i variables, is nothing but

$$\max_x \left\{ -\langle c, x \rangle_E : \begin{array}{l} A_i x - b_i \in \mathbf{K}_i,\ 1 \leq i \leq m \\ Rx = r \end{array} \right\},$$

which is equivalent to (P). □

7.3.2 Conic duality theorem

A conic program (P) is called *strictly feasible*, if it admits a *strictly feasible* solution, that is, a feasible solution $\bar{x}$ such that $A_i\bar{x} - b_i \in \operatorname{int} \mathbf{K}_i$, $i = 1, \ldots, m$.

Conic Duality Theorem is the following statement that resembles very much the Linear Programming Duality Theorem:

Theorem 7.1 (Conic Duality Theorem). *Consider a primal–dual pair of conic programs* (P), (D):

$$\mathrm{Opt}(P) = \min_x \left\{ \langle c, x\rangle_E : \begin{array}{l} A_i x - b_i \in \mathbf{K}_i,\ i = 1, \dots, m, \\ Rx = r \end{array} \right\}, \quad (P)$$

$$\mathrm{Opt}(D) = \max_{z,\{y_i\}} \left\{ \langle z, r\rangle_E + \sum_i \langle y_i, b_i\rangle_{F_i} : \begin{array}{l} y_i \in \mathbf{K}_i^*, 1 \le i \le m \\ R^* z + \sum_i A_i^* y_i = c \end{array} \right\} \quad (D)$$

Then

(i) [Weak Duality] *One has* $\mathrm{Opt}(D) \le \mathrm{Opt}(P)$.
(ii) [Symmetry] *The duality is symmetric:* (D) *is a conic program, and the program dual to* (D) *is (equivalent to)* (P).
(iii) [Strong Duality] *If one of the programs* (P), (D) *is strictly feasible and bounded, then the other program is solvable, and* $\mathrm{Opt}(P) = \mathrm{Opt}(D)$.

If both the programs are strictly feasible, then both are solvable with equal optimal values.

In addition, if one of the problems is strictly feasible, then the optimal values in both the problems are equal to each other.

Proof. We have already verified Weak Duality and Symmetry. Let us prove the first claim in Strong Duality. By Symmetry, we can restrict ourselves to the case when the strictly feasible and bounded program is (P).

Consider the following two sets in the Euclidean space $G = \mathbf{R} \times F \times F_1 \times \cdots \times F_m$:

$$\begin{aligned} T &= \{[t; z; y_1; ...; y_m] : \exists x : t = \langle c, x\rangle_E; y_i = A_i x - b_i, 1 \le i \le m; \\ &\qquad z = Rx - r\}, \\ S &= \{[t; z; y_1; ...; y_m] : t < \mathrm{Opt}(P), y_1 \in \mathbf{K}_1, \dots, y_m \in \mathbf{K}_m, z = 0\}. \end{aligned}$$

The sets T and S are clearly convex and nonempty; observe that they do not intersect. Indeed, assuming that $[t; z; y_1; ...; y_m] \in S \cap T$, we should have $t < \mathrm{Opt}(P)$, and $y_i \in \mathbf{K}_i$, $z = 0$ (since the point is in S), and at the same time for certain $x \in E$ we should have $t = \langle c, x\rangle_E$ and $A_i x - b_i = y_i \in \mathbf{K}_i$, $Rx - r = z = 0$, meaning that there exists a feasible solution to (P) with the value of the objective $< \mathrm{Opt}(P)$, which is impossible. Since the convex and nonempty sets S and T do not intersect, they can be separated by a linear form (Theorem 2.14): there exists $[\tau; \zeta; \eta_1; ...; \eta_m] \in G = \mathbf{R} \times F \times F_1 \times \cdots \times F_m$ such that

$$\begin{aligned}
&\sup_{[t;z;y_1;...;y_m]\in S} \langle[\tau;\zeta;\eta_1;...;\eta_m],[t;z;y_1;...;y_m]\rangle_G && \text{(a)}\\
&\quad\leq \inf_{[t;z;y_1;...;y_m]\in T} \langle[\tau;\zeta;\eta_1;...;\eta_m],[t;z;y_1;...;y_m]\rangle_G, &&\\
&\inf_{[t;z;y_1;...;y_m]\in S} \langle[\tau;\zeta;\eta_1;...;\eta_m],[t;z;y_1;...;y_m]\rangle_G && \text{(b)}\\
&\quad< \sup_{[t;z;y_1;...;y_m]\in T} \langle[\tau;\zeta;\eta_1;...;\eta_m],[t;z;y_1;...;y_m]\rangle_G, &&
\end{aligned}$$

or, which is the same,

$$\begin{aligned}
&\sup_{t<\mathrm{Opt}(P), y_i\in\mathbf{K}_i} \left[\tau t + \sum_i \langle \eta_i, y_i\rangle_{F_i}\right] && \text{(a)}\\
&\quad\leq \inf_{x\in E}\left[\tau\langle c,x\rangle_E + \langle\zeta, Rx-r\rangle_F + \sum_i\langle\eta_i, A_i x - b_i\rangle_{F_i}\right], &&\\
&\inf_{t<\mathrm{Opt}(P), y_i\in\mathbf{K}_i} \left[\tau t + \sum_i \langle \eta_i, y_i\rangle_{F_i}\right] && \text{(b)}\\
&\quad< \sup_{x\in E}\left[\tau\langle c,x\rangle + \langle\zeta, Rx-r\rangle_F + \sum_i\langle\eta_i, A_i x - b_i\rangle_{F_i}\right]. &&
\end{aligned} \tag{7.2}$$

Since the left-hand side in (7.2(a)) is finite, we have

$$\tau \geq 0,\ -\eta_i \in \mathbf{K}_i^*, 1 \leq i \leq m, \tag{7.3}$$

whence the left-hand side in (7.2(a)) is equal to $\tau\mathrm{Opt}(P)$. Since the right-hand side in (7.2(a)) is finite, we have

$$R^*\zeta + \sum_i A_i^*\eta_i + \tau c = 0, \tag{7.4}$$

and the right-hand side in (a) is $\langle -\zeta, r\rangle_F - \sum_i \langle \eta_i, b_i\rangle_{F_i}$, so that (7.2(a)) reads

$$\tau \mathrm{Opt}(P) \le \langle -\zeta, r\rangle_F - \sum_i \langle \eta_i, b_i\rangle_{F_i}. \tag{7.5}$$

We claim that $\tau > 0$. Believing in our claim, let us extract Strong Duality from it. Indeed, setting $y_i = -\eta_i/\tau$, $z = -\zeta/\tau$, (7.3), (7.4) say that $z, \{y_i\}$ is a feasible solution for (D), and by (7.5), the value of the dual objective at this dual feasible solution is $\ge \mathrm{Opt}(P)$. By Weak Duality, this value cannot be larger than $\mathrm{Opt}(P)$, and we conclude that our solution to the dual is in fact an optimal one, and that $\mathrm{Opt}(P) = \mathrm{Opt}(D)$, as claimed.

It remains to prove that $\tau > 0$. Assume this is not the case; then $\tau = 0$ by (7.3). Now, let $\bar{x}$ be a strictly feasible solution to (P). Taking the inner product of both sides in (7.4) with $\bar{x}$, we have

$$\langle \zeta, R\bar{x}\rangle_F + \sum_i \langle \eta_i, A_i\bar{x}\rangle_{F_i} = 0,$$

while (7.5) reads

$$-\langle \zeta, r\rangle_F - \sum_i \langle \eta_i, b_i\rangle_{F_i} \ge 0.$$

Summing up the resulting inequalities and taking into account that $\bar{x}$ is feasible for (P), we get

$$\sum_i \langle \eta_i, A_i\bar{x} - b_i\rangle \ge 0.$$

Since $A_i\bar{x} - b_i \in \operatorname{int} \mathbf{K}_i$ and $\eta_i \in -\mathbf{K}_i^*$, the inner products on the left-hand side of the latter inequality are nonpositive, and ith of them is zero iff $\eta_i = 0$; thus, the inequality says that $\eta_i = 0$ for all i. Adding this observation to $\tau = 0$ and looking at (7.4), we see that $R^*\zeta = 0$, whence $\langle \zeta, Rx\rangle_F = 0$ for all x and, in particular, $\langle \zeta, r\rangle_F = 0$ due to $r = R\bar{x}$. The bottom line is that $\langle \zeta, Rx - r\rangle_F = 0$ for all x. Now, let us look at (7.2(b)). Since $\tau = 0$, $\eta_i = 0$ for all i and $\langle \zeta, Rx - r\rangle_F = 0$ for all x, both sides in this inequality are equal to 0, which is impossible. We arrive at a desired contradiction.

We have proved the first claim in Strong Duality. The second claim there is immediate: if both (P), (D) are strictly feasible, then both programs are bounded as well by Weak Duality, and thus are solvable with equal optimal values by the already proved part of Strong Duality.

Finally, the "In addition" part of Strong Duality can be justified exactly as in the case of LP duality: by primal–dual symmetry we can assume that the strictly feasible problem is the primal one. If this problem is also bounded, then $\mathrm{Opt}(P) = \mathrm{Opt}(D)$ by the already proved part of the Strong Duality claim. And if (P) is unbounded, then $\mathrm{Opt}(P) = -\infty$ and (D) is infeasible by Weak Duality, so that $\mathrm{Opt}(D) = -\infty = \mathrm{Opt}(P)$. □

Remark. The Conic Duality Theorem is a bit weaker than its LO counterpart: where in the LO case plain feasibility was enough, now strong feasibility is required. It can be easily demonstrated by examples that this difference stems from the essence of the matter rather than being a shortcoming of our proofs. Indeed, it can be easily demonstrated by examples that in the case of nonpolyhedral cones various "pathologies" can take place, e.g.,

- (P) can be strictly feasible and below bounded while being unsolvable;
- both (P) and (D) can be solvable, but with different optimal values, etc.

Importance of strong feasibility is the main reason for our chosen way to represent constraints of a conic program as conic inequality/inequalities augmented by a system of linear equality constraints. In principle, we could write a conic problem (P) without equality constrains, as follows:

$$\min_x \left\{ c^T x : A_i x - b_i \in \mathbf{K}_i, 1 \leq i \leq m, Rx - r \in \mathbf{R}^k_+, \; r - Rx \in \mathbf{R}^k_+ \right\} \quad [k = \dim r]$$

— the possibility we used, to save notation, in LO. Now, it would be unwise to treat equality constraints via pairs of opposite inequalities.

— the resulting problem would definitely *not* be strictly feasible.[5]

7.3.2.1 *Refinement*

We can slightly refine the Conic Duality Theorem, extending the "special treatment" from linear equality constraints to *scalar* linear inequalities. Specifically, consider problem (P) and assume that one of the conic constraints in the problem, say, the first one, is just $A_1x - b_1 \geq 0$, that is, $F_1 = \mathbf{R}^\mu$ with the standard inner product, and $\mathbf{K}_1$ is the corresponding nonnegative orthant. Thus, our primal problem is

$$\mathrm{Opt}(P) = \min_x \left\{ \langle c, x \rangle_E : \begin{array}{c} A_1x - b_1 \geq 0 \\ A_ix - b_i \in \mathbf{K}_i,\ 2 \leq i \leq m \\ Rx = r \end{array} \right\}, \qquad (P)$$

so that the dual (D) is

$$\mathrm{Opt}(D) = \max_{z,\{y_i\}_{i=1}^m} \left\{ \langle r, z \rangle_F + b_1^T y_1 + \sum_{i=2}^m \langle b_i, y_i \rangle_{F_i} : \begin{array}{c} y_1 \geq 0, \\ y_i \in \mathbf{K}_i^*,\ 2 \leq i \leq m, \\ R^*z + \sum_{i=1}^m A_i^* y_i = c \end{array} \right\}. \qquad (D)$$

Essentially strict feasibility: Note that the structure of problem (D) is completely similar to the one of (P) — the variables, let them be called ξ, are subject to finitely many *scalar* linear equalities and inequalities and, furthermore, finitely many conic inequalities $P_i\xi - p_i \in \mathbf{L}_i$, $i \in I$, where $\mathbf{L}_i$ are regular cones in Euclidean spaces. Let us call such a conic problem *essentially strictly feasible*, if it admits a feasible solution $\bar{\xi}$ at which the conic inequalities are satisfied strictly: $P_i\bar{\xi} - p_i \in \mathrm{int}\,\mathbf{L}_i$, $i \in I$. It turns out that the

[5]Another way to eliminate equality constraints, which is free of the outlined shortcoming, could be to use the equality constraints to express part of the variables as linear functions of the remaining variables, thus reducing the design dimension of the problem and getting rid of equality constraints.

Conic Duality Theorem 7.1 remains valid when one replaces in it "strict feasibility" with "essentially strict feasibility," which is some progress: a strictly feasible conic problem clearly is essentially strictly feasible, but not necessarily vice versa. Thus, we intend to prove

Theorem 7.2 (Refined Conic Duality Theorem). *Consider a primal–dual pair of conic programs* (P), (D):

$$\mathrm{Opt}(P) = \min_x \left\{ \langle c, x\rangle_E : \begin{array}{lr} A_1 x - b_1 \geq 0 & (a) \\ A_i x - b_i \in \mathbf{K}_i,\ 2 \leq i \leq m & (b) \\ Rx = r & (c) \end{array} \right\}, \quad (P)$$

$$\mathrm{Opt}(D) = \max_{z,\{y_i\}_{i=1}^m} \left\{ \begin{array}{l} \langle r, z\rangle_F + b_1^T y_1 + \sum_{i=2}^m \langle b_i, y_i\rangle_{F_i} : \\ \left. \begin{array}{l} y_1 \geq 0, \\ y_i \in \mathbf{K}_i^*,\ 2 \leq i \leq m, \\ R^* z + \sum_{i=1}^m A_i^* y_i = c \end{array} \right\} \end{array} \right. . \quad (D)$$

Then

(i) [Weak Duality] *One has* $\mathrm{Opt}(D) \leq \mathrm{Opt}(P)$.
(ii) [Symmetry] *The duality is symmetric:* (D) *is a conic program, and the program dual to* (D) *is (equivalent to)* (P).
(iii) [Refined Strong Duality] *If one of the programs* (P), (D) *is essentially strictly feasible and bounded, then the other program is solvable, and* $\mathrm{Opt}(P) = \mathrm{Opt}(D)$.

If both the programs are essentially strictly feasible, then both are solvable with equal optimal values.

In addition, if one of the problems is essentially strictly feasible, then the optimal values in both the problems are equal to each other.

Note that the Refined Conic Duality Theorem covers the usual Linear Programming Duality Theorem: the latter is the particular case $m = 1$ of the former.

Proof of Theorem 7.2. With the Conic Duality Theorem at our disposal, all we should take care of now is the refined strong duality. In other words, invoking primal–dual symmetry, all we need is to prove that

> (!) *If* (P) *is essentially strictly feasible and bounded, then* (D) *is solvable, and* $\mathrm{Opt}(P) = \mathrm{Opt}(D)$.

Thus, assume that (P) is essentially strictly feasible and bounded. Recall that we are in the situation when F_1 is $\mathbf{R}^\mu$ with the standard inner product; we lose nothing when assuming that the remaining Euclidean spaces $E, F, F_2, \ldots, F_m$ are the spaces of column vectors of appropriate dimensions equipped with certain inner products, not necessarily the standard ones. As a result, all linear mappings participating in (P) and (D) can be thought of as matrices.

Let

$$X = \{x : A_1 x \geq b_1, Rx = r\}.$$

This set is nonempty. Let $\Pi = \{x : Sx = s \in \mathbf{R}^\nu\}$ be the affine span of X, so that

$$X = \{x \in \Pi : A_1 x \geq b_1\}, \tag{!}$$

and let $\widehat{x}$ be a point from the relative interior of X, see Section 2.1.3.3. Let us look one by one at the inequalities of the system $A_1^T x \geq b$ cutting X off Π according to (!). When such an inequality is active at $\widehat{x}$, it clearly is satisfied as equality on the entire Π, and we can drop it from the constraints in (!). After looking through all these constraints, we either (case (a)) drop all of them, meaning that $X = \Pi$, or (case (b)) end up with a nonempty system of inequalities $C_1 x \geq d_1$ such that

$$\begin{aligned} \widehat{x} \in X &:= \{x : A_1 x \geq b_1, Rx = r\} \\ &= \{x : Sx = s, C_1 x \geq d_1\} \;\&\; C_1 \widehat{x} > d_1. \end{aligned} \tag{7.6}$$

Note that we lose nothing when assuming that (7.6) takes place in the case of (a) as well — it suffices to set $C_1 x = [0; \ldots; 0]^T x$ and $d_1 = -1$. Thus, from now on we have at our disposal (7.6); to save words, we assume also that the system of equations $Sx = s$ in (7.6) is nonempty (this is automatically so when $\mathrm{Aff}(X) \neq \mathbf{R}^n$, otherwise we can set $Sx = [0; \ldots; 0]^T x$ and $s = 0$).

2^0. Let us pass from the original problem (P) to the *equivalent* problem

$$\mathrm{Opt}(P) = \min_x \left\{ \langle c, x\rangle_E : \begin{array}{ll} C_1 x - d_1 \geq 0 & (a) \\ A_i x - b_i \in \mathbf{K}_i,\ i = 2, 3, \ldots, m, & (b) \\ Sx = s & (c) \end{array} \right\}. \qquad (\bar{P})$$

The equivalence of (P) and $(\bar{P})$ is an immediate corollary of the fact that the set X of x's satisfying $(P(a))$ and $(P(c))$ is, by (7.6), exactly the same as the set of x's satisfying $(\bar{P}(a))$ and $(\bar{P}(c))$. Our next observation is that $(\bar{P})$ is strictly feasible. Indeed, let $\bar{x}$ be a feasible solution to (P) such that $A_i\bar{x} - b_i \in \operatorname{int} \mathbf{K}_i$, $i = 2, \ldots, m$; existence of such a solution is given by the essentially strict feasibility of (P). For every $\lambda \in (0, 1)$, the point $x_\lambda = (1 - \lambda)\bar{x} + \lambda\widehat{x}$ belongs to X and thus satisfies $(P(c))$, same as it satisfies the strict version $C_1 x_\lambda > d_1$ of $(\bar{P}(a))$. For small positive λ x_λ clearly satisfies also the inclusions $A_i x_\lambda - b_i \in \operatorname{int} \mathbf{K}_i$, $i = 2, 3, \ldots, m$ and therefore is a strictly feasible solution to $(\bar{P})$. Thus, $(\bar{P})$ is strictly feasible (and bounded along with (P)), so that by the Conic Duality Theorem the dual to $(\bar{P})$ — the problem

$$\mathrm{Opt}(\bar{D}) = \max_{w, \eta, \{y_i\}_{i=2}^m} \left\{ s^T w + d_1^T \eta + \sum_{i=2}^m \langle y_i, b_i\rangle_{F_i} : \begin{array}{l} \eta \geq 0 \\ y_i \in \mathbf{K}_i^*,\ 2 \leq i \leq m \\ S^* w + C_1^* \eta + \sum_{i=2}^m A_i^* y_i = c \end{array} \right\} \qquad (\bar{D})$$

is solvable with the optimal value $\mathrm{Opt}(P)$. Let $w^*, \eta^*, y_2^*, \ldots, y_m^*$ be an optimal solution to $(\bar{D})$. All we need to prove is that this solution can be converted to a feasible solution to (D) with the value of the objective of (D) at this feasible solution being *at least* $\mathrm{Opt}(\bar{D})$ (since we have seen that $\mathrm{Opt}(\bar{D}) = \mathrm{Opt}(P)$, by Weak Duality the resulting solution will be optimal for (D) with the value of the objective equal to $\mathrm{Opt}(P)$, which is all we need).

To convert $(w^*, \eta^*, y_2^*, \ldots, y_m^*)$ into a feasible solution to (D), let us act as follows. To simplify the notation, we may assume w.l.o.g. that $w^* \geq 0$. Indeed, we can multiply by (-1) the equations in the system $Sx = s$ which correspond to negative entries in w^*, replacing simultaneously these entries with their magnitudes; in our context, this clearly changes nothing.

Now, by (7.6) the system of linear inequalities

$$C_1 x \geq d_1 \ \& \ Sx \geq s$$

is satisfied everywhere on the *nonempty* solution set X of the system

$$A_1 x \geq b_1 \ \& \ Rx = r.$$

Consequently, by Inhomogeneous Farkas Lemma, there exist entrywise nonnegative matrices G, H and matrices U, V of appropriate sizes such that

$$\begin{aligned} C_1 &= GA_1 + UR, && \text{(a.1)} \\ Gb_1 + Ur &\geq d_1; && \text{(a.2)} \\ S &= HA_1 + VR, && \text{(b.1)} \\ Hb_1 + Vr &\geq s. && \text{(b.2)} \end{aligned} \tag{7.7}$$

Now, consider the candidate solution $\bar{z}, \bar{y}_1, \ldots, \bar{y}_m$ to (D) as follows:

$$\begin{aligned} \bar{z} &= U^*\eta^* + V^*w^*, \\ \bar{y}_1 &= G^*\eta^* + H^*w^*, \\ \bar{y}_i &= y_i^*, \ i = 2, \ldots, m. \end{aligned}$$

This indeed is a feasible solution to (D); all we need to verify is that $\bar{y}_1 \geq 0$ (this is true due to $\eta^* \geq 0$, $w^* \geq 0$ and to entrywise nonnegativity of G, H) and that $R^*\bar{z} + \sum_{i=1}^m A_i^*\bar{y}_i = c$. The latter is immediate:

$$\begin{aligned} c &= S^*w^* + C_1^*\eta^* + \sum_{i=2}^m A_i^* y_i^* \quad [\text{constraints in } (\bar{D})] \\ &= [HA_1 + VR]^*w^* + [GA_1 + UR]^*\eta^* \\ &\quad + \sum_{i=2}^m A_i^* y_i^* \ [\text{by } (7.7(\text{a.1})), (7.7(\text{b.1}))] \\ &= R^*[U^*\eta^* + V^*w^*] + A_1^*[G^*\eta^* + H^*w^*] + \sum_{i=2}^m A_i^* y_i^* \\ &= R^*\bar{z} + A_1^*\bar{y}_1 + \sum_{i=2}^m A_i^*\bar{y}_i^* \quad [\text{definition of } \bar{z}, \bar{y}_i]. \end{aligned}$$

Further, we have

$$
\begin{aligned}
[\mathrm{Opt}(P) =]\,\mathrm{Opt}(\bar{D}) &= s^T w^* + d_1^T \eta^* + \sum_{i=2}^{m} \langle y_i^*, b_i\rangle_{F_i} \\
&\le [Hb_1 + Vr]^T w^* + [Gb_1 + Ur]^T \eta^* + \sum_{i=2}^{m} \langle y_i^*, b_i\rangle_{F_i} \\
&= \langle r, U^*\eta^* + V^* w^*\rangle_F + b_1^T[H^* w^* + G^*\eta^*] + \sum_{i=2}^{m} \langle y_i^*, b_i\rangle_{F_i} \\
&= \langle r, \bar{z}\rangle_F + \sum_{i=1}^{m} \langle \bar{y}_i, b_i\rangle_{F_i},
\end{aligned}
$$

where the first inequality is due to $w^* \ge 0, \eta^* \ge 0$ and (7.7(a.2), 7.7(b.2)), and the last equality is due to the definition of $\bar{z}, \bar{y}_1, \ldots, \bar{y}_m$. The resulting inequality, as it was already explained, implies that $\bar{z}, \bar{y}_1, \ldots, \bar{y}_m$ form an optimal solution to (D) and that $\mathrm{Opt}(P) = \mathrm{Opt}(D)$, which is all we need. □

We have proved all the claims of the theorem, except for its "In addition" conclusion. The latter can be proved in exactly the same fashion as in the case of the plain Conic Duality Theorem. □

7.3.3 *Consequences of conic duality theorem*

7.3.3.1 *Optimality conditions in conic programming*

Optimality conditions in Conic Programming are given by the following statement:

Theorem 7.3. *Consider a primal–dual pair*

$$
\mathrm{Opt}(P) = \min_x \left\{ \langle c, x\rangle_E : \begin{array}{lr} A_1 x - b_1 \ge 0 & (a) \\ A_i x - b_i \in \mathbf{K}_i,\ 2 \le i \le m & (b) \\ Rx = r & (c) \end{array} \right\} \quad (P)
$$

$$
\mathrm{Opt}(D) = \max_{z, \{y_i\}_{i=1}^m} \left\{ \begin{array}{l} \langle r, z\rangle_F + b_1^T y_1 + \sum_{i=2}^{m} \langle b_i, y_i\rangle_{F_i} : \\ y_1 \ge 0, \\ y_i \in \mathbf{K}_i^*,\ 2 \le i \le m, \\ R^* z + \sum_{i=1}^{m} A_i^* y_i = c \end{array} \right\} \quad (D)
$$

of conic programs, and let both programs be essentially strictly feasible. A pair $(x, \xi \equiv [z; y_1; ...; y_m])$ of feasible solutions to (P) and (D) comprises optimal solutions to the respective programs iff

(i) [Zero duality gap]: *One has*

$$\begin{aligned}\mathrm{DualityGap}(x;\xi) &:= \langle c, x\rangle_E - \left[\langle z, r\rangle_F + \sum_i \langle b_i, y_i\rangle_{F_i}\right]\\ &= 0,\end{aligned}$$

same as iff.

(ii) [Complementary slackness]:

$$\forall i : \langle y_i, A_i x_i - b_i\rangle_{F_i} = 0.$$

Proof. By Refined Conic Duality Theorem, we are in the situation when $\mathrm{Opt}(P) = \mathrm{Opt}(D) \in \mathbf{R}$. Therefore,

$$\begin{aligned}\mathrm{DualityGap}(x;\xi) = &\underbrace{[\langle c, x\rangle_E - \mathrm{Opt}(P)]}_{a}\\ &+ \underbrace{\left[\mathrm{Opt}(D) - \left[\langle z, b\rangle_F + \sum_i \langle b_i, y_i\rangle_{F_i}\right]\right]}_{b}.\end{aligned}$$

Since x and ξ are feasible for the respective programs, the duality gap is nonnegative and it can vanish iff $a = b = 0$, that is, iff x and ξ are optimal solutions to the respective programs, as claimed in (i). To prove (ii), note that since x is feasible, we have

$$Rx = r,\ A_i x - b_i \in \mathbf{K}_i,\ c = R^* z + \sum_i A_i^* y_i,\ y_i \in \mathbf{K}_i^*,$$

whence

$$\begin{aligned}\mathrm{DualityGap}(x;\xi) &= \langle c, x\rangle_E - \left[\langle z, r\rangle_F + \sum_i \langle b_i, y_i\rangle_{F_i}\right]\\ &= \langle R^* z + \sum_i A_i^* y_i, x\rangle_E - \left[\langle z, r\rangle_F + \sum_i \langle b_i, y_i\rangle_{F_i}\right]\\ &= \underbrace{\langle z, Rx - r\rangle_F}_{=0} + \sum_i \underbrace{\langle y_i, A_i x - b_i\rangle_{F_i}}_{\geq 0},\end{aligned}$$

where the nonnegativity of the terms in the last $\sum_i$ follows from $y_i \in \mathbf{K}_i^*$, $A_i x_i - b_i \in \mathbf{K}_i$. We see that the duality gap, as evaluated at a pair of primal–dual feasible solutions, vanishes iff the complementary slackness holds true, and thus (ii) is readily given by (i). □

7.3.3.2 *A surrogate of GTA*

The following statement is a slightly weakened form of the Inhomogeneous Farkas Lemma (which is equivalent to GTA):

Proposition 7.2 (Conic Inhomogeneous Farkas Lemma). *Let* $\mathbf{K}$ *be a regular cone. A scalar linear inequality*

$$p^T x \geq q \tag{$*$}$$

is a consequence of an essentially strictly feasible system

$$Ax - b \in \mathbf{K}, \quad Rx = r \tag{!}$$

composed of a conic inequality and a system of linear equations[6] *iff* $(*)$ *is a "linear consequence" of* (!), *i.e., iff there exists* λ, μ *such that*

$$\lambda \in \mathbf{K}^*, \quad A^*\lambda + R^*\mu = p, \ \langle b, \lambda \rangle + \langle r, \mu \rangle \geq q. \tag{7.8}$$

Proof. Let $(*)$ be a consequence of (!). Then the (essentially strictly feasible!) conic program

$$\min_x \left\{ p^T x : Ax - b \in \mathbf{K}, Rx = r \right\}$$

is below bounded with optimal value $\geq q$. Applying the Refined Conic Duality Theorem, the dual program has a feasible solution with the value of the dual objective $\geq q$, which is nothing but the solvability of (7.8) (look at the dual!). Vice versa, if λ, μ solve (7.8) and x solves (!), then

$$\begin{aligned} 0 &\leq \langle \lambda, Ax - b \rangle + \langle \mu, Rx - r \rangle = [A^*\lambda + R^*\mu]^T x - \langle b, \lambda \rangle - \langle r, \mu \rangle \\ &= p^T x - \langle b, \lambda \rangle - \langle r, \mu \rangle \leq p^T x - q, \end{aligned}$$

so that $(*)$ is indeed a consequence of (!); note that to get the latter conclusion, no assumption of essentially strict feasibility (and even feasibility) of (!) is needed. □

---

[6]Essentially strict feasibility of (!) is defined completely similar to essentially strict feasibility of a conic problem; it means that $\mathbf{K}$ is the direct product of a nonnegative orthant (perhaps of dimension 0) and a regular cone $\mathbf{K}'$, and (!) has a feasible solution $\bar{x}$ with the $\mathbf{K}'$-component belonging to the interior of $\mathbf{K}'$.

7.3.4 *Sensitivity analysis*

The results we are about to present resemble those of Sensitivity Analysis for LO. Consider a primal–dual pair of conic programs in the single-cone form

$$\min_x \{\langle c, x\rangle : Ax - b \in \mathbf{K}, Rx = r\} \qquad (\mathcal{P})$$

$$\max_{y,z} \{\langle b, y\rangle + \langle r, z\rangle : y \in \mathbf{K}^*, A^*y + R^*z = c\} \qquad (\mathcal{D})$$

In what follows, we treat the part of the data $A, b, \mathbf{K}, R$ as fixed, and b, r, c as varying, so that it makes sense to refer to $(\mathcal{P})$ as $(\mathcal{P}[b, r; c])$, and to its dual $(\mathcal{D})$ as to $(\mathcal{D}[b, r; c])$, and to denote the optimal value of $(\mathcal{P}[b, r; c])$ as $\mathrm{Opt}(b, r; c)$. Our goal is to explore the structure of the *cost function* $\mathrm{Opt}(b, r; c)$ as a function of (b, r), c being fixed, and of c, (b, r) being fixed.

7.3.4.1 *The cost function as a function of c*

Let (b, r) be fixed at certain value $(\bar{b}, \bar{r})$ such that $(\mathcal{P}[\bar{b}, \bar{r}; c])$ is feasible (this fact is independent of the value of c). An immediate observation is that in this case, the function $\mathrm{Opt}_{b,r}(c) = \mathrm{Opt}(b, r; c)$ is a concave function of c. Indeed, this is the infimum of the nonempty family $\{f_x(c) = c^T x : Ax - b \in \mathbf{K}, Rx = r\}$ of linear (and thus concave) functions of c. A less trivial observation is as follows:

Proposition 7.3. *Let $\bar{c}$ be such that $(\mathcal{P}[\bar{b}, \bar{r}, \bar{c}])$ is solvable, and $\bar{x}$ be the corresponding optimal solution. Then $\bar{x}$ is a supergradient of $\mathrm{Opt}_{\bar{b},\bar{r}}(\cdot)$ at $\bar{c}$, meaning that*

$$\forall c : \mathrm{Opt}_{\bar{b},\bar{r}}(c) \le \mathrm{Opt}_{\bar{b},\bar{r}}(\bar{c}) + \langle \bar{x}, c - \bar{c}\rangle.$$

Geometrically: the graph of $\mathrm{Opt}_{\bar{b},\bar{r}}(c)$ never goes above the graph of the affine function $\ell(c) = \mathrm{Opt}_{\bar{b},\bar{r}}(\bar{c}) + \langle \bar{x}, c - \bar{c}\rangle$ and touches this graph at the point $[\bar{c}; \mathrm{Opt}_{\bar{b},\bar{r}}(\bar{c})]$ (and perhaps at other points as well).

Proof is immediate. Since $\bar{x}$ is a feasible solution of $(\mathcal{P}[\bar{b}, \bar{r}; c])$ for every c, we have

$$\mathrm{Opt}_{\bar{b},\bar{r}}(c) \le \langle c, \bar{x}\rangle = \langle c - \bar{c}, \bar{x}\rangle + \langle \bar{c}, \bar{x}\rangle = \mathrm{Opt}_{\bar{b},\bar{r}}(\bar{c}) + \langle \bar{x}, c - \bar{c}\rangle. \quad \square$$

Remark. The fact that a function is convex (or concave) implies, in particular, that the function possesses certain "regularity" e.g., the following is true:

Let f be a convex (or concave) function and X be a closed and bounded set belonging to the relative interior of the function's domain. Then f is Lipschitz continuous on X: there exists $L < \infty$ such that

$$\forall(x, y \in X) : |f(x) - f(y)| \leq L\|x - y\|. \qquad \square$$

7.3.4.2 *The cost function as a function of* (b, r)

Now, let c be fixed at certain value $\bar{c}$, and assume that there exists (b', r') such that the problem $(\mathcal{P}[b', r'; \bar{c}])$ is strictly feasible and bounded. Then the dual problem $(\mathcal{D}[b', r'; \bar{c}])$ is feasible (and even solvable) by the Conic Duality Theorem, meaning that the duals to *all* problems $(\mathcal{D}[b, r; \bar{c})$ are feasible (since the feasible set of the dual is independent of b, r). By Weak Duality, it follows that problems $(P[b, r; \bar{c}])$ are bounded, and thus the cost function $\mathrm{Opt}_{\bar{c}}(b, r) = \mathrm{Opt}(b, r; \bar{c})$ takes only real values and the value $+\infty$. It is easily seen that this function is convex.

Indeed, denoting for short $q = (b, r)$ and suppressing temporarily the subscript $_{\bar{c}}$, we should prove that $\mathrm{Opt}((1-\lambda)q + \lambda q') \leq (1-\lambda)\mathrm{Opt}(q) + \lambda\mathrm{Opt}(q')$ for all q, q' and all $\lambda \in [0, 1]$. There is nothing to prove when $\lambda = 0$ or $\lambda = 1$; when $0 < \lambda < 1$, there is nothing to prove when $\mathrm{Opt}(q)$ or $\mathrm{Opt}(q')$ are infinite. Thus, we can restrict ourselves with the case $q = (b, r) \in \mathrm{Dom}\,\mathrm{Opt}$, $q' = (b', r') \in \mathrm{Dom}\,\mathrm{Opt}$ and $0 < \lambda < 1$. Given $\epsilon > 0$, we can find x and x' such that

$$\begin{aligned}
&Ax - b \in \mathbf{K}, Rx = r, \bar{c}^T x \leq \mathrm{Opt}(q) + \epsilon,\\
&Ax' - b' \in \mathbf{K}, Rx' = r', \bar{c}^T x' \leq \mathrm{Opt}(q') + \epsilon.
\end{aligned}$$

Setting $\tilde{x} = (1-\lambda)x + \lambda x'$, $\tilde{q} = (1-\lambda)q + \lambda q' = (\tilde{b}, \tilde{r})$, we have $A\tilde{x} - \tilde{b} = (1-\lambda)[Ax - b] + \lambda[Ax' - b'] \in \mathbf{K}$, where the inclusion follows from the fact that $\mathbf{K}$ is a cone, and $R\tilde{x} = (1-\lambda)Rx + \lambda Rx' = R\tilde{r}$; thus, $\tilde{x}$ is a feasible solution of $(\mathcal{P}[\tilde{q}, \bar{c}])$. We also have

$$\begin{aligned}
\langle \bar{c}, \tilde{x}\rangle &= (1-\lambda)\langle \bar{c}, x\rangle + \lambda\langle \bar{c}, x'\rangle \leq (1-\lambda)[\mathrm{Opt}(q) + \epsilon] + \lambda[\mathrm{Opt}(q') + \epsilon]\\
&= (1-\lambda)\mathrm{Opt}(q) + \lambda\mathrm{Opt}(q') + \epsilon.
\end{aligned}$$

Since $\tilde{x}$ is feasible for $(\mathcal{P}[\tilde{q}, \bar{c}])$, we have

$$\mathrm{Opt}(\tilde{q}) \leq \langle \bar{c}, \tilde{x}\rangle \leq (1-\lambda)\mathrm{Opt}(q) + \lambda\mathrm{Opt}(q') + \epsilon.$$

The resulting inequality holds true for every $\epsilon > 0$, whence

$$\mathrm{Opt}(\tilde{q}) \leq (1-\lambda)\mathrm{Opt}(q) + \lambda\mathrm{Opt}(q'),$$

which completes the proof of convexity of $\mathrm{Opt}(\cdot)$. $\square$

We have the following analogy of Proposition 7.3:

Proposition 7.4. *Let $\bar{c}$ be such that the problem $(\mathcal{P}[b', r'; \bar{c}])$ is strictly feasible and bounded for some (b', r'), implying, as we already know, that $(\mathcal{D}[b, r; \bar{r}])$ is feasible for all b, r. Let $\bar{b}, \bar{r}$ be such that $(\mathcal{P}[\bar{b}, \bar{r}; \bar{c}])$ is strictly feasible. Then the dual problem $(\mathcal{D}[\bar{b}, \bar{r}; \bar{c}])$ is solvable, and every optimal solution $(\bar{y}, \bar{z})$ to the latter program is a subgradient of the convex function* $\mathrm{Opt}_{\bar{c}}(\cdot)$ *at the point $(\bar{b}, \bar{r})$, meaning that*

$$\forall (b, r) : \mathrm{Opt}_{\bar{c}}(b, r) \geq \mathrm{Opt}_{\bar{c}}(\bar{b}, \bar{r}) + \langle \bar{y}, b - \bar{b} \rangle + \langle \bar{z}, r - \bar{r} \rangle.$$

Geometrically: the graph of $\mathrm{Opt}_{\bar{c}}(b, r)$ *never goes below the graph of the affine function* $\ell(b, r) = \mathrm{Opt}_{\bar{c}}(\bar{b}, \bar{r}) + \langle \bar{y}, b - \bar{b} \rangle + \langle \bar{z}, r - \bar{r} \rangle$ *and touches this graph at the point* $(\bar{b}, \bar{r}; \mathrm{Opt}_{\bar{c}}(\bar{b}, \bar{r}))$ *(and perhaps at other points as well).*

Proof is immediate. As we have already mentioned, our choice of $\bar{c}$ ensures that $(\mathcal{D}[\bar{b}, \bar{r}; \bar{c}])$ is feasible, and thus the program $(\mathcal{P}[\bar{b}, \bar{r}; \bar{c}])$ is bounded; since by assumption the latter program is strictly feasible, the Conic Duality Theorem says that $(\mathcal{D}[\bar{b}, \bar{r}; \bar{c}])$ is solvable, which is the first claim in the proposition. Now, let $(\bar{y}, \bar{r})$ be an optimal solution to $(\mathcal{D}[\bar{b}, \bar{r}; \bar{c}])$. Then for every feasible solution x to $(\mathcal{P}[b, r; \bar{c}])$, we have

$$\begin{aligned} \langle \bar{c}, x \rangle &= \langle A^* \bar{y} + R^* \bar{z}, x \rangle = \langle \bar{y}, Ax - b \rangle + \langle \bar{z}, Rx \rangle + \langle \bar{y}, b \rangle \\ &\geq \langle \bar{y}, b \rangle + \langle \bar{z}, r \rangle = \langle \bar{y}, b - \bar{b} \rangle + \langle \bar{z}, r - \bar{r} \rangle + \underbrace{\langle \bar{y}, \bar{b} \rangle + \langle \bar{z}, \bar{r} \rangle}_{= \mathrm{Opt}_{\bar{c}}(\bar{b}, \bar{r})}. \end{aligned}$$

Since the resulting inequality is valid for all feasible solutions x to $(\mathcal{P}[b, r; \bar{c}])$, we conclude that

$$\mathrm{Opt}_{\bar{c}}(b, r) \geq \mathrm{Opt}_{\bar{c}}(\bar{b}, \bar{r}) + \langle \bar{y}, b - \bar{b} \rangle + \langle \bar{z}, r - \bar{r} \rangle. \qquad \square$$

7.3.5 *Geometry of primal–dual pair of conic problems*

We are about to derive geometric interpretation of a primal–dual pair (P), (D) of conic programs completely similar to the interpretation of a primal–dual pair of LO programs (Section 3.3.2). As was explained in the beginning of Section 7.2, we lose nothing when assuming that the primal program is a single-cone one and that the space E of the primal decision vectors is $\mathbf{R}^n$, so that the primal program reads

$$\mathrm{Opt}(P) = \min_x \left\{c^T x : Ax - b \in \mathbf{K}, Rx = r\right\}, \qquad (P)$$

where $x \mapsto Ax$ is a linear mapping from $\mathbf{R}^n$ to Euclidean space F, R is $k \times n$ matrix and $\mathbf{K}$ is a regular cone in F. The dual program now reads

$$\mathrm{Opt}(D) = \max_{z,y} \left\{r^T z + \langle b, y\rangle : y \in \mathbf{K}^*, A^* y + R^T z = c\right\} \qquad (D)$$

($\langle\cdot,\cdot\rangle$ stands for $\langle\cdot,\cdot\rangle_F$). Assume that the systems of linear equality constraints in (P) and in (D) are solvable, and let $\bar{x}$ and $[\bar{y};\bar{z}]$ be solutions to these systems:

$$\begin{aligned} R\bar{x} &= r, && \text{(a)} \\ A^*\bar{y} + R^T\bar{z} &= c. && \text{(b)} \end{aligned} \qquad (7.9)$$

Let us express (P) in terms of the primal slack $\xi = Ax - b \in F$. The constraints of (P) say that this vector should belong to the intersection of $\mathbf{K}$ and the *primal feasible plane* $\mathcal{M}_P$, which is the image of the affine plane $\{x : Rx = r\}$ in the x-space under the affine mapping $x \mapsto Ax - b$. The linear subspace $\mathcal{L}_P$ in F which is parallel to $\mathcal{M}_P$ is $\mathcal{L}_P = \{\xi = Ax : Rx = 0\}$, and we can take the point $A\bar{x} - b := -\bar{\xi}$ as the shift vector for $\mathcal{M}_P$. Thus,

$$\mathcal{M}_P = \mathcal{L}_P - \bar{\xi}, \; \bar{\xi} = b - A\bar{x}, \; \mathcal{L}_P = \{\xi = Ax : Rx = 0\}. \qquad (7.10)$$

Now, let us express the primal objective in terms of the primal slack. Given x satisfying the equality constraints in (P), we have

$$\begin{aligned} c^T x &= [A^*\bar{y} + R^T\bar{z}]^T x = [A^*\bar{y}]^T x + [R^T\bar{z}]^T x = \langle\bar{y}, Ax\rangle + \bar{z}^T Rx \\ &= \left\langle \bar{y}, \underbrace{Ax - b}_{\xi} \right\rangle + \mathrm{const}_P, \; \mathrm{const}_P = \langle\bar{y}, b\rangle + \bar{z}^T r. \end{aligned}$$

We have arrived at the following intermediate conclusion:

Program (P) can be reduced to the program

$$\mathrm{Opt}(\mathcal{P}) = \min_{\xi \in F} \{\langle\bar{y}, \xi\rangle : \xi \in \mathbf{K} \cap \mathcal{M}_P\}$$
$$\left[\begin{array}{c} \mathcal{M}_P = \mathcal{L}_P - \bar{\xi}, \\ \mathcal{L}_P = \{\xi = Ax : Rx = 0\} \\ \bar{\xi} = b - A\bar{x} \\ \mathrm{Opt}(P) = \mathrm{Opt}(\mathcal{P}) + \langle\bar{y}, b\rangle + \bar{z}^T r. \end{array}\right] \qquad (\mathcal{P})$$

Now, let us process in a similar fashion the dual program (D), specifically, express it in terms of the vector y. The constraints of (D) say that this vector should belong to the intersection of $\mathbf{K}^*$ and the *dual feasible plane* $\mathcal{M}_D = \{y : \exists z : A^*y + R^T z = c\}$. This plane is parallel to the linear subspace $\mathcal{L}_D = \{y : \exists z : A^*y + R^T z = 0\}$, and as a shift vector for $\mathcal{M}_D$ we can take $\bar{y} \in \mathcal{M}_D$. It remains to express the dual objective in terms of y. To this end, note that if $[y; z]$ satisfies the linear equality constraints of (D), then

$$\begin{aligned} r^T z + \langle b, y\rangle &= \bar{x}^T R^T z + \langle b, y\rangle = \bar{x}^T[c - A^*y] + \langle b, y\rangle \\ &= \bar{x}^T c + \langle b - A\bar{x}, y\rangle = \langle \bar{\xi}, y\rangle + \mathrm{const}_D, \ \mathrm{const}_D = c^T\bar{x}. \end{aligned}$$

We have arrived at the following conclusion:

Program (D) can be reduced to the program

$$\begin{aligned} &\mathrm{Opt}(\mathcal{D}) = \max_{y\in F}\left\{\langle\bar{\xi}, y\rangle : y \in \mathbf{K}^* \cap \mathcal{M}_D\right\} \\ &\left[\begin{array}{l} \mathcal{M}_D = \mathcal{L}_D + \bar{y} \\ \quad \mathcal{L}_D = \{y : \exists z : A^*y + R^*z = 0\} \\ \mathrm{Opt}(D) = \mathrm{Opt}(\mathcal{D}) + c^T\bar{x}. \end{array}\right] \end{aligned} \qquad (\mathcal{D})$$

Now, same as in the LO case, $\mathcal{L}_D$ is just the orthogonal complement of $\mathcal{L}_P$. Indeed, $h \in (\mathcal{L}_P)^\perp$ iff $\langle h, Ax\rangle = 0$ whenever $Rx = 0$, that is, iff the linear equation $x^T[A^*h] = 0$ in variables x is a consequence of the linear system $Rx = 0$, which is the case iff $A^*h = R^T w$ for some w, which, after substitution $w = -z$, is nothing but the characterization of $\mathcal{L}_D$.

Finally, let us compute the duality gap at a pair $(x, [y, z])$ of candidate solutions satisfying the equality constraints in (P) and (D) as follows:

$$\begin{aligned} c^T x - \langle b, y\rangle - r^T z &= [A^*y + R^T z]^T x - \langle b, y\rangle - r^T z \\ &= \langle Ax - b, y\rangle + [Rx - r]^T z = \langle Ax - b, y\rangle. \end{aligned}$$

Putting things together, we arrive at a perfectly symmetric, purely geometric description of (P), (D):

Assume that the systems of linear constraints in (P) and (D) are solvable. Then the primal–dual pair (P), (D) of conic problems reduces to the following geometric problem. We are given

- *two cones dual to each other, $\mathbf{K}$, $\mathbf{K}^*$, in a Euclidean space F,*

- *a pair of linear subspaces* $\mathcal{L}_P$, $\mathcal{L}_D$ *in* F *which are orthogonal complements to each other, and*
- *a pair of shift vectors* $\bar{\xi}, \bar{y}$ *in* F.

These geometric data define affine subspaces $\mathcal{M}_P = \mathcal{L}_P - \bar{\xi}$, $\mathcal{M}_D = \mathcal{L}_D + \bar{y}$.

The primal problem (P) *reduces to minimizing the linear form* $\langle \bar{y}, \cdot \rangle$ *over the intersection of the primal feasible plane* $\mathcal{M}_P$ *and the cone* $\mathbf{K}$, *which is the primal feasible set; the dual problem* (D) *reduces to maximizing the linear form* $\langle \bar{\xi}, \cdot \rangle$ *over the intersection of the dual feasible plane* $\mathcal{M}_D$ *and the cone* $\mathbf{K}^*$, *which is the dual feasible set. Given feasible solutions* ξ, y *to these geometric problems, the corresponding duality gap is the inner product of the solutions.*

Strict feasibility of a problem from our primal–dual pair means that the corresponding feasible plane intersects the interior of the corresponding cone. Whenever both problems are strictly feasible, the minimal value of the duality gap is zero, and the duality gap, as evaluated at a pair of primal and dual feasible solutions, is the sum of their nonoptimalities, in terms of the objectives of the respective problems. Under the same assumption of primal–dual strict feasibility, pairs of optimal solutions to the respective problems are exactly the pairs of primal and dual feasible solutions orthogonal to each other, and these pairs do exist.

We see that geometrically, a primal–dual pair of conic problems looks completely similar to a pair of primal–dual LO programs: in both situations (in the second — under additional assumption that both problems are strictly feasible) we are looking for pairs of vectors orthogonal to each other with one member of the pair belonging to the intersection of "primal" affine plane and "primal" cone, and the other member belonging to the intersection of the "dual" affine plane and the "dual" cone.

The pair of primal and dual affine planes cannot be arbitrary: they should be shifts of linear subspaces which are orthogonal complements to each other. Similarly, the pair of cones in question are "rigidly connected" to each other — they are duals of each other. In LO, the underlying cone is the nonnegative orthant and thus is *self-dual*, this is why in our LO investigations (Section 3.3.2) we did not see *two* cones, just one of them.

We complete this section by mentioning that, same as in the LO case, the choice of the shift vectors for $\mathcal{M}_P$, $\mathcal{M}_D$ (or, which is the same, the objectives in $(\mathcal{P})$ and $(\mathcal{D})$) is immaterial: when replacing the above $\bar{\xi}$ with any other vector from the minus primal feasible plane $[-\mathcal{M}_P]$, the primal problem $(\mathcal{P})$ clearly remains intact, and

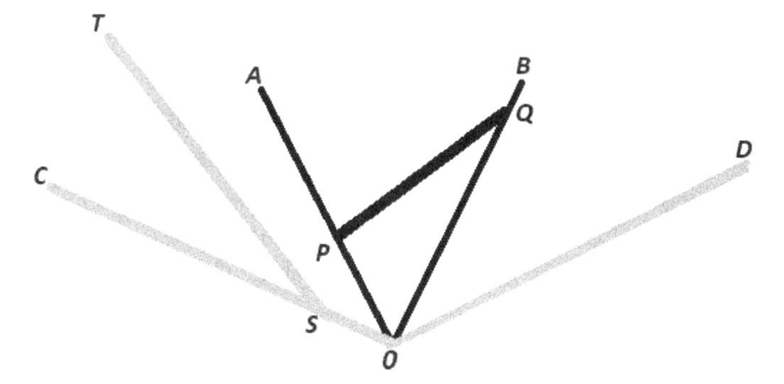

Fig. 7.1. Geometry of primal–dual conic pair.

Notes:

- $\angle AOB$ — cone $\mathbf{K}$; $\angle COD$ — cone $\mathbf{K}^*$
- segment $[P, Q]$ — feasible set of $(\mathcal{P})$; ray $[ST)$ — feasible set of $(\mathcal{D})$.
- Q is the primal, and S is the dual optimal solution. Pay attention to the orthogonality of $\overline{PQ}$ to $\overline{ST}$ and of $\overline{OQ}$ to $\overline{OS}$.

the dual objective $\langle\bar{\xi}, \cdot\rangle$, *restricted on the dual feasible plane*, changes by an additive constant, which affects nothing but the optimal value $\text{Opt}(\mathcal{D})$. By similar reasons, replacing $\bar{y}$ with any other vector from $\mathcal{M}_D$ keeps $(\mathcal{D})$ intact and changes by additive constant the restriction of the primal objective $\langle\bar{y}, \cdot\rangle$ on the primal feasible plane.

Our geometric findings are illustrated in Fig. 7.1.

7.4 Conic representations of sets and functions

It is easily seen that every convex program $\min_{x\in X} f(x)$ ($f : \mathbf{R}^n \to \mathbf{R}$ is convex, $X \subset \mathbf{R}^n$ is convex and closed) can be equivalently reformulated as a conic program. This fact is of no actual use, since a general-type cone is not simpler than a general-type closed convex set. What indeed is important is to recognize when a given convex program can be posed as a conic program *from a given family*, primarily — when it can be posed as an LO/CQO/SDO program. To this end, we can develop an approach completely similar to the one we used in Section 1.3.2. Specifically, assume we are given a family $\mathcal{K}$ of regular cones, every one of them "living" in its own Euclidean space. It makes sense to assume also that the family contains the nonnegative ray and is closed w.r.t. taking finite direct products and to passing from a cone to its dual cone. The examples most important for us are as follows:

- The family $\mathcal{LO}$ of nonnegative orthants; this family underlies LO.
- The family $\mathcal{CQ}$ of finite direct products of Lorentz cones; this family underlies CQO (Conic Quadratic Optimization).

  Note that $\mathcal{CQ}$ contains $\mathbf{R}_+ = \mathbf{L}^1$; the fact that all other requirements are satisfied is evident (recall that the Lorentz cones are self-dual).
- The family $\mathcal{SD}$ of finite direct products of semidefinite cones; this family underlies SDO (Semidefinite Optimization).

  Note that $\mathcal{SD}$ contains $\mathbf{R}_+ = \mathbf{S}^1_+$, and satisfies all other requirements by exactly the same reasons as $\mathcal{CQ}$.

Now, given a family $\mathcal{K}$, we call a set $X \subset \mathbf{R}^n$ *$\mathcal{K}$-representable* if it can be represented in the form

$$X = \{x \in \mathbf{R}^n : \exists w : Px + Qw + r \in \mathbf{K}\}, \qquad (!)$$

where $\mathbf{K}$ is a cone from $\mathcal{K}$; corresponding data $(P, Q, r, \mathbf{K})$, same as the representation itself, are called *$\mathcal{K}$-representation* of X ($\mathcal{K}$-r. of X for short). Note that this definition mirrors the definition of a polyhedral representation of a set (which in our now language becomes $\mathcal{LO}$-representation). Completely similar to the polyhedral case, given a $\mathcal{K}$-representation (!) of X, we can immediately rewrite the problem of minimizing a linear objective $c^T x$ over X as a conic program on the cone from the family $\mathcal{K}$, specifically, the program

$$\min_{x,w} \left\{c^T x : Px + Qw + r \in \mathbf{K}\right\}.$$

Bearing in mind this observation, we understand why it is important to build a calculus of $\mathcal{K}$-representable sets *and functions.* A $\mathcal{K}$-r. of a function f is, by definition, the same as $\mathcal{K}$-r. of its epigraph, and a function is called $\mathcal{K}$-representable if it admits a $\mathcal{K}$-r. Same as in the polyhedral case, a $\mathcal{K}$-r.

$$\{[x; \tau] : \tau \geq f(x)\} = \{[x; \tau] : \exists w : Px + \tau p + Qw + r \in \mathbf{K}\}$$

of a function f implies $\mathcal{K}$-r.'s of the sublevel sets of the function:

$$\{x : a \geq f(x)\} = \{x : \exists w : Px + ap + Qw + r \in \mathbf{K}\}.$$

We are about to present "executive summary" of $\mathcal{K}$-representability of sets and functions; for a self-contained, in-depth exposition of this topic, see Ben-Tal and Nemirovski (2001, 2022).

7.4.1 *Calculus of $\mathcal{K}$-representability: Calculus rules*

The "elementary" calculus rules from Section 1.3.3 extend word by word from the polyhedral representability to $\mathcal{K}$-representability. In particular,

- *Finite intersections, direct products, arithmetic sums, affine images and inverse affine images of $\mathcal{K}$-r. sets are $\mathcal{K}$-r., with $\mathcal{K}$-r. of the result readily given by those of the operands.*

  For example, given $\mathcal{K}$-r.'s $X_i = \{x \in \mathbf{R}^n : \exists w^i : P_i x + Q_i w^i + r^i \in \mathbf{K}_i\}$, $\mathbf{K}_i \in \mathcal{K}$, of sets $X_i \subset \mathbf{R}^n$, $i = 1, \ldots, m$, their intersection admits the $\mathcal{K}$-r.

$$\bigcap_{i=1}^{m} X_i = \{x \in \mathbf{R}^m : \exists w = [w^1; \ldots; w^m] : \\ [P_1 x + Q_1 w^1 + r^1; \ldots; P_m x + Q_m w^m + r^m], \\ \in \mathbf{K} := \mathbf{K}_1 \times \cdots \times \mathbf{K}_m\},$$

  and $\mathbf{K} \in \mathcal{K}$, since $\mathcal{K}$ is closed w.r.t. taking direct products.

- *Taking finite linear combinations with nonnegative coefficients, direct summation, taking finite maxima, affine substitutions of variables, and restrictions onto $\mathcal{K}$-r. sets, as applied to $\mathcal{K}$-r. functions, produce $\mathcal{K}$-r. results, with $\mathcal{K}$-r. of the results readily given by those of the operands.*

  In fact, all these rules are immediate consequences of the following observation (cf. Theorem on superposition, item F.5 in Section 1.3.3):

  Assume we are given functions $f_i : \mathbf{R}^n \to \mathbf{R} \cup \{+\infty\}$, $1 \le i \le m$, integer $k \in [0, m]$, function $F : \mathbf{R}^m \to \mathbf{R} \cup \{+\infty\}$ and a set $Y \subset \mathbf{R}^m$ such that

  - *F, f_i, $i \le k$, F, and Y are $\mathcal{K}$-r:*

$$\begin{aligned} \tau \ge F(y) &\Leftrightarrow \exists w^0 : P_0 y + \tau p_0 + Q_0 w^0 + r^0 \in \mathbf{K}_0 \in \mathcal{K} \\ \tau \ge f_i(x) &\Leftrightarrow \exists w^i : P_i x + \tau p_i + Q_i w^i + r^i \in \mathbf{K}_i \in \mathcal{K}, 1 \le i \le k \\ y \in Y &\Leftrightarrow \exists z : z : Sy + Tz + s \in \mathbf{M} \in \mathcal{K} \end{aligned}$$

    and functions $f_{k+1}, f_{k+2}, \ldots, f_m$ are affine;

- *whenever* $x \in \mathbf{R}^n$ *is such that* $f_i(x) \in \mathbf{R}$, $i \leq m$, *the vector* $f(x) = [f_1(x); ...; f_m(x)]$ *belongs to* Y;
- *restricted on* Y, *function* F *possesses the following monotonicity property: whenever* $y, y' \in Y$ *are such that* $y_i \geq y'_i$ *for* $i \leq k$ *and* $y_i = y'_i$ *for* $k+1 \leq i \leq m$, *one has* $F(y) \geq F(y')$.

Then the superposition

$$g(x) = \begin{cases} F(f_1(x, \ldots, f_m(x)), & f_i(x) \in \mathbf{R}, i \leq m, \\ +\infty, & \textit{otherwise}, \end{cases}$$

is $\mathcal{K}$*-r. as follows:*

$$\begin{aligned}
\tau \geq g(x) &\Leftrightarrow \exists y : \begin{cases} y \in Y \\ F(y) \leq \tau \\ y_i \geq f_i(x), i \leq k \\ y_i = f_i(x), k+1 \leq i \leq m, \end{cases} \\
&\Leftrightarrow \exists y = [y_1; ...; y_m], z, w^0, w^1, \ldots, w^m : \\
&\begin{cases} Sy + Tz + u \in \mathbf{M} & \text{“says” that } y \in Y \\ P_0 y + \tau p_0 + Q_0 w^0 + r^0 \in \mathbf{K}_0 & \text{“says” that } F(y) \leq \tau \\ P_i x + y_i p_i + Q_i w^i + r^i \in \mathbf{K}_i, 1 \leq i \leq k & \text{“says” that } y_i \geq f_i(x) \\ f_i(x) - y_i = 0, k+1 \leq i \leq m \end{cases} \\
&\Leftrightarrow \exists u = [y; z; w^0; w^1; ...; w^k] : \\
&Ax + Bu + c := \begin{bmatrix} Sy + Tz + s \\ P_0 y + \tau p_0 + Q_0 w^0 + r^0 \\ P_1 x + y_1 p_1 + Q_1 w^1 + r^1 \\ \vdots \\ P_k x + y_k p_k + Q_k w^k + r^k \\ [f_{k+1}(x); ...; f_m(x)] - [y_{k+1}; ...; y_m] \\ [y_{k+1}; ...; y_m] - [f_{k+1}(x); ...; f_m(x)] \end{bmatrix} \in \mathbf{K},
\end{aligned}$$

with

$$\mathbf{K} = \mathbf{M} \times \mathbf{K}_0 \times \mathbf{K}_1 \times \cdots \times \mathbf{K}_k \times \mathbf{R}_+^{m-k} \times \mathbf{R}_+^{m-k} \in \mathcal{K}.$$

Extending more advanced rules of calculus of polyhedral representability to the case of $\mathcal{K}$-representability requires certain care. For example,

(1) Assume that an LO program $\min_x\{c^T x : Ax \geq b, Rx = r\}$ is feasible and bounded for some value of $[b; r]$; as we remember from

Section 3.3.4,[7] under this assumption the function $\text{Opt}([b;r]) = \min_x\{c^T x : Ax \geq b, Rx = r\}$ is convex and polyhedrally representable:

$$\{[b;r;\tau] : \tau \geq \text{Opt}([b;r])\} = \left\{[b;r;\tau] : \exists x : Ax - b \geq 0, Bx = r, c^T x \leq \tau\right\}. \tag{!}$$

Now, let us pass from the optimal value in an LO program to the one in a $\mathcal{K}$-conic program:

$$\text{Opt}([b;r]) = \inf\left\{c^T x : Ax - b \in \mathbf{K}, Rx = r\right\}. \tag{7.11}$$

Assume that the program is strictly feasible and bounded for some value of $[b;r]$. Then the dual program is feasible, and since the latter fact is independent of $[b;r]$, we conclude from Weak Duality that (7.11) is bounded for all values of $[b;r]$, so that the cost function $\text{Opt}([b;r])$ takes only real values and perhaps the value $+\infty$. You can easily verify that the cost function is convex. Now, the "literal analogy" of (!) would be

$$\begin{aligned}\{[b;r;\tau] : \tau \geq \text{Opt}([b;r])\} = \{[b;r;\tau] : \exists x : Ax - b \in \mathbf{K},\\ Rx = r, c^T x \leq \tau\},\end{aligned}$$

but this relation is not necessarily true: $\text{Opt}([b;r])$ can be finite and nonachievable, meaning that the fact that the point $[b;r;\tau]$ with $\tau = \text{Opt}([b;r])$ is in the epigraph of Opt cannot be "certified" by any x.

The correct version of (!) is as follows:

Let (7.11) *be bounded and feasible for some value of* $[b;r]$. *Then the cost function* $\text{Opt}([b;r])$ *is a convex function, and the* $\mathcal{K}$-*r. set*

$$\mathcal{G} = \left\{[b;r;\tau] : \exists x : Ax - b \in \mathbf{K}, Rx = r, c^T x \leq \tau\right\}$$

is in-between the epigraph

$$\text{Epi}(\text{Opt}(\cdot)) = \{[b;r;\tau] : \tau \geq \text{Opt}([b;r])\}$$

[7] Take into account that in Section 3.3.4 we were speaking about the optimal value of a *maximization* problem, while now we are speaking about optimal value of a minimization one; as a result, what used to be concave now becomes convex.

of the cost function and the "strictly upper part"

$$\mathrm{Epi}^+(\mathrm{Opt}(\cdot)) = \{[b;r;\tau] : \tau > \mathrm{Opt}([b;r])\}$$

of this epigraph:

$$\mathrm{Epi}^+(\mathrm{Opt}(\cdot)) \subset \mathcal{G} \subset \mathrm{Epi}(\mathrm{Opt}(\cdot)).$$

(2) The closed perspective transform

$$\overline{X} = \mathrm{cl}\{[x;t] : t > 0, t^{-1}x \in X\}$$

of a nonempty polyhedral set $X = \{x : \exists w : Px + Qw + r \geq 0\}$ admits the polyhedral representation

$$\overline{X} = \{[x;t] : \exists w : Px + Qw + tr \geq 0 \ \&\ t \geq 0\}$$

(Section 2.4.4.2). The "nonpolyhedral" version of this rule reads as follows:

Let $X = \{x \in \mathbf{R}^n : \exists w : Px + Qw + r \in \mathbf{K}\}$ *be a nonempty* $\mathcal{K}$*-r. set, and let*

$$\widehat{X} = \{[x;t] : \exists w : Px + Qw + tr \in \mathbf{K} \ \&\ t \geq 0\}.$$

The $\mathcal{K}$*-r. set* $\widehat{X}$ *is in-between the "incomplete" perspective transform*

$$X^+ = \{[x;t] : t > 0, t^{-1}x \in X\}$$

and the closed perspective transform of X*:*

$$X^+ \subset \widehat{X} \subset \overline{X} := \mathrm{cl}X^+.$$

(3) The support function of a nonempty polyhedral set is polyhedrally representable, with a p.r. readily given by a p.r. of the set (Section 3.3.4). To get a similar result in the general conic case, we need essentially strict feasibility of the representation of the set. The precise statement reads as follows:

(!) *Let X be a nonempty $\mathcal{K}$-representable set given by a $\mathcal{K}$-representation*

$$X = \{x : \exists w : Px + Qw - b \in \mathbf{K},\ Rx + Sw \geq r\} \qquad [\mathbf{K} \in \mathcal{K}],$$

which is essentially strictly feasible, meaning that there exists $\bar{x}, \bar{w}$ such that

$$P\bar{x} + Q\bar{w} - b \in \text{int}\,\mathbf{K},\ R\bar{x} + S\bar{w} \geq r.$$

Then the support function

$$\text{Supp}(\xi) = \sup_{x \in X} \xi^T x$$

of the set X admits the explicit $\mathcal{K}$-representation

$$\begin{aligned} &\{[\xi;\tau] : \tau \geq \text{Supp}(\xi)\} \\ &= \left\{ [\xi;\tau] : \exists \lambda, \mu : \begin{array}{l} \lambda \in \mathbf{K}^*, \mu \geq 0, P^*\lambda + R^*\mu + \xi = 0, Q^*\lambda + S^*\mu = 0, \\ \langle b, \lambda \rangle + \langle r, \mu \rangle + \tau \geq 0 \end{array} \right\}. \end{aligned} \tag{7.12}$$

Note that (7.12) is indeed a $\mathcal{K}$-representation of the support function, since $\mathcal{K}$ is closed w.r.t. passing from a cone to its dual.

Proof of (!). Observe that $-\text{Supp}(\xi)$ is the optimal value in the $\mathcal{K}$-conic program

$$\min_{x,w} \left\{ -\xi^T x : Px + Qw - b \in \mathbf{K},\ Rx + Sw \geq r \right\}. \tag{$*$}$$

The latter problem is essentially strictly feasible; thus, for a real τ, we have $\tau \geq \text{Supp}(\xi)$ iff $(*)$ is bounded with the optimal value $\geq -\tau$, which, by Conic Duality Theorem, is the case iff the conic dual of $(*)$ admits a feasible solution with the optimal value $\geq -\tau$, that is, iff

$$\begin{aligned} &\exists \lambda \in \mathbf{K}^*, \mu \geq 0 : P^*\lambda + R^*\mu = -\xi, Q^*\lambda + S^*\mu = 0, \\ &\quad \langle b, \lambda \rangle + \langle r, \mu \rangle \geq -\tau, \end{aligned}$$

and (7.12) follows. □

Remark. Similar to what was done in Section 3.3.4, the possibility to point out a $\mathcal{K}$-representation of the support function of a (nonempty) $\mathcal{K}$-r. set implies, along with other consequences, the following result:

Let the (nonempty) uncertainty set of an uncertain LO problem be given by an essentially strictly feasible $\mathcal{K}$-representation. Then the Robust Counterpart of the problem can be straightforwardly converted into an explicit $\mathcal{K}$-conic program.

7.4.2 Calculus of $\mathcal{K}$-representability: Expressive abilities of CQO and SDO

We have seen that the "rule" part of the calculus of $\mathcal{K}$-representable sets and functions remains intact (and in its "advanced" parts is even slightly weaker) than the calculus of polyhedral representability. What extends *dramatically* when passing from LO to CQO and especially SDO is the spectrum of "raw materials" of the calculus, that is, "elementary" $\mathcal{CQ}$- (Conic Quadratic-) and $\mathcal{SD}$ (Semidefinite)-representable functions and sets. With slight exaggeration, one can say that "for all practical purposes," *all* computationally tractable convex sets and functions arising in applications are $\mathcal{SD}$-representable, so that all "real-life" convex problems are within the grasp of SDO.[8] In our LO-oriented book, we omit the list of "raw materials" for the calculus of $\mathcal{CQ}$- and $\mathcal{SD}$-representable functions/sets; such a list can be found in Ben-Tal and Nemirovski (2001, 2022). Here, we restrict ourselves with a "short list" as follows:

- *Every polyhedrally representable set/function is $\mathcal{CQ}$-representable, and every $\mathcal{CQ}$-representable set/function is $\mathcal{SD}$-representable, with $\mathcal{CQ}$ (resp., $\mathcal{SD}$) representation readily given by the initial polyhedral (resp., $\mathcal{CQ}$) representation.*

  This is an immediate consequence of Section 7.2.1, where, in particular, we have seen that the Lorentz cone $\mathbf{L}^n$ admits the $\mathcal{SD}$-representation

$$\mathbf{L}^n = \left\{ x \in \mathbf{R}^n : \left[\begin{array}{c|ccc} x_n & x_1 & \cdots & x_{n-1} \\ \hline x_1 & x_n & & \\ \vdots & & \ddots & \\ x_{n-1} & & & x_n \end{array}\right] \succeq 0 \right\}.$$

[8] Of course, whatever be a family $\mathcal{K}$ of cones, the $\mathcal{K}$-representable sets and functions are convex (why?), so that Conic Programming stays within the boundaries of Convex Optimization.

- *The following sets and functions admit explicit $\mathcal{CQ}$-r.'s:*
  - *"Rotated" Lorentz cone* $X = \{[x;u;v] \in \mathbf{R}^m \times \mathbf{R}_+ \times \mathbf{R}_+ : \|x\|_2^2 \leq 4uv\}$:

$$X = \{x;u;v] : [x;u-v;u+v] \in \mathbf{L}^{n+2}\}$$

  - *Univariate convex power functions* $\begin{cases} x^p & ,x \geq 0 \\ 0 & ,x \leq 0 \end{cases}$ *with rational* $p \geq 1$, $\begin{cases} -x^p & ,x \geq 0 \\ +\infty & ,x < 0 \end{cases}$ *with rational* $p \in [0,1]$, $\begin{cases} x^{-p} & ,x > 0 \\ +\infty & ,x \leq 0 \end{cases}$ *with rational* $p > 0$;
  - *Convex quadratic form* $f(x) = x^T Q^T Q x + b^T x + c : \mathbf{R}^n \to \mathbf{R}$:

$$\tau \geq f(x) \Leftrightarrow \left[2Qx; \tau + b^T x + c; \tau - b^T x - c\right] \in \mathbf{L}^{m+2} \qquad [Q : m \times n].$$

  As a consequence, ellipsoids and elliptic cylinders — sublevel sets of convex quadratic forms — admit explicit $\mathcal{CQ}$-r.'s;
  - *The hypograph of "geometric mean"-type monomial — the set*

$$\{[x;t] \in \mathbf{R}_x^n \times \mathbf{R}_t : x \geq 0, t \leq x_1^{p_1} x_2^{p_2} \cdots x_n^{p_n}\}$$

  with rational $p_i > 0$ *such that* $\sum_i p_i \leq 1$;
  - *The epigraph of algebraic "inverse-type" monomial — the set*

$$\{[x;t] : x > 0, t \geq x_1^{-p_1} \cdots x_n^{-p_n}\}$$

  with rational $p_i > 0$;
  - *The* p*-norm* $\|\cdot\|_p$ *with rational* $p \in [1,\infty)$ *or with* $p = \infty$.

Next,

- *When* $f(x) : \mathbf{R}^n \to \mathbf{R} \cup \{+\infty\}$ *is $\mathcal{SD}$-r. function symmetric w.r.t. permutations of entries in the argument:* $f(Px) = f(x)$ *for every* x *and every permutation* $n \times n$ *matrix* P *— then the function*

$$F(X) = f(\lambda(X)) : \mathbf{S}^n \to \mathbf{R} \cup \{+\infty\}$$

  ($\lambda(X)$ is the vector of eigenvalues, taken with their multiplicities and written down in the nonascending order, of symmetric matrix X) is $\mathcal{SD}$-representable with $\mathcal{SD}$-representation readily given by $\mathcal{SD}$-r. of f.

For example,

- *The sum $S_k(X)$ of $k \leq n$ largest eigenvalues of $X \in \mathbf{S}^n$ is $\mathcal{SD}$-r. with the representation*

$$\tau \geq S_k(X) \Leftrightarrow \exists Z \in \mathbf{S}^n, s : X \preceq Z + sI_n, \, Z \succeq 0, \mathrm{Tr}(Z) + ks \leq \tau$$

  (cf. the polyhedral representation (3.47) of the sum $s_k(x)$ of the k largest entries in $x \in \mathbf{R}^n$);
- *The Shatten norm $|X|_p = \|\lambda(X)\|_p : \mathbf{S}^n \to \mathbf{R}_+$ of symmetric matrix, where $p \in [1, \infty]$ is rational or ∞, admits explicit $\mathcal{SD}$-r. In particular, the simplest $\mathcal{SD}$-r. of the spectral norm $|X|_\infty = \|\lambda(X)\|_\infty = \max_{\xi:\|\xi\|_2 \leq 1} \|X\xi\|_2$ on $\mathbf{S}^n$ is*

$$\tau \geq \|X\|_\infty \Leftrightarrow -\tau I_n \preceq X \preceq \tau I_n;$$

- The set $\{[X; t] \in \mathbf{S}^n \times \mathbf{R} : X \succeq 0, t \leq \sqrt[n]{\mathrm{Det}\,(X)}\}$ admits explicit $\mathcal{SD}$-r.

Similarly,

- *When $f(x) : \mathbf{R}^m_+ \to \mathbf{R} \cup \{+\infty\}$ is $\mathcal{SD}$-r. function symmetric w.r.t. permutations of entries in the argument and nondecreasing in every one of the components of the argument, the function $F(X) = f(\sigma(X)) : \mathbf{R}^{m\times n} \to \mathbf{R} \cup \{+\infty\}$ ($\sigma(X) \in \mathbf{R}^m$ is the vector of singular values of $X \in \mathbf{R}^{m\times n}$) is $\mathcal{SD}$-representable with $\mathcal{SD}$-r. readily given by the one of f.*

  For example,

  - *The sum $\Sigma_k(X)$ of $k \leq m$ largest singular values of $X \in \mathbf{R}^{m\times n}$ admits explicit $\mathcal{SD}$-r.;*
  - *The Shatten norm $|X|_p = \|\sigma(X)\|_p : \mathbf{R}^{m\times n} \to \mathbf{R}_+$ of $m \times n$ matrix, where $p \in [1, \infty]$ is rational or ∞, admits explicit $\mathcal{SD}$-r. In particular, the simplest $\mathcal{SD}$-r. of the spectral norm $|X|_\infty = \|\sigma(X)\|_\infty = \max_{\xi:\|\xi\|_2 \leq 1} \|X\xi\|_2$ on $\mathbf{R}^{m\times n}$ is*

$$\tau \geq \|X\|_\infty \Leftrightarrow \left[\begin{array}{c|c} \tau I_m & X \\ \hline X^T & \tau I_n \end{array}\right] \succeq 0.$$

Here is an "advertising example:" the messy and highly nonlinear optimization program

	minimize $\sum_{\ell=1}^{n} x_\ell^2$
(a)	$x \geq 0;$
(b)	$a_\ell^T \; x \leq b_\ell, \; \ell = 1, \ldots, n;$
(c)	$\|Px - p\|_2 \leq c^T \; x + d;$
(d)	$x_\ell^{\frac{\ell+1}{\ell}} \leq e_\ell^T x + f_\ell, \; \ell = 1, \ldots, n;$
(e)	$x_\ell^{\frac{\ell}{\ell+3}} x_{\ell+1}^{\frac{1}{\ell+3}} \geq g_\ell^T x + h_\ell, \; \ell = 1, \ldots, n-1;$
(f)	$\mathrm{Det} \begin{bmatrix} x_1 & x_2 & x_3 & \cdots & x_n \\ x_2 & x_1 & x_2 & \cdots & x_{n-1} \\ x_3 & x_2 & x_1 & \cdots & x_{n-2} \\ \vdots & \vdots & \vdots & \ddots & \vdots \\ x_n & x_{n-1} & x_{n-2} & \cdots & x_1 \end{bmatrix} \geq 1;$
(g)	$1 \leq \sum_{\ell=1}^{n} x_\ell \cos(\ell\omega) \leq 1 + \sin^2(5\omega) \; \forall \omega \in \left[-\frac{\pi}{7}, 1.3\right]$

can be converted *in a systematic way* into a semidefinite program; omitting the constraints (f) and (g), the problem can be systematically converted into a $\mathcal{CQ}$-program (and thus solving it within accuracy ϵ can be reduced in polynomial time to a similar task for an LO program, see Section 1.4).

Remark. In the case of polyhedral representability, ignoring the "compactness" of a representation, we can always avoid slack variables: if a set in $\mathbf{R}^n$ admits a polyhedral representation, it is always polyhedral, i.e., can be represented as the solution set of a finite system of nonstrict linear inequalities in the "original" variables — the coordinates of a vector running through $\mathbf{R}^n$. In the nonpolyhedral case, using slack variables in representations of sets and functions is a must. For example, take the epigraph of the univariate function x^4; this set is conic quadratic representable:

$$
\begin{aligned}
G &:= \{[x, \tau] \in \mathbf{R}^2 : \tau \geq x^4\} \\
&= \{[x; \tau] : \exists w \in \mathbf{R} : \underbrace{\|[2x; w-1]\|_2 \leq w+1}_{\text{(a)}}, \underbrace{\|[2w; \tau-1]\|_2 \leq \tau+1}_{\text{(b)}}\}.
\end{aligned}
$$

Indeed, (a) says that $w \geq x^2$, and (b) says that $\tau \geq w^2$; what these inequalities say about τ, x is exactly $\tau \geq x^4$. On the other hand, assume that we managed to find a conic quadratic representation of the same set without slack variables:

$$
G = \{[x; \tau] : \|xa_i + \tau b_i + c_i\|_2 \leq \alpha_i x + \beta_i \tau + \gamma_i,\ 1 \leq i \leq m, xp + \tau q = r\}. \quad (!)
$$

Observe, first, that the system of linear equations should be trivial: $p = q = r = 0$. Indeed, otherwise these equations would cut off the 2D plane of x and τ a line or a point containing G, which is clearly impossible. Now, the sets of the form $\{[x; \tau] : \|xa_i + \tau b_i + c_i\|_2 \leq \alpha_i x + \beta_i \tau + \gamma_i\}$ are convex sets representable as intersections of solutions sets of quadratic inequalities

$$
\|xa_i + \tau b_i + \gamma_i\|_2^2 \leq [\alpha_i x + \beta_i \tau + \gamma_i]^2
$$

with half-planes (or the entire 2D planes) $\{[x; \tau] : \alpha_i x + \beta_i \tau + \gamma_i \geq 0\}$. A set of this type is bounded by finitely many "arcs," every one of them being either a line segment (including rays and entire lines), or parts of ellipses/parabolas/hyperbolas, and thus the right-hand side set in (!) is bounded by finitely many arcs of the same type. But such an arc, as it is easily seen, can intersect the true boundary of G — the curve given by $t = x^4$ — only in finitely many points, so that a finite number of the arcs cannot cover the curve. The conclusion is that G cannot be represented by conic quadratic inequalities in variables x, τ only.

7.5 Exercise for Chapters 6, 7

Exercise 7.1 (Computational study: Stabilizing LTIS). LTIS Background. Dynamics of *Linear Time-Invariant System* (LTIS) is given by the recurrence

$$x_0 = z$$
$$x_{t+1} = Ax_t + Bu_t + d_t,\ t = 0, 1, \ldots$$
$$\left[\begin{array}{l} \bullet\ \ x_t \in \mathbf{R}^{n_x}\text{: state at time } t \\ \bullet\ \ u_t \in \mathbf{R}^{n_u}\text{: control at time } t \\ \bullet\ \ d_t \in \mathbf{R}^{n_x}\text{: external disturbance at time } t \end{array}\right].$$

The conceptually simplest control policy is the *state-based linear feedback* given by

$$u_t = Kx_t$$
$$[K \in \mathbf{R}^{n_u \times n_x}\text{: controller}].$$

Equipped with feedback control policy, the system exhibits *closed-loop* dynamics given by

$$x_0 = z$$
$$x_{t+1} = [A + BK]x_t + d_t,\ t = 0, 1, \ldots$$

The closed-loop dynamics is called *stable* if in the absence of external disturbances — in the case when $d_t \equiv 0$ — all trajectories go to 0 as $t \to \infty$. It is known that closed-loop dynamics is stable iff the moduli of all eigenvalues of the matrix $A+BK$ are < 1, same as iff this matrix admits *Lyapunov Stability Certificate* (LSC) X — positive definite $n \times n$ matrix X such that for some *decay rate* $\gamma \in [0, 1)$ it holds

$$[A + BK]^T X^{-1}[A + BK] \preceq \gamma X^{-1}. \qquad (!)$$

As for the necessity of this condition, we refer the reader to a whatever standard control textbook. Sufficiency is immediate: Assuming (!), and setting $Z = X^{-1/2}[A + BK]X^{1/2}$, (!) says that

$$\|Zy\|_2 \leq \sqrt{\gamma}\|y\|_2 \quad \forall y.$$

On the other hand, passing from states x to states $y = X^{-1/2}x$, closed-loop dynamics becomes

$$y_{t+1} = Zy_t + X^{-1/2}d_t, \quad t = 1, 2, \ldots$$

implying that when $d_t \equiv 0$, $\|y_t\|_2 \le \gamma^{t/2}\|y_0\|_2$ holds, whence $\|x_t\|_2 \le \|X^{1/2}\|_{2,2}\|X^{-1/2}x_0\|_2\gamma^{t/2}$; here

$$\|Q\|_{2,2} = \max_{\xi:\|\xi\|_2\le 1} \|Q\xi\|_2$$

is the spectral norm of matrix Q. It follows that

$$\|x_t\|_2 \le \sqrt{\lambda_{\max}(X)/\lambda_{\min}(X)}\gamma^{t/2}\|x_0\|_2,\ t = 0, 1, \ldots \tag{$*$}$$

where $\lambda_{\max}(X)$ and $\lambda_{\min}(X)$ are the largest and the smallest eigenvalues of $X \succ 0$, respectively.

Building stabilizing controller: One of the standard problems related to LTIS is to specify, given A and B, a feedback matrix K resulting in stable closed-loop dynamics. We intend to consider this problem in a slightly simplified setting, where the decay rate $\gamma \in [0, 1)$ is given in advance. Observe that (!) is homogeneous in X, so that the constraint $X \succ 0$ can be w.l.o.g. replaced with $X \succeq I_{n_x}$. Passing from variables X and K to X and $F = KX$ and multiplying both sides in (!) by X from the left and from the right, (!) reduces to solving the system

$$X \succeq I_{n_x},\ [AX + BF]^T X^{-1}[AX + BF] \preceq \gamma X$$

of matrix inequalities in variables $X \in \mathbf{S}^{n_x}$ and $F \in \mathbf{R}^{n_u\times n_x}$. Invoking the Schur Complement Lemma (Lemma 7.2), this system is equivalent to the system of *Linear* Matrix Inequalities

$$X \succeq I_{n_x},\ \left[\begin{array}{c|c} \gamma X & AX + BF \\ \hline [AX + BF]^T & X \end{array}\right] \succeq 0 \tag{\#}$$

in variables X,F, which is an SDO feasibility problem.

Under the circumstances, the natural objective to be minimized is the maximal eigenvalue of X, resulting in the optimization problem

$$\mathrm{Opt} = \min_{X,\lambda}\left\{\lambda : \lambda I \succeq X \succeq I_{n_x},\ \left[\begin{array}{c|c} \gamma X & AX + BF \\ \hline [AX + BF]^T & X \end{array}\right] \succeq 0\right\}, \tag{D}$$

the rationale being the desire to minimize the right-hand side in $(*)$.

Now, for your task:

(1) Implement in software the design of stabilizing controller via solving (D) and carry out numerical experimentation.

Recommended platform: `MATLAB`,
Recommended setup:

$$A = \left[\begin{array}{c|c} \cos(\theta) & \sin(\theta) \\ \hline -\sin(\theta) & \cos(\theta) \end{array}\right] \text{ with } \theta = \frac{\pi}{32},\ B = \begin{bmatrix} 0 \\ 1 \end{bmatrix}.$$

You could use as a solver, either the Ellipsoid Algorithm, or an IPM like `SDPT3` or `Mosek` invoked via `cvx`.[9]

Recommended experimentation is as follows. With no control (i.e., with zero K), the closed-look dynamics given by the above A is what is called "neutral" — with zero external disturbances, the subsequent 2D states are clockwise rotations of each other by angle $\pi/32$, so that the trajectory neither goes to 0, nor runs to ∞. However, it is easily seen that nonzero disturbances, even approaching zero as time grows, with zero controls may push the trajectory to ∞. In contrast, with stable dynamics, bounded sequence of external disturbances results in bounded trajectories, and converging to zero as $t \to \infty$ disturbances result in converging to 0 as $t \to \infty$ trajectories.

You are recommended to observe these phenomena in numerical simulations, by making the sequence of disturbances a linear combination of harmonic oscillations: $d_t = \sum_{i=1}^{k} \cos(\omega_i t) f^i$, $t = 1, 2, \ldots$.

[9] "CVX: Matlab Software for Disciplined Convex Programming," Michael Grant and Stephen Boyd, http://cvxr.com/cvx/.

Chapter 8

Interior Point Methods for LO and Semidefinite Optimization

In this chapter, we present the basic theory of IPMs for LO and SDO. In this presentation, same as everywhere else in this book, we intend to stick to the "grand scheme of things," without going into algorithmic and implementation details. This allows for unified treatment of IPMs for LO and SDO. Our exposition of the (by itself nowadays standard) material to be presented here follows the one in Ben-Tal and Nemirovski (2001a, Chapter 6).

8.1 SDO program and its dual

In what follows, the problem to be solved is a semidefinite program

$$\min_{x\in\mathbf{R}^n}\left\{c^Tx:\sum_{j=1}^n x_jA_i^j-B_i\succeq 0,\ i=1,\ldots,m\right\},\qquad (*)$$

where B_i and A_i^j are symmetric $\nu_i\times\nu_i$ matrices.

Note that for notational convenience, we have omitted linear equality constraints which used to be a component of the standard form of a conic program. This can be done without loss of generality. Indeed, we can use the linear equality constraints, if any, to express part of the decision variables as affine functions of the remaining "independent" variables. Substituting these representations in the conic constraints and discarding the linear equality constraints, we end up with an equivalent "equalities-free" reformulation of the original problem.

It is convenient to think about the data matrices B_i, $1 \leq i \leq m$, as about diagonal blocks in a block-diagonal matrix B, and about matrices A_i^j, $1 \leq i \leq m$, as about diagonal blocks in block-diagonal matrices A^j. Since a block-diagonal symmetric matrix is positive semidefinite iff its diagonal blocks are symmetric and positive semidefinite, we can rewrite $(*)$ equivalently as

$$\mathrm{Opt}(P) = \min_x \left\{c^T x : \mathcal{A}x - B \in \mathbf{S}_+^\nu\right\}, \tag{P}$$

where

- $\mathcal{A}$ denotes the linear mapping

$$x \mapsto \mathcal{A}x = \sum_{j=1}^N x_j A^j \tag{$\mathcal{A}$}$$

  from the space $\mathbf{R}^N$ of the design variables x to the space $\mathbf{S}^\nu$ of block-diagonal symmetric matrices of the block-diagonal structure $\nu = [\nu_1; ...; \nu_m]$, that is, symmetric block-diagonal matrices with m diagonal blocks of sizes $\nu_1, \ldots, \nu_m$.
- $\mathbf{S}_+^\nu$ is the cone of positive semidefinite matrices from $\mathbf{S}^\nu$.

Note that $\mathbf{S}^\nu$ clearly is a linear subspace in the space $\mathbf{S}^n$ of $n \times n$ symmetric matrices, where

$$n = |\nu| := \sum_{i=1}^m \nu_i. \tag{8.1}$$

$\mathbf{S}^\nu$ is equipped with the Frobenius inner product

$$\langle X, S\rangle = \mathrm{Tr}(XS) = \sum_{i,j=1}^n X_{ij}S_{ij} \qquad [X, S \in \mathbf{S}^\nu]$$

inherited from $\mathbf{S}^n$. Note that the cone $\mathbf{S}_+^\nu$ is regular and self-dual (why?).

Notational conventions.

I. We are in the situation when the cone associated with the conic problem we want to solve "lives" in the space $\mathbf{S}^\nu$ of matrices. As a result, we need a special notation for linear mappings with $\mathbf{S}^\nu$ as the argument or the image space, to distinguish these mappings from matrices "living" in $\mathbf{S}^\nu$. To this end, we will use script letters $\mathcal{A}, \mathcal{B}, \ldots$ as in $(\mathcal{A})$; matrices from $\mathbf{S}^\nu$, as usual will be denoted by capital Roman letters.

II. As always, we write $A \succeq B$ ($\Leftrightarrow B \preceq A$) to express that A, B are symmetric matrices of the same size such that $A - B$ is positive semidefinite; in particular, $A \succeq 0$ means that A is symmetric positive semidefinite. We write $A \succ B$ ($\Leftrightarrow B \prec A$) to express that A, B are symmetric matrices of the same size such that $A - B$ is positive definite; in particular, $A \succ 0$ means that A is symmetric positive definite. The set of all positive definite matrices from $\mathbf{S}^\nu$ is exactly the interior $\operatorname{int} \mathbf{S}^\nu_+$ of the cone $\mathbf{S}^\nu_+$.

III. We shall denote the Euclidean norm associated with the Frobenius inner product by $\|\cdot\|_{\mathrm{Fr}}$ as follows:

$$\|A\|_{\mathrm{Fr}} = \sqrt{\langle A, A\rangle} = \sqrt{\operatorname{Tr}(A^2)} = \sqrt{\sum_{i,j} A_{ij}^2} \qquad [A \in \mathbf{S}^\nu]$$

Assumption A. From now on we assume that

A. *The linear mapping $\mathcal{A}$ has the trivial kernel*, or, which is the same, *the matrices $A^1, \ldots, A^N$ are linearly independent.*

To justify this assumption, note that when $\mathcal{A}$ has a nontrivial kernel and c is not orthogonal to this kernel, problem (P) definitely is bad — either infeasible, or unbounded. And if c is orthogonal to $\operatorname{Ker}\mathcal{A}$, we lose nothing by restricting x to reside in the orthogonal complement L to $\operatorname{Ker}\mathcal{A}$, which amounts to passing from (P) to an equivalent program of the same structure and with linear mapping possessing a trivial kernel.

8.1.1 *The problem dual to (P)*

The problem dual to the conic problem (P), according to our general theory, is

$$\operatorname{Opt}(D) = \max_{S \in \mathbf{S}^\nu} \left\{ \langle B, S\rangle : \mathcal{A}^* S = c, S \in (\mathbf{S}^\nu_+)^* = \mathbf{S}^\nu_+ \right\}, \qquad (D)$$

where $\mathcal{A}^* : \mathbf{S}^\nu \to \mathbf{R}^N$ is the linear mapping conjugate to the mapping $\mathcal{A} : \mathbf{R}^N \to \mathbf{S}^\nu$. Let us compute $\mathcal{A}^*$. By definition, $\mathcal{A}^* S$ is characterized by the identity

$$\forall x \in \mathbf{R}^N : \langle S, \mathcal{A}x\rangle = x^T[\mathcal{A}^* S].$$

Invoking $(\mathcal{A})$, this reads

$$\forall x = [x_1; ...; x_N] : \sum_{j=1}^{N} x_j \langle S, A^j \rangle = \sum_{j=1}^{N} x_j [\mathcal{A}^* S]_j,$$

whence, recalling what $\langle \cdot, \cdot \rangle$ is,

$$\mathcal{A}^* S = [\mathrm{Tr}(A^1 S); ...; \mathrm{Tr}(A^N S)]. \qquad (\mathcal{A}^*)$$

In other words, the equality constraints in (D) read $\mathrm{Tr}(A^j S) = c_j$, $1 \leq j \leq N$.

Assumption B. In what follows, we, in addition to **A**, make the following crucial assumption:

B. *The primal–dual pair* (P), (D) *is primal–dual strictly feasible: there exist* $\bar{x}$ *and* $\bar{S}$ *such that*

$$\begin{aligned} &\mathcal{A}\bar{x} - B \succ 0 \\ &\bar{S} \succ 0 \ \& \ \mathcal{A}^* \bar{S} = c. \end{aligned} \qquad (8.2)$$

By Conic Duality Theorem, **B** implies that both (P) and (D) are solvable with equal optimal values.

8.1.2 *Geometric form of the primal–dual pair* (P), (D)

By elementary Linear Algebra, assumption **A** ensures solvability of the system of linear constraints in (D). Denoting by $C \in \mathbf{S}^\nu$ a solution to this system and recalling the construction from Section 7.3.5, we can pass from (P), (D) to geometric reformulations of these problems, specifically, to the pair

$$\min_X \left\{ \langle C, X \rangle : X \in \mathcal{M}_P \cap \mathbf{S}^\nu_+ \right\}$$
$$\left[C : \mathcal{A}^* C = c, \mathcal{M}_P = \mathcal{L}_P - B, \ \mathcal{L}_P = \mathrm{Im}\mathcal{A} := \{\mathcal{A}x : x \in \mathbf{R}^N\}\right], \qquad (\mathcal{P})$$

$$\max_S \left\{ \langle B, S \rangle : S \in \mathcal{M}_D \cap \mathbf{S}^\nu_+ \right\}$$
$$\left[\mathcal{M}_D = \mathcal{L}_D + C, \ \mathcal{L}_D = \mathrm{Ker}\,\mathcal{A}^* := \{S \in \mathbf{S}^\nu : \mathcal{A}^* S = 0\} = \mathcal{L}_P^\perp\right]. \qquad (\mathcal{D})$$

Recall that

- x is feasible for (P) iff the corresponding primal slack $X(x) = \mathcal{A}x - B$ is feasible for $(\mathcal{P})$, and $c^T x = \langle C, X(x)\rangle + \langle C, B\rangle$;
- $(\mathcal{D})$ is nothing but (D) written in the geometric form;
- When x is feasible for (P) and S is feasible for (D), the corresponding duality gap satisfies

$$\text{DualityGap}(x, S) := c^T x - \langle B, S\rangle = \langle X(x), S\rangle; \tag{8.3}$$

- A pair of optimal solutions to $(\mathcal{P})$, $(\mathcal{D})$ is exactly a pair of orthogonal to each other feasible solutions to the respective problems. A pair x, S of optimal solutions to (P), (D) is exactly a pair of feasible solutions to the respective problems such that $X(x)$ is orthogonal to S.

8.2 Path-following interior point methods for (P), (D): Preliminaries

8.2.1 *Log-det barrier*

In what follows, the central role is played by the *log-det barrier* for the cone $\mathbf{S}^\nu_+$ — the function

$$\Phi(X) = -\ln \text{Det}\, X : \text{int}\, \mathbf{S}^\nu_+ \to \mathbf{R}. \tag{8.4}$$

This function is clearly well defined and smooth (has derivatives of all orders) on the interior int $\mathbf{S}^\nu_+$ of the positive semidefinite cone; we extend Φ from this interior to the entire $\mathbf{S}^\nu$ by setting $\Phi(X) = +\infty$ when $X \notin \text{Dom}\,\Phi := \text{int}\, \mathbf{S}^\nu_+$, obtaining, on the closest inspection, a convex function (it was shown in Section 3.3.8.2). We now list the properties of this barrier that are most important in our context. It will be very instructive for a reader to "translate" these properties to the case of LO, which is nothing but the case of the simple block-diagonal structure $\nu = [1; ...; 1]$, where $\mathbf{S}^\nu$ is just the space of $n \times n$ diagonal matrices, or, identifying such a matrix with the vector of its diagonal entries, just $\mathbf{R}^n$. With this identification, $\mathbf{S}^\nu_+$ is nothing but the nonnegative orthant $\mathbf{R}^n_+$, and the log-det barrier becomes the log barrier

$$F(x) = -\sum_{i=1}^{n} \ln(x_i) : \text{int}\, \mathbf{R}^n_+ \to \mathbf{R}.$$

I. **Barrier property:** *Whenever* $\{X_i\}$ *is a sequence of points from* $\mathrm{Dom}\,\Phi = \mathrm{int}\,\mathbf{S}^n_+$ (i.e., a sequence of positive definite matrices) *converging to a matrix* $X \in \partial\,\mathrm{Dom}\,\Phi$ (i.e., converging to a positive semidefinite and singular matrix), *we have* $\Phi(X_i) \to +\infty$ *as* $i \to \infty$.

This property is evident: the vector of eigenvalues $\lambda(X)$ of X (see Section C.2.C) is the limit of the vectors $\lambda(X_i)$, and since X is a positive semidefinite but not a positive definite matrix, some of $\lambda_j(X)$ are zeros. Thus, some of the n sequences $\{\lambda_j(X_i)\}_{i=1}^{\infty}$, $1 \le j \le n$, have zero limits, while the remaining have positive limits, whence

$$\Phi(X_i) = -\log \mathrm{Det}\,(X_i) = -\sum_{j=1}^{n} \ln(\lambda_j(X_i)) \to +\infty,\ i \to \infty. \qquad \square$$

II. **Derivatives of Φ:** *Let* $X \succ 0$. *Then the gradient* $\nabla\Phi(X)$ *of* Φ *at* X *is*

$$\nabla\Phi(X) = -X^{-1} \tag{8.5}$$

and the Hessian $\mathcal{H}(X) = \nabla\Phi'(X)$ (which is a self-conjugate linear mapping acting from $\mathbf{S}^\nu$ to $\mathbf{S}^\nu$, see Example 3 in Section B.3.9) *is given by*

$$\mathcal{H}(X)H = X^{-1}HX^{-1}, \quad H \in \mathbf{S}^\nu. \tag{8.6}$$

For derivation of these formulas, see Examples 3 in Sections B.3.6 and B.3.9.

III. **Strong convexity of Φ on its domain:** *For every* $X \in \mathrm{Dom}\,\Phi$, *we have*

$$\forall H \in \mathbf{S}^\nu : \langle H, \mathcal{H}(X)H\rangle = \|X^{-1/2}HX^{-1/2}\|_{Fr}^2 \ge 0,^1 \tag{8.7}$$

with the concluding inequality being strict unless $H = 0$.

Indeed, we have

$$\begin{aligned}\langle H, \mathcal{H}(X)H\rangle &= \mathrm{Tr}(HX^{-1}HX^{-1}) \underbrace{=}_{(a)} \mathrm{Tr}([HX^{-1}HX^{-1/2}]X^{-1/2})\\ &= \mathrm{Tr}(X^{-1/2}[HX^{-1}HX^{-1/2}]) = \mathrm{Tr}([X^{-1/2}HX^{-1/2}]^2),\end{aligned}$$

where (a) is given by the simple and important fact we have already mentioned (see C.4.B.4): *whenever* A *and* B *are such that* AB *is well defined and is square, we have* $\mathrm{Tr}(AB) = \mathrm{Tr}(BA)$.

---

[1]Here and it what follows for $A \succeq 0$, $A^{1/2} \succeq 0$ is the matrix square root of A as defined in C.4.A.1. When $A \succ 0$, one has $A^{1/2} \succ 0$, and we set $A^{-1/2} = [A^{1/2}]^{-1} = [A^{-1}]^{1/2}$.

IV. **"Self-duality" of Φ and basic identities:** We have

- ["Self-duality"] The mapping $X \mapsto -\nabla\Phi(X) = X^{-1}$ is a one-to-one mapping of int $\mathbf{S}^\nu_+$ onto itself, which is self-inverse: $-\nabla\Phi(-\nabla\Phi(X)) = (X^{-1})^{-1} = X$ for all $X \succ 0$.
- ["Basic identities"] For $X \in \mathrm{Dom}\,\Phi$, we have

$$\begin{array}{lll} \langle X, \nabla\Phi(X)\rangle & = -n,\ n := \sum_{i=1}^m \nu_i; & \text{(a)} \\ \nabla\Phi(X) & = -\mathcal{H}(X)X; & \text{(b)} \\ \nabla\Phi(tX) & = t^{-1}\nabla\Phi(X). & \text{(c)} \end{array} \tag{8.8}$$

Indeed, by (8.5), $\langle X, \nabla\Phi(X)\rangle = \mathrm{Tr}(X[-X^{-1}]) = \mathrm{Tr}(-I_n) = -n$, while by (8.6) and (8.5), $\nabla\Phi(X) = -X^{-1} = -X^{-1}XX^{-1} = -\mathcal{H}(X)X$. We have proved (a) and (b); (c) is evident due to $\nabla\Phi(X) = -X^{-1}$.

8.2.2 *Path-following scheme: The idea*

The idea underlying the IPMs we are about to consider is as follows. Let $\mu > 0$. Consider the *penalized primal objective, the penalty parameter being* μ — the function

$$P_\mu(X) := \langle C, X\rangle + \mu\Phi(X) : \mathcal{M}_P^+ := \mathcal{M}_P \cap \mathrm{int}\ \mathbf{S}^\nu_+ \to \mathbf{R}.$$

This is a smooth convex function on the strictly feasible part $\mathcal{M}^+$ of the feasible set of $(\mathcal{P})$ which has the *barrier property* — it blows up to $+\infty$ along every sequence of strictly feasible solutions to $(\mathcal{P})$ converging to a boundary point of $\mathcal{M}_P^+$ (i.e., to a feasible solution which is a degenerate matrix; in the LO case, this is a feasible solution with not all the coordinates strictly positive). Assume that this function attains its minimum over $\mathcal{M}_P^+$ at a unique point $X = X_*(\mu)$ (as we shall see in a while, this is indeed the case). We arrive at a curve $X_*(\mu)$, $\mu > 0$, in the space of strictly feasible solutions to $(\mathcal{P})$, called the *primal central path.* Now, what happens when $\mu \to +0$? When μ is small, the function $P_\mu(X)$ is close to the primal objective $\langle C, X\rangle$ in "almost entire" set $\mathcal{M}_P^+$ of strictly feasible solutions to $(\mathcal{P})$, that is, everywhere in the set $\mathcal{M}_P^+$ except for a narrow "strip" along the relative boundary $\partial\mathcal{M}_P^+ = \{X \in \mathcal{M}_P \cap \mathbf{S}^\nu_+ : \mathrm{Det}\,(X) = 0\}$ of the set where $\Phi(X)$ is so large that the penalty term $\mu\Phi(X)$ is "nonnegligible." As $\mu \to +0$, this stripe becomes more and more thin, so that $P_\mu(X)$ becomes more and more close to the primal objective in a larger and larger part of the set $\mathcal{M}_P^+$ of strictly feasible primal solutions. As a result, it is natural to expect (and indeed is true) that *the primal central path approaches the primal optimal set as* $\mu \to 0$ — it is a kind of Ariadne's thread leading to the primal optimal set, see Fig. 8.1. In the path-following IPM, *we move along this path, staying "close" to it* — exactly as Theseus used Ariadne's thread — *and thus approach the optimal set.* How we "move along the path staying close to it" will be explained later. What is worth noting right now is that the dual problem $(\mathcal{D})$, which is in no sense inferior to $(\mathcal{P})$, also defines a path, called the *dual central path*, which leads to the dual optimal set. It turns out that it makes full sense to follow both the paths *simultaneously* (these two processes "help each other"), thus approaching both primal and dual optimal solutions.

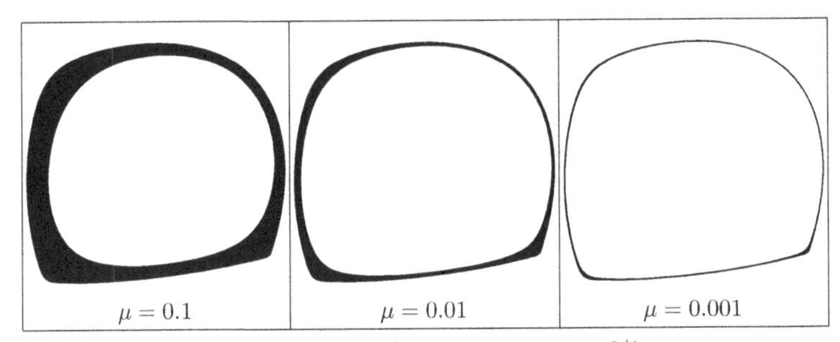

Black: part of $\mathcal{M}_P^+$ where $|\mu\Phi(X)| > \mu^{3/4}$

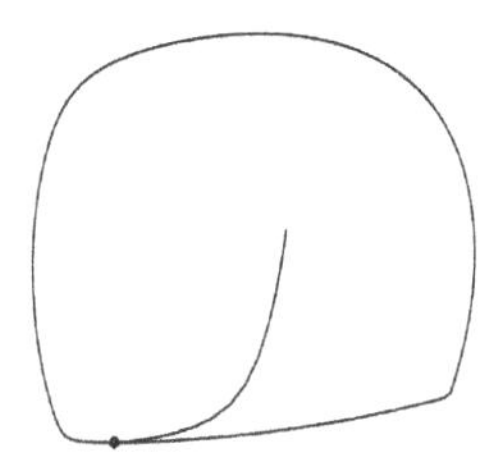

Central path and optimal solution
(the lowest point of the feasible domain)

Fig. 8.1. 2D primal feasible set $\mathcal{M}_P \cap \mathbf{S}^6_+$ and the primal central path.

8.3 Path-following interior point methods for (P), (D): Constructions & results

8.3.1 *Central path: Existence and characterization*

Proposition 8.1. *Under Assumption* **B** *of strict primal–dual feasibility, for every $\mu > 0$, the functions*

$$P_\mu(X) = \langle C, X\rangle + \mu\Phi(X) : \mathcal{M}_P^+ := \mathcal{M}_P \cap \text{int }\mathbf{S}_+^\nu \to \mathbf{R},$$
$$D_\mu(X) = -\langle B, S\rangle + \mu\Phi(S) : \mathcal{M}_D^+ := \mathcal{M}_D \cap \text{int }\mathbf{S}_+^\nu \to \mathbf{R},$$

attain their minima on their domains, the respective minimizers $X_*(\mu)$ and $S_*(\mu)$ being unique. These minimizers $X_* = X_*(\mu)$ and $S_* = S_*(\mu)$ are fully characterized by the following property: X_* is a strictly feasible solution to $(\mathcal{P})$, S_* is a strictly feasible solution to $(\mathcal{D})$ and the following four equivalent-to-each-other relations take place:

$$\begin{aligned} & S_* = \mu X_*^{-1} \quad [\Leftrightarrow S_* = -\mu\nabla\Phi(X_*)], && \text{(a)} \\ \Leftrightarrow\ & X_* = \mu S_*^{-1} \quad [\Leftrightarrow X_* = -\mu\nabla\Phi(S_*)], && \text{(b)} \\ \Leftrightarrow\ & X_*S_* = \mu I, && \text{(c)} \\ \Leftrightarrow\ & S_*X_* = \mu I, && \text{(d)} \\ \Leftrightarrow\ & X_*S_* + S_*X_* = 2\mu I. && \text{(e)} \end{aligned} \tag{8.9}$$

Proof.
1^0. Let us prove first that $P_\mu(\cdot)$ achieves its minimum at $\mathcal{M}_P^+$. Indeed, let X_i be a minimizing sequence for P_μ, that is, $X_i \in \mathcal{M}_P^+$ and

$$P_\mu(X_i) \to \inf_{X\in\mathcal{M}_P^+} P_\mu(X), \quad i \to \infty.$$

All we need is to prove that the sequence $\{X_i\}$ is bounded. Indeed, taking this fact for granted, we could extract from the sequence a converging subsequence; w.l.o.g., let it be the sequence itself: $\exists\bar{X} = \lim_{i\to\infty} X_i$. We claim that $\bar{X} \in \mathcal{M}_P^+$, which would imply that $\bar{X}$ is a desired minimizer of $\Phi_\mu(\cdot)$ on $\mathcal{M}_P^+$. Indeed, the only alternative to $\bar{X} \in \mathcal{M}_P^+$ is that $\bar{X}$ is a boundary point of $\mathcal{S}_+^\nu$; but in this case, $\Phi(X_i) \to \infty$ as $i \to \infty$ due to the barrier property of Φ, while the sequence $\langle B, X_i \rangle$ is bounded since the sequence $\{X_i\}$ is so. It would follow that $P_\mu(X_i) \to +\infty$ as $i \to \infty$, which contradicts the origin of the sequence $\{X_i\}$.

It remains to verify our claim that the sequence $\{X_i\}$ is bounded. To this end, note that, as we remember from Section 7.3.5, when we replace C (which by its origin belongs to the dual feasible plane $\mathcal{M}_D = \{S : \mathcal{A}^* S = c\}$, see $(\mathcal{P})$, $(\mathcal{D})$) with any other point $C' \in \mathcal{M}_D$, the primal objective, restricted to the primal feasible plane $\mathcal{M}_P$, is shifted by a constant, meaning that $\{X_i\}$ is a minimizing sequence for the function $P'_\mu(X) = \langle C', X \rangle + \mu\Phi(X)$ on $\mathcal{M}_P^+$. Since $(\mathcal{D})$ is strictly feasible, we can choose C' to be $\succ 0$, which, by Proposition 7.1 and due to the fact that $\mathbf{S}_+^\nu$ is self-dual, implies that $\|X\|_{\mathrm{Fr}} \leq \theta\langle C', X\rangle$ for some θ and all $X \in \mathbf{S}_+^\nu$. Assuming that $\|X_i\|_{\mathrm{Fr}} \to \infty$ as $i \to \infty$ and taking into account that $\Phi(X) \geq -n \ln(\|X\|_{\mathrm{Fr}})$ when $X \in \operatorname{int} \mathbf{S}_+^\nu$ (why?), we would conclude that

$$\begin{aligned} P'_\mu(X_i) &= \langle C', X_i \rangle + \mu\Phi(X_i) \\ &\geq \theta^{-1}\|X_i\|_{\mathrm{Fr}} - \mu n \ln(\|X_i\|_{\mathrm{Fr}}) \to \infty, \quad i \to \infty, \end{aligned}$$

which contradicts the fact that $\{X_i\}$ is a minimizing sequence for $P'_\mu(\cdot)$ on $\mathcal{M}_P^+$. This contradiction shows that the sequence $\{X_i\}$ is bounded, as claimed.

2^0. Now, let us prove that the minimizer of $P_\mu(\cdot)$ on $\mathcal{M}_P^+$ is unique. This is immediately given by the fact that $\mathcal{M}_P^+$ is convex and $P_\mu(\cdot)$ is strongly convex along with Φ.

Since we do not assume previous knowledge of continuous optimization, here is the verification. Assume that X', X'' are two distinct minimizers of $P_\mu(\cdot)$ on $\mathcal{M}_P^+$, and let us lead this assumption to a contradiction. To this end let $H = X'' - X'$, $X_t = X' + tH$ and $\phi(t) = P_\mu(X_t)$, $0 \leq t \leq 1$. Computing $\phi''(t)$, we get $\phi''(t) = \langle H, \mathcal{H}(X_t)H \rangle > 0$, see item III in Section 8.2. At the same time, ϕ attains its minimum on $[0, 1]$ at 0 and at 1 due to the origin of X' and X''. Since $\phi(0) \leq \phi(t)$ for $t \in [0, 1]$, $\phi'(0) \geq 0$, and since $\phi(t) \geq \phi(1)$ for $t \in [0, 1]$, $\phi'(1) \leq 0$. But this is impossible, since $\phi'(1) = \phi'(0) + \int_0^1 \phi''(t)dt > \phi'(0)$.

3^0. We have proved that the minimizer X_* of $P_\mu(\cdot)$ on $\mathcal{M}_P^+$ exists and is unique. By exactly the same argument but applied to $(\mathcal{D})$ instead of $(\mathcal{P})$, the minimizer $S_* = S_*(\mu)$ of $D_\mu(\cdot)$ on $\mathcal{M}_D^+$ exists and is unique. Now, for every $H \in \mathcal{L}_P$ a segment $\{X_* + tH, -\delta \leq t \leq \delta\}$ is contained in $\mathcal{M}_P^+$ provided $\delta > 0$ is small enough; indeed, this segment clearly is contained in $\mathcal{M}_P$ for all δ and comprises positive definite matrices for small enough $\delta > 0$ due to $X_* \succ 0$. Applying the Fermat rule, we see that the directional derivative $DP_\mu(X_*)[H] = \frac{d}{dt}\big|_{t=0} P_\mu(X_* + tH) = \langle C - \mu\nabla\Phi(X_*), H\rangle$ of $P_\mu(X_*)$ taken at X_* along the direction H should be 0, and this should be so for every $H \in \mathcal{L}_P$, meaning that the gradient $\nabla P_\mu(X_*)$ should be orthogonal to $\mathcal{L}_P$ and thus should belong to $\mathcal{L}_D = \mathcal{L}_P^\perp$. Vice versa, if $\bar{X} \in \mathcal{M}_P^+$ is such that $\nabla P_\mu(\bar{X}) \in \mathcal{L}_P^\perp$, then, taking into account that $P_\mu(\cdot)$ is convex on $\mathcal{M}_P^+$ along with $\Phi(\cdot)$ (see the beginning of Section 8.2) and applying Gradient inequality, we conclude that $\bar{X}$ is a minimizer of $P_\mu(\cdot)$ over $\mathcal{M}_P^+$. The bottom line is as follows: X_* *is fully characterized by the fact that it is a strictly feasible solution to* $(\mathcal{P})$ *such that* $C + \mu\nabla\Phi(X_*) = C - \mu X_*^{-1} \in \mathcal{L}_P^\perp = \mathcal{L}_D$. Since $X_*^{-1} \succ 0$ due to $X_* \succ 0$, recalling the definition of the dual feasible plane, this conclusion can be reformulated as follows:

> (!) $X_* = X_*(\mu)$ *is fully characterized by the fact that it is a strictly feasible solution to* $(\mathcal{P})$ *such that* $\mu X_*^{-1} \equiv -\mu\nabla\Phi(X_*)$ *is a strictly feasible solution to* $(\mathcal{D})$.

By similar argument,

> (!!) $S_* = S_*(\mu)$ *is fully characterized by the fact that it is a strictly feasible solution to* $(\mathcal{D})$ *such that* $\mu S_*^{-1} \equiv -\mu\nabla\Phi(S_*)$ *is a strictly feasible solution to* $(\mathcal{P})$.

Now, note that the "strictly feasible solution to $(\mathcal{D})$" mentioned in (!) is nothing but S_*. Indeed, setting temporary $\widehat{S} = \mu X_*^{-1}$, we get a strictly feasible solution to $(\mathcal{D})$ such that $\mu\widehat{S}^{-1} = X_*$ is a strictly feasible solution to $(\mathcal{P})$; but by (!!), these two facts fully characterize S_*, and we conclude that $S_* = \mu X_*^{-1}$.

4^0. We have proved that $X_* = X_*(\mu)$, $S_* = S_*(\mu)$ are fully characterized by the facts that they are strictly feasible solutions to the respective problems $(\mathcal{P})$, $(\mathcal{D})$ and are linked to each other by any one of the first four (clearly equivalent to each other) relations in (8.9). It remains to verify that these four relations, taken along with the fact that X_* and S_* are positive definite, are equivalent to the fifth relation in (8.9). To this end, it clearly suffices to verify that if X, S are positive semidefinite symmetric matrices such that $SX + XS = 2tI$ with certain real t, then the matrices commute: $SX = SX$, so that $SX = XS = tI$. Indeed, from $SX + XS = 2tI$

it follows that $X^2S = 2tX - XSX$ is a symmetric matrix, that is, $X^2S = (X^2S)^T = SX^2$, i.e., the symmetric matrices X^2 and S commute. By the theorem on simultaneous diagonalization of commuting symmetric matrices (Section C.2.E), there exists an orthogonal matrix U such that both the matrices $USU^T = P$ and $UX^2U^T = Q$ are diagonal (and positive semidefinite along with S, X). Setting $R = Q^{1/2}$, consider the matrices $S = U^TPU$ and $\bar{X} = U^TRU$. These matrices clearly commute, so that all that we need to prove is that in fact $\bar{X} = X$. But this is evident: by construction, $\bar{X} \succeq 0$ and $\bar{X}^2 = U^T[Q^{1/2}]^2U = X^2$, that is, $\bar{X} = (X^2)^{1/2}$; but the latter matrix is nothing but X due to $X \succeq 0$. □

Primal–dual central path: We have proved that the primal and the dual central paths $X_*(\mu)$, $S_*(\mu)$ are well defined. In the sequel, we refer to $(X_*(\cdot), S_*(\cdot))$ as to the *primal–dual central path* of the strictly primal–dual feasible primal–dual pair $(\mathcal{P})$, $(\mathcal{D})$ of semidefinite programs. By the origin of $(\mathcal{P})$, $X_*(\mu) = \mathcal{A}x_*(\mu) - B$ for uniquely defined (by assumption A) path $x_*(\mu)$ of feasible solutions to (P).

Augmented complementary slackness: The relations equivalent to each other (8.9(c–e)) are called *augmented complementary slackness conditions* The reason is clear: the necessary and sufficient condition for a pair of primal–dual feasible solutions (X, S) to $(\mathcal{P})$, $(\mathcal{D})$ to be composed of optimal solutions is $\mathrm{Tr}(XS) = 0$, or, which is the same for symmetric positive semidefinite matrices, $XS = SX = 0$.[2] Relations (8.9(c–e)) replace $XS = 0$ with a close, for small $\mu > 0$, relation $XS = \mu I$, which allows X and S to be positive definite.

---

[2]The equivalence is shown as follows: Let $X \succeq 0$, $S \succeq 0$; we should prove that $\mathrm{Tr}(XS) = 0$ iff XS and iff $SX = 0$. In one direction this is evident: if $XS = 0$, then also $SX = (XS)^T = 0$, and vice versa, and of course in this case $\mathrm{Tr}(XS) = 0$. In the opposite direction: assume that $\mathrm{Tr}(XS) = 0$. Then $0 = \mathrm{Tr}(XS) = \mathrm{Tr}(X^{1/2}SX^{1/2}) = \mathrm{Tr}((X^{1/2}S^{1/2})(X^{1/2}S^{1/2})^T) = \|X^{1/2}S^{1/2}\|_{\mathrm{Fr}}^2$, whence $X^{1/2}S^{1/2} = 0$, so that $S^{1/2}X^{1/2} = (X^{1/2}S^{1/2})^T = 0$ as well. We see that the matrices $X^{1/2}$ and $S^{1/2}$ commute and their product is 0, whence also $XS = (X^{1/2})^2(S^{1/2})^2 = X^{1/2}S^{1/2}S^{1/2}X^{1/2} = 0$.

8.3.1.1 *Duality gap along the primal–dual central path and around it*

We start with the following immediate.

Observation 8.1. The duality gap along the primal–dual central path is equal to μn, $n = \sum_i \nu_i$:

$$\forall \mu > 0 : \text{DualityGap}(X_*(\mu), S_*(\mu)) := c^T x_*(\mu) - \langle B, S_*(\mu) \rangle \\ = \langle X_*(\mu), S_*(\mu) \rangle = \mu n. \tag{8.10}$$

Indeed, the first equality in (8.10) is explained in Section 7.3.5, and the second equality is due to the definition of the Frobenius inner product and (8.9).

Observation 8.1 shows that if we were able to move along the primal–dual central path pushing μ to $+0$, we were approaching primal–dual optimality at the rate depending solely on the rate at which μ approaches 0. Unfortunately, the primal–dual central path is a curve, and there is no possibility to stay at it all the time. What we intend to do is to trace this path, staying close to it. The immediate related questions are as follows:

A. What kind of closeness is appropriate for us?
B. How to trace the path staying "appropriately close" to it?

We answer question A here; question B, which is the essence of the matter, will be answered in the next section.

Proximity measure: A good, by far not evident in advance, notion of closeness to the path is offered by the proximity measure as follows:

Given a target value $\mu > 0$ of the path parameter and a pair $Z = (X, S)$ of strictly feasible solutions to $(\mathcal{P})$, $(\mathcal{D})$, we quantify the closeness of (X, S) to $Z_*(\mu) = (X_*(\mu), S_*(\mu))$ by the quantity

$$\begin{aligned} \text{dist}(Z, Z_*(\mu)) &= \sqrt{\langle [\mu^{-1}S - X^{-1}], [\mathcal{H}(X)]^{-1}[\mu^{-1}S - X^{-1}] \rangle} \\ &= \sqrt{\text{Tr}(X[\mu^{-1}S - X^{-1}]X[\mu^{-1}S - X^{-1}])}. \end{aligned} \tag{8.11}$$

Observe that this "strange" proximity measure is well defined: $\mathcal{H}(X)$, due to its origin, is a symmetric positive definite mapping of $\mathcal{S}^\nu$ onto itself, so that the quantity under the square root is always nonnegative and is 0 iff $S = \mu X^{-1}$ — the equality which, for strongly feasible solutions to $(\mathcal{P})$, $(\mathcal{D})$, characterizes the pair $X_*(\mu)$, $S_*(\mu)$. Note also that the above "distance," which looks asymmetric w.r.t. X and S, is in fact perfectly symmetric:

$$\begin{aligned}
&\mathrm{Tr}(X[\mu^{-1}S - X^{-1}]X[\mu^{-1}S - X^{-1}]) \\
&\quad = \mathrm{Tr}([\mu^{-1}XS - I][\mu^{-1}XS - I]) \\
&\quad = \mathrm{Tr}(\mu^{-2}XSXS - 2\mu^{-1}XS + I) \\
&\quad = \mathrm{Tr}(\mu^{-2}SXSX - 2\mu^{-1}SX + I) \\
&\quad = \mathrm{Tr}([\mu^{-1}SX - I][\mu^{-1}SX - I]) \\
&\quad = \mathrm{Tr}(S[\mu^{-1}X - S^{-1}]S[\mu^{-1}X - S^{-1}]).
\end{aligned}$$

Besides recovering the symmetry w.r.t. X, S, this computation shows also that

$$\begin{aligned}
\mathrm{dist}^2((X,s), Z_*(\mu)) &= \mathrm{Tr}(\mu^{-2}XSXS - 2\mu^{-1}XS + I) \\
&= \mathrm{Tr}(\mu^{-2}X^{1/2}SXSX^{1/2} - 2\mu X^{1/2}SX^{1/2} + I) \\
&= \mathrm{Tr}([\mu^{-1}X^{1/2}SX^{1/2} - I]^2) \\
&= \|\mu^{-1}X^{1/2}SX^{1/2} - I\|_{\mathrm{Fr}}^2;
\end{aligned}$$

taking into account the symmetry, we arrive at

$$\begin{aligned}
\mathrm{dist}^2((X,S), Z_*(\mu)) &= \|\mu^{-1}X^{1/2}SX^{1/2} - I\|_{\mathrm{Fr}}^2 \\
&= \|\mu^{-1}S^{1/2}XS^{1/2} - I\|_{\mathrm{Fr}}^2. \qquad (8.12)
\end{aligned}$$

We extract from these relations an important

Corollary 8.1. *Let $Z = (X,S)$ be a pair of strictly feasible solutions to $(\mathcal{P})$, $(\mathcal{D})$. Then*

$$\begin{aligned}
&\mathrm{DualityGap}(X,S) \\
&\quad \le \mu n(1 + \rho/\sqrt{n}), \quad \rho = \mathrm{dist}((X,S), (X_*(\mu), S_*(\mu))). \quad (8.13)
\end{aligned}$$

Indeed, the matrix $R = \mu^{-1}X^{1/2}SX^{1/2} - I$ is symmetric, and $\rho^2 = \sum_{i,j} R_{ij}^2 \geq \sum_i R_{ii}^2$, whence $\sum_i R_{ii} \leq \rho\sqrt{n}$. In other words,

$$\begin{aligned}\text{DualityGap}(X,S) &= \text{Tr}(XS) = \text{Tr}(X^{1/2}SX^{1/2}) \\ &= \mu\sum_{i=1}^{n}(1+R_{ii}) \leq \mu n + \mu\rho\sqrt{n}.\end{aligned}$$ □

In the sequel, we shall say that a pair (X,S) *is close to* $Z_*(\mu) = (X_*(\mu), S_*(\mu))$, if X is a strictly feasible solution to $(\mathcal{P})$, S is a strictly feasible solution to $(\mathcal{D})$, and

$$\text{dist}((X,S), Z_*(\mu) \leq 0.1.$$

Corollary 8.1 says that as far as the duality gap is concerned, a pair of primal–dual strictly feasible solutions (X,S), which is close to $(X_*(\mu), S_*(\mu))$, is essentially as good as the latter pair itself:

$$\text{DualityGap}(X,S) \leq 1.1\mu n.$$

Thus, it is wise to *trace* the path as $\mu \to +0$ — to build a sequence of triples (X_i, S_i, μ_i) with (X_i, S_i) close to $Z_*(\mu_i)$ and $\mu_i \to 0$ as $i \to \infty$. The rate of convergence of such a scheme depends solely on the rate at which we can push μ_i to 0.

8.3.2 *Conceptual path-following scheme*

Assume we are given a current value $\bar{\mu} > 0$ of the path parameter and a current iterate $(\bar{X}, \bar{S})$ close to $Z_*(\bar{\mu})$. How could we decrease the value of μ to a smaller value $\mu_+ > 0$ and to update $(\bar{X}, \bar{S})$ into a new iterate (X_+, S_+) close to $Z_*(\mu_+)$? If we knew an answer to this question, we could update the iteration $(\bar{X}, \bar{S}, \bar{\mu}) \mapsto (X_+, S_+, \mu_+)$, thus hopefully obtaining a converging algorithm.

Assume that we have somehow chosen $\mu_+ > 0$. Denoting $\Delta\bar{X} = X_+ - \bar{X}$, $\Delta S = S_+ - \bar{S}$, these two matrices should satisfy the following restrictions:

$$\begin{array}{lll} \Delta X \in \mathcal{L}_P & \text{(a)} & \\ \bar{X} + \Delta X \succ 0 & \text{(a}'\text{)} & \\ \Delta S \in \mathcal{L}_D = \mathcal{L}_P^\perp & \text{(b)} & (8.14) \\ \bar{S} + \Delta S \succ 0 & \text{(b}'\text{)} & \\ G_{\mu_+}(\bar{X} + \Delta X, \bar{S} + \Delta S) \approx 0. & \text{(c)} & \end{array}$$

Here (a), (a′) ensure that $X_+ = \bar{X} + \Delta X$ is strictly primal feasible, (b), (b′) ensure that $S_+ = \bar{S} + \Delta S$ is strictly dual feasible, and $G_{\mu_+}(X, S)$ represents equivalently the augmented complementary slackness condition $\mu_+^{-1} S - X^{-1} = 0$, so that (c) is responsible for closeness of (X_+, S_+) to $Z_*(\mu_+)$.

Now, (a) and (b) are just linear equality constraints, and we know from Linear Algebra how to handle them. Since (a′) and (b′) are *strict* conic inequalities which are satisfied when $\Delta X = 0$, $\Delta S = 0$, they will be automatically satisfied when ΔX and ΔS are small enough, which we can hope to ensure by decreasing μ "not too aggressively" — by choosing $\mu_+ < \bar{\mu}$ close enough to $\bar{\mu}$. What is an actual troublemaker is (c) — this is a system of *nonlinear* (in)equalities. Well, in Computational Mathematics there is a standard way to cope with nonlinearity of systems of equations we want to solve — *linearization.* Given current iterate, we linearize the equations of the system at this iterate and solve the resulting system of *linear equations*, treating the resulting solution as our new iterate. The rationale behind this approach, called the *Newton method* (it indeed goes back to Newton) is that if we already are close to a solution to the true system, then the linearized system will be "very close" to the true one, so that the new iterate will be "much closer" to the actual solution than the previous one. When replacing (8.15) with the linearization of the augmented complementary slackness condition taken at "previous iterate" $\Delta X = 0$, $\Delta S = 0$, we end up with the system of linear equations in variables ΔX, ΔS

$$\begin{array}{ll} \Delta X \in \mathcal{L}_P, & \text{(a)} \\ \Delta S \in \mathcal{L}_D = \mathcal{L}_P^\perp, & \text{(b)} \\ G_{\mu_+}(\bar{X}, \bar{S}) + \dfrac{\partial G_{\mu_+}(\bar{X}, \bar{S})}{\partial X} \Delta X + \dfrac{\partial G_{\mu_+}(\bar{X}, \bar{S})}{\partial S} \Delta S = 0 & \text{(c)} \end{array} \qquad (8.15)$$

augmented by $\succ$-inequalities

$$\bar{x} + \Delta X \succ 0, \bar{S} + \Delta S \succ 0. \tag{8.16}$$

At a step of a "simple" path-following method, given current value $\bar{\mu} > 0$ of the path parameter along with current strictly primal–dual feasible iterate $(\bar{X}, \bar{S})$, we

- choose a new value $\mu_+ > 0$ of the path parameter,
- form and solve the system (8.15) of linear equations in matrix variables ΔX, ΔS, and
- update the iterate according to $(\bar{X}, \bar{S}) \mapsto (X_+ = \bar{X} + \Delta X, S_+ = \bar{S} + \Delta S)$,

and go to the next step (i,.e., replace $\bar{X}, \bar{S}, \bar{\mu}$ with X_+, S_+, μ_+ and repeat the above actions). In such a method, the care on $\succ$-inequalities (8.16) comes from the construction of the method — it should be such that these inequalities are automatically satisfied. In more advanced methods, the solution ΔX, ΔS to (8.15) plays the role of *search direction* rather than the actual shift in the iterates; the new iterates are given by

$$X_+ = \bar{X} + \alpha \Delta X, \quad S_+ = \bar{S} + \alpha \Delta S, \tag{!}$$

where $\alpha > 0$ is a stepsize chosen according to rules (different in different methods) which include, in particular, the requirement that (!) preserves positive definiteness. With this modification, the restrictions (8.16) become redundant and are eliminated.

8.3.2.1 *Primal path-following method*

Let us look at what happens when we use the augmented complementary slackness condition in its initial form in (8.15)

$$G_{\mu_+}(\bar{X} + \Delta X, \bar{S} + \Delta S) := \mu_+^{-1}[\bar{S} + \Delta S] - [\bar{X} + \Delta X]^{-1} = 0. \tag{$*$}$$

In this case, recalling that $-(\bar{X}+\Delta X)^{-1} = \nabla\Phi(\bar{X}+\Delta X) \approx \nabla\Phi(\bar{X}) + \mathcal{H}(\bar{X})\Delta X$ and taking into account item II in Section 8.2, relation

(8.15(c)) reads

$$\mu_+^{-1}(\bar{S}+\Delta S) - \bar{X}^{-1} + \bar{X}^{-1}\Delta X \bar{X}^{-1} = 0,$$

so that (8.15) becomes the system

$$\begin{array}{lr}
\Delta X \in \mathcal{L}_P \qquad\qquad \Leftrightarrow \Delta X = \mathcal{A}\delta x, \quad [\delta x \in \mathbf{R}^N] & \text{(a)}\\
\Delta S \in \mathcal{L}_D \qquad\qquad \Leftrightarrow \mathcal{A}^*\Delta S = 0, & \text{(b)}\\
\mu_+^{-1}[\bar{S}+\Delta S] - \bar{X}^{-1} + \bar{X}^{-1}\Delta X \bar{X}^{-1} = 0 & \text{(c)}\\
\Leftrightarrow \mu_+^{-1}[\bar{S}+\Delta S] + \nabla\Phi(\bar{X}) + \mathcal{H}(\bar{X})\Delta X = 0, & \\
 & \text{(Nwt)}
\end{array}$$

(we have used the description of $\mathcal{L}_P$ and $\mathcal{L}_D$ as given in $(\mathcal{P})$, $(\mathcal{D})$). To process the system, we act as follows. We set

$$K(x) = \Phi(X(x)) \equiv \Phi(\mathcal{A}x - B), \qquad [X(x) = \mathcal{A}x - B]$$

thus getting a barrier for the feasible domain of (P); note that for a strictly feasible solution x to (P), we have

$$\nabla K(x) = \mathcal{A}^*\nabla\Phi(X(x)), \nabla^2 K(x)h = \mathcal{A}^*\mathcal{H}(X(x))\mathcal{A}h.$$

Let also $\bar{x}$ be the strictly feasible solution to (P) corresponding to the strictly feasible solution $\bar{X}$ to $(\mathcal{P})$:

$$\bar{X} = X(\bar{x}) = \mathcal{A}\bar{x} - B.$$

Multiplying both sides in (Nwt(c)) by $\mathcal{A}^*$ and taking into account that $\mathcal{A}^*(\bar{S}+\Delta S) = c$ (since $\bar{S}$ is dual feasible and ΔS should satisfy $\mathcal{A}^*\Delta S = 0$), we get the equation

$$\nabla^2 K(\bar{x})\delta x + \nabla K(\bar{x}) + \mu_+^{-1}c = 0 \qquad (*)$$

in variable $\delta x \in \mathbf{R}^N$. It is immediately seen that assumption **A** and strong convexity of Φ (item III in Section 8.2) ensure that $\nabla^2 K(\bar{x})$ is positive definite and thus nonsingular, and we can solve $(*)$ thus getting

$$\begin{array}{rl}
 & \delta x = -\left[\nabla^2 K(\bar{x})\right]^{-1}\left[\nabla K(\bar{x}) + \mu_+^{-1}c\right]\\
 & X_+ = X(\bar{x}+\delta x) = \mathcal{A}[\bar{x}+\delta x] - B\\
\Rightarrow & S_+ = \bar{S} + \Delta S = \mu_+\left[\bar{X}^{-1} - \mathcal{H}(\bar{X})\mathcal{A}\delta x\right].
\end{array}$$

We see that this process can be expressed in terms of the original decision variables $x \in \mathbf{R}^N$ and the barrier $K(x)$ for the feasible domain of (P). Iterating this process, we arrive at the *primal path-following method* where the iterates x_i in the space of original decision variables and the values μ_i of the path parameter are updated according to

$$\begin{aligned} &\mu_i \mapsto \mu_{i+1} > 0, && \text{(a)} \\ &x_i \mapsto x_{i+1} = x_i - \left[\nabla^2 K(x_i)\right]^{-1} [\nabla K(x_i) + \mu_{i+1}^{-1} c], && \text{(b)} \end{aligned} \tag{8.17}$$

and this process is accompanied by generating primal slacks X_{i+1} and dual solutions S_{i+1} according to

$$\begin{aligned} X_{i+1} &= X(x_{i+1}) = \mathcal{A}x_{i+1} - B, \\ S_{i+1} &= \mu_{i+1}\left[X_i^{-1} - X_i^{-1}[\mathcal{A}[x_{i+1} - x_i]]X_i^{-1}\right]. \end{aligned} \tag{8.18}$$

Note that the relation (8.17(b)) is quite transparent. Indeed, the original design variables $x \in \mathbf{R}^N$ affinely parameterize the primal feasible plane $\mathcal{M}_P$ in $(\mathcal{P})$ according to $X = X(x) = \mathcal{A}x - B$, and the objective $\langle C, X\rangle$ of $(\mathcal{P})$ in this parameterization, up to an irrelevant additive constant, is the original objective $c^T x$. With this parameterization, the set $\mathcal{M}_P^+$ of strictly feasible solutions to $(\mathcal{P})$ is parameterized by the interior int $\mathcal{X}$ of the feasible domain $\mathcal{X} = \{X : \mathcal{A}x - B \in \mathcal{M}_P^+\}$ of (P), and the primal central path $X_*(\mu)$ is parameterized by the primal central path $x_*(\mu) \in \text{int } \mathcal{X}$ in the space of x-variables. Now, $X_*(\mu)$ minimizes $\langle C, X\rangle + \mu\Phi(X)$ over $X \in \mathcal{M}_P^+$, meaning that $x_*(\mu)$ minimizes $c^T x + \mu\Phi(\mathcal{A}x - B) = c^T x + \mu K(x)$ over $x \in \text{int } \mathcal{X}$. In other words, $x_*(\mu)$ is given by the Fermat equation

$$\nabla K(x) + \mu^{-1} c = 0.$$

Linearizing this equation (where μ is set to μ_{i+1}) at ith iterate $x_i \in \text{int } \mathcal{X}$, we get the Newton equation for x_{i+1}:

$$\nabla K(x_i) + \nabla^2 K(x_{i+1})(x_{i+1} - x_i) + \mu_{i+1}^{-1} c = 0.$$

Solving this equation with respect to x_{i+1}, we arrive at the recurrence (8.17(b)).

Note: With the augmented complementary slackness relation $G_{\mu_+}(X, S) = 0$ rewritten in the form $\mu_+^{-1} X - S^{-1} = 0$, the construction completely similar to one we have just presented leads to the *dual path-following method* which, essentially, is the primal path-following algorithm as applied to the "swapped" pair $(\mathcal{D})$, $(\mathcal{P})$ of problems.

The complexity analysis of the primal path-following method can be summarized in the following statement (we omit its proof):

Theorem 8.1. *Assume that we are given a starting point* (x_0, S_0, μ_0) *such that* x_0 *is a strictly feasible solution to* (P), S_0 *is a strictly feasible solution to* $(\mathcal{D})$, $\mu_0 > 0$ *and the pair*

$$(X_0 = \mathcal{A}x_0 - B, S_0)$$

is close to $Z_*(\mu_0)$:

$$\mathrm{dist}((X_0, S_0), Z_*(\mu_0)) \leq 0.1.$$

Starting with (μ_0, x_0, X_0, S_0), *let us iterate process* (8.17)–(8.18) *equipped with the updating policy for* μ_i *given by*

$$\mu_{i+1} = \left(1 + \frac{0.1}{\sqrt{n}}\right)^{-1} \mu_i \quad \left[n = \sum_{i=1}^{m} \nu_i\right]. \tag{8.19}$$

The resulting process is well defined and generates strictly primal–dual feasible pairs (X_i, S_i) *such that* (X_i, S_i) *stay close to the points* $Z_*(\mu_i)$ *on the primal–dual central path:*

$$\mathrm{dist}((X_i, S_i), Z_*(\mu_i)) \leq 0.1, \quad i = 1, 2, \ldots \tag{8.20}$$

The theorem says that getting close to the primal–dual central path once, we can trace it by the primal path-following method, keeping the iterates close to the path (see (8.20)) and decreasing the penalty parameter by an absolute constant factor every $O(1)\sqrt{n}$ steps. Taking into account Corollary 8.1, we conclude that

$$\mathrm{DualityGap}(X_i, S_i) \leq 1.1 n\mu_0 \exp\{-O(1)i/\sqrt{n}\}, \tag{8.21}$$

that is,

> (!) *Every* $O(1)\sqrt{n}$ *iterations of the method reduce the* (upper bound on the) *duality gap by an absolute constant factor*, say, by factor 10.

This fact is extremely important theoretically; in particular, it underlies the so far best-known RACM-polynomial time complexity bounds for LO, CQO and SDO. As a practical tool, the primal and the dual path-following algorithms, at least in their short-step form presented above, are not that attractive. The computational power of the methods can be improved by passing to appropriate large-step versions of the algorithms, but even these versions are thought of to be inferior as compared to "true" primal–dual path-following methods (those which "indeed work with both $(\mathcal{P})$ and $(\mathcal{D})$," see what follows). There are, however, cases when the primal or the dual path-following scheme seems to be unavoidable; these are, essentially, the situations where the pair $(\mathcal{P})$, $(\mathcal{D})$ is "highly asymmetric," e.g., $(\mathcal{P})$ and $(\mathcal{D})$ have different by orders of magnitude design dimensions $\dim \mathcal{L}_P$, $\dim \mathcal{L}_D$. Here, it becomes too expensive computationally to treat $(\mathcal{P})$, $(\mathcal{D})$ in a "nearly symmetric way," and it is better to focus solely on the problem with smaller design dimensions.

8.3.3 *Primal–dual path-following methods*

8.3.3.1 *Monteiro–Zhang's family of path-following IPMs*

The augmented complementary slackness condition can be represented in many different ways, for example, as $\mu^{-1}S - X^{-1} = 0$, $\mu^{-1}X - S^{-1} = 0$, or $XS + SX = 2\mu I$. Different representations of the primal–dual central path as the solution set of a system of nonlinear equations result in different path-following algorithms — these algorithms operate with *linearizations* of the equations outside of the path, and these linearizations for different representations of the path differ from each other. We are about to consider the *Zhang's family* of representations of the augmented complementary slackness condition — representations by the matrix equation

$$QXSQ^{-1} + Q^{-1}SXQ = 2\mu I. \tag{8.22}$$

in symmetric matrix variables $X, S \in \mathbf{S}^\nu$, where the "parameter" Q is a positive definite symmetric matrix. We claim that *positive definite* solutions X, S to this equation are exactly positive definite X, S satisfying the augmented complementary slackness condition $XS = \mu I$.

Indeed, if $X \succ 0, S \succ 0$ satisfy $XS = \mu I$, then, of course, X, S are positive definite solutions to (8.22). In the opposite direction: let $X \succ 0$, $S \succ 0$ solve (8.22). To prove that $XS = \mu I$, set $\widehat{X} = QXQ$, $\widetilde{S} = Q^{-1}SQ^{-1}$, so that $\widehat{X}$ and $\widetilde{S}$ are symmetric positive definite along with

Q, X, S. We have

$$\widehat{X}\widetilde{S} + \widetilde{S}\widehat{X} = [QXQ][Q^{-1}SQ^{-1}] + [Q^{-1}SQ^{-1}][QXQ] \\ = QXSQ^{-1} + Q^{-1}SXQ = 2\mu I,$$

where the last equality is due to (8.22). From the proof of Proposition 8.1 we know that the relations $\widehat{X} \succ 0$, $\widetilde{S} \succ 0$, $\widehat{X}\widetilde{S} + \widetilde{S}\widehat{X} = 2\mu I$ imply $\widehat{X}\widetilde{S} = \mu I$, that is, $QXSQ^{-1} = \mu I$. Thus, $XS = \mu I$, as required.

We are about to consider the *Monteiro-Zhang family* of path-following methods, where the augmented complementary slackness condition is represented by (8.22) with varying from step to step scaling matrices $Q \succ 0$ resulting in *commutative scalings*, that is, in commuting with each other scaled matrices

$$\widetilde{S} = Q^{-1}\bar{S}Q^{-1}, \quad \widehat{X} = Q\bar{X}Q,$$

where $\bar{X}, \bar{S}$ are the iterates to be updated; this commutativity dramatically simplifies the convergence analysis. Examples of commutative scalings are

(1) $Q = \bar{S}^{1/2}$ ($\widetilde{S} = I, \widehat{X} = \bar{S}^{1/2}\bar{X}\bar{S}^{1/2}$) (the "$XS$" method);
(2) $Q = \bar{X}^{-1/2}$ ($\widetilde{S} = \bar{X}^{1/2}\bar{S}\bar{X}^{1/2}$, $\widehat{X} = I$) (the "SX" method);
(3) Q is such that $\widetilde{S} = \widehat{X}$ (the NT (Nesterov-Todd) method, extremely attractive and deep)

If $\bar{X}$ and $\bar{S}$ were just positive reals, the formula for Q would be simple: $Q = \left(\frac{\bar{S}}{\bar{X}}\right)^{1/4}$. In the matrix case this simple formula becomes a bit more complicated (to simplify notation, below we write X instead of $\bar{X}$ and S instead of $\bar{S}$):

$$Q = R^{1/2}, \quad R = X^{-1/2}(X^{1/2}SX^{1/2})^{-1/2}X^{1/2}S.$$

We should verify that (a) R is symmetric positive definite, and thus Q is well-defined, and that (b) $Q^{-1}SQ^{-1} = QXQ$.

(a): Verification of the symmetry of R is as follows:

$$\begin{array}{rl} & X^{-1/2}(X^{1/2}SX^{1/2})X^{-1/2}S^{-1} = I \\ \Rightarrow & X^{-1/2}(X^{1/2}SX^{1/2})^{-1/2}(X^{1/2}SX^{1/2})(X^{1/2}SX^{1/2})^{1/2}X^{-1/2}S^{-1} = I \\ \Rightarrow & \left(X^{-1/2}(X^{1/2}SX^{1/2})^{-1/2}X^{1/2}S\right)\left(X^{1/2}(X^{1/2}SX^{1/2})^{1/2}X^{-1/2}S^{-1}\right) = I \\ \Rightarrow & X^{-1/2}(X^{1/2}SX^{1/2})^{-1/2}X^{1/2}S = SX^{1/2}(X^{1/2}SX^{1/2})^{-1/2}X^{-1/2} \\ \Rightarrow & R = R^T. \end{array}$$

To verify that $R \succ 0$, recall that by Linear Algebra the spectrum (the set of distinct eigenvalues) of the product AB of two square matrices remains intact when swapping the factors.[3] Thus, denoting the spectrum of A by $\sigma(A)$, it holds

$$\begin{aligned}\sigma(R) &= \sigma\left(X^{-1/2}(X^{1/2}SX^{1/2})^{-1/2}X^{1/2}S\right)\\ &= \sigma\left((X^{1/2}SX^{1/2})^{-1/2}X^{1/2}SX^{-1/2}\right)\\ &= \sigma\left((X^{1/2}SX^{1/2})^{-1/2}(X^{1/2}SX^{1/2})X^{-1}\right)\\ &= \sigma\left((X^{1/2}SX^{1/2})^{1/2}X^{-1}\right)\\ &= \sigma(\underbrace{X^{-1/2}(X^{1/2}SX^{1/2})^{1/2}X^{-1/2}}_{P}),\end{aligned}$$

and matrix P clearly is symmetric positive definite, so that all eigenvalues of $R = R^T$ are positive, and thus $R \succ 0$.

(b): We want to prove that $QXQ = Q^{-1}SQ^{-1}$, or, which is the same, that $R^{1/2}XR^{1/2} = R^{-1/2}SR^{-1/2}$, or, which again is the same, that $RXR = S$. Here is the computation:

$$\begin{aligned}RXR &= \left(X^{-1/2}(X^{1/2}SX^{1/2})^{-1/2}X^{1/2}S\right)X\left(X^{-1/2}(X^{1/2}SX^{1/2})^{-1/2}X^{1/2}S\right)\\ &= X^{-1/2}(X^{1/2}SX^{1/2})^{-1/2}(X^{1/2}SX^{1/2})(X^{1/2}SX^{1/2})^{-1/2}X^{1/2}S\\ &= X^{-1/2}X^{1/2}S\\ &= S.\end{aligned}$$

We remark that Nesterov and Todd did not *guess* their scaling; they have developed a deep theory, applicable to the general LO-CQO-SDO case rather than just to SDO, which, in particular, guarantees that the scaling resulting in $\widehat{X} = \widetilde{S}$ does exist (and is unique); with this fact at hand, deriving formula for Q becomes relatively easy.

8.3.3.2 *Primal–dual short-step path-following methods based on commutative scalings*

Path-following methods we are about to consider trace the primal–dual central path of $(\mathcal{P})$, $(\mathcal{D})$, staying close to it. The path is traced by iterating the following updating:

[3] Here is the verification: If $\lambda = 0$ is an eigenvalue of AB, then the matrix AB is singular, that is, $\mathrm{Det}\,(AB) = 0$, whence $\mathrm{Det}\,(BA) = 0$ and thus BA is singular, meaning that $\lambda = 0$ is an eigenvalue of BA as well. Now let $\lambda \neq 0$ be an eigenvalue of AB. Assume that λ is not an eigenvalue of BA, and let us lead this assumption to a contradiction. Setting $R = I + \lambda^{-1}A[I - \lambda^{-1}BA]^{-1}B$, we have $R[I - \lambda^{-1}AB] = [I - \lambda^{-1}AB] + \lambda^{-1}A[I - \lambda^{-1}BA]^{-1}(B[I - \lambda^{-1}AB]) = [I - \lambda^{-1}AB] + \lambda^{-1}A[I - \lambda^{-1}BA]^{-1}[I - \lambda^{-1}BA]B = [I - \lambda^{-1}AB] + \lambda^{-1}AB = I$, that is, $[I - \lambda^{-1}AB]$ is invertible, which is impossible since λ is an eigenvalue of AB.

(U): Given a current pair of strictly feasible primal and dual solutions $(\bar{X}, \bar{S})$ such that the triple

$$(\bar{\mu} = n^{-1}\operatorname{Tr}(\bar{X}\bar{S}), \bar{X}, \bar{S}) \tag{8.23}$$

satisfies $\operatorname{dist}((\bar{X}, \bar{S}), Z_*(\bar{\mu})) \leq \kappa \leq 0.1$, or, equivalently, (see (8.12))

$$\|\bar{\mu}^{-1}\bar{X}^{1/2}\bar{S}\bar{X}^{1/2} - I\|_{\mathrm{Fr}} \leq \kappa, \tag{8.24}$$

we

(1) Choose the new value μ_+ of the path parameter according to

$$\mu_+ = \left(1 - \frac{\chi}{\sqrt{n}}\right)\bar{\mu},\ n = \sum_{i=1}^{m} \nu_i \tag{8.25}$$

($\chi \in (0, \kappa]$ is control parameter);

(2) Choose somehow the scaling matrix $Q \succ 0$ such that the matrices $\widehat{X} = Q\bar{X}Q$ and $\widetilde{S} = Q^{-1}\bar{S}Q^{-1}$ commute with each other;

(3) Linearize the equation

$$QXSQ^{-1} + Q^{-1}SXQ = 2\mu_+ I$$

at the point $(\bar{X}, \bar{S})$, thus coming to the equation

$$\begin{aligned} Q[\Delta XS + X\Delta S]Q^{-1} + Q^{-1}[\Delta SX + S\Delta X]Q \\ = 2\mu_+ I - [Q\bar{X}\bar{S}Q^{-1} + Q^{-1}\bar{S}\bar{X}Q]; \end{aligned} \tag{8.26}$$

(4) Add to (8.26) the linear equations

$$\begin{aligned} \Delta X &\in \mathcal{L}_P, \\ \Delta S &\in \mathcal{L}_D = \mathcal{L}_P^{\perp}; \end{aligned} \tag{8.27}$$

(5) Solve system (8.26), (8.27), thus getting "primal–dual search direction" $(\Delta X, \Delta S)$;

(6) Update current primal–dual solutions $(\bar{X}, \bar{S})$ into a new pair (X_+, S_+) according to

$$X_+ = \bar{X} + \Delta X, \quad S_+ = \bar{S} + \Delta S.$$

The rationale for (U) has already been explained; the only "novelty" for us is maintaining the relation

$$\bar{\mu} = n^{-1}\operatorname{Tr}(\bar{X}\bar{S}). \tag{8.28}$$

between the current value $\bar{\mu}$ of the path parameter and current strictly feasible primal and dual solutions $\bar{X}$, $\bar{S}$ which we used to consider as three "independent" of each other entities. Note that on the path, where $X = X_*(\mu)$ and $S = S_*(\mu)$, we have $XS = \mu I$ and therefore do have $\mathrm{Tr}(XS) = \mu n$; thus, (8.28) is a natural way to reduce the number of independent entities we are operating with.

The major element of the complexity analysis of path-following polynomial time methods for LO and SDO is as follows:

Theorem 8.2. *Let $0 < \chi \leq \kappa \leq 0.1$. Let, further, $(\bar{X}, \bar{S})$ be a pair of strictly feasible primal and dual solutions to $(\mathcal{P})$, $(\mathcal{D})$ such that the triple (8.23) satisfies (8.24). Then the updated pair (X_+, S_+) is well defined (i.e., system (8.26), (8.27) is solvable with a unique solution), X_+, S_+ are strictly feasible solutions to $(\mathcal{P})$, $(\mathcal{D})$, respectively,*

$$\mu_+ = n^{-1}\,\mathrm{Tr}(X_+ S_+)$$

and the triple (μ_+, X_+, S_+) is close to the path:

$$\mathrm{dist}((X_+, S_+), Z_*(\mu_+)) \leq \kappa.$$

Discussion. The theorem says that updating (U) converts a close-to-the primal–dual-central-path, in the sense of (8.24), strictly primal–dual feasible iterate $(\bar{X}, \bar{S})$ into a new strictly primal–dual feasible iterate with the same closeness-to-the-path property and smaller, by factor $(1 - \chi n^{-1/2})$, value of the path parameter. Thus, after we once get close to the path, we are able to trace this path, staying close to it and decreasing the path parameter by absolute constant factor in $O(1)\sqrt{n}$ steps. According to Corollary 8.1, this means that every $O(1)\sqrt{n}$ step decreases the (upper bound on the) duality gap by an absolute constant factor.

Note that path-following methods implemented in software work *more or less* according to the outlined scheme, up to the fact that they decrease the path parameter more aggressively and move in an essentially larger neighborhood of the central path than the short-step methods we have described. From the theoretical worst-case-oriented viewpoint, this "aggressive behavior" is dangerous and can result in decreasing the duality gap by an absolute constant factor at the cost of $O(n)$ iterations, rather than of $O(\sqrt{n})$ of them. In actual computations, however, this aggressive policy outperforms significantly the worst-case-oriented "safe short-step policy."

8.3.3.3 *Proof of Theorem 8.2

We are about to carry out the complexity analysis of the primal–dual path-following methods based on commutative scalings. This analysis, originally due to Monteiro and Zhang (1998) is reproducing Ben-Tal and Nemirovski (2001a, Chapter 6); although not that difficult, it is more technical than all else in our book, and a noninterested reader may skip it without any harm.

Scalings: Let Q be a nonsingular matrix of the same size and block-diagonal structure as those of the matrices from $\mathbf{S}^\nu$. We can associate with Q a one-to-one linear transformation on $\mathbf{S}^\nu$ ("scaling by Q") given by the formula

$$H \mapsto \mathcal{Q}[H] = QHQ^T. \tag{Scl}$$

It is immediately seen that (Scl) is a symmetry of the semidefinite cone $\mathbf{S}^\nu_+$, meaning that it maps the cone onto itself, same as it maps onto itself the interior of the cone. This family of symmetries is quite rich: for every pair of points A, B from the interior of the semidefinite cone, there exists a scaling which maps A onto B, e.g., the scaling

$$H \mapsto (\underbrace{B^{1/2}A^{-1/2}}_{Q})H(\underbrace{A^{-1/2}B^{1/2}}_{Q^T}).$$

Essentially, this is exactly the existence of that rich family of symmetries of the underlying cones which makes SDO (same as LO and CQO, where the cones also are "perfectly symmetric") especially well suited for IP methods.

In what follows, we will be interested in scalings associated with *positive definite* scaling matrices from $\mathbf{S}^\nu$. The scaling given by such a matrix Q $(X, S, \ldots)$ will be denoted by $\mathcal{Q}$ (resp., $\mathcal{X}, \mathcal{S}, \ldots$):

$$\mathcal{Q}[H] = QHQ.$$

Given a problem of interest (P) and a scaling matrix $Q \succ 0$, we can *scale* the problem, i.e., pass from it to the problem

$$\min_x \left\{c^T x : \mathcal{Q}\left[\mathcal{A}x - B\right] \succeq 0\right\}, \tag{$\mathcal{Q}(P)$}$$

which, of course, is equivalent to (P) (since $\mathcal{Q}[H]$ is positive semidefinite iff H is so). In terms of "geometric reformulation" $(\mathcal{P})$ of (P), this transformation is nothing but the substitution of variables

$$QXQ = Y \quad \Leftrightarrow \quad X = Q^{-1}YQ^{-1};$$

with respect to Y-variables, $(\mathcal{P})$ is the problem

$$\min_Y \left\{ \mathrm{Tr}(C[Q^{-1}YQ^{-1}]) : Y \in \widehat{\mathcal{M}}_P := \mathcal{Q}[\mathcal{L}_P] - \mathcal{Q}[B],\ Y \in \mathbf{S}^\nu_+ \right\},$$

i.e., the problem

$$\begin{array}{c} \min_Y \left\{ \mathrm{Tr}(\widetilde{C}Y) : Y \in \widehat{\mathcal{M}}_P := \widehat{\mathcal{L}}_P - \widehat{B},\ Y \in \mathbf{S}^\nu \right\} \\ \left[\begin{array}{l} \widetilde{C} = Q^{-1}CQ^{-1}, \widehat{B} = QBQ, \widehat{\mathcal{L}}_P = \mathrm{Im}(\mathcal{Q} \cdot \mathcal{A}) = \mathcal{Q}[\mathcal{L}_P] \\ \quad := \{\mathcal{Q}[H] : H \in \mathcal{L}_P\} \end{array} \right] \end{array} \tag{$\widehat{\mathcal{P}}$}$$

The problem dual to $(\widehat{\mathcal{P}})$ is

$$\max_Z \left\{ \mathrm{Tr}(\widehat{B}Z) : Z \in \widehat{\mathcal{L}}_P^\perp + \widetilde{C},\ Z \in \mathbf{S}^\nu_+ \right\}. \tag{$\widetilde{\mathcal{D}}$}$$

It is immediate to realize what is $\widehat{\mathcal{L}}_P^\perp$:

$$\langle Z, QXQ \rangle = \mathrm{Tr}(ZQXQ) = \mathrm{Tr}(QZQX) = \langle QZQ, X \rangle;$$

thus, Z is orthogonal to every matrix from $\widehat{\mathcal{L}}_P$, i.e., to every matrix of the form QXQ with $X \in \mathcal{L}_P$ iff the matrix QZQ is orthogonal to every matrix from $\mathcal{L}_P$, i.e., iff $QZQ \in \mathcal{L}_P^\perp$. It follows that

$$\widehat{\mathcal{L}}_P^\perp = \mathcal{Q}^{-1}[\mathcal{L}_P^\perp].$$

Thus, when acting on the primal–dual pair $(\mathcal{P})$, $(\mathcal{D})$ of SDO programs, a scaling, given by a matrix $Q \succ 0$, converts it into another primal–dual pair of problems, and this new pair is as follows:

- The "primal" geometric data — the subspace $\mathcal{L}_P$ and the primal shift B (which has a part-time job to be the dual objective as well) — are replaced with their images under the mapping $\mathcal{Q}$;
- The "dual" geometric data — the subspace $\mathcal{L}_D = \mathcal{L}_P^\perp$ and the dual shift C (it is the primal objective as well) — are replaced with their images under the mapping $\mathcal{Q}^{-1}$ *inverse* to $\mathcal{Q}$; this inverse mapping again is a scaling, the scaling matrix being Q^{-1}.

We see that it makes sense to speak about *primal–dual scaling* which acts on both the primal and the dual variables and maps a primal variable X onto QXQ, and a dual variable S onto $Q^{-1}SQ^{-1}$. Formally speaking, the primal–dual scaling associated with a matrix $Q \succ 0$ is the linear transformation $(X, S) \mapsto (QXQ, Q^{-1}SQ^{-1})$ of the direct product of two copies of $\mathbf{S}^\nu$ (the "primal" and the "dual" ones). A primal–dual scaling acts naturally on different entities associated with a primal–dual pair $(\mathcal{P})$, $(\mathcal{D})$, in particular, at the following:

- the pair $(\mathcal{P})$, $(\mathcal{D})$ itself — it is converted into another primal–dual pair of problems $(\widehat{\mathcal{P}})$, $(\widetilde{\mathcal{D}})$;
- a primal–dual feasible pair (X, S) of solutions to $(\mathcal{P})$, $(\mathcal{D})$ — it is converted to the pair $(\widehat{X} = QXQ, \widetilde{S} = Q^{-1}SQ^{-1})$, which, as it is immediately seen, is a pair of feasible solutions to $(\widehat{\mathcal{P}})$, $(\widetilde{\mathcal{D}})$. Note that the primal–dual scaling preserves strict feasibility and the duality gap:
$$\begin{aligned}\mathrm{DualityGap}_{\mathcal{P},\mathcal{D}}(X, S) &= \mathrm{Tr}(XS) = \mathrm{Tr}(QXSQ^{-1}) = \mathrm{Tr}(\widehat{X}\widetilde{S})\\ &= \mathrm{DualityGap}_{\widehat{\mathcal{P}},\widetilde{\mathcal{D}}}(\widehat{X}, \widetilde{S});\end{aligned}$$
- the primal–dual central path $(X_*(\cdot), S_*(\cdot))$ of $(\mathcal{P})$, $(\mathcal{D})$; it is converted into the curve $(\widehat{X}_*(\mu) = QX_*(\mu)Q, \widetilde{S}_*(\mu) = Q^{-1}S_*(\mu)Q^{-1})$, which is nothing but the primal–dual central path $\overline{Z}(\mu)$ of the primal–dual pair $(\widehat{\mathcal{P}})$, $(\widetilde{\mathcal{D}})$.

  The latter fact can be easily derived from the characterization of the primal–dual central path; a more instructive derivation is based on the fact that our "hero" — the barrier $\Phi(\cdot)$ — is "semi-invariant" w.r.t. scaling:
$$\begin{aligned}\Phi(\mathcal{Q}[X]) &= -\ln\mathrm{Det}\,(QXQ)\\ &= -\ln\mathrm{Det}\,(X) - 2\ln\mathrm{Det}\,(Q) = \Phi(X) + \mathrm{const}(Q).\end{aligned}$$
  Now, a point on the primal central path of the problem $(\widehat{\mathcal{P}})$ associated with path parameter μ — let this point be temporarily denoted by $Y(\mu)$ — is the unique minimizer of the aggregate
$$P_\mu(Y) = \langle Q^{-1}CQ^{-1}, Y\rangle + \mu\Phi(Y) \equiv \mathrm{Tr}(Q^{-1}CQ^{-1}Y) + \mu\Phi(Y)$$
  over the set $\widehat{M_P^+}$ of strictly feasible solutions of $(\widehat{\mathcal{P}})$. The latter set is exactly the image of the set $\mathcal{M}_P^+$ of strictly feasible solutions to

$(\mathcal{P})$ under the transformation $\mathcal{Q}$, so that $Y(\mu)$ is the image, under the same transformation, of the point — let it be called $X(\mu)$ — which minimizes the aggregate

$$\mathrm{Tr}((Q^{-1}CQ^{-1})(QXQ)) + \mu\Phi(QXQ) = \mathrm{Tr}(CX) + \mu\Phi(X) + \mathrm{const}(Q),$$

over the set $\mathcal{M}_P^+$ of strictly feasible solutions to $(\mathcal{P})$. We see that $X(\mu)$ is exactly the point $X_*(\mu)$ on the primal central path associated with problem $(\mathcal{P})$. Thus, the point $Y(\mu)$ of the primal central path associated with $(\widehat{\mathcal{P}})$ is nothing but $\widehat{X}_*(\mu) = QX_*(\mu)Q$. Similarly, the point of the central path associated with the problem $(\widetilde{\mathcal{D}})$ is exactly $\widetilde{S}_*(\mu) = Q^{-1}S_*(\mu)Q^{-1}$.

- The neighborhood of the primal–dual central path,

$$\mathcal{N}_\kappa = \{(X \in \mathcal{M}_P^+, s \in \mathcal{M}_D^+) : \mathrm{dist}((X,S), Z_*(\mathrm{Tr}(XS))) \le \kappa\},$$

associated with the pair of problems (P), $(\mathcal{D})$. As you can guess, the image of $\mathcal{N}_\kappa$ is exactly the neighborhood $\overline{\mathcal{N}}_\kappa$ of the primal–dual central path $\overline{Z}(\cdot)$ of $(\widehat{\mathcal{P}})$, $(\widetilde{\mathcal{D}})$.

The latter fact is immediate: for a pair (X,S) of strictly feasible primal and dual solutions to $(\mathcal{P})$, $(\mathcal{D})$, and a $\mu > 0$, we have (see (8.11)):

$$\begin{aligned}
&\mathrm{dist}^2((\widehat{X},\widetilde{S}),\overline{Z}_*(\mu))\\
&\quad = \mathrm{Tr}\left([QXQ](\mu^{-1}Q^{-1}SQ^{-1} - [QXQ]^{-1})[QXQ](\mu^{-1}Q^{-1}SQ^{-1} - [QXQ]^{-1})\right)\\
&\quad = \mathrm{Tr}\left(QX(\mu^{-1}S - X^{-1})X(\mu^{-1}S - X^{-1})Q^{-1}\right)\\
&\quad = \mathrm{Tr}\left(X(\mu^{-1}S - X^{-1})X(\mu^{-1}S - X^{-1})\right)\\
&\quad = \mathrm{dist}^2((X,S),Z_*(\mu)),
\end{aligned}$$

and

$$\mathrm{Tr}(\widehat{X}\widetilde{S}) = \mathrm{Tr}([QXQ][Q^{-1}SQ^{-1}]) = \mathrm{Tr}(QXSQ^{-1}) = \mathrm{Tr}(XS).$$

Proof of Theorem 8.2. The proof to follow reproduces, with some modifications, the proof of Ben-Tal and Nemirovski (2001a, Theorem 6.5.2).

1^0. Observe, first (this observation is crucial!) that it suffices to prove our theorem in the particular case when $\bar{X}, \bar{S}$ commute with each other and $Q = I$. Indeed, it is immediately seen that the updating (U) can be represented as follows:

(1) We first scale by Q the "input data" of (U) — the primal–dual pair of problems $(\mathcal{P})$, $(\mathcal{D})$ and the strictly feasible pair $\bar{X}, \bar{S}$ of primal and dual solutions to these problems, as explained in section "Scaling." Note that the resulting entities — a pair of primal–dual problems and a strictly feasible pair of primal–dual solutions to these problems — are linked with each other exactly in the same fashion as the original entities, due to scaling invariance of the duality gap and the neighborhood $\mathcal{N}_\kappa$. In addition, the scaled primal and dual solutions commute;
(2) We apply to the "scaled input data" yielded by the previous step the updating $(\widehat{\mathrm{U}})$ completely similar to (U), but using the unit matrix in the role of Q;
(3) We "scale back" the result of the previous step, i.e., subject this result to the scaling associated with Q^{-1}, thus obtaining the updated iterate (X_+, S_+).

Given that the second step of this procedure preserves primal–dual strict feasibility, w.r.t. the scaled primal–dual pair of problems, of the iterate and keeps the iterate in the neighborhood $\mathcal{N}_\kappa$ of the corresponding central path, we could use once again the "scaling invariance" reasoning to assert that the result (X_+, S_+) of (U) is well defined, is strictly feasible for $(\mathcal{P})$, $(\mathcal{D})$ and is close to the original central path, as claimed in the theorem. Thus, all we need is to justify the above "Given," and this is exactly the same as to prove the theorem in the particular case of $Q = I$ and commuting $\bar{X}$, $\bar{S}$. In the rest of the proof we assume that $Q = I$ and that the matrices $\bar{X}, \bar{S}$ commute with each other. Due to the latter property, $\bar{X}, \bar{S}$ are diagonal in a properly chosen orthonormal basis, meaning that there exists a block-diagonal, with block-diagonal structure ν, orthogonal matrix U such that $U\bar{X}U^T$ and $U\bar{S}U^T$ are diagonal. Representing all matrices from $\mathbf{S}^\nu$ in this basis (i.e., passing from a matrix A to the matrix UAU^T), we can reduce the situation to the case when $\bar{X}$ and $\bar{S}$ are diagonal. Thus, we may (and do) assume in the sequel that $\bar{X}$ and $\bar{S}$ are diagonal, with diagonal entries x_i, s_i, $i = 1, \ldots, n$, respectively, and that $Q = I$. Finally, to simplify the notation, we write μ, X, S instead of $\bar{\mu}$, $\bar{X}$, $\bar{S}$, respectively.

2^0. Our situation and goals now are as follows. We are given affine planes $\mathcal{L}_P - B$, $\mathcal{L}_D + C$, $\mathcal{L}_D = \mathcal{L}_P^\perp$, in $\mathbf{S}^\nu$ and two positive definite diagonal matrices $X = \mathrm{Diag}(\{x_i\}) \in \mathcal{L}_P - B$, $S = \mathrm{Diag}(\{s_i\}) \in \mathcal{L}_D + C$. We set

$$\mu = n^{-1}\,\mathrm{Tr}(XS),$$

and know that

$$\|\mu^{-1}X^{1/2}SX^{1/2} - I\|_{\mathrm{Fr}} \leq \kappa.$$

We further set

$$\mu_+ = (1 - \chi n^{-1/2})\mu, \tag{8.29}$$

and consider the system of equations w.r.t. unknown symmetric matrices $\Delta X, \Delta S$:

$$\begin{aligned} \Delta X &\in \mathcal{L}_P, & \text{(a)}\\ \Delta S &\in \mathcal{L}_D = \mathcal{L}_P^{\perp}, & \text{(b)}\\ \Delta XS + X\Delta S + \Delta SX + S\Delta X &= 2\mu_+ I - 2XS. & \text{(c)} \end{aligned} \tag{8.30}$$

We should prove that the system has a unique solution, and for this solution the matrices

$$X_+ = X + \Delta X, \quad S_+ = S + \Delta S$$

are

(i) Positive definite;
(ii) Belong, respectively, to $\mathcal{L}_P - B$, $\mathcal{L}_P^{\perp} + C$ and satisfy the relation

$$\mathrm{Tr}(X_+S_+) = \mu_+ n; \quad \text{and} \tag{8.31}$$

(iii) Satisfy the relation

$$\Omega \equiv \|\mu_+^{-1}X_+^{1/2}S_+X_+^{1/2} - I\|_{\mathrm{Fr}} \leq \kappa. \tag{8.32}$$

Observe that the situation can be reduced to the one with $\mu = 1$. Indeed, let us pass from the matrices $X, S, \Delta X, \Delta S, X_+, S_+$ to $X, S' = \mu^{-1}S, \Delta X, \Delta S' = \mu^{-1}\Delta S, X_+, S'_+ = \mu^{-1}S_+$. Now, the "we are given" part of our situation becomes as follows: we are given two diagonal positive definite matrices X, S' such that $X \in \mathcal{L}_P - B$, $S' \in \mathcal{L}_P^{\perp} + C'$, $C' = \mu^{-1}C$,

$$\mathrm{Tr}(XS') = n \times 1,$$

and

$$\|X^{1/2}S'X^{1/2} - I\|_{\mathrm{Fr}} = \|\mu^{-1}X^{1/2}SX^{1/2} - I\|_{\mathrm{Fr}} \leq \kappa.$$

The "we should prove" part becomes: to verify that the system of equations

$$\begin{aligned} \Delta X &\in \mathcal{L}_P, && \text{(a)}\\ \Delta S' &\in \mathcal{L}_P^{\perp}, && \text{(b)}\\ \Delta X S' + X\Delta S' + \Delta S' X + S'\Delta X &= 2(1-\chi n^{-1/2})I - 2XS', && \text{(c)} \end{aligned}$$

has a unique solution and that the matrices $X_+ = X + \Delta X$, $S'_+ = S' + \Delta S'_+$ are positive definite, are contained in $\mathcal{L}_P - B$, respectively, $\mathcal{L}_P^{\perp} + C'$ and satisfy the relations

$$\mathrm{Tr}(X_+ S'_+) = \frac{\mu_+}{\mu} = 1 - \chi n^{-1/2},$$

and

$$\|(1-\chi n^{-1/2})^{-1} X_+^{1/2} S'_+ X_+^{1/2} - I\|_{\mathrm{Fr}} \leq \kappa.$$

Thus, the general situation indeed can be reduced to the one with $\mu = 1$, $\mu_+ = 1 - \chi n^{-1/2}$, and we loose nothing assuming, in addition to what was already postulated, that

$$\mu \equiv n^{-1}\,\mathrm{Tr}(XS) = 1, \quad \mu_+ = 1 - \chi n^{-1/2},$$

whence

$$[\mathrm{Tr}(XS) =] \quad \sum_{i=1}^{n} x_i s_i = n, \tag{8.33}$$

and

$$[\|\mu^{-1}X^{1/2}SX^{1/2} - I\|_{\mathrm{Fr}}^2 \equiv] \quad \sum_{i=1}^{n} (x_i s_i - 1)^2 \leq \kappa^2. \tag{8.34}$$

3^0. We start with proving that (8.30) indeed has a unique solution. It is convenient to pass in (8.30) from the unknowns ΔX, ΔS to the unknowns

$$\begin{aligned} \delta X = X^{-1/2}\Delta X X^{-1/2} &\Leftrightarrow \Delta X = X^{1/2}\delta X X^{1/2}, \\ \delta S = X^{1/2}\Delta S X^{1/2} &\Leftrightarrow \Delta S = X^{-1/2}\delta S X^{-1/2}. \end{aligned} \tag{8.35}$$

With respect to the new unknowns, (8.30) becomes

$$\begin{array}{rr} X^{1/2}\delta X X^{1/2} \in \mathcal{L}_P, & \text{(a)} \\ X^{-1/2}\delta S X^{-1/2} \in \mathcal{L}_P^\perp, & \text{(b)} \\ X^{1/2}\delta X X^{1/2} S + X^{1/2}\delta S X^{-1/2} + X^{-1/2}\delta S X^{1/2} + S X^{1/2}\delta X X^{1/2} & \text{(c)} \\ = 2\mu_+ I - 2XS & \\ \Updownarrow & \\ L(\delta X, \delta S) \equiv \left[\underbrace{\sqrt{x_i x_j}(s_i + s_j)}_{\phi_{ij}}(\delta X)_{ij} + \underbrace{\left(\sqrt{\frac{x_i}{x_j}} + \sqrt{\frac{x_j}{x_i}}\right)}_{\psi_{ij}}(\delta S)_{ij} \right]_{(i,j)\in\mathcal{IJ}}, & \text{(d)} \\ = 2\left[(\mu_+ - x_i s_i)\delta_{ij}\right]_{(i,j)\in\mathcal{IJ}} & \end{array} \tag{8.36}$$

where $\delta_{ij} = \begin{cases} 0, & i \neq j \\ 1, & i = j \end{cases}$ are the Kronecker symbols and $\mathcal{IJ}$ is the set of pairs (i,j) of indexes of diagonal $(i = j)$ and under-diagonal $(i > j)$ cells in the diagonal blocks of block-diagonal matrices from $\mathbf{S}^\nu$.

We first claim that (8.36), regarded as a system with unknown matrices $\delta X \in \mathbf{S}^\nu$, $\delta S \in \mathbf{S}^\nu$, has a unique solution. Observe that (8.36) is a system with $2\dim \mathbf{S}^\nu \equiv 2K$, $K = \operatorname{Card}\mathcal{IJ}$, scalar unknowns and $2K$ scalar linear equations. Indeed, (8.36(a)) is a system of $K' \equiv K - \dim \mathcal{L}_P$ linear equations, (8.36(b)) is a system of $K'' = K - \dim \mathcal{L}_P^\perp = \dim \mathcal{L}_P$ linear equations, and (8.36(c)) has K equations, so that the total # of linear equations in our system is $K' + K'' + K = (K - \dim \mathcal{L}_P) + \dim \mathcal{L}_P + K = 2K$. Now, to verify that the *square* system of linear equations (8.36) has exactly one solution, it suffices to prove that the homogeneous system

$$X^{1/2}\delta X X^{1/2} \in \mathcal{L}_P, \quad X^{-1/2}\delta S X^{-1/2} \in \mathcal{L}_P^\perp, \quad L(\delta X, \delta S) = 0$$

has only a trivial solution. Let $(\delta X, \delta S)$ be a solution to the homogeneous system. Relation $L(\delta X, \delta S) = 0$ taken together with

$\delta X, \delta S \in \mathbf{S}^\nu$ means that

$$(\delta X)_{ij} = -\frac{\psi_{ij}}{\phi_{ij}}(\delta S)_{ij}, \tag{8.37}$$

whence

$$\mathrm{Tr}(\delta X \delta S) = -\sum_{i,j} \frac{\psi_{ij}}{\phi_{ij}}(\delta S)_{ij}^2. \tag{8.38}$$

Representing $\delta X, \delta S$ via $\Delta X, \Delta S$ according to (8.35), we get

$$\begin{aligned}\mathrm{Tr}(\delta X \delta S) &= \mathrm{Tr}(X^{-1/2}\Delta X X^{-1/2} X^{1/2} \Delta S X^{1/2})\\ &= \mathrm{Tr}(X^{-1/2}\Delta X \Delta S X^{1/2})\\ &= \mathrm{Tr}(\Delta X \Delta S),\end{aligned}$$

and the latter quantity is 0 due to $\Delta X = X^{1/2}\delta X X^{1/2} \in \mathcal{L}_P$ and $\Delta S = X^{-1/2}\delta S X^{-1/2} \in \mathcal{L}_P^{\perp}$. Thus, the left-hand side in (8.38) is 0; since $\phi_{ij} > 0$, $\psi_{ij} > 0$, (8.38) implies that $\delta S = 0$. But then $\delta X = 0$ in view of (8.37). Thus, the homogeneous version of (8.36) has the trivial solution only, so that (8.36) is solvable with a unique solution.

4^0. Let $\delta X, \delta S$ be the unique solution to (8.36), and let ΔX, ΔS be linked to $\delta X, \delta S$ according to (8.35). Our local goal is to bound from above the Frobenius norms of δX and δS.

From (8.36(d)), it follows that

$$\begin{aligned}(\delta X)_{ij} &= -\frac{\psi_{ij}}{\phi_{ij}}(\delta S)_{ij} + 2\frac{\mu_+ - x_i s_i}{\phi_{ii}}\delta_{ij}, \quad i,j = 1,\ldots,n; \quad &\text{(a)}\\ (\delta S)_{ij} &= -\frac{\phi_{ij}}{\psi_{ij}}(\delta X)_{ij} + 2\frac{\mu_+ - x_i s_i}{\psi_{ii}}\delta_{ij}, \quad i,j = 1,\ldots,n. \quad &\text{(b)}\end{aligned} \tag{8.39}$$

Same as in the concluding part of 3^0, relations (8.36(a–b)) imply that

$$\mathrm{Tr}(\Delta X \Delta S) = \mathrm{Tr}(\delta X \delta S) = \sum_{i,j}(\delta X)_{ij}(\delta S)_{ij} = 0. \tag{8.40}$$

Multiplying (8.39(a)) by $(\delta S)_{ij}$ and taking the sum over i, j, we get, in view of (8.40), the relation

$$\sum_{i,j} \frac{\psi_{ij}}{\phi_{ij}}(\delta S)_{ij}^2 = 2\sum_i \frac{\mu_+ - x_i s_i}{\phi_{ii}}(\delta S)_{ii}; \tag{8.41}$$

by "symmetric" reasoning, we get

$$\sum_{i,j} \frac{\phi_{ij}}{\psi_{ij}} (\delta X)_{ij}^2 = 2 \sum_i \frac{\mu_+ - x_i s_i}{\psi_{ii}} (\delta X)_{ii}. \tag{8.42}$$

Now, let

$$\theta_i = x_i s_i, \tag{8.43}$$

so that in view of (8.33) and (8.34), one has

$$\begin{aligned} \sum_i \theta_i &= n, \qquad &\text{(a)} \\ \sum_i (\theta_i - 1)^2 &\le \kappa^2. \qquad &\text{(b)} \end{aligned} \tag{8.44}$$

Observe that

$$\phi_{ij} = \sqrt{x_i x_j}(s_i + s_j) = \sqrt{x_i x_j}\left(\frac{\theta_i}{x_i} + \frac{\theta_j}{x_j}\right) = \theta_j \sqrt{\frac{x_i}{x_j}} + \theta_i \sqrt{\frac{x_j}{x_i}}.$$

Thus,

$$\begin{aligned} \phi_{ij} &= \theta_j \sqrt{\frac{x_i}{x_j}} + \theta_i \sqrt{\frac{x_j}{x_i}}, \\ \psi_{ij} &= \sqrt{\frac{x_i}{x_j}} + \sqrt{\frac{x_j}{x_i}}; \end{aligned} \tag{8.45}$$

since $1 - \kappa \le \theta_i \le 1 + \kappa$ by (8.44(b)), we get

$$1 - \kappa \le \frac{\phi_{ij}}{\psi_{ij}} \le 1 + \kappa. \tag{8.46}$$

By the geometric-arithmetic mean inequality, we have $\psi_{ij} \ge 2$, whence in view of (8.46)

$$\phi_{ij} \ge (1 - \kappa)\psi_{ij} \ge 2(1 - \kappa) \quad \forall i, j. \tag{8.47}$$

We now have

$$
\begin{aligned}
(1-\kappa)\sum_{i,j}(\delta X)_{ij}^2 &\le \sum_{i,j}\frac{\phi_{ij}}{\psi_{ij}}(\delta X)_{ij}^2 \quad \text{[see (8.46)]}\\
&\le 2\sum_i \frac{\mu_+ - x_i s_i}{\psi_{ii}}(\delta X)_{ii} \quad \text{[see (8.42)]}\\
&\le 2\sqrt{\sum_i(\mu_+ - x_i s_i)^2}\sqrt{\sum_i \psi_{ii}^{-2}(\delta X)_{ii}^2}\\
&\le \sqrt{\sum_i((1-\theta_i)^2 - 2\chi n^{-1/2}(1-\theta_i) + \chi^2 n^{-1})}\sqrt{\sum_{i,j}(\delta X)_{ij}^2} \quad \text{[see (8.47)]}\\
&\le \sqrt{\chi^2 + \sum_i(1-\theta_i)^2}\sqrt{\sum_{i,j}(\delta X)_{ij}^2} \quad \left[\text{since } \sum_i(1-\theta_i) = 0 \text{ by (8.44(a))}\right]\\
&\le \sqrt{\chi^2+\kappa^2}\sqrt{\sum_{i,j}(\delta X)_{ij}^2} \quad \text{[see (8.44(b))]},
\end{aligned}
$$

and from the resulting inequality it follows that

$$
\|\delta X\|_{\mathrm{Fr}} \le \rho \equiv \frac{\sqrt{\chi^2+\kappa^2}}{1-\kappa}. \tag{8.48}
$$

Similarly,

$$
\begin{aligned}
(1+\kappa)^{-1}\sum_{i,j}(\delta S)_{ij}^2 &\le \sum_{i,j}\frac{\psi_{ij}}{\phi_{ij}}(\delta S)_{ij}^2 \quad \text{[see (8.46)]}\\
&\le 2\sum_i \frac{\mu_+ - x_i s_i}{\phi_{ii}}(\delta S)_{ii} \quad \text{[see (8.41)]}\\
&\le 2\sqrt{\sum_i(\mu_+ - x_i s_i)^2}\sqrt{\sum_i \phi_{ii}^{-2}(\delta S)_{ii}^2}\\
&\le (1-\kappa)^{-1}\sqrt{\sum_i(\mu_+ - \theta_i)^2}\sqrt{\sum_{i,j}(\delta S)_{ij}^2} \quad \text{[see (8.47)]}\\
&\le (1-\kappa)^{-1}\sqrt{\chi^2+\kappa^2}\sqrt{\sum_{i,j}(\delta S)_{ij}^2} \quad \text{[same as above]}
\end{aligned}
$$

and from the resulting inequality, it follows that

$$\|\delta S\|_{\rm Fr} \le \frac{(1+\kappa)\sqrt{\chi^2+\kappa^2}}{1-\kappa} = (1+\kappa)\rho. \tag{8.49}$$

5^0. We are ready to prove 2^0 (i–ii). We have

$$X_+ = X + \Delta X = X^{1/2}(I+\delta X)X^{1/2},$$

and the matrix $I+\delta X$ is positive definite due to (8.48) (indeed, the right-hand side in (8.48) is $\rho < 1$, whence the Frobenius norm (and therefore the maximum of moduli of eigenvalues) of δX is less than 1). Note that by the just indicated reasons $I+\delta X \preceq (1+\rho)I$, whence

$$X_+ \preceq (1+\rho)X. \tag{8.50}$$

Similarly, the matrix

$$S_+ = S + \Delta S = X^{-1/2}(X^{1/2}SX^{1/2} + \delta S)X^{-1/2}$$

is positive definite. Indeed, the eigenvalues of the matrix $X^{1/2}SX^{1/2}$ are $\ge \min_i \theta_i \ge 1-\kappa$, while the moduli of eigenvalues of δS, by (8.49), do not exceed $\frac{(1+\kappa)\sqrt{\chi^2+\kappa^2}}{1-\kappa} < 1-\kappa$. Thus, the matrix $X^{1/2}SX^{1/2}+\delta S$ is positive definite, whence S_+ also is so. We have proved 2^0(i).

2^0(ii) is easy to verify. First, by (8.36), we have $\Delta X \in \mathcal{L}_P$, $\Delta S \in \mathcal{L}_P^\perp$, and since $X \in \mathcal{L}_P - B$, $S \in \mathcal{L}_P^\perp + C$, we have $X_+ \in \mathcal{L}_P - B$, $S_+ \in \mathcal{L}_P^\perp + C$. Second, we have

$$\begin{aligned}
{\rm Tr}(X_+S_+) &= {\rm Tr}(XS + X\Delta S + \Delta X S + \Delta X \Delta S)\\
&= {\rm Tr}(XS + X\Delta S + \Delta X S) \quad [\text{since } {\rm Tr}(\Delta X\Delta S)\\
&\qquad = 0 \text{ due to } \Delta X \in \mathcal{L}_P,\ \Delta S \in \mathcal{L}_P^\perp]\\
&= \mu_+ n \quad [\text{take the trace of both sides in (8.30(c))}]
\end{aligned}$$

2^0(ii) is proved.

6^0. It remains to verify 2^0(iii). We should bound from above the quantity

$$\Omega = \left\| \mu_+^{-1} X_+^{1/2} S_+ X_+^{1/2} - I \right\|_{\mathrm{Fr}} = \left\| X_+^{1/2} (\mu_+^{-1} S_+ - X_+^{-1}) X_+^{1/2} \right\|_{\mathrm{Fr}},$$

and our plan is first to bound from above the "close" quantity

$$\widehat{\Omega} = \left\| X^{1/2} (\mu_+^{-1} S_+ - X_+^{-1}) X^{1/2} \right\|_{\mathrm{Fr}} = \mu_+^{-1} \|Z\|_{\mathrm{Fr}},$$
$$Z = X^{1/2}(S_+ - \mu_+ X_+^{-1}) X^{1/2}, \tag{8.51}$$

and then to bound Ω in terms of $\widehat{\Omega}$.

$6^0.1$. *Bounding* $\widehat{\Omega}$. We have

$$\begin{aligned}
Z &= X^{1/2}(S_+ - \mu_+ X_+^{-1}) X^{1/2} \\
&= X^{1/2}(S + \Delta S) X^{1/2} - \mu_+ X^{1/2} [X + \Delta X]^{-1} X^{1/2} \\
&= XS + \delta S - \mu_+ X^{1/2} [X^{1/2}(I + \delta X) X^{1/2}]^{-1} X^{1/2} \\
& \qquad\qquad\qquad\qquad\qquad\qquad\qquad\qquad [\text{see } (8.35)] \\
&= XS + \delta S - \mu_+ (I + \delta X)^{-1} \\
&= XS + \delta S - \mu_+ (I - \delta X) - \mu_+ [(I + \delta X)^{-1} - I + \delta X] \\
&= \underbrace{XS + \delta S + \delta X - \mu_+ I}_{Z^1} + \underbrace{(\mu_+ - 1)\delta X}_{Z^2} + \underbrace{\mu_+ [I - \delta X - (I + \delta X)^{-1}]}_{Z^3},
\end{aligned}$$

so that

$$\|Z\|_{\mathrm{Fr}} \le \|Z^1\|_{\mathrm{Fr}} + \|Z^2\|_{\mathrm{Fr}} + \|Z^3\|_{\mathrm{Fr}}. \tag{8.52}$$

We are about to bound separately all 3 terms in the right-hand side of the latter inequality.

Bounding $\|Z^2\|_{Fr}$: We have

$$\|Z^2\|_{\mathrm{Fr}} = |\mu_+ - 1| \|\delta X\|_{\mathrm{Fr}} \le \chi n^{-1/2} \rho \tag{8.53}$$

(see (8.48) and take into account that $\mu_+ - 1 = -\chi n^{-1/2}$).

Bounding $\|Z^3\|_{\text{Fr}}$: Let λ_i be the eigenvalues of δX. We have

$$\begin{aligned}
\|Z^3\|_{\text{Fr}} &= \|\mu_+[(I+\delta X)^{-1} - I + \delta X]\|_{\text{Fr}} \\
&\le \|(I+\delta X)^{-1} - I + \delta X\|_{\text{Fr}} \quad [\text{since } |\mu_+| \le 1] \\
&= \sqrt{\sum_i \left(\frac{1}{1+\lambda_i} - 1 + \lambda_i\right)^2} \\
&\qquad [\text{pass to the orthonormal eigenbasis of } \delta X] \\
&= \sqrt{\sum_i \frac{\lambda_i^4}{(1+\lambda_i)^2}} \\
&\le \sqrt{\sum_i \frac{\rho^2\lambda_i^2}{(1-\rho)^2}} \le \frac{\rho^2}{1-\rho} \\
&\qquad \left[\text{see (8.48) and note that } \sum_i \lambda_i^2 = \|\delta X\|_{\text{Fr}}^2 \le \rho^2\right]
\end{aligned} \tag{8.54}$$

Bounding $\|Z^1\|_{\text{Fr}}$: This is a bit more involving. We have

$$\begin{aligned}
Z^1_{ij} &= (XS)_{ij} + (\delta S)_{ij} + (\delta X)_{ij} - \mu_+\delta_{ij} \\
&= (\delta X)_{ij} + (\delta S)_{ij} + (x_i s_i - \mu_+)\delta_{ij} \\
&= (\delta X)_{ij}\left[1 - \frac{\phi_{ij}}{\psi_{ij}}\right] + \left[2\frac{\mu_+ - x_i s_i}{\psi_{ii}} + x_i s_i - \mu_+\right]\delta_{ij} \\
&\qquad [\text{we have used (8.39(b))}] \\
&= (\delta X)_{ij}\left[1 - \frac{\phi_{ij}}{\psi_{ij}}\right] \\
&\qquad [\text{since } \psi_{ii} = 2, \text{ see (8.45)}]
\end{aligned}$$

whence, in view of (8.46),

$$|Z^1_{ij}| \le \kappa|(\delta X)_{ij}|,$$

so that

$$\|Z^1\|_{\text{Fr}} \le \kappa\|\delta X\|_{\text{Fr}} \le \kappa\rho \tag{8.55}$$

(the concluding inequality is given by (8.48)).

Assembling (8.53), (8.54), (8.55) and (8.52), we come to

$$\|Z\|_{\rm Fr} \le \rho\left[\frac{\chi}{\sqrt{n}} + \frac{\rho}{1-\rho} + \kappa\right],$$

whence, by (8.51),

$$\widehat{\Omega} \le \frac{\rho}{1-\chi n^{-1/2}}\left[\frac{\chi}{\sqrt{n}} + \frac{\rho}{1-\rho} + \kappa\right]. \tag{8.56}$$

$6^0.2$. *Bounding* Ω. We have

$$\begin{aligned}
\Omega^2 &= \|\mu_+^{-1}X_+^{1/2}S_+X_+^{1/2} - I\|_{\rm Fr}^2\\
&= \left\| X_+^{1/2}\underbrace{[\mu_+^{-1}S_+ - X_+^{-1}]}_{\Theta=\Theta^T} X_+^{1/2}\right\|_{\rm Fr}^2\\
&= {\rm Tr}\left(X_+^{1/2}\Theta X_+\Theta X_+^{1/2}\right)\\
&\le (1+\rho)\,{\rm Tr}\left(X_+^{1/2}\Theta X\Theta X_+^{1/2}\right) \qquad \text{[see (8.50)]}\\
&= (1+\rho)\,{\rm Tr}\left(X_+^{1/2}\Theta X^{1/2}X^{1/2}\Theta X_+^{1/2}\right)\\
&= (1+\rho)\,{\rm Tr}\left(X^{1/2}\Theta X_+^{1/2}X_+^{1/2}\Theta X^{1/2}\right)\\
&= (1+\rho)\,{\rm Tr}\left(X^{1/2}\Theta X_+\Theta X^{1/2}\right)\\
&\le (1+\rho)^2\,{\rm Tr}\left(X^{1/2}\Theta X\Theta X^{1/2}\right) \qquad \text{[the same (8.50)]}\\
&= (1+\rho)^2\|X^{1/2}\Theta X^{1/2}\|_{\rm Fr}^2\\
&= (1+\rho)^2\|X^{1/2}[\mu_+^{-1}S_+ - X_+^{-1}]X^{1/2}\|_{\rm Fr}^2\\
&= (1+\rho)^2\widehat{\Omega}^2 \qquad \text{[see (8.51)]}
\end{aligned}$$

so that

$$\begin{aligned}
\Omega \le (1+\rho)\widehat{\Omega} &= \frac{\rho(1+\rho)}{1-\chi n^{-1/2}}\left[\frac{\chi}{\sqrt{n}} + \frac{\rho}{1-\rho} + \kappa\right],\\
\rho &= \frac{\sqrt{\chi^2+\kappa^2}}{1-\kappa}.
\end{aligned} \tag{8.57}$$

(see (8.56) and (8.48)).

It is immediately seen that if $0 < \chi \le \kappa \le 0.1$, the right-hand side in the resulting bound for Ω is $\le \kappa$, as required in 2^0(iii). □

8.4 How to start path-tracing

So far, we have explained "how to travel along the highway" — how to trace the primal–dual central path after we get close to it. The question which still remains open is "how to reach the highway" — how to come close to the primal–dual central path. It turns out that this can be achieved with the same path-following technique as applied to an appropriate auxiliary problem. We consider two schemes of the latter type — the first which requires *a priori* knowledge of a strictly feasible primal solution, and the second which does not require such knowledge.

8.4.1 *Tracing auxiliary path*

Assume that the feasible set of (P) is bounded, and that we have at our disposal a strictly feasible solution $\bar{x}$ to (P). In this case, we can act as follows. Let us look at the primal central path $x_*(\mu)$ in the space of the original design variables:

$$x_*(\mu) = \operatorname*{argmin}_x \left\{c^T x + \mu\Phi(\mathcal{A}x - B)\right\}.$$

We know that as $\mu \to +0$, this path approaches the optimal set of (P). But what happens when $\mu \to \infty$? Rewriting the relation defining the path equivalently as

$$x_*(\mu) = \operatorname*{argmin}_x \left\{\mu^{-1}c^T x + \Phi(\mathcal{A}x - B)\right\},$$

the answer becomes clear: *as $\mu \to +\infty$, $x_*(\mu)$ converges to the unique minimizer x_K of the function $K(x) = \Phi(\mathcal{A}x - B)$ on* int $\mathcal{X}$, *where $\mathcal{X}$ is the feasible domain of (P); this minimizer is called the analytic center of $\mathcal{X}$.*

The fact that x_K is well defined and is unique follows immediately from the fact that $K(x)$ is a strongly convex function on the *bounded* domain int $\mathcal{X}$ and that this function possesses the barrier property: $K(x_i) \to +\infty$ along every sequence of points $x_i \in \mathcal{X}$ converging to a boundary point of $\mathcal{X}$, where strong convexity of K follows from item III, Section 8.2, and the fact that $\mathcal{A}$ has a trivial kernel.

We see that the "starting point" of $x_*(\cdot)$ — limit of the path $x_*(\mu)$ as $\mu \to \infty$ — is independent of c. On the other hand, let us look at the central path $\widetilde{x}(\mu)$ associated with the objective $d^T x$, where $d = -\nabla K(\bar{x}) = \mathcal{A}^*(-\nabla\Phi(\mathcal{A}\bar{x} - B))$. Now, let (P') be the problem obtained from (P) by replacing the objective with $d^T x$. Setting $D = -\nabla\Phi(\mathcal{A}\bar{x} - B)$, we see that $D \in \text{int}\, \mathbf{S}^\nu_+$, meaning that the objective D in the "primal slack" reformulation

$$\begin{gathered} \min_X \left\{\text{Tr}(DX) : X \in [\mathcal{L}_P - B] \bigcap \mathbf{S}^\nu_+\right\} \\ [\mathcal{L}_P = \text{Im}\mathcal{A}] \end{gathered} \tag{$\mathcal{P}'$}$$

of (P') is $\succ 0$ and thus is a strictly feasible solution to the geometric dual $(\mathcal{D}')$ of (P'). Since $(\mathcal{P}')$ is strictly feasible along with (P), the primal–dual pair $(\mathcal{P}')$, $(\mathcal{D}')$ generates a primal–dual central path, so that $\widetilde{x}(\cdot)$ is well defined. Now, observe that by construction, $\nabla[d^T x + \Phi(\mathcal{A}x - B)]$ vanishes when $x = \bar{x}$, that is, we know that $\widetilde{x}(1) = \bar{x}$. As a result, we can trace the primal–dual central path $\widetilde{Z}(\cdot)$ of $(\mathcal{P}')$, $(\mathcal{D}')$, since we know that the pair $(\mathcal{A}\bar{x} - B, D)$ is exactly on this path, the value of the path parameter being 1. Let us trace this path "backward in time," that is, increasing the path parameter instead of decreasing it. It is immediately seen that the result of Theorem 8.2 remains intact when the updating rule (8.25) in (U) is replaced with $\mu_+ = (1 + \chi n^{-1/2})\mu$, that is, we can trace $\widetilde{Z}(\mu)$, increasing the value of μ by absolute constant factor every $O(1)\sqrt{n}$ steps and staying all the time in the neighborhood $\mathcal{N}_{0.09}$ of this path. It is not difficult to see that checking in this "back tracing" of the auxiliary path the proximity to the path of interest, we eventually will discover that this proximity is ≤ 0.1. At this moment we already "are on the highway" and can switch to tracing the path of interest.

A shortcoming of the construction we have just presented is in the necessity to know an initial strictly feasible solution to (P) and to have the feasible domain of the problem bounded (otherwise the analytic center does not exist). We are about to present another scheme, where we trace an *infeasible* path and approach optimality and feasibility simultaneously.

8.4.2 ***Infeasible start path-following method*

What follows originates from Nesterov *et al.* (1996), see also Frenk *et al.* (2010, Chapter 4) and references therein; our exposition reproduces, with some modifications, Ben-Tal and Nemirovski (2001a, Exercises 6.18–6.23).

Situation and goals: For the sake of a reader, we start with reiterating our assumptions and goals. We are interested to solve an SDO program

$$\min_x \left\{ c^T x : X \equiv \mathcal{A}x - B \in \mathbf{S}^\nu_+ \right\}. \tag{P}$$

The corresponding primal–dual pair, in its geometric form, is

$$\begin{array}{ll}
\min\limits_X \left\{ \langle C, X \rangle : X \in (\mathcal{M}_P := \mathcal{L}_P - B) \cap \mathbf{S}^\nu_+ \right\} & (\mathcal{P}) \\
\max\limits_S \left\{ \langle B, S \rangle : S \in (\mathcal{M}_D := \mathcal{L}_P^\perp + C) \cap \mathbf{S}^\nu_+ \right\} & (\mathcal{D}) \\
\quad \left[\mathcal{L}_P = \operatorname{Im} \mathcal{A}, \ \mathcal{L}_D = \mathcal{L}_P^\perp = \operatorname{Ker} \mathcal{A}^*, \mathcal{A}^* C = c \right] &
\end{array} \tag{8.58}$$

where $\mathcal{A}$ has a trivial kernel and $(\mathcal{P})$, $(\mathcal{D})$ are strictly feasible.

To proceed, it is convenient to "normalize" the data as follows: when we shift B, the shift belonging to $\mathcal{L}_P$, $(\mathcal{P})$ remains intact, while $(\mathcal{D})$ is replaced with an equivalent problem (since when shifting B by a shift along $\mathcal{L}_P$, the dual objective, restricted to the dual feasible plane, gets a constant additive term). Similarly, when we shift C, the shift belonging to $\mathcal{L}_D$, the dual problem $(\mathcal{D})$ remains intact, and the primal $(\mathcal{P})$ is replaced with an equivalent problem. Thus, we can shift B by a shift from $\mathcal{L}_P$ and C by a shift from $\mathcal{L}_D = \mathcal{L}_P^\perp$, while not varying the primal–dual pair $(\mathcal{P})$, $(\mathcal{D})$ (or, better to say, converting it to an equivalent primal–dual pair). With appropriate shifts of this type we can ensure that $B \in \mathcal{L}_P^\perp$ and $C \in \mathcal{L}_P$. We lose nothing when assuming that the data from the very beginning are normalized by the requirements

$$B \in \mathcal{L}_D = \mathcal{L}_P^\perp, \quad C \in \mathcal{L}_P, \tag{Nrm}$$

which, in particular, implies that $\langle C, B \rangle = 0$, so that the duality gap at a pair (X, S) of primal–dual feasible solutions becomes

$$\operatorname{DualityGap}(X, S) = \langle X, S \rangle = \langle C, X \rangle - \langle B, S \rangle; \tag{8.59}$$

to see it, it suffices to open parentheses in the equality

$$0 = \langle \underbrace{X + B}_{\in \mathcal{L}_P}, \underbrace{S - C}_{\in \mathcal{L}_P^\perp} \rangle.$$

Our goal is rather ambitious:

to develop an interior point method for solving $(\mathcal{P})$, $(\mathcal{D})$ which requires neither a priori knowledge of a primal–dual strictly feasible pair of solutions, nor a specific initialization phase.

The scheme: The construction we are about to present achieves the announced goal as follows.

(1) We write down the following system of conic constraints in variables $X, S \in \mathbf{S}^\nu$ and additional scalar variables τ, σ:

$$\begin{aligned} X + \tau B - P &\in \mathcal{L}_P; && \text{(a)} \\ S - \tau C - D &\in \mathcal{L}_P^\perp; && \text{(b)} \\ \langle C, X\rangle - \langle B, S\rangle + \sigma - d &= 0; && \text{(c)} \\ \mathrm{Diag}\{X, S, \sigma, \tau\} &\in \mathbf{S}_+^\nu \times \mathbf{S}_+^\nu \times \mathbf{R}_+ \times \mathbf{R}_+. && \text{(d)} \end{aligned} \qquad \text{(C)}$$

Here $P, D \in \mathbf{S}^\nu, d \in \mathbf{R}$ are certain fixed entities which we choose in such a way that

(i) We can easily point out a strictly feasible solution $\widehat{Y} = \mathrm{Diag}\{\widehat{X}, \widehat{S}, \widehat{\sigma}, \widehat{\tau} = 1\}$ to the system;

(ii) The solution set $\mathcal{Y}$ of (C) is unbounded; moreover, whenever $Y_i = (X_i, S_i, \sigma_i, \tau_i) \in \mathcal{Y}$ is an unbounded sequence, we have $\tau_i \to \infty$.

(2) Imagine that we have a mechanism which allows us to "run away to ∞ along $\mathcal{Y}$," i.e., to generate a sequence of points $Y_i = (X_i, S_i, \sigma_i, \tau_i) \in \mathcal{Y}$ such that $\|Y_i\| \equiv \sqrt{\|X_i\|_{\mathrm{Fr}}^2 + \|S_i\|_{\mathrm{Fr}}^2 + \sigma_i^2 + \tau_i^2} \to \infty$. In this case, by (ii) $\tau_i \to \infty$, $i \to \infty$. Let us define the normalizations

$$\widetilde{X}_i = \tau_i^{-1} X_i, \quad \widetilde{S}_i = \tau_i^{-1} S_i,$$

of X_i, S_i. Since $(X_i, S_i, \sigma_i, \tau_i)$ is a solution to (C), these normalizations satisfy the relations

$$\begin{aligned} \widetilde{X}_i &\in (\mathcal{L}_P - B + \tau_i^{-1} P) \cap \mathbf{S}_+^\nu; && \text{(a)} \\ \widetilde{S}_i &\in (\mathcal{L}_P^\perp + C + \tau_i^{-1} D) \cap \mathbf{S}_+^\nu; && \text{(b)} \\ \langle C, \widetilde{X}_i\rangle - \langle B, \widetilde{S}_i\rangle &\le \tau_i^{-1} d. && \text{(c)} \end{aligned} \qquad \text{(C}')$$

Since $\tau_i \to \infty$, relations (C$'$) say that as $i \to \infty$, the normalizations $\widetilde{X}_i, \widetilde{S}_i$ simultaneously approach primal–dual feasibility for (P), (D) (see (C$'$(a–b))) and primal–dual optimality (see (C$'$(c)) and recall that the duality gap, with our normalization $\langle C, B\rangle = 0$, is $\langle C, X\rangle - \langle B, S\rangle$).

(3) *The* issue, of course, is how to build a mechanism which allows to run away to ∞ along $\mathcal{Y}$. The mechanism we intend to use is as follows. (C) can be rewritten in the form

$$Y \equiv \mathrm{Diag}\{X, S, \sigma, \tau\} \in (\mathcal{E} + R) \cap \mathbf{S}^{\widetilde{\nu}}_+ \tag{G}$$

where

- $\widetilde{\nu} = [\nu; \nu; 1; 1]$, that is, $\mathbf{S}^{\widetilde{\nu}} = \mathbf{S}^\nu \times \mathbf{S}^\nu \times \mathbf{R} \times \mathbf{R}$;
- $\mathcal{E}$ is the linear subspace in $\mathbf{S}^{\widetilde{\nu}}$ given by

$$\mathcal{E} = \Big\{ \mathrm{Diag}\{U, V, s, r\} : U + rB \in \mathcal{L}_P, V - rC \in \mathcal{L}_P^\perp, \\ \langle C, U\rangle - \langle B, V\rangle + s = 0 \Big\};$$

- The point $R \in \mathbf{S}^{\widetilde{\nu}}$ is given by

$$R = \mathrm{Diag}\{P, D, d - \langle C, P\rangle + \langle B, D\rangle, 0\}.$$

Let

$$\widehat{Y} = \mathrm{Diag}\{\widehat{X}, \widehat{S}, \widehat{\sigma}, \widehat{\tau} = 1\}$$

be the strictly feasible solution to (G) given by 1(i), and let

$$\widetilde{C} = -\nabla\Phi(\widehat{Y}). \qquad [\Phi(Y) = -\ln \mathrm{Det}\,(Y) : \mathrm{int}\, \mathbf{S}^{\widetilde{\nu}}_+ \to \mathbf{R}]$$

Consider the auxiliary problem

$$\min_Y \left\{ \langle \widetilde{C}, Y\rangle : Y \in (\mathcal{E} + R) \cap \mathbf{S}^{\widetilde{\nu}}_+ \right\}. \tag{Aux}$$

This problem is strictly feasible, a strictly feasible solution being $\widehat{Y}$. Besides this, $\widetilde{C} \succ 0$, so that the semidefinite dual of (Aux) is strictly feasible as well, a strictly feasible solution being $\widetilde{C}$. By Proposition 8.1 the primal central path $\widetilde{Y}_*(\mu)$ of problem (Aux) is well defined, and by the origin of $\widetilde{C}$, the point $\widehat{Y}$ belongs to the primal central path $\widetilde{Y}_*(\mu)$ of the problem:

$$\widehat{Y} = \widetilde{Y}_*(1).$$

Let us trace the primal central path $\widetilde{Y}_*(\cdot)$, but increasing the value μ of the path parameter instead of decreasing it, thus enforcing μ to go to $+\infty$. What will happen in this process? Recall that the point $\widetilde{Y}_*(\mu)$ of the primal central path of (Aux) minimizes the aggregate

$$\langle \widetilde{C}, Y \rangle + \mu\Phi(Y)$$

over $Y \in \mathcal{Y}^+ = \{Y \in (\mathcal{E} + R) \cap \operatorname{int} \mathbf{S}^{\widetilde{\nu}}_+\}$. When μ is large, we, essentially, are trying to minimize over $\mathcal{Y}^+$ just $\Phi(Y)$. But *the log-det barrier restricted to an unbounded intersection of an affine plane and the interior of the associated semidefinite cone is not bounded below on this intersection.* Taking this fact for granted for the moment, let us look at the consequences. If we were minimizing the barrier Φ over $\mathcal{Y}^+$, the minimum "would be achieved at infinity"; it is natural to guess (and this is indeed true) that when minimizing a slightly perturbed barrier, the minimum will run away to infinity as the level of perturbations goes to 0. Thus, we may expect (and again it is indeed true) that $\|\widetilde{Y}_*(\mu)\|_{\mathrm{Fr}} \to \infty$ as $\mu \to +\infty$, so that when tracing the path $\widetilde{Y}(\mu)$ as $\mu \to 0$, we are achieving our goal of running away to infinity along $\mathcal{Y}$.

Here is the justification of the claim we took for granted. By 1(i–ii), $\mathcal{Y}$ intersects int $\mathbf{S}^{\widetilde{\nu}}_+$ and is unbounded; since $\mathcal{Y}$ is a closed and unbounded convex set, it has a nonzero recessive direction H[4]: $Y + tH \in \mathcal{Y}$ whenever $Y \in \mathcal{Y}$ and $t \geq 0$. Since $\mathcal{Y} \subset \mathbf{S}^{\widetilde{\nu}}_+$, H is a positive semidefinite matrix. Now, let us look at what happens to Φ on the ray $\{\widehat{Y} + tH : t \geq 0\} \subset \mathcal{Y}$. The matrix $\widehat{Y}$ by assumption is positive definite, while H is positive semidefinite and nonzero. It follows that as $t \to \infty$, the eigenvalues of $\widehat{Y} + tH$ are positive, remain bounded away from zero and some of them tend to $+\infty$, meaning that $\Phi(\widehat{Y} + tH) = -\sum_{i=1}^{2n+2} \ln(\lambda_i(\widehat{X} + tH)) \to -\infty$ as $t \to \infty$, so that Φ is not bounded from below on $\mathcal{Y}^+$.

[4]While we know this fact only for polyhedral sets, it is true for all closed convex sets.

Now, let us implement the outlined approach.

Specifying P, D, d: Given the data of (P), let us choose $P, D \in \mathbf{S}^{\nu}$ in such a way that $P \succ B$, $D \succ -C$, $\widehat{\sigma} > 0$ and set

$$d = \langle C, P\rangle - \langle B, D\rangle + \widehat{\sigma}.$$

Invoking (Nrm), it is immediately seen that with this setup, the point

$$\widehat{Y} = \mathrm{Diag}\{\widehat{X} = P - B, \widehat{S} = C + D, \widehat{\sigma}, \widehat{\tau} = 1\} \qquad (8.60)$$

is a strictly feasible solution to (Aux). Thus, our setup ensures 1(i).

Verifying 1(ii). This step is crucial:

Lemma 8.1. *Let*

$$\max_{U}\left\{-\langle R, U\rangle : U \in \mathbf{S}^{\widetilde{\nu}}_{+} \cap \left(\mathcal{E}^{\perp} + \tilde{C}\right)\right\} \qquad (\mathrm{Aux}')$$

be the problem dual to (Aux). *Then* (Aux), (Aux′) *is a strictly primal–dual feasible pair of problems, and, in addition, the feasible set* $\mathcal{Y}$ *of* (Aux) *is unbounded.*

Proof. By construction, $\widehat{Y}$ is a strictly feasible solution to (Aux), whence (Aux) is strictly feasible. Also by construction, $\widetilde{C} = -\nabla\Phi(\widehat{Y}) \succ 0$, and since, $\widetilde{C}$ clearly belongs to the feasible plane of (Aux′), $\widetilde{C}$ is a strictly feasible solution to (Aux′).

Now, $(\mathcal{P})$, $(\mathcal{D})$ are strictly feasible, which by Conic Duality Theorem implies that both problems are solvable with equal optimal values. Let X_* be an optimal solution to $(\mathcal{P})$, S_* be an optimal solution to $(\mathcal{D})$, meaning that $X_* + B \in \mathcal{L}_P$ and $S_* - C \in \mathcal{L}_P^{\perp}$. Besides this, $0 = \mathrm{DualityGap}(X_*, S_*) = \langle C, X_*\rangle - \langle B, S_*\rangle$ (see (8.59)). Now, consider the direction $\Delta = \mathrm{Diag}\{X_*, S_*, 0, 1\}$. Looking at the definition of $\mathcal{E}$, we see that $\Delta \in \mathcal{E}$, and since $X_* \succeq 0$, $S_* \succeq 0$, $\Delta \in \mathbf{S}^{\widetilde{\nu}}_{+}$ as well. It follows that Δ is a recessive direction of $\mathcal{Y}$, and since Δ is nonzero, $\mathcal{Y}$ is unbounded. □

To complete the verification of 1(ii), we need the following simple fact:

Lemma 8.2. *Let* $\bar{X}$, $\bar{S}$ *be a strictly feasible pair of primal–dual solutions to* $(\mathcal{P})$, $(\mathcal{D})$, *so that by Proposition* 7.1 *there exists* $\gamma \in (0, 1)$

such that

$$\begin{aligned} \gamma\|X\|_{Fr} &\le \langle \bar{S}, X\rangle \quad \forall X \in \mathbf{S}^{\nu}_{+}, \\ \gamma\|S\|_{Fr} &\le \langle \bar{X}, S\rangle \quad \forall S \in \mathbf{S}^{\nu}_{+}. \end{aligned} \tag{!}$$

Then for every feasible solution $Y = \mathrm{Diag}\{X, S, \sigma, \tau\}$ *to* (Aux) *one has*

$$\begin{aligned} &\|Y\|_{Fr} \le \alpha\tau + \beta, \\ &\qquad \alpha = \gamma^{-1}\left[\langle \bar{X}, C\rangle - \langle \bar{S}, B\rangle\right] + 1, \\ &\qquad \beta = \gamma^{-1}\left[\langle \bar{X} + B, D\rangle + \langle \bar{S} - C, P\rangle + d\right]. \end{aligned} \tag{8.61}$$

Note that (8.61) clearly implies that whenever $Y_i = \mathrm{Diag}\{..., \tau_i\} \in \mathcal{Y}$ are such that $\|Y_i\|_{\mathrm{Fr}} \to \infty$ as $i \to \infty$, then $\tau_i \to \infty$ as $i \to \infty$ as well.

Proof of Lemma 8.2. Feasible solution Y to (Aux), by construction, is a feasible solution to (C). By (C.a), we have $X + \tau B - P \in \mathcal{L}_P$, whence, since $\bar{S} - C \in \mathcal{L}_P^{\perp}$,

$$0 = \langle X + \tau B - P, \bar{S} - C\rangle \Rightarrow \langle X, \bar{S}\rangle \le \tau\langle B, C - \bar{S}\rangle + \langle P, \bar{S} - C\rangle + \langle C, X\rangle;$$

similarly, by (C.b) and due to $\bar{X} + B \in \mathcal{L}_P$, we have

$$0 = \langle S - \tau C - D, \bar{X} + B\rangle \Rightarrow \langle S, \bar{X}\rangle \le \tau\langle C, \bar{X} + B\rangle + \langle D, \bar{X} + B\rangle - \langle B, S\rangle.$$

Summing up the resulting inequalities, we get

$$\begin{aligned} \langle X, \bar{S}\rangle + \langle S, \bar{X}\rangle &\le \tau\left[\langle B, C - \bar{S}\rangle + \langle C, \bar{X} + B\rangle\right] \\ &\quad + \underbrace{\left[\langle C, X\rangle - \langle B, S\rangle\right]}_{\le d - \sigma \text{ by (C.c)}} + \left[\langle P, \bar{S} - C\rangle + \langle D, \bar{X} + B\rangle\right], \end{aligned}$$

whence, taking into account that $X \succeq 0$, $S \succeq 0$ by (C.d) and invoking (!),

$$\begin{aligned} \gamma[\|X\|_{\mathrm{Fr}} + \|S\|_{\mathrm{Fr}}] + \sigma &\le \tau\left[\langle B, C - \bar{S}\rangle + \langle C, \bar{X} + B\rangle\right] \\ &\qquad + \left[\langle P, \bar{S} - C\rangle + \langle D, \bar{X} + B\rangle\right] + d; \end{aligned}$$

since $\sigma \ge 0$ by (C.d) and $\langle B, C\rangle = 0$, (8.61) follows. $\square$

Tracing the path $\widetilde{Y}_*(\mu)$ as $\mu \to +\infty$: The path $\widetilde{Y}_*(\mu)$ is the primal central path of the strictly primal–dual feasible primal–dual pair of problems (Aux), (Aux$'$) which are in the form of $(\mathcal{P})$, $(\mathcal{P}')$. The only difference with the situation discussed in previous sections is that now we are interested in tracing the path as $\mu \to +\infty$, starting the process from the point $\widehat{Y} = \widetilde{Y}_*(1)$ given by 1(i), rather than tracing the path as $\mu \to +0$. It turns out that we have exactly the same possibilities of tracing the path $\widetilde{Y}_*(\mu)$ as the path parameter goes to $+\infty$ as when tracing the path as $\mu \to +0$; in particular, we can use short-step primal and primal–dual path-following methods with stepsize policies "opposite" to those mentioned, respectively, in Theorems 8.1 and 8.2 ("opposite" means that instead of decreasing μ at each iteration in certain ratio, we increase it in exactly the same ratio). It can be straightforwardly verified that the results of Theorems 8.1 and 8.2 remain valid in this new situation as well. Thus, in order to generate a triple (μ, Y, U) such that $\mu \in (1, \infty)$, Y is strictly feasible for (Aux), U is strictly feasible for the problem (Aux$'$) dual to (Aux), and $\mathrm{dist}((Y, U), \widetilde{Z}_*(\mu)) \le \kappa \le 0.1$, it suffices to carry out

$$\mathcal{N}(\mu) = O(1)\sqrt{2n+2}\ln(2\mu)$$

steps of the path-following method; here $\widetilde{Z}_*(\cdot)$ is the primal–dual central path of the primal–dual pair (Aux), (Aux$'$), and dist from now on is the distance to this path, as defined in (8.11) (the latter definition should, of course, be applied to (Aux), (Aux$'$)). Thus, we understand what is the cost of arriving at a close-to-the-path triple (μ, Y, U) with a desired value $\mu \in (1, \infty)$ of the path parameter. Further, our original scheme explains how to convert the Y-component of such a triple into a pair X_μ, S_μ of approximate solutions to $(\mathcal{P})$, $(\mathcal{D})$:

$$X_\mu = \frac{1}{\tau[Y]} X[Y]; \quad S_\mu = \frac{1}{\tau[Y]} S[Y],$$

where

$$Y = \mathrm{Diag}\{X[Y], S[Y], \sigma[Y], \tau[Y]\}.$$

What we do *not* know for the moment is

(?) *What is the quality of the resulting pair (X_μ, S_μ) of approximate solutions to $(\mathcal{P})$, $(\mathcal{D})$ as a function of μ?*

Looking at (C′), we see that (?) is, essentially, the question of how rapidly the component $\tau[Y]$ of our "close-to-the-path triple (μ, Y, U)" blows up when μ goes to ∞. In view of the bound (8.61), the latter question, in turn, becomes "how large is $\|Y\|_{\rm Fr}$ when μ is large." The answers to all these questions will be obtained in the three steps to follow.

Step 1. We start with associating with $V \in \operatorname{int} \mathbf{S}^{\widetilde{\nu}}_+$ a norm on the space $\mathbf{S}^{\widetilde{\nu}}$, specifically, the Euclidean norm

$$\begin{aligned}\|H\|_V = \|V^{-1/2}HV^{-1/2}\|_{\rm Fr} &= \sqrt{{\rm Tr}([V^{-1/2}HV^{-1/2}]^2)}\\ &= \sqrt{{\rm Tr}(V^{-1}HV^{-1}H)}\\ &= \sqrt{\langle H, \mathcal{H}(V)H\rangle}.\end{aligned}$$

Lemma 8.3. *Let $V \in \operatorname{int} \mathbf{S}^{\widetilde{\nu}}_+$. Then the Dikin ellipsoid of V — the set*

$$E(V) = \{W \in \mathbf{S}^{\widetilde{\nu}} : \|W - V\|_V \le 1\}$$

— is contained in $\mathbf{S}^{\widetilde{\nu}}_+$.

Besides this, if $H \in \mathbf{S}^{\widetilde{\nu}}_+$, then

$$\langle \nabla\Phi(V), -H\rangle \ge \|H\|_V. \tag{8.62}$$

Proof. For $W \in \mathbf{S}^{\widetilde{\nu}}$, we have

$$\begin{aligned}\|W - V\|_V = \|V^{-1/2}[W - V]V^{-1/2}\|_{\rm Fr} &= \|[V^{-1/2}WV^{-1/2} - I\|_{\rm Fr}\\ &= \sqrt{\sum_i [\lambda_i(V^{-1/2}WV^{-1/2}) - 1]^2}\end{aligned}$$

If now $\|W - V\|_V \le 1$, then from the latter equality it follows that $\lambda_i(V^{-1/2}WV^{-1/2}) \ge 0$, that is, the matrix $V^{-1/2}WV^{-1/2}$ is positive semidefinite, whence W is so as well. We have proved the first statement of the lemma.

[4]We have used the fact that for a symmetric matrix A, $\|A\|^2_{\rm Fr} = {\rm Tr}(A^2) = \sum_i \lambda_i^2(A)$. The latter equality is readily given by the eigenvalue decomposition: $A = U\,{\rm Diag}\{\lambda(A)\}U^T$ with an orthogonal U, whence ${\rm Tr}(A^2) = {\rm Tr}(U[{\rm Diag}\{\lambda(A)\}]^2U^T) = {\rm Tr}([{\rm Diag}\{\lambda(A)\}]^2) = \sum_i \lambda_i^2(A)$.

The verification of the second statement is immediate:

$$\begin{aligned}\langle \nabla\Phi(V), -H\rangle &= \langle V^{-1}, H\rangle = \mathrm{Tr}(V^{-1}H) = \mathrm{Tr}(V^{-1/2}HV^{-1/2})\\ &= \sum_i \lambda_i(V^{-1/2}HV^{-1/2}) = \|\lambda(V^{-1/2}HV^{-1/2})\|_1,\end{aligned}$$

where the last equality is given by the fact that $\lambda(V^{-1/2}HV^{-1/2}) \geq 0$ due to $H \succeq 0$. On the other hand, $\|H\|_V = \sqrt{\mathrm{Tr}([V^{-1/2}HV^{-1/2}]^2)} = \|\lambda(V^{-1/2}HV^{-1/2})\|_2$. Since $\|h\|_2 \leq \|h\|_1$ for all h, we conclude that $\|H\|_V \leq \langle \nabla\Phi(H), -H\rangle$. □

Step 2. We have the following.

Lemma 8.4. *Let* (μ, Y, U) *be a "close-to-the-path" triple, so that* $\mu > 0$, Y *is strictly feasible for* (Aux), U *is strictly feasible for the dual to* (Aux) *problem* (Aux′) *and*

$$\mathrm{dist}((Y, U), \widetilde{Z}_*(\mu)) \leq \kappa \leq 0.1.$$

Then

$$\begin{aligned}&\max\{\langle -\nabla\Phi(Y), H\rangle : H \in \mathcal{E},\ \|H\|_Y \leq 1\} \geq 1, && \text{(a)}\\ &\max\{\left|\langle \mu^{-1}\widetilde{C} + \nabla\Phi(Y), H\rangle\right| : H \in \mathcal{E},\ \|H\|_Y \leq 1\} \leq \kappa \leq 0.1, && \text{(b)}\end{aligned} \tag{8.63}$$

whence also

$$\max\{\langle -\mu^{-1}\widetilde{C}, H\rangle : H \in \mathcal{E},\ \|H\|_Y \leq 1\} \geq 0.9. \tag{8.64}$$

Proof. Since the feasible set $\mathcal{Y}$ of (Aux) is unbounded, this set admits a nonzero recessive direction H; clearly, $H \in \mathcal{E}$ and $H \succeq 0$. Multiplying H by appropriate positive real, we can ensure also that $\|H\|_Y = 1$. By Lemma 8.3 applied with Y in the role of V, we have $\langle -\nabla\Phi(Y), H\rangle \geq \|H\|_Y = 1$, and (8.63(a)) follows.

To prove (8.63(b)), note that the feasible plane in (Aux′) is $\mathcal{E}^{\perp} + \widetilde{C}$. Recalling what U is, we conclude that $U - \widetilde{C} \in \mathcal{E}^{\perp}$, that is, whenever

$H \in \mathcal{E}$, we have

$$\begin{aligned}\langle \mu^{-1}\widetilde{C} + \nabla\Phi(Y), H\rangle &= \langle \mu^{-1}U + \nabla\Phi(Y), H\rangle = \langle \mu^{-1}U - Y^{-1}, H\rangle \\ &= \mathrm{Tr}([\mu^{-1}U - Y^{-1}]H) = \mathrm{Tr}(Y^{1/2}[\mu^{-1}U - Y^{-1}]HY^{-1/2}) \\ &= \mathrm{Tr}(\left[Y^{1/2}[\mu^{-1}U - Y^{-1}]Y^{1/2}\right]\left[Y^{-1/2}HY^{-1/2}\right]),\end{aligned}$$

whence by the Cauchy inequality

$$\begin{aligned}&|\langle \mu^{-1}\widetilde{C} + \nabla\Phi(Y), H\rangle| \\ &\quad\le \sqrt{\mathrm{Tr}(\left[Y^{1/2}[\mu^{-1}U - Y^{-1}]Y^{1/2}\right]^2)}\sqrt{\mathrm{Tr}([Y^{-1/2}HY^{-1/2}]^2)} \\ &\quad= \sqrt{\mathrm{Tr}(Y^{1/2}[\mu^{-1}U - Y^{-1}]Y[\mu^{-1}U - Y^{-1}]Y^{1/2})}\|H\|_Y \\ &\quad= \sqrt{\mathrm{Tr}(Y[\mu^{-1}U - Y^{-1}]Y[\mu^{-1}U - Y^{-1}])}\|H\|_Y \\ &\quad\underbrace{=}_{(*)} \mathrm{dist}((Y,U); \widetilde{Z}_*(\mu))\|H\|_Y \le \kappa\|H\|_Y \le 0.1\|H\|_Y.\end{aligned}$$

Where $(*)$ is due to (8.11). The resulting inequality holds true for all $H \in \mathcal{E}$, and (8.63(b)) follows.

Relation (8.64) is an immediate consequence of (8.63). □

Step 3. Now, consider the following geometric construction. Given a triple (μ, Y, U) satisfying the premise of Lemma 8.4, let us denote by W^1 the intersection of the Dikin ellipsoid of $\widehat{Y}$ with the feasible plane of (Aux), and by W^μ the intersection of the Dikin ellipsoid of Y with the same feasible plane. Let us also extend the line segment $[\widehat{Y}, Y]$ to the left of $\widehat{Y}$ until it crosses the boundary of W^1 at certain point P. Further, let us choose $H \in \mathcal{E}$ such that $\|H\|_Y = 1$ and

$$\langle -\mu^{-1}\widetilde{C}, H\rangle \ge 0.9,$$

(such an H exists in view of (8.64)) and set

$$M = Y + H; \quad N = \widehat{Y} + \omega H, \quad \omega = \frac{\|\widehat{Y} - P\|_{\mathrm{Fr}}}{\|Y - P\|_{\mathrm{Fr}}}.$$

The cross-section of the entities involved by the 2D plane passing through P, Y, M looks as shown in Fig. 8.2.

A. We claim, first, that *the points* P, M, N *belong to the feasible set* $\mathcal{Y}$ *of* (Aux).

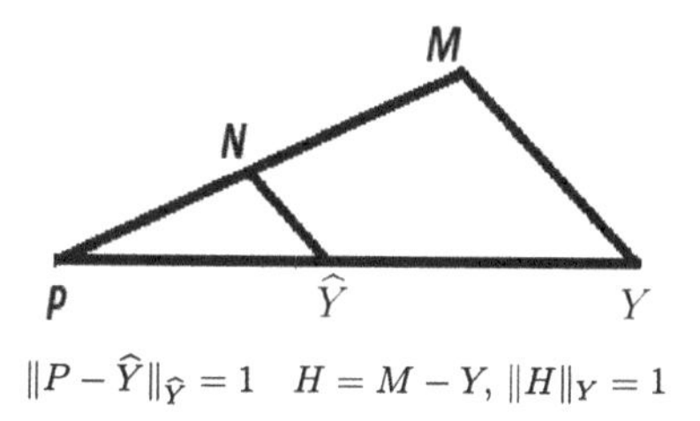

Fig. 8.2. The entities related to Step 3.

Indeed, since the direction $\widehat{Y} - Y$ belongs to $\mathcal{E}$, P belongs to the feasible plane of (Aux), and since by construction P belongs to the Dikin ellipsoid of $\widehat{Y}$, P belongs to $\mathbf{S}_+^{\widetilde{\nu}}$ by Lemma 8.3. Thus, P belongs to $\mathcal{Y}$. By similar reasons, with Y in the role of $\widehat{Y}$, we have $M \in \mathcal{Y}$. Since N is a convex combination of P and M, $N \in \mathcal{Y}$ as well.

B. We further claim that

$$\langle \nabla\Phi(\widehat{Y}), N - \widehat{Y} \rangle = \mu\omega\langle -\mu^{-1}\widetilde{C}, H \rangle \geq 0.9\omega\mu. \tag{8.65}$$

Indeed, the equality in (8.65) is evident, since by construction $\widetilde{C} = -\nabla\Phi(\widehat{Y})$; the inequality is given by the choice of H.

C. We make the following simple and extremely important.

Observation 8.2. Let $X \in \operatorname{int} \mathbf{S}_+^k$, $W \in \mathbf{S}_+^k$ and let $\Psi(\cdot)$ be the log-det barrier for S_+^k. Then

$$\langle \nabla\Psi(X), W - X \rangle \leq k.$$

Verification is immediate:

$$\begin{aligned} \langle \nabla\Psi(X), W - X \rangle &= \langle X^{-1}, X - W \rangle = \operatorname{Tr}(X^{-1}[X - W]) \\ &= \operatorname{Tr}(I_k - X^{-1}W) = k - \langle X^{-1}, W \rangle \leq k, \end{aligned}$$

where the concluding $\leq$ is due to $X^{-1}, W \in \mathbf{S}_+^k$ and the fact that the latter cone is self-dual. □

From A, B, C, it follows that

$$\omega \le \frac{2n+2}{0.9\mu}. \tag{8.66}$$

Now, let

$$\Omega = \frac{0.9 \min\limits_D [\|D\|_{\rm Fr} : \|D\|_{\widehat{Y}} = 1]}{2n+2}, \quad \Omega' = \max_D [\|D\|_{\rm Fr} : \|D - \widehat{Y}\|_{\widehat{Y}} = 1].$$

Observe that Ω and Ω' are positive quantities depending on our "starting point" $\widehat{Y}$ and completely independent of μ. The role of these quantities in our context is clear from the following.

Observation 8.3. Under the premise of Lemma 8.4, one has

$$\|Y\|_{\rm Fr} \ge \Omega\mu - \Omega'. \tag{8.67}$$

Indeed, by definition of ω, we have $\|Y - P\|_{\rm Fr} = \|\widehat{Y} - P\|_{\rm Fr}\omega^{-1}$, whence, by (8.66), $\|Y - P\|_{\rm Fr} \ge 0.9\mu\frac{\|\widehat{Y}-P\|_{\rm Fr}}{2n+2}$, and by the Triangle inequality,

$$\|Y\|_{\rm Fr} \ge 0.9\mu\frac{\|\widehat{Y} - P\|_{\rm Fr}}{2n+2} - \|P\|_{\rm Fr}.$$

It remains to note that by construction $\|\widehat{Y} - P\|_{\widehat{Y}} = 1$, whence $0.9\frac{\|\widehat{Y}-P\|_{\rm Fr}}{2n+2} \ge \Omega$ and $\|P\|_{\rm Fr} \le \Omega'$. □

The result: Combining Observation 8.3 and (8.61), we arrive at the following.

Theorem 8.3. *Whenever a triple (μ, Y, U) is "close-to-the-path" (i.e., satisfies the premise of Lemma 8.4) and $Y = \mathrm{Diag}\{X, S, \sigma, \tau\}$, one has*

$$\tau \ge \Theta\mu - \Theta^{-1} \tag{8.68}$$

with $\Theta > 0$ depending solely on the data of $(\mathcal{P})$. Consequently, when $\mu \ge 2\Theta^{-2}$, the pair $(X_\tau = \tau^{-1}X, S_\tau = \tau^{-1}S)$ satisfies the relations (cf. (C′))

$$\begin{array}{ll} X_\tau \in \mathbf{S}^\nu_+ \cap (\mathcal{L}_P - B + 2\mu^{-1}\Theta^{-1}P) & [\text{"primal } O(1/\mu)\text{-feasibility"}], \\ S_\tau \in \mathbf{S}^\nu_+ \cap (\mathcal{L}_P^\perp + C + 2\mu^{-1}\Theta^{-1}D) & [\text{"dual } O(1/\mu)\text{-feasibility"}], \\ \langle C, X_\tau\rangle - \langle B, S_\tau\rangle \le 2\mu^{-1}\Theta^{-1}d & [\text{"}O(1/\mu)\text{-duality gap"}]. \end{array} \tag{8.69}$$

Discussion. Theorem 8.3 says that in order to get an "ϵ-primal–dual feasible ϵ-optimal" solution to $(\mathcal{P})$, $(\mathcal{D})$, see (8.58), it suffices to trace the primal central path of (Aux), starting at the point $\widehat{Y}$ (path parameter equals 1) until a close-to-the-path point with path parameter $O(1/\epsilon)$ is reached, which requires $O(1)\sqrt{2n+2}\ln\left(\frac{1}{O(\epsilon)}\right)$ iterations. Thus, we arrive at a process with the same complexity characteristics as for the path-following methods described by Theorems 8.1 and 8.2; note, however, that *now we have absolutely no trouble with how to start tracing the path.*

At this point, a careful reader should protest: relations (8.69) do say that when μ is large, X_τ is nearly feasible for $(\mathcal{P})$ and S_τ is nearly feasible for $(\mathcal{D})$; but why do we know that X_τ, S_τ are nearly optimal for the respective problems? What pretends the latter property, is the "$O(\mu^{-1})$-duality gap" relation in (8.69), and indeed, the left-hand side of this inequality looks as the duality gap, while the right-hand side is $O(\mu^{-1})$. But in fact the relation

$$\begin{aligned}\text{DualityGap}(X,S) &\equiv [\langle C,X\rangle - \text{Opt(P)}] + [\text{Opt}(\mathcal{D}) - \langle B,S\rangle] \\ &= \langle C,X\rangle - \langle B,S\rangle^5\end{aligned}$$

is valid only for *primal–dual feasible pairs* (X,S), while our X_τ, S_τ are only $O(\mu^{-1})$-feasible.

Here is the missing element:

Proposition 8.2. *Let the primal–dual pair of problems* $(\mathcal{P})$, $(\mathcal{D})$, *see* (8.58), *be strictly primal–dual feasible and be normalized by* $\langle C,B\rangle = 0$, *let* (X_*, S_*) *be a primal–dual optimal solution to the pair, and let* X, S *"ϵ-satisfy" the feasibility and optimality conditions for* $(\mathcal{P})$, $(\mathcal{D})$, *i.e.,*

$$\begin{aligned} &X \in \mathbf{S}^\nu_+ \cap (\mathcal{L}_P - B + \Delta X),\ \|\Delta X\|_{\text{Fr}} \le \epsilon, && \text{(a)}\\ &S \in \mathbf{S}^\nu_+ \cap (\mathcal{L}_P^\perp + C + \Delta S),\ \|\Delta S\|_{\text{Fr}} \le \epsilon, && \text{(b)}\\ &\langle C,X\rangle - \langle B,S\rangle \le \epsilon. && \text{(c)}\end{aligned}$$

Then

$$\begin{aligned}\langle C,X\rangle - \text{Opt}(\mathcal{P}) &\le \epsilon(1 + \|X_* + B\|_{\text{Fr}}),\\ \text{Opt}(\mathcal{D}) - \langle B,S\rangle &\le \epsilon(1 + \|S_* - C\|_{\text{Fr}}).\end{aligned}$$

Proof. We have $S - C - \Delta S \in \mathcal{L}_P^\perp$, $X_* + B \in \mathcal{L}_P$, whence

$$\begin{aligned} 0 &= \langle S - C - \Delta S, X_* + B \rangle \\ &= \underbrace{\langle S, X_* \rangle}_{\geq 0: X_*, S \in \mathbf{S}^\nu_+ = [\mathbf{S}^\nu_+]^*} - \mathrm{Opt}(\mathcal{P}) + \langle S, B \rangle + \langle -\Delta S, X_* + B \rangle \end{aligned}$$

$$\Rightarrow -\mathrm{Opt}(\mathcal{P}) \leq -\langle S, B \rangle + \langle \Delta S, X_* + B \rangle \leq -\langle S, B \rangle + \epsilon \|X_* + B\|_{\mathrm{Fr}}.$$

Combining the resulting inequality and (c), we get the first of the inequalities to be proved; the second of them is given by "symmetric" reasoning. □

Appendix A

Prerequisites from Linear Algebra

Regarded as mathematical entities, the objective and the constraints in a mathematical programming problem are functions of several real variables; therefore before entering the optimization theory and methods, we need to recall several basic notions and facts about the spaces $\mathbf{R}^n$ where these functions live, same as about the functions themselves. The reader is supposed to know most of the facts to follow, so he/she should not be surprised by a "cooking book" style, which we intend to use in the following.

A.1 Space $\mathbf{R}^n$: Algebraic structure

Basically, all events and constructions to be considered will take place in the *space* $\mathbf{R}^n$ *of* n*-dimensional real vectors*. This space can be described as follows.

A.1.1 *A point in* $\mathbf{R}^n$

A point in $\mathbf{R}^n$ (also called an n*-dimensional vector*) is an ordered collection $x = [x_1; \ldots; x_n]$ of n reals, called the *coordinates*, or *components*, or *entries* of vector x; the space $\mathbf{R}^n$ itself is the set of all collections of this type.

A.1.2 *Linear operations*

$\mathbf{R}^n$ is equipped with two *basic operations*:

- *Addition of vectors*: This operation takes on input two vectors $x = [x_1; \ldots; x_n]$ and $y = [y_1; \ldots; y_n]$ and produces from them a new vector

$$x + y = [x_1 + y_1; \dots ; x_n + y_n]$$

with entries which are sums of the corresponding entries in x and in y.

- *Multiplication of vectors by reals*: This operation takes on input a real λ and an n-dimensional vector $x = [x_1; \dots ; x_n]$ and produces from them a new vector

$$\lambda x = [\lambda x_1; \dots ; \lambda x_n]$$

with entries which are λ times the entries of x.

As far as addition and multiplication by reals are concerned, the arithmetic of $\mathbf{R}^n$ inherits most of the common rules of real arithmetic, such as $x + y = y + x$, $(x + y) + z = x + (y + z)$, $(\lambda + \mu)(x + y) = \lambda x + \mu x + \lambda y + \mu y$, and $\lambda(\mu x) = (\lambda\mu)x$.

A.1.3 *Linear subspaces*

Linear subspaces in $\mathbf{R}^n$ are, by definition, nonempty subsets of $\mathbf{R}^n$ which are closed with respect to addition of vectors and multiplication of vectors by reals:

$$L \subset \mathbf{R}^n \text{ is a linear subspace } \Leftrightarrow \begin{cases} L \neq \emptyset; \\ x, y \in L \Rightarrow x + y \in L; \\ x \in L, \lambda \in \mathbf{R} \Rightarrow \lambda x \in L. \end{cases}$$

A.1.3.A. Examples of linear subspaces

Following are some examples of linear subspaces:

(1) the entire $\mathbf{R}^n$;
(2) the *trivial* subspace containing the single zero vector $0 = [0; \dots ; 0]$[1] (this vector/point is called also *the origin*);

[1]Pay attention to the notation: We use the same symbol 0 to denote the real zero and the n-dimensional vector with all coordinates equal to zero; these two zeros are not the same, and one should understand from the context (it is always very easy) which zero is meant.

(3) the set $\{x \in \mathbf{R}^n : x_1 = 0\}$ of all vectors x with the first coordinate equal to zero;
the latter example admits a natural extension;

(4) the set of all solutions to a *homogeneous* (i.e., with zero right-hand side) system of linear equations

$$\left\{ x \in \mathbf{R}^n : \begin{array}{c} a_{11}x_1 + \cdots + a_{1n}x_n = 0 \\ a_{21}x_1 + \cdots + a_{2n}x_n = 0 \\ \cdots \\ a_{m1}x_1 + \cdots + a_{mn}x_n = 0 \end{array} \right\} \tag{A.1}$$

is always a linear subspace in $\mathbf{R}^n$. This example is "generic," that is, *every* linear subspace in $\mathbf{R}^n$ is the solution set of a (finite) system of homogeneous linear equations, see Proposition A.6;

(5) *Linear span of a set of vectors*: Given a nonempty set X of vectors, one can form a linear subspace $\mathrm{Lin}(X)$, called the *linear span* of X; this subspace consists of all vectors x which can be represented as *linear combinations* $\sum_{i=1}^{N} \lambda_i x_i$ of vectors from X (in $\sum_{i=1}^{N} \lambda_i x_i$, N is an arbitrary positive integer, λ_i are reals and x_i belong to X). Note that

> $\mathrm{Lin}(X)$ *is the smallest w.r.t. inclusion linear subspace which contains* X*: If* L *is a linear subspace such that* $L \supset X$*, then* $L \supset L(X)$ (why?).

The "linear span" example is also generic:

> *Every linear subspace in* $\mathbf{R}^n$ *is the linear span of an appropriately chosen finite set of vectors from* $\mathbf{R}^n$.

(see Theorem A.2(i)).

By definition, the linear span of the empty set of vectors is the trivial linear subspace $\{0\}$: $\mathrm{Lin}(\emptyset) = \{0\}$. This definition is in full accordance with the claim "$\mathrm{Lin}(X)$ is the smallest w.r.t. inclusion linear subspace containing X," same as is in accordance with the standard convention "empty (with empty set of indices) sum of vectors from $\mathbf{R}^n$ has value, namely, 0."

A.1.3.B. Sums and intersections of linear subspaces

Let $\{L_\alpha\}_{\alpha \in I}$ be a family (finite or infinite) of linear subspaces of $\mathbf{R}^n$. From this family, one can build two sets:

(1) *the sum* $\sum_\alpha L_\alpha$ of the subspaces L_α which consists of all vectors which can be represented as finite sums of vectors taken each from its own subspace of the family;
(2) *the intersection* $\bigcap_\alpha L_\alpha$ of the subspaces from the family.

Theorem A.1. *Let $\{L_\alpha\}_{\alpha\in I}$ be a family of linear subspaces of $\mathbf{R}^n$. Then the following holds true:*

(i) *The sum $\sum_\alpha L_\alpha$ of the subspaces from the family is itself a linear subspace of $\mathbf{R}^n$; it is the smallest w.r.t. inclusion of those subspaces of $\mathbf{R}^n$ which contain every subspace from the family.*
(ii) *The intersection $\bigcap_\alpha L_\alpha$ of the subspaces from the family is itself a linear subspace of $\mathbf{R}^n$; it is the largest w.r.t. inclusion of those subspaces of $\mathbf{R}^n$ which are contained in every subspace from the family.*

A.1.4 *Linear independence, bases, dimensions*

A collection $X = \{x^1; \dots; x^N\}$ of vectors from $\mathbf{R}^n$ is called *linearly independent* if no nontrivial (i.e., with at least one nonzero coefficient) linear combination of vectors from X is zero.

Example of linearly independent set: The collection of n *standard basic orths* $e_1 = [1; 0; \dots; 0]$, $e_2 = [0; 1; 0; \dots; 0], \dots, e_n = [0; \dots; 0; 1]$.

Examples of linearly dependent sets: (1) $X = \{0\}$; (2) $X = \{e_1, e_1\}$; (3) $X = \{e_1, e_2, e_1 + e_2\}$.

A collection of vectors $f^1, \dots, f^m$ is called a *basis* in $\mathbf{R}^n$ if

(1) the collection is linearly independent;
(2) every vector from $\mathbf{R}^n$ is a linear combination of vectors from the collection (i.e., $\text{Lin}\{f^1, \dots, f^m\} = \mathbf{R}^n$).

Example of a basis: The collection of standard basic orths $e_1, \dots, e_n$ is a basis in $\mathbf{R}^n$.

Examples of non-bases: (1) The collection $\{e_2, \dots, e_n\}$: This collection is linearly independent, but not every vector is a linear combination of the vectors from the collection. (2) The collection $\{e_1, e_1, e_2, \dots, e_n\}$: Every vector is a linear combination of vectors form the collection, but the collection is not linearly independent.

Besides the bases of the entire $\mathbf{R}^n$, one can speak about the bases of linear subspaces:

A collection $\{f^1, \ldots, f^m\}$ of vectors is called a *basis of a linear subspace* L if

(1) the collection is linearly independent;
(2) $L = \text{Lin}\{f^1, \ldots, f^m\}$, i.e., all vectors f^i belong to L, and every vector from L is a linear combination of the vectors $f^1, \ldots, f^m$.

In order to avoid trivial remarks, it makes sense to agree once for ever the following:

An empty set of vectors is linearly independent, and an empty linear combination of vectors $\sum_{i\in\emptyset} \lambda_i x_i$ equals to zero.

With this convention, the trivial linear subspace $L = \{0\}$ also has a basis, specifically, an empty set of vectors.

Theorem A.2. (i) *Let L be a linear subspace of $\mathbf{R}^n$. Then L admits a (finite) basis, and all bases of L are composed of the same number of vectors; this number is called the dimension of L and is denoted by* $\dim(L)$.

We have seen that $\mathbf{R}^n$ admits a basis composed of n elements (the standard basic orths). From (i), it follows that *every* basis of $\mathbf{R}^n$ contains exactly n vectors, and the dimension of $\mathbf{R}^n$ is n.

(ii) *The larger a linear subspace of $\mathbf{R}^n$, the larger is its dimension: If $L \subset L'$ are linear subspaces of $\mathbf{R}^n$, then $\dim(L) \le \dim(L')$, and the equality takes place if and only if $L = L'$.*

We have seen that the dimension of $\mathbf{R}^n$ is n; according to the above convention, the trivial linear subspace $\{0\}$ of $\mathbf{R}^n$ admits an empty basis, so that its dimension is 0. Since $\{0\} \subset L \subset \mathbf{R}^n$ for every linear subspace L of $\mathbf{R}^n$, it follows from (ii) that the dimension of a linear subspace in $\mathbf{R}^n$ is an integer between 0 and n.

(iii) *Let L be a linear subspace in $\mathbf{R}^n$. Then the following holds true:*

(iii.1) *every linearly independent subset of vectors from L can be extended to a basis of L;*
(iii.2) *From every spanning subset X for L, i.e., a set X such that $\text{Lin}(X) = L$, one can extract a basis of L.*

It follows from (iii) that

- every linearly independent subset of L contains at most $\dim(L)$ vectors, and if it contains exactly $\dim(L)$ vectors, it is a basis of L;

- every spanning set for L contains at least $\dim(L)$ vectors, and if it contains exactly $\dim(L)$ vectors, it is a basis of L.

(iv) *Let L be a linear subspace in $\mathbf{R}^n$, and $f^1, \ldots, f^m$ be a basis in L. Then every vector $x \in L$ admits exactly one representation*

$$x = \sum_{i=1}^{m} \lambda_i(x) f^i$$

as a linear combination of vectors from the basis, and the mapping

$$x \mapsto (\lambda_1(x), \ldots, \lambda_m(x)) : L \to \mathbf{R}^m$$

is a one-to-one mapping of L onto $\mathbf{R}^m$ which is linear, i.e., for every $i = 1, \ldots, m$, one has

$$\begin{aligned} \lambda_i(x+y) &= \lambda_i(x) + \lambda_i(y) \quad \forall (x, y \in L); \\ \lambda_i(\nu x) &= \nu \lambda_i(x) \quad \forall (x \in L, \nu \in \mathbf{R}). \end{aligned} \tag{A.2}$$

The reals $\lambda_i(x)$, $i = 1, \ldots, m$, are called the coordinates of $x \in L$ in the basis $f^1, \ldots, f^m$.

For example, the coordinates of a vector $x \in \mathbf{R}^n$ in the *standard basis* $e_1, \ldots, e_n$ of $\mathbf{R}^n$ — the one composed of the standard basic orths — are exactly the entries of x.

(v) [Dimension formula]: *Let L_1, L_2 be linear subspaces of $\mathbf{R}^n$. Then*

$$\dim(L_1 \cap L_2) + \dim(L_1 + L_2) = \dim(L_1) + \dim(L_2).$$

A.1.5 *Linear mappings and matrices*

A function $\mathcal{A}(x)$ (another name — *mapping*) defined on $\mathbf{R}^n$ and taking values in $\mathbf{R}^m$ is called *linear* if it preserves linear operations:

$$\begin{aligned} \mathcal{A}(x+y) &= \mathcal{A}(x) + \mathcal{A}(y) \quad \forall (x, y \in \mathbf{R}^n); \\ \mathcal{A}(\lambda x) &= \lambda \mathcal{A}(x) \quad \forall (x \in \mathbf{R}^n, \lambda \in \mathbf{R}). \end{aligned}$$

It is immediately seen that a linear mapping from $\mathbf{R}^n$ to $\mathbf{R}^m$ can be represented as multiplication by an $m \times n$ matrix:

$$\mathcal{A}(x) = Ax,$$

and this matrix is uniquely defined by the mapping: the columns A_j of A are just the images of the standard basic orths e_j under the mapping $\mathcal{A}$:

$$A_j = \mathcal{A}(e_j).$$

Linear mappings from $\mathbf{R}^n$ into $\mathbf{R}^m$ can be added to each other:

$$(\mathcal{A} + \mathcal{B})(x) = \mathcal{A}(x) + \mathcal{B}(x)$$

and multiplied by reals:

$$(\lambda \mathcal{A})(x) = \lambda \mathcal{A}(x),$$

and the results of these operations are again linear mappings from $\mathbf{R}^n$ to $\mathbf{R}^m$. The addition of linear mappings and multiplication of these mappings by reals correspond to the same operations with the matrices representing the mappings: adding/multiplying by reals mappings, we add, respectively, multiply by reals the corresponding matrices.

Given two linear mappings $\mathcal{A}(x) : \mathbf{R}^n \to \mathbf{R}^m$ and $\mathcal{B}(y) : \mathbf{R}^m \to \mathbf{R}^k$, we can build their superposition

$$\mathcal{C}(x) \equiv \mathcal{B}(\mathcal{A}(x)) : \mathbf{R}^n \to \mathbf{R}^k,$$

which is again a linear mapping, now from $\mathbf{R}^n$ to $\mathbf{R}^k$. In the language of matrices representing the mappings, the superposition corresponds to matrix multiplication: the $k \times n$ matrix C representing the mapping $\mathcal{C}$ is the product of the matrices representing $\mathcal{A}$ and $\mathcal{B}$:

$$\mathcal{A}(x) = Ax, \ \mathcal{B}(y) = By \Rightarrow \mathcal{C}(x) \equiv \mathcal{B}(\mathcal{A}(x)) = B \cdot (Ax) = (BA)x.$$

Important convention: When speaking about adding n-dimensional vectors and multiplying them by reals, it is absolutely unimportant whether we treat the vectors as the column ones, or the row ones, or write down the entries in rectangular tables, or something else. However, when matrix operations (matrix–vector multiplication, transposition, etc.) become involved, it is important whether we treat our vectors as columns, as rows, or as something else. For the sake of definiteness, *from now on, we treat all vectors as column ones*, independent of how we refer to them in the text. For example, when saying for the first time what a vector is, we wrote $x = [x_1; \dots; x_n]$, which might suggest that we were speaking about row vectors. We stress that it is *not* the case, and the only reason for using the "MATLAB notation" $x = [x_1; \dots; x_n]$ (bracketed row of entries separated by semicolons) instead of the "correct" one $x = \begin{bmatrix} x_1 \\ \vdots \\ x_n \end{bmatrix}$ is to save space and to avoid ugly formulas like $f\left(\begin{bmatrix} x_1 \\ \vdots \\ x_n \end{bmatrix}\right)$ when speaking about functions with vector arguments. After we have agreed that *there is no such thing as a row vector in this book*, we can use (and do use) without any harm whatever notation we want.

A.1.6 *Determinant and rank*

We are about to remind the basic facts about ranks and determinants of matrices.

A.1.6.1 *Determinant*

Let $A = [a_{ij}]_{\substack{1 \le i \le n, \\ 1 \le j \le n}}$ be a square matrix. A *diagonal* in A is the collection of n cells with indices $(1, j_1), (2, j_2), \dots, (n, j_n)$, where $j_1, j_2, \dots, j_n$ are distinct from each other, so that the mapping $\sigma : \{1, \dots, n\} \to \{1, \dots, n\}$ given by $\sigma(i) = j_i$, $1 \le i \le n$, is a *permutation* — a one-to-one mapping of the set $\{1, \dots, n\}$ onto itself. There are $n!$ different permutations of $\{1, \dots, n\}$, and they form group Σ_n with group operation — the product – $(\sigma^1 \sigma^2)(i) = \sigma^1(\sigma^2(i))$, $1 \le i \le n$. Permutations $\sigma \in \Sigma_n$ can be assigned *signs* $\mathrm{sign}(\sigma) \in \{\pm 1\}$ in such a way that the sign of the identity permutation $\sigma(i) \equiv i$ is $+1$, the sign of the product of two permutations is the product of their signs: $\mathrm{sign}(\sigma^1 \sigma^2) = \mathrm{sign}(\sigma^1)\,\mathrm{sign}(\sigma^2)$ for all σ^1, σ^2, and the sign of a *transposition* — permutation swapping two distinct indexes and keeping intact all other indexes — is (-1); these properties specify the sign in a unique fashion.

Now, let $A = [a_{ij}]_{1 \le i,j \le n}$ be an $n \times n$ real or complex matrix. The quantity

$$\mathrm{Det}\,(A) = \sum_{\sigma \in \Sigma_n} \mathrm{sign}(\sigma) \prod_{i=1}^{n} a_{i\sigma(i)}$$

is called the *determinant* of A. The main properties of determinant are as follows:

(1) $\text{Det}\,(A) = \text{Det}\,(A^T)$.
(2) [Polylinearity] $\text{Det}\,(A)$ is linear in rows of A: when all but one rows are fixed, $\text{Det}\,(A)$ is a linear function of the remaining row and similarly for columns.
(3) [Antisymmetry] When swapping rows with two distinct indexes, the determinant is multiplied by (-1) and similarly for columns.
(4) $\text{Det}\,(I_n) = 1$.
Note: The last three properties uniquely define $\text{Det}\,(\cdot)$.
(5) [Multiplicativity] For two $n \times n$ matrices A, B, one has $\text{Det}\,(AB) = \text{Det}\,(A)\,\text{Det}\,(B)$.
(6) An $n \times n$ matrix A is nonsingular, that is, $AB = I_n$ for properly selected B if and only if $\text{Det}\,(A) \neq 0$.
(7) [Cramer's rule] For a nonsingular $n \times n$ matrix A, linear system $Ax = b$ in variables x has unique solution with entries given by

$$x_i = \frac{\text{Det}_i}{\text{Det}},\ 1 \leq i \leq n,$$

where $\text{Det} = \text{Det}\,(A)$ and Det_i is the determinant of the matrix obtained from A by replacing the ith column with the right-hand side vector b.
(8) [Decomposition] Let A be an $n \times n$ matrix, $n > 1$. Then for every $i \in \{1, \ldots, n\}$, one has

$$\text{Det}\,(A) = \sum_{j=1}^{n} a_{ij} C^{ij},$$

where the *algebraic complement* C^{ij} of (i,j)th entry in A is $(-1)^{i+j}\,\text{Det}\,(A^{ij})$, A^{ij} being the $(n-1) \times (n-1)$ matrix obtained from A by eliminating ith row and jth column.
(9) An affine mapping $x \mapsto Ax + b : \mathbf{R}^n \to \mathbf{R}^n$ multiplies n-dimensional volumes by $|\,\text{Det}\,(A)|$. In particular, for a real $n \times n$ matrix A, the quantity $|\,\text{Det}\,(A)|$ is the n-dimensional volume of the parallelotope

$$X = \left\{ x = \sum_j s_j A_j : 0 \leq s_j \leq 1, j \leq n \right\}$$

spanned by the columns $A_1, \ldots, A_n$ of A.

A.1.6.2 *Rank*

Let A be an $m \times n$ matrix.

Let $R_k = \{i_1 < i_2 < \cdots < i_p\}$ be a collection of $p \geq 1$ distinct indexes of rows of A, and $C_q = \{j_1 < j_2 < \cdots < j_q\}$ be a collection of $q \geq 1$ distinct indexes of columns of A. The $p \times q$ matrix with entries $B_{k\ell} = A_{i_k, j_\ell}$, $1 \leq k \leq p, 1 \leq \ell \leq q$ is called submatrix of A with row indexes R_p and column indexes C_q; this is what you get from 2D array A in the intersection of rows with indexes from R_p and columns with indexes from C_q.

The *rank* Rank(A) of A is, by definition, the largest of the row, or, which is the same, the column sizes of *nonsingular* square submatrices of A; when no such submatrices exist, that is, when A is the zero matrix, the rank of A by definition is 0.

The main properties of rank are as follows:

(1) Rank(A) is the dimension of the *image space* ImA — the linear span of columns of A or, which is the same, the subspace of the destination space $\mathbf{R}^m$ composed of vectors representable as Ax, $x \in \mathbf{R}^n$; equivalently, Rank(A) is the maximum of cardinalities of linearly independent collections of columns of A. Moreover, a collection of columns of A with Rank(A) distinct indexes is linearly independent if and only if its intersection with the collection of Rank(A) properly selected rows is a nonsingular submatrix of A.

(2) Rank(A) = Rank(A^T), so that Rank(A) is the maximum of cardinalities of linearly independent collections of rows of A. Rank(A) is the codimension of the kernel (a.k.a. nullspace) Ker $A = \{x : Ax = 0\}$ of A:

$$\dim \text{Ker}\, A = n - \text{Rank}(A) \qquad [A : m \times n].$$

A collection of rows of A with Rank(A) distinct indexes is linearly independent if and only if its intersection with the collection of Rank(A) properly selected columns is a nonsingular submatrix of A.

Note: Ker(A) is the orthogonal complement of ImA^T.

(3) Whenever product AB of two matrices makes sense, one has

$$\text{Rank}(AB) \leq \min[\text{Rank}(A), \text{Rank}(B)].$$

For two matrices A, B of the same sizes, one has

$$|\,\mathrm{Rank}(A) - \mathrm{Rank}(B)| \leq \mathrm{Rank}(A + B) \leq \mathrm{Rank}(A) + \mathrm{Rank}(B).$$

(4) An $n \times n$ matrix B is nonsingular if and only if $\mathrm{Rank}(B) = n$.

A.2 Space $\mathbf{R}^n$: Euclidean structure

So far, we were interested solely in the algebraic structure of $\mathbf{R}^n$ or, which is the same, in the properties of the *linear* operations (addition of vectors and multiplication of vectors by scalars) the space is endowed with. Now, let us consider another structure on $\mathbf{R}^n$ — the *standard Euclidean structure* — which allows us to speak about distances, angles, convergence, etc., and thus makes the space $\mathbf{R}^n$ a much richer mathematical entity.

A.2.1 *Euclidean structure*

The standard Euclidean structure on $\mathbf{R}^n$ is given by the *standard inner product* — an operation which takes on input two vectors x, y and produces from them a real, specifically, the real

$$\langle x, y \rangle \equiv x^T y = \sum_{i=1}^{n} x_i y_i.$$

The basic properties of the inner product are as follows:

(1) [Bi-linearity]: The real-valued function $\langle x, y \rangle$ of two vector arguments $x, y \in \mathbf{R}^n$ is linear with respect to every one of the arguments, the other argument being fixed:

$$\langle \lambda u + \mu v, y \rangle = \lambda \langle u, y \rangle + \mu \langle v, y \rangle \quad \forall (u, v, y \in \mathbf{R}^n, \lambda, \mu \in \mathbf{R}),$$
$$\langle x, \lambda u + \mu v \rangle = \lambda \langle x, u \rangle + \mu \langle x, v \rangle \quad \forall (x, u, v \in \mathbf{R}^n, \lambda, \mu \in \mathbf{R}).$$

(2) [Symmetry]: The function $\langle x, y \rangle$ is symmetric:

$$\langle x, y \rangle = \langle y, x \rangle \quad \forall (x, y \in \mathbf{R}^n).$$

(3) [Positive definiteness]: The quantity $\langle x, x \rangle$ always is nonnegative, and it is zero if and only if x is zero.

Remark A.1. The three outlined properties — bi-linearity, symmetry and positive definiteness — form a definition of an Euclidean inner product, and there are infinitely many different ways to satisfy these properties; in other words, there are infinitely many different Euclidean inner products on $\mathbf{R}^n$. The standard inner product $\langle x, y\rangle = x^T y$ is just a particular case of this general notion. Although in the sequel, we normally work with the standard inner product, the reader should remember that the facts we are about to recall are valid for all Euclidean inner products and not only for the standard one.

The notion of an inner product underlies a number of purely algebraic constructions, in particular, those of *inner product representation of linear forms* and of *orthogonal complement.*

A.2.2 *Inner product representation of linear forms on* $\mathbf{R}^n$

A *linear form* on $\mathbf{R}^n$ is a real-valued function $f(x)$ on $\mathbf{R}^n$ which is additive ($f(x+y) = f(x)+f(y)$) and homogeneous ($f(\lambda x) = \lambda f(x)$)

Example of linear form: $f(x) = \sum_{i=1}^n ix_i$.
Examples of nonlinear functions: (1) $f(x) = x_1 + 1$; (2) $f(x) = x_1^2 - x_2^2$; (3) $f(x) = \sin(x_1)$.

When adding/multiplying by reals linear forms, we again get linear forms (scientifically speaking, "linear forms on $\mathbf{R}^n$ form a linear space"). *Euclidean structure allows us to identify linear forms on* $\mathbf{R}^n$ *with vectors from* $\mathbf{R}^n$.

Theorem A.3. *Let* $\langle\cdot,\cdot\rangle$ *be a Euclidean inner product on* $\mathbf{R}^n$.

(i) *Let* $f(x)$ *be a linear form on* $\mathbf{R}^n$. *Then there exists a uniquely defined vector* $f \in \mathbf{R}^n$ *such that the form is just the inner product with* f :

$$f(x) = \langle f, x\rangle \quad \forall x.$$

(ii) *Vice versa, every vector* $f \in \mathbf{R}^n$ *defines, via the formula*

$$f(x) \equiv \langle f, x\rangle,$$

a linear form on $\mathbf{R}^n$.

(iii) *The above one-to-one correspondence between the linear forms and vectors on* $\mathbf{R}^n$ *is linear: adding linear forms (or multiplying a linear form by a real), we add (respectively, multiply by the real) the vector(s) representing the form(s).*

A.2.3 *Orthogonal complement*

An Euclidean structure allows us to associate with a linear subspace $L \subset \mathbf{R}^n$ another linear subspace $L^\perp$ — the *orthogonal complement* (or the *annulator*) of L; by definition, $L^\perp$ consists of all vectors which are orthogonal to every vector from L:

$$L^\perp = \{f : \langle f, x \rangle = 0 \quad \forall x \in L\}.$$

Theorem A.4. (i) *Twice taken, orthogonal complement recovers the original subspace: whenever* L *is a linear subspace of* $\mathbf{R}^n$, *one has*

$$(L^\perp)^\perp = L.$$

(ii) *The larger a linear subspace* L, *the smaller is its orthogonal complement: If* $L_1 \subset L_2$ *are linear subspaces of* $\mathbf{R}^n$, *then* $L_1^\perp \supset L_2^\perp$.

(iii) *The intersection of a subspace and its orthogonal complement is trivial, and the sum of these subspaces is the entire* $\mathbf{R}^n$:

$$L \cap L^\perp = \{0\}, \quad L + L^\perp = \mathbf{R}^n.$$

Remark A.2. From Theorem A.4(iii) and the dimension formula (Theorem A.2(v)), it follows, first, that for every subspace L in $\mathbf{R}^n$, one has

$$\dim(L) + \dim(L^\perp) = n.$$

Second, every vector $x \in \mathbf{R}^n$ admits a unique decomposition

$$x = x_L + x_{L^\perp}$$

into a sum of two vectors: the first of them, x_L, belongs to L, and the second, $x_{L^\perp}$, belongs to $L^\perp$. This decomposition is called the *orthogonal decomposition* of x *taken with respect to* $L, L^\perp$; x_L is called the *orthogonal projection* of x onto L, and $x_{L^\perp}$ — the orthogonal projection of x onto the orthogonal complement of L. Both projections depend on x linearly, for example,

$$(x+y)_L = x_L + y_L, \quad (\lambda x)_L = \lambda x_L.$$

The mapping $x \mapsto x_L$ is called the *orthogonal projector* onto L.

A.2.4 *Orthonormal bases*

A collection of vectors $f^1, \ldots, f^m$ is called *orthonormal* w.r.t. Euclidean inner product $\langle \cdot, \cdot \rangle$ if distinct vector from the collection are orthogonal to each other:

$$i \neq j \Rightarrow \langle f^i, f^j \rangle = 0$$

and inner product of every vector f^i with itself is unit:

$$\langle f^i, f^i \rangle = 1, \quad i = 1, \ldots, m.$$

Theorem A.5. (i) *An orthonormal collection $f^1, \ldots, f^m$ is always linearly independent and is therefore a basis of its linear span* $L = \mathrm{Lin}(f^1, \ldots, f^m)$ (such a basis in a linear subspace is called *orthonormal*). *The coordinates of a vector $x \in L$ w.r.t. an orthonormal basis $f^1, \ldots, f^m$ of L are given by explicit formulas:*

$$x = \sum_{i=1}^{m} \lambda_i(x) f^i \quad \Leftrightarrow \quad \lambda_i(x) = \langle x, f^i \rangle.$$

Example of an orthonormal basis in $\mathbf{R}^n$: The standard basis $\{e_1, \ldots, e_n\}$ is orthonormal *with respect to the standard inner product* $\langle x, y \rangle = x^T y$ on $\mathbf{R}^n$ (but is not orthonormal w.r.t. other Euclidean inner products on $\mathbf{R}^n$).

Proof of (i). Taking inner product of both sides in the equality

$$x = \sum_j \lambda_j(x) f^j$$

with f^i, we get

$$\begin{aligned}
\langle x, f_i \rangle &= \left\langle \sum_j \lambda_j(x) f^j, f^i \right\rangle \\
&= \sum_j \lambda_j(x) \langle f^j, f^i \rangle \quad \text{[bilinearity of inner product]} \\
&= \lambda_i(x) \quad \text{[orthonormality of } \{f^i\}\text{]}.
\end{aligned}$$

Similar computation demonstrates that if 0 is represented as a linear combination of f^i with certain coefficients λ_i, then $\lambda_i = \langle 0, f^i \rangle = 0$, i.e., all the coefficients are zero; this means that an orthonormal system is linearly independent.

(ii) *If $f^1, \ldots, f^m$ is an orthonormal basis in a linear subspace L, then the inner product of two vectors $x, y \in L$ in the coordinates $\lambda_i(\cdot)$ w.r.t. this basis is given by the standard formula*

$$\langle x, y \rangle = \sum_{i=1}^{m} \lambda_i(x)\lambda_i(y).$$

Proof.

$$\begin{aligned} x &= \sum_i \lambda_i(x) f^i, \ y = \sum_i \lambda_i(y) f^i \\ \Rightarrow \langle x, y \rangle &= \left\langle \sum_i \lambda_i(x) f^i, \sum_i \lambda_i(y) f^i \right\rangle \\ &= \sum_{i,j} \lambda_i(x)\lambda_j(y)\langle f^i, f^j \rangle \qquad \text{[bilinearity of inner product]} \\ &= \sum_i \lambda_i(x)\lambda_i(y) \qquad \text{[orthonormality of } \{f^i\}\text{].} \end{aligned}$$

(iii) *Every linear subspace L of $\mathbf{R}^n$ admits an orthonormal basis; moreover, every orthonormal system $f^1, \ldots, f^m$ of vectors from L can be extended to an orthonormal basis in L.*

Important corollary: *All Euclidean spaces of the same dimension are "the same." Specifically, if L is an m-dimensional space in a space $\mathbf{R}^n$ equipped with an Euclidean inner product $\langle \cdot, \cdot \rangle$, then there exists a one-to-one mapping $x \mapsto A(x)$ of L onto $\mathbf{R}^m$ such that*

- *the mapping preserves linear operations:*

$$A(x+y) = A(x) + A(y) \quad \forall (x, y \in L); A(\lambda x) = \lambda A(x) \quad \forall (x \in L, \lambda \in \mathbf{R});$$

- *the mapping converts the $\langle \cdot, \cdot \rangle$ inner product on L into the standard inner product on $\mathbf{R}^m$:*

$$\langle x, y \rangle = (A(x))^T A(y) \quad \forall x, y \in L.$$

Indeed, by (iii), L admits an orthonormal basis $f^1, \ldots, f^m$; using (ii), one can immediately check that the mapping

$$x \mapsto A(x) = [\lambda_1(x); \ldots; \lambda_m(x)],$$

which maps $x \in L$ into the m-dimensional vector composed of the coordinates of x in the basis $f^1, \ldots, f^m$, meets all the requirements.

Proof of (iii) is given by important by its own right Gram–Schmidt orthogonalization process as follows. We start with an arbitrary basis $h^1, \ldots, h^m$ *in* L *and step by step convert it into an orthonormal basis* $f^1, \ldots, f^m$*. At the beginning of a step* t *of the construction, we already have an orthonormal collection* $f^1, \ldots, f^{t-1}$ *such that* $\mathrm{Lin}\{f^1, \ldots, f^{t-1}\} = \mathrm{Lin}\{h^1, \ldots, h^{t-1}\}$*. At a step* t*, we note the following:*

(1) *Build the vector*

$$g^t = h^t - \sum_{j=1}^{t-1} \langle h^t, f^j \rangle f^j.$$

It is easily seen (check it!) that

(a) *one has*

$$\mathrm{Lin}\{f^1, \ldots, f^{t-1}, g^t\} = \mathrm{Lin}\{h^1, \ldots, h^t\}; \tag{A.3}$$

(b) $g^t \neq 0$ *(derive this fact from* (A.3) *and the linear independence of the collection* $h^1, \ldots, h^m$*);*

(c) g^t *is orthogonal to* $f^1, \ldots, f^{t-1}$*.*

(2) *Since* $g^t \neq 0$*, the quantity* $\langle g^t, g^t \rangle$ *is positive (positive definiteness of the inner product), so that the vector*

$$f^t = \frac{1}{\sqrt{\langle g^t, g^t \rangle}} g^t$$

is well defined. It is immediately seen (check it!) that the collection $f^1, \ldots, f^t$ *is orthonormal and*

$$\mathrm{Lin}\{f^1, \ldots, f^t\} = \mathrm{Lin}\{f^1, \ldots, f^{t-1}, g^t\} = \mathrm{Lin}\{h^1, \ldots, h^t\}.$$

Step t *of the orthogonalization process is completed.*

After m *steps of the optimization process, we end up with an orthonormal system* $f^1, \ldots, f^m$ *of vectors from* L *such that*

$$\mathrm{Lin}\{f^1, \ldots, f^m\} = \mathrm{Lin}\{h^1, \ldots, h^m\} = L,$$

so that $f^1, \ldots, f^m$ *is an orthonormal basis in* L*.*

The construction can be easily modified (do it!) to extend a given orthonormal system of vectors from L *to an orthonormal basis of* L*.*

A.3 Affine subspaces in $\mathbf{R}^n$

Many of the events to come will take place not in the entire $\mathbf{R}^n$, but in its *affine subspaces* which, geometrically, are planes of different dimensions in $\mathbf{R}^n$. Let us become acquainted with these subspaces.

A.3.1 *Affine subspaces and affine hulls*

Definition of an affine subspace. Geometrically, a linear subspace L of $\mathbf{R}^n$ is a special plane — the one passing through the origin of the space (i.e., containing the zero vector). To get an arbitrary plane M, it suffices to subject an appropriate special plane L to a translation — to add to all points from L a fixed *shifting vector* a. This geometric intuition leads to the following.

Definition A.1 (Affine subspace). An affine subspace (a plane) in $\mathbf{R}^n$ is a set of the form

$$M = a + L := \{y = a + x : x \in L\}, \tag{A.4}$$

where L is a linear subspace in $\mathbf{R}^n$ and a is a vector from $\mathbf{R}^n$, e.g., shifting the linear subspace L composed of vectors with zero first entry by a vector $a = [a_1; \ldots ; a_n]$, we get the set $M = a + L$ of all vectors x with $x_1 = a_1$; according to our terminology, this is an affine subspace.

Immediate question about the notion of an affine subspace is: What are the "degrees of freedom" in decomposition (A.4) — how "strict" M determines a and L? The answer is as follows:

Proposition A.1. *The linear subspace L in decomposition* (A.4) *is uniquely defined by M and is the set of all differences of the vectors from M:*

$$L = M - M := \{x - y : x, y \in M\}. \tag{A.5}$$

The shifting vector a is not uniquely defined by M and can be chosen as an arbitrary vector from M.

A.3.2 *Intersections of affine subspaces, affine combinations and affine hulls*

An immediate conclusion of Proposition A.1 is as follows.

Corollary A.1. *Let $\{M_\alpha\}$ be an arbitrary family of affine subspaces in $\mathbf{R}^n$, and assume that the set $M = \cap_\alpha M_\alpha$ is nonempty. Then M_α is an affine subspace.*

From Corollary A.1, it immediately follows that for every nonempty subset Y of $\mathbf{R}^n$, there exists the smallest w.r.t. inclusion affine subspace containing Y — the intersection of all affine subspaces containing Y. This smallest affine subspace containing Y is called the *affine hull* of Y (notation: $\mathrm{Aff}(Y)$).

All this resembles the story about linear spans a lot. Can we further extend this analogy and get a description of the affine hull $\mathrm{Aff}(Y)$ in terms of elements of Y similar to the one of the linear span ("linear span of X is the set of all linear combinations of vectors from X")? Sure we can!

Let us choose somehow a point $y_0 \in Y$, and consider the set

$$X = Y - y_0 := \{y - y_0 : y \in Y\}.$$

All affine subspaces containing Y should also contain y_0 and therefore, by Proposition A.1, can be represented as $M = y_0 + L$, L being a linear subspace. It is absolutely evident that an affine subspace $M = y_0 + L$ contains Y if and only if the subspace L contains X and that the larger the L, the larger is the M:

$$L \subset L' \Rightarrow M = y_0 + L \subset M' = y_0 + L'.$$

Thus, to find the smallest among *affine subspaces containing* Y, it suffices to find the smallest among the *linear subspaces containing* X and to translate the latter space by y_0:

$$\mathrm{Aff}(Y) = y_0 + \mathrm{Lin}(X) = y_0 + \mathrm{Lin}(Y - y_0). \tag{A.6}$$

Now, we know what $\mathrm{Lin}(Y - y_0)$ is — this is a set of all linear combinations of vectors from $Y - y_0$, so that a generic element of $\mathrm{Lin}(Y - y_0)$ is

$$x = \sum_{i=1}^{k} \mu_i (y_i - y_0) \quad [k \text{ may depend of } x],$$

with $y_i \in Y$ and real coefficients μ_i. It follows that the generic element of $\mathrm{Aff}(Y)$ is

$$y = y_0 + \sum_{i=1}^{k} \mu_i (y_i - y_0) = \sum_{i=0}^{k} \lambda_i y_i,$$

where

$$\lambda_0 = 1 - \sum_i \mu_i, \ \lambda_i = \mu_i, \ i \geq 1.$$

We see that a generic element of $\mathrm{Aff}(Y)$ is a linear combination of vectors from Y. Note, however, that the coefficients λ_i

in this combination are not completely arbitrary: their sum is equal to 1. Linear combinations of this type — with the unit sum of coefficients — have a special name: they are called *affine combinations.*

We have seen that every vector from $\text{Aff}(Y)$ is an affine combination of vectors of Y. Is the inverse true, i.e., does $\text{Aff}(Y)$ contain all affine combinations of vectors from Y? The answer is positive. Indeed, if

$$y = \sum_{i=1}^{k} \lambda_i y_i$$

is an affine combination of vectors from Y, then, using the equality $\sum_i \lambda_i = 1$, we can write it also as

$$y = y_0 + \sum_{i=1}^{k} \lambda_i (y_i - y_0),$$

y_0 being the "marked" vector we used in our previous reasoning, and the vector of this form, as we already know, belongs to $\text{Aff}(Y)$. Thus, we come to the following.

Proposition A.2 (Structure of affine hull).

$$\text{Aff}(Y) = \{\textit{the set of all affine combinations of vectors from } Y\}.$$

When Y itself is an affine subspace, it, of course, coincides with its affine hull, and the above proposition leads to the following.

Corollary A.2. *An affine subspace M is closed with respect to taking affine combinations of its members — every combination of this type is a vector from M. Vice versa, a nonempty set which is closed with respect to taking affine combinations of its members is an affine subspace.*

A.3.3 *Affinely spanning sets, affinely independent sets, affine dimension*

Affine subspaces are closely related to linear subspaces, and the basic notions associated with linear subspaces have natural and useful affine analogies. Here, we introduce these notions and discuss their basic properties.

Affinely spanning sets: Let $M = a + L$ be an affine subspace. We say that a subset Y of M is *affinely spanning* for M (we say also that Y spans M affinely or that M is affinely spanned by Y), if $M = \text{Aff}(Y)$ or, which is the same, due to Proposition A.3.2, if every point of M is an affine combination of points from Y. An immediate consequence of the reasoning of the previous section is as follows.

Proposition A.3. *Let $M = a + L$ be an affine subspace and Y be a subset of M, and let $y_0 \in Y$. The set Y affinely spans M – $M = \text{Aff}(Y)$ – if and only if the set*

$$X = Y - y_0$$

spans the linear subspace L: $L = \text{Lin}(X)$.

Affinely independent sets: A linearly independent set $x_1, \ldots, x_k$ is a set such that no nontrivial linear combination of $x_1, \ldots, x_k$ equals to zero. An equivalent definition is given by Theorem A.2(iv): $x_1, \ldots, x_k$ are linearly independent if the coefficients in a linear combination

$$x = \sum_{i=1}^{k} \lambda_i x_i$$

are uniquely defined by the value x of the combination. This equivalent form reflects the essence of the matter — what we indeed need is the uniqueness of the coefficients in expansions. Accordingly, this equivalent form is the prototype for the notion of an affinely independent set: We want to introduce this notion in such a way that the coefficients λ_i in an *affine* combination

$$y = \sum_{i=0}^{k} \lambda_i y_i$$

of "affinely independent" set of vectors $y_0, \ldots, y_k$ would be uniquely defined by y. Nonuniqueness would mean that

$$\sum_{i=0}^{k} \lambda_i y_i = \sum_{i=0}^{k} \lambda'_i y_i$$

for two different collections of coefficients λ_i and λ'_i with unit sums of coefficients; if it is the case, then

$$\sum_{i=0}^{m}(\lambda_i - \lambda'_i)y_i = 0,$$

so that y_i's are linearly dependent and, moreover, there exists a nontrivial zero combination of them with *zero sum of coefficients* (since $\sum_i(\lambda_i - \lambda'_i) = \sum_i \lambda_i - \sum_i \lambda'_i = 1 - 1 = 0$). Our reasoning can be inverted — if there exists a nontrivial linear combination of y_i's with zero sum of coefficients which is zero, then the coefficients in the representation of a vector as an affine combination of y_i's are not uniquely defined. Thus, in order to get uniqueness, we should for sure forbid relations

$$\sum_{i=0}^{k} \mu_i y_i = 0,$$

with nontrivial zero sum coefficients μ_i. Thus, we have motivated the following.

Definition A.2 (Affinely independent set). A collection $y_0, \dots, y_k$ of n-dimensional vectors is called affinely independent if no nontrivial linear combination of the vectors with zero sum of coefficients is zero:

$$\sum_{i=0}^{k} \lambda_i y_i = 0, \ \sum_{i=0}^{k} \lambda_i = 0 \Rightarrow \lambda_0 = \lambda_1 = \cdots = \lambda_k = 0.$$

With this definition, we get the result completely similar to the one of Theorem A.2(iv).

Corollary A.3. *Let $y_0, \dots, y_k$ be affinely independent. Then the coefficients λ_i in an affine combination*

$$y = \sum_{i=0}^{k} \lambda_i y_i \quad \left[\sum_i \lambda_i = 1\right]$$

of the vectors $y_0, \dots, y_k$ are uniquely defined by the value y of the combination.

Verification of affine independence of a collection can be immediately reduced to verification of linear independence of closely related collection.

Proposition A.4. *$k+1$ vectors $y_0, \ldots, y_k$ are affinely independent if and only if the k vectors $(y_1 - y_0), (y_2 - y_0), \ldots, (y_k - y_0)$ are linearly independent.*

From the latter proposition, it follows, e.g., that the collection $0, e_1, \ldots, e_n$ composed of the origin and the standard basic orths is affinely independent. Note that this collection is linearly dependent (as every collection containing zero). You should definitely know the difference between the two notions of independence we deal with: linear independence means that no nontrivial linear combination of the vectors can be zero, while affine independence means that no nontrivial linear combination *from certain restricted class of them* (with zero sum of coefficients) can be zero. Therefore, there are more affinely independent sets than the linearly independent ones: a linearly independent set is for sure affinely independent, but not vice versa.

Affine bases and affine dimension: Propositions A.3.2 and A.3.3 reduce the notions of affine spanning/affine independent sets to the notions of spanning/linearly independent ones. Combined with Theorem A.2, they result in the following analogies of the latter two statements.

Proposition A.5 (Affine dimension). *Let $M = a + L$ be an affine subspace in $\mathbf{R}^n$. Then the following two quantities are finite integers which are equal to each other:*

(i) *minimal # of elements in the subsets of M which affinely span M;*
(ii) *maximal # of elements in affine independent subsets of M.*
The common value of these two integers is by 1 more than the dimension $\dim L$ *of L.*

By definition, the affine dimension of an affine subspace $M = a + L$ is the dimension $\dim L$ *of L. Thus, if M is of affine dimension k, then the minimal cardinality of sets affinely spanning M, same* *as the maximal cardinality of affine independent subsets of M, is $k+1$.*

Theorem A.6 (Affine bases). *Let $M = a + L$ be an affine subspace in $\mathbf{R}^n$.*

A. *Let $Y \subset M$. The following three properties of Y are equivalent:*

   (i) *Y is an affine independent set which affinely spans M;*
   (ii) *Y is affine independent and contains $1 + \dim L$ elements;*
   (iii) *Y affinely spans M and contains $1 + \dim L$ elements.*

   A subset Y of M possessing the indicated properties equivalent to each other is called an ***affine basis*** *of M. Affine bases in M are exactly the collections $y_0, \ldots, y_{\dim L}$ such that $y_0 \in M$ and $(y_1 - y_0), \ldots, (y_{\dim L} - y_0)$ is a basis in L.*

B. *Every affinely independent collection of vectors of M either itself is an affine basis of M or can be extended to such a basis by adding new* **vectors***. In particular, there exists affine basis of M.*

C. *Given a set Y which affinely spans M, you can always extract from this set an affine basis of M.*

We already know that the standard basic orths $e_1, \ldots, e_n$ form a basis of the entire space $\mathbf{R}^n$. What about affine bases in $\mathbf{R}^n$? According to Theorem A.6.A, you can choose as such a basis a collection $e_0, e_0 + e_1, \ldots, e_0 + e_n$, e_0 being an arbitrary vector.

Barycentric coordinates: Let M be an affine subspace, and let $y_0, \ldots, y_k$ be an affine basis of M. Since the basis, by definition, affinely spans M, every vector y from M is an affine combination of the vectors of the basis:

$$y = \sum_{i=0}^{k} \lambda_i y_i \quad \left[\sum_{i=0}^{k} \lambda_i = 1\right],$$

and since the vectors of the affine basis are affinely independent, the coefficients of this combination are uniquely defined by y (Corollary A.3). These coefficients are called *barycentric coordinates* of y with respect to the affine basis in question. In contrast to the usual coordinates with respect to a (linear) basis, the barycentric coordinates could not be quite arbitrary: their sum should be equal to 1.

A.3.4 *Dual description of linear subspaces and affine subspaces*

To the moment, we have introduced the notions of linear subspace and affine subspace and have presented a scheme of generating these entities: to obtain, e.g., a linear subspace, you start from an arbitrary nonempty set $X \subset \mathbf{R}^n$ and add to it all linear combinations of the vectors from X. When replacing linear combinations with the affine ones, you get a way to generate affine subspaces.

The just indicated way of generating linear subspaces/affine subspaces resembles the approach of a worker building a house: he starts with the base and then adds to it new elements until the house is ready. There exists, anyhow, an approach of an artist creating a sculpture: he takes something large and then deletes extra parts of it. Is there something like "artist's way" to represent linear subspaces and affine subspaces? The answer is positive and very instructive.

A.3.4.1 *Affine subspaces and systems of linear equations*

Let L be a linear subspace. According to Theorem A.4(i), it is an orthogonal complement, namely, the orthogonal complement to the linear subspace $L^\perp$. Now, let $a_1, \dots, a_m$ be a finite spanning set in $L^\perp$. A vector x which is orthogonal to $a_1, \dots, a_m$ is orthogonal to the entire $L^\perp$ (since every vector from $L^\perp$ is a linear combination of $a_1, \dots, a_m$ and the inner product is bilinear); and of course vice versa, a vector orthogonal to the entire $L^\perp$ is orthogonal to $a_1, \dots, a_m$. We see that

$$L = (L^\perp)^\perp = \left\{x : a_i^T x = 0,\, i = 1, \dots, k\right\}. \tag{A.7}$$

Thus, we get a very important, although simple, proposition.

Proposition A.6 ("Outer" description of a linear subspace). *Every linear subspace L in $\mathbf{R}^n$ is a set of solutions to a homogeneous linear system of equations*

$$a_i^T x = 0, \quad i = 1, \dots, m, \tag{A.8}$$

given by properly chosen m and vectors $a_1, \dots, a_m$.

Proposition A.6 is an "if and only if" statement: as we remember from Example A.1.3A(4), solution set to a homogeneous system of linear equations with n variables is always a linear subspace in $\mathbf{R}^n$.

From Proposition A.6 and the facts we know about the dimension, we can easily derive several important consequences:

- Systems (A.8) which define a given linear subspace L are exactly the systems given by the vectors $a_1, \ldots, a_m$ which span $L^\perp$.[2]
- The smallest possible number m of equations in (A.8) is the dimension of $L^\perp$, i.e., by Remark A.2, is codim $L \equiv n - \dim L$.[3]

Now, an affine subspace M is, by definition, a translation of a linear subspace: $M = a + L$. As we know, vectors x from L are exactly the solutions of certain *homogeneous* system of linear equations

$$a_i^T x = 0, \quad i = 1, \ldots, m.$$

It is absolutely clear that adding to these vectors a fixed vector a, we get exactly the set of solutions to the *inhomogeneous* solvable linear system

$$a_i^T x = b_i \equiv a_i^T a, \quad i = 1, \ldots, m.$$

Vice versa, the set of solutions to a *solvable* system of linear equations

$$a_i^T x = b_i, \quad i = 1, \ldots, m$$

with n variables is the sum of a particular solution to the system and the solution set to the corresponding homogeneous system (the latter set, as we already know, is a linear subspace in $\mathbf{R}^n$), i.e., is an affine subspace. Thus, we get the following.

Proposition A.7 ("Outer" description of an affine subspace). *Every affine subspace $M = a + L$ in $\mathbf{R}^n$ is a set of solutions to a solvable linear system of equations*

$$a_i^T x = b_i, \quad i = 1, \ldots, m, \tag{A.9}$$

given by properly chosen m and vectors $a_1, \ldots, a_m$.

---

[2]The reasoning which led us to Proposition A.6 says that $[a_1, \ldots, a_m$ span $L^\perp]$ $\Rightarrow$ [(A.8) defines L]; now, we claim that the inverse is also true.

[3]To make this statement also true in the extreme case when $L = \mathbf{R}^n$ (i.e., when codim $L = 0$), we from now on make a convention that an *empty* set of equations or inequalities defines, as the solution set, the entire space.

Vice versa, the set of all solutions to a solvable system of linear equations with n variables is an affine subspace in $\mathbf{R}^n$.

The linear subspace L associated with M is exactly the set of solutions of the homogeneous (with the right-hand side set to 0) *version of system* (A.9).

Comment: The "outer" description of a linear subspace/affine subspace — the "artist's" one — is in many cases much more useful than the "inner" description via linear/affine combinations (the "worker's" one), e.g., with the outer description, it is very easy to check whether a given vector belongs or does not belong to a given linear subspace/affine subspace, which is not that easy with the inner one.[4] In fact, both descriptions are "complementary" to each other and work perfectly well in parallel: What is difficult to see with one of them is clear with another. The idea of using "inner" and "outer" descriptions of the entities we meet with — linear subspaces, affine subspaces, convex sets, optimization problems — the general idea of *duality* — is, I would say, the main driving force of convex analysis and optimization, and in the sequel, we would all the time meet with different implementations of this fundamental idea.

A.3.5 *Structure of the simplest affine subspaces*

This small section deals mainly with terminology. According to their dimensions, affine subspaces in $\mathbf{R}^n$ are named as follows:

- *Subspaces of dimension 0*: These subspaces are translations of the only zero-dimensional linear subspace $\{0\}$, i.e., are singleton sets — vectors from $\mathbf{R}^n$. These subspaces are called *points*; a point is a solution to a square system of linear equations with nonsingular matrix.
- *Subspaces of dimension 1 (lines)*: These subspaces are translations of one-dimensional linear subspaces of $\mathbf{R}^n$. A one-dimensional linear subspace has a single-element basis given by a nonzero vector d and composed of all multiples of this vector. Consequently, line

[4]In principle, it is not difficult to certify that a given point belongs to, say, a linear subspace given as the linear span of some set — it suffices to point out a representation of the point as a linear combination of vectors from the set. But how could you certify that the point does *not* belong to the subspace?

is a set of the form

$$\{y = a + td : t \in \mathbf{R}\},$$

given by a pair of vectors a (the origin of the line) and d (the direction of the line), $d \neq 0$. The origin of the line and its direction are not uniquely defined by the line; you can choose as origin any point on the line and multiply a particular direction by nonzero reals.

In the barycentric coordinates, a line is described as follows:

$$l = \{\lambda_0 y_0 + \lambda_1 y_1 : \lambda_0 + \lambda_1 = 1\} = \{\lambda y_0 + (1 - \lambda) y_1 : \lambda \in \mathbf{R}\},$$

where y_0, y_1 is an affine basis of l; you can choose as such a basis any pair of distinct points on the line.

The "outer" description of a line is as follows: It is the set of solutions to a linear system with n variables and $n - 1$ linearly independent equations.

- Subspaces of dimension > 2 and $< n - 1$ have no special names; sometimes, they are called affine planes of such and such dimension.
- Affine subspaces of dimension $n - 1$, due to important role they play in convex analysis, have a special name — they are called *hyperplanes*. The outer description of a hyperplane is that a hyperplane is the solution set of a *single* linear equation

$$a^T x = b,$$

with nontrivial left-hand side ($a \neq 0$). In other words, a hyperplane is the level set $a(x) = \text{const}$ of a nonconstant linear form $a(x) = a^T x$.
- The "largest possible" affine subspace — the one of dimension n — is unique and is the entire $\mathbf{R}^n$. This subspace is given by an empty system of linear equations.

Appendix B

Prerequisites from Real Analysis

B.1 Space $\mathbf{R}^n$: Metric structure and topology

Euclidean structure on the space $\mathbf{R}^n$ gives rise to a number of extremely important *metric* notions — distances, convergence, etc. For the sake of definiteness, we associate these notions with the standard inner product $x^T y$.

B.1.1 *Euclidean norm and distances*

By positive definiteness, the quantity $x^T x$ is always nonnegative, so that the quantity

$$|x| \equiv \|x\|_2 = \sqrt{x^T x} = \sqrt{x_1^2 + x_2^2 + \cdots + x_n^2}$$

is well defined; this quantity is called the (standard) *Euclidean norm* of vector x (or simply the norm of x) and is treated as the distance from the origin to x. The distance between two arbitrary points $x, y \in \mathbf{R}^n$ is, by definition, the norm $|x - y|$ of the difference $x - y$. The notions we have defined satisfy all basic requirements on the general notions of a norm and distance, specifically the following:

(1) *Positivity of norm: The norm of a vector is always nonnegative; it is zero if and only if the vector is zero:*

$$|x| \geq 0 \quad \forall x; \quad |x| = 0 \Leftrightarrow x = 0.$$

(2) *Homogeneity of norm: When a vector is multiplied by a real, its norm is multiplied by the absolute value of the real:*

$$|\lambda x| = |\lambda| \cdot |x| \quad \forall (x \in \mathbf{R}^n, \lambda \in \mathbf{R}).$$

(3) *Triangle inequality: Norm of the sum of two vectors is $\leq$ the sum of their norms:*

$$|x+y| \leq |x| + |y| \quad \forall (x, y \in \mathbf{R}^n).$$

In contrast to the properties of positivity and homogeneity, which are absolutely evident, the triangle inequality is not trivial and definitely requires a proof. The proof goes through a fact which is extremely important by its own right — the *Cauchy Inequality*, which perhaps is the most frequently used inequality in mathematics:

Theorem B.1 (Cauchy's Inequality). *The absolute value of the inner product of two vectors does not exceed the product of their norms:*

$$|x^T y| \leq |x||y| \quad \forall (x, y \in \mathbf{R}^n)$$

and is equal to the product of the norms if and only if one of the vectors is proportional to the other one:

$$|x^T y| = |x||y| \Leftrightarrow \{\exists \alpha : x = \alpha y \text{ or } \exists \beta : y = \beta x\}.$$

Proof is immediate. We may assume that both x and y are nonzero (otherwise, the Cauchy inequality is clearly equality, and one of the vectors is constant times (specifically, zero times) the other one, as announced in the theorem). Assuming $x, y \neq 0$, consider the function

$$f(\lambda) = (x - \lambda y)^T (x - \lambda y) = x^T x - 2\lambda x^T y + \lambda^2 y^T y.$$

By positive definiteness of the inner product, this function — which is a second-order polynomial — is nonnegative on the entire axis, whence the discriminant of the polynomial

$$(x^T y)^2 - (x^T x)(y^T y)$$

is nonpositive:

$$(x^T y)^2 \leq (x^T x)(y^T y).$$

Taking square roots of both sides, we arrive at the Cauchy inequality. We also see that the inequality is equality if and only if the discriminant of the second-order polynomial $f(\lambda)$ is zero, i.e., if and only if the polynomial has a (multiple) real root; but due to positive definiteness of inner product, $f(\cdot)$ has a root λ if and only if $x = \lambda y$, which proves the second part of theorem. $\square$

From Cauchy's Inequality to the Triangle Inequality: Let $x, y \in \mathbf{R}^n$. Then

$$\begin{aligned}
|x+y|^2 &= (x+y)^T(x+y) && \text{[definition of norm]} \\
&= x^Tx + y^Ty + 2x^Ty && \text{[opening parentheses]} \\
&\le \underbrace{x^Tx}_{|x|^2} + \underbrace{y^Ty}_{|y|^2} + 2|x||y| && \text{[Cauchy's Inequality]} \\
&= (|x|+|y|)^2 \\
\Rightarrow |x+y| &\le |x|+|y|. && \square
\end{aligned}$$

The properties of norm (i.e., of the distance to the origin) we have established induce properties of the distances between pairs of arbitrary points in $\mathbf{R}^n$, specifically the following:

(1) *Positivity of distances*: The distance $|x-y|$ between two points is positive, except for the case when the points coincide ($x=y$), when the distance between x and y is zero.
(2) *Symmetry of distances*: The distance from x to y is the same as the distance from y to x:

$$|x-y| = |y-x|.$$

(3) *Triangle inequality for distances*: For every three points x, y, z, the distance from x to z does not exceed the sum of distances between x and y and between y and z:

$$|z-x| \le |y-x| + |z-y| \quad \forall (x, y, z \in \mathbf{R}^n).$$

B.1.2 *Convergence*

Equipped with distances, we can define the fundamental notion of *convergence of a sequence of vectors*. Specifically, we say that *a sequence $x^1, x^2, \ldots$ of vectors from $\mathbf{R}^n$ converges to a vector $\bar{x}$*, or, equivalently, that *$\bar{x}$ is the limit of the sequence $\{x^i\}$* (notation: $\bar{x} = \lim_{i\to\infty} x^i$) if the distances from $\bar{x}$ to x^i go to 0 as $i \to \infty$:

$$\bar{x} = \lim_{i\to\infty} x^i \Leftrightarrow |\bar{x} - x^i| \to 0, i \to \infty,$$

or, which is the same, for every $\epsilon > 0$, there exists $i = i(\epsilon)$ such that the distance between every point x^i, $i \ge i(\epsilon)$, and $\bar{x}$ does not exceed ϵ:

$$\left\{|\bar{x} - x^i| \to 0, i \to \infty\right\} \Leftrightarrow \left\{\forall \epsilon > 0 \exists i(\epsilon) : i \ge i(\epsilon) \Rightarrow |\bar{x} - x^i| \le \epsilon\right\}.$$

B.1.3 *Closed and open sets*

After we have in our disposal distance and convergence, we can speak about *closed* and *open* sets:

- A set $X \subset \mathbf{R}^n$ is called *closed* if it contains limits of all converging sequences of elements of X:

$$\left\{x^i \in X, x = \lim_{i\to\infty} x^i\right\} \Rightarrow x \in X.$$

- A set $X \subset \mathbf{R}^n$ is called *open* if whenever x belongs to X, all points close enough to x also belong to X:

$$\forall(x \in X)\exists(\delta > 0) : |x' - x| < \delta \Rightarrow x' \in X.$$

  An open set containing a point x is called a *neighbourhood* of x.

  Examples of closed sets: (1) $\mathbf{R}^n$; (2) $\emptyset$; (3) the sequence $x^i = (i, 0, ..., 0)$, $i = 1, 2, 3, ...$; (4) $\{x \in \mathbf{R}^n : \sum_{i=1}^n a_{ij}x_j = 0, i = 1, ..., m\}$ (in other words, a linear subspace in $\mathbf{R}^n$ is always closed, see Proposition A.6); (5) $\{x \in \mathbf{R}^n : \sum_{i=1}^n a_{ij}x_j = b_i, i = 1, ..., m\}$ (in other words, an affine subset of $\mathbf{R}^n$ is always closed, see Proposition A.7); (6) any finite subset of $\mathbf{R}^n$.

  Examples of non-closed sets: (1) $\mathbf{R}^n\backslash\{0\}$; (2) the sequence $x^i = (1/i, 0, ..., 0)$, $i = 1, 2, 3, ...$; (3) $\{x \in \mathbf{R}^n : x_j > 0, j = 1, ..., n\}$; (4) $\{x \in \mathbf{R}^n : \sum_{i=1}^n x_j > 5\}$.

  Examples of open sets: (1) $\mathbf{R}^n$; (2) $\emptyset$; (3) $\{x \in \mathbf{R}^n : \sum_{j=1}^n a_{ij}x_j > b_j$, $i = 1, ..., m\}$; (4) complement of a finite set.

  Examples of non-open sets: (1) A nonempty finite set; (2) the sequence $x^i = (1/i, 0, ..., 0)$, $i = 1, 2, 3, ...$, and the sequence $x^i = (i, 0, 0, ..., 0)$, $i = 1, 2, 3, ...$; (3) $\{x \in \mathbf{R}^n : x_j \geq 0, j = 1, ..., n\}$; (4) $\{x \in \mathbf{R}^n : \sum_{i=1}^n x_j \geq 5\}$.

B.1.4 *Local compactness of* $\mathbf{R}^n$

A fundamental fact about convergence in $\mathbf{R}^n$, which in certain sense is characteristic for this series of spaces, is the following.

Theorem B.2. *From every bounded sequence $\{x^i\}_{i=1}^\infty$ of points from $\mathbf{R}^n$, one can extract a converging subsequence $\{x^{i_j}\}_{j=1}^\infty$. Equivalently, a closed and bounded subset X of $\mathbf{R}^n$ is compact, i.e., a set possessing the following two properties equivalent to each other:*

(i) *From every sequence of elements of X, one can extract a subsequence which converges to certain point of X.*

(ii) *From every open covering of X (i.e., a family $\{U_\alpha\}_{\alpha \in A}$ of open sets such that $X \subset \bigcup_{\alpha \in A} U_\alpha$), one can extract a finite sub-covering, i.e., a finite subset of indices $\alpha_1, \ldots, \alpha_N$ such that $X \subset \bigcup_{i=1}^{N} U_{\alpha_i}$.*

B.2 Continuous functions on $\mathbf{R}^n$

B.2.1 *Continuity of a function*

Let $X \subset \mathbf{R}^n$ and $f(x) : X \to \mathbf{R}^m$ be a function (another name — mapping) defined on X and taking values in $\mathbf{R}^m$.

(1) f is called *continuous at a point* $\bar{x} \in X$ if for every sequence x^i of points of X converging to $\bar{x}$, the sequence $f(x^i)$ converges to $f(\bar{x})$. Equivalent definition:

$f : X \to \mathbf{R}^m$ is continuous at $\bar{x} \in X$ if for every $\epsilon > 0$, there exists $\delta > 0$ such that

$$x \in X, |x - \bar{x}| < \delta \Rightarrow |f(x) - f(\bar{x})| < \epsilon.$$

(2) f is called *continuous on* X if f is continuous at every point from X. Equivalent definition: f preserves convergence: whenever a sequence of points $x^i \in X$ converges to a point $x \in X$, the sequence $f(x^i)$ converges to $f(x)$.

Examples of continuous mappings:

(1) An *affine* mapping

$$f(x) = \begin{bmatrix} \sum_{j=1}^{m} A_{1j} x_j + b_1 \\ \vdots \\ \sum_{j=1}^{m} A_{mj} x_j + b_m \end{bmatrix} \equiv Ax + b : \mathbf{R}^n \to \mathbf{R}^m$$

is continuous on the entire $\mathbf{R}^n$ (and thus on every subset of $\mathbf{R}^n$) (check it!).

(2) The norm $|x|$ is a continuous on $\mathbf{R}^n$ (and thus on every subset of $\mathbf{R}^n$) real-valued function (check it!).

Task B.1.

- Consider the function

$$f(x_1,x_2) = \begin{cases} \frac{x_1^2-x_2^2}{x_1^2+x_2^2}, & (x_1,x_2) \neq 0, \\ 0, & x_1 = x_2 = 0 \end{cases} : \mathbf{R}^2 \to \mathbf{R}.$$

  Check whether this function is continuous on the following sets:
  (1) $\mathbf{R}^2$;
  (2) $\mathbf{R}^2 \backslash \{0\}$;
  (3) $\{x \in \mathbf{R}^2 : x_1 = 0\}$;
  (4) $\{x \in \mathbf{R}^2 : x_2 = 0\}$;
  (5) $\{x \in \mathbf{R}^2 : x_1 + x_2 = 0\}$;
  (6) $\{x \in \mathbf{R}^2 : x_1 - x_2 = 0\}$;
  (7) $\{x \in \mathbf{R}^2 : |x_1 - x_2| \leq x_1^4 + x_2^4\}$.
- Let $f : \mathbf{R}^n \to \mathbf{R}^m$ be a continuous mapping. Mark those of the following statements which always are true:
  (1) If U is an open set in $\mathbf{R}^m$, then so is the set $f^{-1}(U) = \{x : f(x) \in U\}$.
  (2) If U is an open set in $\mathbf{R}^n$, then so is the set $f(U) = \{f(x) : x \in U\}$.
  (3) If F is a closed set in $\mathbf{R}^m$, then so is the set $f^{-1}(F) = \{x : f(x) \in F\}$.
  (4) If F is a closed set in $\mathbf{R}^n$, then so is the set $f(F) = \{f(x) : x \in F\}$.

B.2.2 *Elementary continuity-preserving operations*

All "elementary" operations with mappings preserve continuity. Specifically, we have the following:

Theorem B.3. *Let X be a subset in $\mathbf{R}^n$.*

(i) [Stability of continuity w.r.t. linear operations]: *If $f_1(x), f_2(x)$ are continuous functions on X taking values in $\mathbf{R}^m$ and $\lambda_1(x)$, $\lambda_2(x)$ are continuous real-valued functions on X, then the function*

$$f(x) = \lambda_1(x) f_1(x) + \lambda_2(x) f_2(x) : X \to \mathbf{R}^m$$

is continuous on X;

(ii) [Stability of continuity w.r.t. superposition]: *Let*

- $X \subset \mathbf{R}^n$, $Y \subset \mathbf{R}^m$;
- $f : X \to \mathbf{R}^m$ *be a continuous mapping such that* $f(x) \in Y$ *for every* $x \in X$;
- $g : Y \to \mathbf{R}^k$ *be a continuous mapping.*

Then the composite mapping

$$h(x) = g(f(x)) : X \to \mathbf{R}^k$$

is continuous on X.

B.2.3 *Basic properties of continuous functions on* $\mathbf{R}^n$

The basic properties of continuous functions on $\mathbf{R}^n$ can be summarized as follows.

Theorem B.4. *Let* X *be a nonempty closed and bounded subset of* $\mathbf{R}^n$.

(i) *If a mapping* $f : X \to \mathbf{R}^m$ *is continuous on* X, *it is bounded on* X: *there exists* $C < \infty$ *such that* $|f(x)| \leq C$ *for all* $x \in X$.

Proof. Assume, on the contrary to what should be proved, that f is unbounded, so that for every i, there exists a point $x^i \in X$ such that $|f(x^i)| > i$. By Theorem B.2, we can extract from the sequence $\{x^i\}$ a subsequence $\{x^{i_j}\}_{j=1}^{\infty}$ which converges to a point $\bar{x} \in X$. The real-valued function $g(x) = |f(x)|$ is continuous (as the superposition of two continuous mappings, see Theorem B.3(ii)), and therefore, its values at the points x^{i_j} should converge, as $j \to \infty$, to its value at $\bar{x}$; on the other hand, $g(x^{i_j}) \geq i_j \to \infty$ as $j \to \infty$, and we get the desired contradiction.

(ii) *If a mapping* $f : X \to \mathbf{R}^m$ *is continuous on* X, *it is uniformly continuous: for every* $\epsilon > 0$, *there exists* $\delta > 0$ *such that*

$$x, y \in X, |x - y| < \delta \quad \Rightarrow \quad |f(x) - f(y)| < \epsilon.$$

Proof. Assume, on the contrary to what should be proved, that there exists $\epsilon > 0$ such that for every $\delta > 0$, one can find a pair of points x, y in X such that $|x - y| < \delta$ and $|f(x) - f(y)| \geq \epsilon$. In particular, for every $i = 1, 2, \ldots$, we can find two points x^i, y^i in X such that $|x^i - y^i| \leq 1/i$ and $|f(x^i) - f(y^i)| \geq \epsilon$. By Theorem B.2, we can extract from the sequence $\{x^i\}$ a subsequence $\{x^{i_j}\}_{j=1}^{\infty}$ which converges to certain point $\bar{x} \in X$. Since $|y^{i_j} - x^{i_j}| \leq 1/i_j \to 0$ as $j \to \infty$, the sequence $\{y^{i_j}\}_{j=1}^{\infty}$

converges to the same point $\bar{x}$ as the sequence $\{x^{i_j}\}_{j=1}^{\infty}$ (why?) Since f is continuous, we have

$$\lim_{j\to\infty} f(y^{i_j}) = f(\bar{x}) = \lim_{j\to\infty} f(x^{i_j}),$$

whence $\lim_{j\to\infty}(f(x^{i_j}) - f(y^{i_j})) = 0$, which contradicts the fact that $|f(x^{i_j}) - f(y^{i_j})| \geq \epsilon > 0$ for all j.

(iii) *Let f bc a real valued continuous function on X. The f attains its minimum on X :*

$$\operatorname*{Argmin}_X f \equiv \{x \in X : f(x) = \inf_{y\in X} f(y)\} \neq \emptyset,$$

same as f attains its maximum at certain points of X :

$$\operatorname*{Argmax}_X f \equiv \{x \in X : f(x) = \sup_{y\in X} f(y)\} \neq \emptyset.$$

Proof. Let us prove that f attains its maximum on X (the proof for minimum is completely similar). Since f is bounded on X by (i), the quantity

$$f^* = \sup_{x\in X} f(x)$$

is finite; of course, we can find a sequence $\{x^i\}$ of points from X such that $f^* = \lim_{i\to\infty} f(x^i)$. By Theorem B.2, we can extract from the sequence $\{x^i\}$ a subsequence $\{x^{i_j}\}_{j=1}^{\infty}$ which converges to certain point $\bar{x} \in X$. Since f is continuous on X, we have

$$f(\bar{x}) = \lim_{j\to\infty} f(x^{i_j}) = \lim_{i\to\infty} f(x^i) = f^*,$$

so that the maximum of f on X indeed is achieved (e.g., at the point $\bar{x}$).

Task B.2. Prove that in general *no one* of the three statements in Theorem B.4 remains valid when X is closed, but not bounded, same as when X is bounded, but not closed.

B.3 Differentiable functions on $\mathbf{R}^n$

B.3.1 *The derivative*

The reader is definitely familiar with the notion of derivative of a real-valued function $f(x)$ of real variable x:

$$f'(x) = \lim_{\Delta x\to 0} \frac{f(x+\Delta x) - f(x)}{\Delta x}$$

This definition does not work when we pass from functions of single real variable to functions of several real variables or, which is the same, to functions with vector arguments. Indeed, in this case, the shift in the argument Δx should be a vector, and we do not know how to *divide* by a vector.

A proper way to extend the notion of the derivative to real- and vector-valued functions of vector argument is to realize what in fact is the meaning of the derivative in the univariate case. What $f'(x)$ says to us is *how to approximate f in a neighbourhood of x by a linear function.* Specifically, *if $f'(x)$ exists, then the linear function $f'(x)\Delta x$ of Δx approximates the change $f(x + \Delta x) - f(x)$ in f up to a remainder which is of highest order as compared with Δx as $\Delta x \to 0$:*

$$|f(x + \Delta x) - f(x) - f'(x)\Delta x| \leq \bar{o}(|\Delta x|) \text{ as } \Delta x \to 0.$$

In the above formula, we meet with the notation $\bar{o}(|\Delta x|)$, and here is the explanation of this notation:

> $\bar{o}(|\Delta x|)$ *is a common name of all functions $\phi(\Delta x)$ of Δx which are well defined in a neighbourhood of the point $\Delta x = 0$ on the axis, vanish at the point $\Delta x = 0$ and are such that*
>
> $$\frac{\phi(\Delta x)}{|\Delta x|} \to 0 \text{ as } \Delta x \to 0.$$
>
> For example,
>
> (1) $(\Delta x)^2 = \bar{o}(|\Delta x|)$, $\Delta x \to 0$,
> (2) $|\Delta x|^{1.01} = \bar{o}(|\Delta x|)$, $\Delta x \to 0$,
> (3) $\sin^2(\Delta x) = \bar{o}(|\Delta x|)$, $\Delta x \to 0$,
> (4) $\Delta x \neq \bar{o}(|\Delta x|)$, $\Delta x \to 0$.
>
> Later on, we shall meet with the notation "$\bar{o}(|\Delta x|^k)$ as $\Delta x \to 0$," where k is a positive integer. The definition is completely similar to the one for the case of $k = 1$:
>
> $\bar{o}(|\Delta x|^k)$ *is a common name of all functions $\phi(\Delta x)$ of Δx which are well defined in a neighbourhood of the point $\Delta x = 0$ on the axis, vanish at the point $\Delta x = 0$ and are such that*
>
> $$\frac{\phi(\Delta x)}{|\Delta x|^k} \to 0 \quad \text{as } \Delta x \to 0.$$

Note that if $f(\cdot)$ is a function defined in a neighbourhood of a point x on the axis, then there perhaps are many linear functions $a\Delta x$ of

Δx which well approximate $f(x+\Delta x) - f(x)$, in the sense that the remainder in the approximation

$$f(x+\Delta x) - f(x) - a\Delta x$$

tends to 0 as $\Delta x \to 0$; among these approximations, however, there exists *at most one* which approximates $f(x+\Delta x) - f(x)$ "very well," so that the remainder is $\bar{o}(|\Delta x|)$, and not merely tends to 0 as $\Delta x \to 0$. Indeed, if

$$f(x+\Delta x) - f(x) - a\Delta x = \bar{o}(|\Delta x|),$$

then, dividing both sides by Δx, we get

$$\frac{f(x+\Delta x) - f(x)}{\Delta x} - a = \frac{\bar{o}(|\Delta x|)}{\Delta x};$$

by definition of $\bar{o}(\cdot)$, the right-hand side in this equality tends to 0 as $\Delta x \to 0$, whence

$$a = \lim_{\Delta x \to 0} \frac{f(x+\Delta x) - f(x)}{\Delta x} = f'(x).$$

Thus, *if* a linear function $a\Delta x$ of Δx approximates the change $f(x+\Delta x) - f(x)$ in f up to the remainder which is $\bar{o}(|\Delta x|)$ as $\Delta x \to 0$, *then* a is the derivative of f at x. You can easily verify that the inverse statement is also true: *If* the derivative of f at x exists, *then* the linear function $f'(x)\Delta x$ of Δx approximates the change $f(x+\Delta x) - f(x)$ in f up to the remainder, which is $\bar{o}(|\Delta x|)$ as $\Delta x \to 0$.

The advantage of the "$\bar{o}(|\Delta x|)$"-definition of derivative is that it can be naturally extended onto vector-valued functions of vector arguments (you should just replace "axis" with $\mathbf{R}^n$ in the definition of $\bar{o}$) and enlightens the *essence* of the notion of derivative: When it exists, this is exactly *the linear function of Δx which approximates the change $f(x+\Delta x) - f(x)$ in f up to a remainder which is $\bar{o}(|\Delta x|)$*. The precise definition is as follows.

Definition B.1 (Frechet differentiability). Let f be a function which is well defined in a neighbourhood of a point $x \in \mathbf{R}^n$ and takes values in $\mathbf{R}^m$. We say that f is differentiable at x if there exists a linear function $Df(x)[\Delta x]$ of $\Delta x \in \mathbf{R}^n$ taking values in $\mathbf{R}^m$, which approximates the change $f(x+\Delta x) - f(x)$ in f up to a remainder which is $\bar{o}(|\Delta x|)$:

$$|f(x+\Delta x) - f(x) - Df(x)[\Delta x]| \leq \bar{o}(|\Delta x|). \tag{B.1}$$

Equivalently, a function f which is well defined in a neighbourhood of a point $x \in \mathbf{R}^n$ and takes values in $\mathbf{R}^m$ is called differentiable at x if there exists a linear function $Df(x)[\Delta x]$ of $\Delta x \in \mathbf{R}^n$ taking values in $\mathbf{R}^m$ such that for every $\epsilon > 0$, there exists $\delta > 0$, satisfying the relation

$$|\Delta x| \leq \delta \quad \Rightarrow \quad |f(x+\Delta x) - f(x) - Df(x)[\Delta x]| \leq \epsilon|\Delta x|.$$

B.3.2 *Derivative and directional derivatives*

We have defined what it means by the fact that a function $f : \mathbf{R}^n \to \mathbf{R}^m$ is differentiable at a point x, but did not say yet what the derivative is. The reader could guess that the derivative is exactly the "linear function $Df(x)[\Delta x]$ of $\Delta x \in \mathbf{R}^n$ taking values in $\mathbf{R}^m$ which approximates the change $f(x+\Delta x) - f(x)$ in f up to a remainder, which is $\leq \bar{o}(|\Delta x|)$" participating in the definition of differentiability. The guess is correct, but we cannot merely call the entity participating in the definition the derivative — why do we know that this entity is unique? Perhaps, there are many different linear functions of Δx approximating the change in f up to a remainder, which is $\bar{o}(|\Delta x|)$. In fact, there is no more than a single linear function with this property due to the following observation:

Let f be differentiable at x and $Df(x)[\Delta x]$ be a linear function participating in the definition of differentiability. Then

$$\forall \Delta x \in \mathbf{R}^n : \quad Df(x)[\Delta x] = \lim_{t \to +0} \frac{f(x+t\Delta x) - f(x)}{t}. \tag{B.2}$$

In particular, the derivative $Df(x)[\cdot]$ is uniquely defined by f and x.

Proof. We have

$$
\begin{array}{ll}
|f(x+t\Delta x) - f(x) - Df(x)[t\Delta x]| \leq \bar{o}(|t\Delta x|) & \\
\Downarrow & \\
\left|\dfrac{f(x+t\Delta x) - f(x)}{t} - \dfrac{Df(x)[t\Delta x]}{t}\right| \leq \dfrac{\bar{o}(|t\Delta x|)}{t} & \\
\Updownarrow & [\text{since } Df(x)[\cdot] \text{ is linear}] \\
\left|\dfrac{f(x+t\Delta x) - f(x)}{t} \quad Df(x)[\Delta x]\right| \leq \dfrac{\bar{o}(|t\Delta x|)}{t} & \\
\Downarrow & \\
Df(x)[\Delta x] = \lim\limits_{t\to+0} \dfrac{f(x+t\Delta x) - f(x)}{t} & \\
 & \left[\begin{array}{l}\text{passing to limit as } t \to +0; \\ \text{note that } \frac{\bar{o}(|t\Delta x|)}{t} \to 0, t \to +0\end{array}\right]
\end{array}
$$

Pay attention to important remarks as follows:

(1) The right-hand side limit in (B.2) is an important entity called the *directional derivative of f taken at x along (a direction)* Δx; note that this quantity is defined in the "purely univariate" fashion — by dividing the change in f by the magnitude of a shift in a direction Δx and passing to limit as the magnitude of the shift approaches 0. Relation (B.2) says that the derivative, if it exists, is, at every Δx, nothing that the directional derivative of f taken at x along Δx. Note, however, that differentiability is much more than the existence of directional derivatives along all directions Δx; differentiability also requires *the directional derivatives to be "well organized" — to depend linearly on the direction* Δx. It is easily seen that just existence of directional derivatives does not imply their "good organization," for example, the Euclidean norm

$$f(x) = |x|$$

at $x = 0$ possesses directional derivatives along all directions:

$$\lim_{t\to+0} \frac{f(0+t\Delta x) - f(0)}{t} = |\Delta x|;$$

these derivatives, however, depend *nonlinearly* on Δx, so that the Euclidean norm is *not* differentiable at the origin (although is differentiable everywhere outside the origin, but this is another story).

(2) It should be stressed that the derivative, if it exists, is what it is: *a linear function of* $\Delta x \in \mathbf{R}^n$ *taking values in* $\mathbf{R}^m$. As we shall see in a while, we can *represent* this function by something "tractable," like a vector or a matrix, and can understand how to compute such a representation; however, an intelligent reader should bear in mind that *a* representation is not exactly the same as *the* represented entity. Sometimes, the difference between derivatives and the entities which represent them is reflected in the terminology: what we call the *derivative* is also called the *differential*, while the word "derivative" is reserved for the vector/matrix representing the differential.

B.3.3 *Representations of the derivative*

By definition, the derivative of a mapping $f : \mathbf{R}^n \to \mathbf{R}^m$ at a point x is a linear function $Df(x)[\Delta x]$ taking values in $\mathbf{R}^m$. How could we represent such a function?

Case of $m = 1$ — The gradient: Let us start with real-valued functions (i.e., with the case of $m = 1$); in this case, the derivative is a *linear* real-valued function on $\mathbf{R}^n$. As we remember, the standard Euclidean structure on $\mathbf{R}^n$ allows us to represent every linear function on $\mathbf{R}^n$ as the inner product of the argument with certain fixed vector. In particular, the derivative $Df(x)[\Delta x]$ of a scalar function can be represented as

$$Df(x)[\Delta x] = [\text{vector}]^T \Delta x;$$

what is denoted "vector" in this relation is called the *gradient* of f at x and is denoted by $\nabla f(x)$:

$$Df(x)[\Delta x] = (\nabla f(x))^T \Delta x. \tag{B.3}$$

How can we compute the gradient? The answer is given by (B.2). Indeed, let us look what (B.3) and (B.2) say when Δx is the ith standard basic orth. According to (B.3), $Df(x)[e_i]$ is the ith coordinate of the vector $\nabla f(x)$; according to (B.2),

$$\left.\begin{array}{r} Df(x)[e_i] = \lim\limits_{t\to+0} \frac{f(x+te_i)-f(x)}{t}, \\ Df(x)[e_i] = -Df(x)[-e_i] = -\lim\limits_{t\to+0} \frac{f(x-te_i)-f(x)}{t} = \lim\limits_{t\to-0} \frac{f(x+te_i)-f(x)}{t} \end{array}\right\}$$
$$\Rightarrow Df(x)[e_i] = \frac{\partial f(x)}{\partial x_i}.$$

Thus, we have the following:

If a real-valued function f is differentiable at x, then the first-order partial derivatives of f at x exist, and the gradient of f at x is just the vector with the coordinates which are the first-order partial derivatives of f taken at x :

$$\nabla f(x) = \begin{bmatrix} \frac{\partial f(x)}{\partial x_1} \\ \vdots \\ \frac{\partial f(x)}{\partial x_n} \end{bmatrix}.$$

The derivative of f, taken at x, is the linear function of Δx given by

$$Df(x)[\Delta x] = (\nabla f(x))^T \Delta x = \sum_{i=1}^{n} \frac{\partial f(x)}{\partial x_i} (\Delta x)_i.$$

General case — The Jacobian: Now, let $f : \mathbf{R}^n \to \mathbf{R}^m$ with $m \geq 1$. In this case, $Df(x)[\Delta x]$, regarded as a function of Δx, is a linear mapping from $\mathbf{R}^n$ to $\mathbf{R}^m$; as we remember, the standard way to represent a linear mapping from $\mathbf{R}^n$ to $\mathbf{R}^m$ is to represent it as the multiplication by $m \times n$ matrix:

$$Df(x)[\Delta x] = [m \times n \text{ matrix}] \cdot \Delta x. \tag{B.4}$$

What is denoted by "matrix" in (B.4) is called *the Jacobian* of f at x and is denoted by $f'(x)$. How can we compute the entries of the Jacobian? Here again, the answer is readily given by (B.2). Indeed, on the one hand, we have

$$Df(x)[\Delta x] = f'(x)\Delta x, \tag{B.5}$$

whence

$$[Df(x)[e_j]]_i = (f'(x))_{ij},\ i = 1, ..., m, j = 1, ..., n.$$

On the other hand, denoting

$$f(x) = \begin{bmatrix} f_1(x) \\ \vdots \\ f_m(x) \end{bmatrix},$$

the same computation as in the case of gradient demonstrates that

$$[Df(x)[e_j]]_i = \frac{\partial f_i(x)}{\partial x_j},$$

and we arrive at the following conclusion:

If a vector-valued function $f(x) = (f_1(x), ..., f_m(x))$ is differentiable at x, then the first-order partial derivatives of all f_i at x exist, and the Jacobian of f at x is just the $m \times n$ matrix with the entries $[\frac{\partial f_i(x)}{\partial x_j}]_{i,j}$ (so that the rows in the Jacobian are $[\nabla f_1(x)]^T, ..., [\nabla f_m(x)]^T$. The derivative of f, taken at x, is the linear vector-valued function of Δx given by

$$Df(x)[\Delta x] = f'(x)\Delta x = \begin{bmatrix} [\nabla f_1(x)]^T \Delta x \\ \vdots \\ [\nabla f_m(x)]^T \Delta x \end{bmatrix}.$$

Remark B.1. Note that for a real-valued function f, we have defined both the gradient $\nabla f(x)$ and the Jacobian $f'(x)$. These two entities are "nearly the same," but not exactly the same: the Jacobian is a vector row, and the gradient is a vector column linked by the relation

$$f'(x) = (\nabla f(x))^T.$$

Of course, both these representations of the derivative of f yield the same linear approximation of the change in f:

$$Df(x)[\Delta x] = (\nabla f(x))^T \Delta x = f'(x)\Delta x.$$

B.3.4 *Existence of the derivative*

We have seen that the existence of the derivative of f at a point implies the existence of the first order partial derivatives of the (components $f_1, ..., f_m$ of) f. The inverse statement is not exactly true — the existence of all first-order partial derivatives $\frac{\partial f_i(x)}{\partial x_j}$ does not necessarily imply the existence of the derivative; we need a bit more.

Theorem B.5 (Sufficient condition for differentiability). *Assume the following:*

(1) *The mapping $f = (f_1, ..., f_m)$: $\mathbf{R}^n \to \mathbf{R}^m$ is well defined in a neighbourhood U of a point $x_0 \in \mathbf{R}^n$.*
(2) *The first-order partial derivatives of the components f_i of f exist everywhere in U.*
(3) *The first-order partial derivatives of the components f_i of f are continuous at the point x_0.*

Then f is differentiable at the point x_0.

B.3.5 *Calculus of derivatives*

The calculus of derivatives is given by the following result.

Theorem B.6. (i) [Differentiability and linear operations]: *Let $f_1(x)$, $f_2(x)$ be mappings defined in a neighbourhood of $x_0 \in \mathbf{R}^n$ and taking values in $\mathbf{R}^m$ and $\lambda_1(x), \lambda_2(x)$ be real-valued functions defined in a neighbourhood of x_0. Assume that $f_1, f_2, \lambda_1, \lambda_2$ are differentiable at x_0. Then so is the function $f(x) = \lambda_1(x)f_1(x) + \lambda_2(x)f_2(x)$, with the derivative at x_0 given by*

$$\begin{aligned} Df(x_0)[\Delta x] &= [D\lambda_1(x_0)[\Delta x]]f_1(x_0) + \lambda_1(x_0)Df_1(x_0)[\Delta x] \\ &\quad + [D\lambda_2(x_0)[\Delta x]]f_2(x_0) + \lambda_2(x_0)Df_2(x_0)[\Delta x], \\ &\Downarrow \\ f'(x_0) &= f_1(x_0)[\nabla\lambda_1(x_0)]^T + \lambda_1(x_0)f_1'(x_0) \\ &\quad + f_2(x_0)[\nabla\lambda_2(x_0)]^T + \lambda_2(x_0)f_2'(x_0). \end{aligned}$$

(ii) (Chain rule): *Let a mapping $f : \mathbf{R}^n \to \mathbf{R}^m$ be differentiable at x_0 and a mapping $g : \mathbf{R}^m \to \mathbf{R}^n$ be differentiable at $y_0 = f(x_0)$. Then the superposition $h(x) = g(f(x))$ is differentiable at x_0, with the derivative at x_0 given by*

$$\begin{aligned} Dh(x_0)[\Delta x] &= Dg(y_0)[Df(x_0)[\Delta x]] \\ &\Downarrow \\ h'(x_0) &= g'(y_0)f'(x_0) \end{aligned}$$

If the outer function g is real-valued, then the latter formula implies that

$$\nabla h(x_0) = [f'(x_0)]^T \nabla g(y_0)$$

(recall that for a real-valued function ϕ, $\phi' = (\nabla\phi)^T$).

B.3.6 *Computing the derivative*

Representations of the derivative via first-order partial derivatives normally allow us to compute it by the standard Calculus rules, in a completely mechanical fashion, not thinking at all of *what* we are computing. The examples to follow (especially the third of them) demonstrate that it often makes sense to bear in mind *what* the derivative is; this sometimes yields the result much faster than blind implementation of Calculus rules.

Example 1: The gradient of an affine function: An *affine* function

$$f(x) = a + \sum_{i=1}^{n} g_i x_i \equiv a + g^T x : \mathbf{R}^n \to \mathbf{R}$$

is differentiable at every point (Theorem B.5) and its gradient, of course, equals g:

$$\begin{aligned}(\nabla f(x))^T \Delta x &= \lim_{t\to+0} t^{-1}\left[f(x+t\Delta x) - f(x)\right] & [(\text{B.2})]\\ &= \lim_{t\to+0} t^{-1}[tg^T\Delta x] & [\text{arithmetics}],\end{aligned}$$

and we arrive at

$$\boxed{\nabla(a + g^T x) = g}.$$

Example 2: The gradient of a quadratic form: For the time being, let us define a homogeneous quadratic form on $\mathbf{R}^n$ as a function

$$f(x) = \sum_{i,j} A_{ij} x_i x_j = x^T A x,$$

where A is an $n \times n$ matrix. Note that the matrices A and A^T define the same quadratic form, and therefore, the *symmetric* matrix $B = \frac{1}{2}(A + A^T)$ also produces the same quadratic form as A and A^T. It follows that we may always assume (and do assume from now on) that the matrix A producing the quadratic form in question is symmetric.

A quadratic form is a simple polynomial and as such is differentiable at every point (Theorem B.5). What is the gradient of f at a point x? Here is the computation:

$$\begin{aligned}(\nabla f(x))^T \Delta x &= Df(x)[\Delta x]\\ &= \lim_{t\to+0} t^{-1}\left[(x+t\Delta x)^T A(x+t\Delta x) - x^T A x\right]\\ & \qquad [(\text{B.2})]\end{aligned}$$

$$
\begin{aligned}
&= \lim_{t\to+0} t^{-1}[x^TAx + t(\Delta x)^TAx \\
&\qquad + tx^TA\Delta x + t^2(\Delta x)^TA\Delta x - x^TAx] \\
&\qquad\qquad [\text{opening parentheses}] \\
&= \lim_{t\to+0} t^{-1}\left[2t(Ax)^T\Delta x + t^2(\Delta x)^TA\Delta x\right] \\
&\qquad\qquad [\text{arithmetics + symmetry of } A] \\
&= 2(Ax)^T\Delta x.
\end{aligned}
$$

We conclude that

$$\boxed{\nabla(x^TAx) = 2Ax}$$

(recall that $A = A^T$).

Example 3: The derivative of the log-det barrier: Let us compute the derivative of the *log-det barrier* (playing an extremely important role in modern optimization)

$$F(X) = \ln \mathrm{Det}\,(X),$$

where X is an $n\times n$ matrix (or, if you prefer, n^2-dimensional vector). Note that $F(X)$ is well defined and differentiable in a neighbourhood of every point $\bar{X}$ with positive determinant (indeed, $\mathrm{Det}(X)$ is a polynomial of the entries of X and thus is everywhere continuous and differentiable with continuous partial derivatives, while the function $\ln(t)$ is continuous and differentiable on the positive ray; by Theorems B.3(ii), B.6(ii), F is differentiable at every X such that $\mathrm{Det}\,(X) > 0$). The reader is kindly asked to try to find the derivative of F by the standard techniques; if the result will not be obtained in, say, 30 minutes, look at the eight-line computation to follow (in this computation, $\mathrm{Det}\,(\bar{X}) > 0$, and $G(X) = \mathrm{Det}\,(X)$):

$$
\begin{aligned}
&DF(\bar{X})[\Delta X] \\
&\quad = D\ln(G(\bar{X}))[DG(\bar{X})[\Delta X]] \quad [\text{chain rule}] \\
&\quad = G^{-1}(\bar{X})DG(\bar{X})[\Delta X] \quad [\ln'(t) = t^{-1}]
\end{aligned}
$$

$$= \mathrm{Det}^{-1}(\bar{X}) \lim_{t\to+0} t^{-1}\left[\mathrm{Det}\,(\bar{X}+t\Delta X) - \mathrm{Det}\,(\bar{X})\right]$$

[definition of G and (B.2)]

$$= \mathrm{Det}^{-1}(\bar{X}) \lim_{t\to+0} t^{-1}\left[\mathrm{Det}\,(\bar{X}(I+t\bar{X}^{-1}\Delta X)) - \mathrm{Det}\,(\bar{X})\right]$$

$$= \mathrm{Det}^{-1}(\bar{X}) \lim_{t\to+0} t^{-1}\left[\mathrm{Det}\,(\bar{X})(\mathrm{Det}\,(I+t\bar{X}^{-1}\Delta X) - 1)\right]$$

[$\mathrm{Det}\,(AB) = \mathrm{Det}\,(A)\,\mathrm{Det}\,(B)$]

$$= \lim_{t\to+0} t^{-1}\left[\mathrm{Det}\,(I+t\bar{X}^{-1}\Delta X) - 1\right]$$

$$= \mathrm{Tr}(\bar{X}^{-1}\Delta X) = \sum_{i,j}[\bar{X}^{-1}]_{ji}(\Delta X)_{ij},$$

where the concluding equality

$$\lim_{t\to+0} t^{-1}[\mathrm{Det}\,(I+tA) - 1] = \mathrm{Tr}(A) \equiv \sum_i A_{ii} \tag{B.6}$$

is immediately given by recalling what the determinant of $I + tA$ is: this is a polynomial of t which is the sum of products, taken along all diagonals of a $n \times n$ matrix and assigned certain signs, of the entries of $I + tA$. At every one of these diagonals, except for the main one, there are at least two cells with the entries proportional to t, so that the corresponding products do not contribute to the constant and the linear in t terms in $\mathrm{Det}\,(I + tA)$ and thus do not affect the limit in (B.6). The only product which does contribute to the linear and the constant terms in $\mathrm{Det}\,(I + tA)$ is the product $(1 + tA_{11})(1 + tA_{22})...(1 + tA_{nn})$ coming from the main diagonal; it is clear that in this product, the constant term is 1, and the linear in t term is $t(A_{11} + \cdots + A_{nn})$, and (B.6) follows.

B.3.7 *Higher-order derivatives*

Let $f : \mathbf{R}^n \to \mathbf{R}^m$ be a mapping which is well defined and differentiable at every point x from an open set U. The Jacobian of this mapping $J(x)$ is a mapping from $\mathbf{R}^n$ to the space $\mathbf{R}^{m\times n}$ matrices, i.e., is a mapping taking values in certain $\mathbf{R}^M$ ($M = mn$). The derivative of this mapping, if it exists, is called the *second derivative* of f; it again is a mapping from $\mathbf{R}^n$ to certain $\mathbf{R}^M$ and as such can be differentiable, and so on, so that we can speak about the second, the third, ... derivatives of a vector-valued function of vector argument. A *sufficient* condition for the existence of k derivatives of f in U is that f is C^k in U, i.e., that all partial derivatives of f of orders $\leq k$ exist and are continuous everywhere in U (cf. Theorem B.5).

We have explained what does it mean that f has k derivatives in U; note, however, that according to the definition, highest order derivatives at a point x are just long vectors; say, the second-order derivative of a scalar function f of 2 variables is the Jacobian of the mapping $x \mapsto f'(x) : \mathbf{R}^2 \to \mathbf{R}^2$, i.e., a mapping from $\mathbf{R}^2$ to $\mathbf{R}^{2\times 2} = \mathbf{R}^4$; the third-order derivative of f is therefore the Jacobian of a mapping from $\mathbf{R}^2$ to $\mathbf{R}^4$, i.e., a mapping from $\mathbf{R}^2$ to $\mathbf{R}^{4\times 2} = \mathbf{R}^8$, and so on. The question which should be addressed now is: *What is a natural and transparent way to represent the highest order derivatives?*

The answer is as follows:

(*) *Let $f : \mathbf{R}^n \to \mathbf{R}^m$ be C^k on an open set $U \subset \mathbf{R}^n$. The derivative of order $\ell \leq k$ of f, taken at a point $x \in U$, can be naturally identified with a function*

$$D^\ell f(x)[\Delta x^1, \Delta x^2, ..., \Delta x^\ell]$$

of ℓ vector arguments $\Delta x^i \in \mathbf{R}^n$, $i = 1, ..., \ell$, and taking values in $\mathbf{R}^m$. This function is linear in every one of the arguments Δx^i, the other arguments being fixed, and is symmetric with respect to permutation of arguments $\Delta x^1, ..., \Delta x^\ell$.

In terms of f, the quantity $D^\ell f(x)[\Delta x^1, \Delta x^2, ..., \Delta x^\ell]$ (full name: "the ℓth derivative (or differential) of f taken at a point x along the directions $\Delta x^1, ..., \Delta x^\ell$") is given by

$$D^\ell f(x)[\Delta x^1, \Delta x^2, \ldots, \Delta x^\ell] = \frac{\partial^\ell}{\partial t_\ell \partial t_{\ell-1} ... \partial t_1}\Big|_{t_1=\ldots=t_\ell=0} f(x + t_1\Delta x^1 + t_2\Delta x^2 + \cdots + t_\ell \Delta x^\ell). \tag{B.7}$$

The explanation to our claims is as follows. Let $f : \mathbf{R}. \to \mathbf{R}^m$ be C^k on an open set $U \subset \mathbf{R}^n$.

(1) When $\ell = 1$, (*) says to us that the first-order derivative of f, taken at x, is a linear function $Df(x)[\Delta x^1]$ of $\Delta x^1 \in \mathbf{R}^n$, taking values in $\mathbf{R}^m$, and that the value of this function at every Δx^1

is given by the relation

$$Df(x)[\Delta x^1] = \tfrac{\partial}{\partial t_1}\Big|_{t_1=0} f(x + t_1\Delta x^1) \tag{B.8}$$

(cf. (B.2)), which is in complete accordance with what we already know about the derivative.

(2) To understand what the second derivative is, let us take the first derivative $Df(x)[\Delta x^1]$, *let us temporarily fix somehow the argument* Δx^1 and treat the derivative as a function of x. As a function of x, Δx^1 being fixed, the quantity $Df(x)[\Delta x^1]$ is again a mapping which maps U into $\mathbf{R}^m$ and is differentiable by Theorem B.5 (provided, of course, that $k \geq 2$). The derivative of this mapping is certain linear function of $\Delta x \equiv \Delta x^2 \in \mathbf{R}^n$, depending on x as on a parameter; of course, it depends on Δx^1 as on a parameter as well. Thus, the derivative of $Df(x)[\Delta x^1]$ in x is certain function

$$D^2 f(x)[\Delta x^1, \Delta x^2]$$

of $x \in U$ and $\Delta x^1, \Delta x^2 \in \mathbf{R}^n$ and taking values in $\mathbf{R}^m$. What we know about this function is that it is linear in Δx^2. In fact, it is also linear in Δx^1, since it is the derivative in x of certain function (namely, of $Df(x)[\Delta x^1]$) *linearly depending on the parameter* Δx^1, so that the derivative of the function *in* x is linear in the parameter Δx^1 as well (differentiation is a linear operation with respect to a function we are differentiating: summing up functions and multiplying them by real constants, we sum up, respectively, multiply by the same constants, the derivatives). Thus, $D^2 f(x)[\Delta x^1, \Delta x^2]$ is linear in Δx^1 when x and Δx^2 are fixed and is linear in Δx^2 when x and Δx^1 are fixed. Moreover, we have

$$\begin{aligned}
D^2 f(x)[\Delta x^1, \Delta x^2] &\\
&= \frac{\partial}{\partial t_2}\Big|_{t_2=0} Df(x + t_2\Delta x^2)[\Delta x^1] && \text{[cf. (B.8)]}\\
&= \frac{\partial}{\partial t_2}\Big|_{t_2=0} \frac{\partial}{\partial t_1}\Big|_{t_1=0} f(x + t_2\Delta x^2 + t_1\Delta x^1) && \text{[by (B.8)]}\\
&= \frac{\partial^2}{\partial t_2 \partial t_1}\Bigg|_{t_1=t_2=0} f(x + t_1\Delta x^1 + t_2\Delta x^2),
\end{aligned} \tag{B.9}$$

as claimed in (B.7) for $\ell = 2$. The only piece of information about the second derivative which is contained in (*) and is not justified yet is that $D^2 f(x)[\Delta x^1, \Delta x^2]$ is symmetric in $\Delta x^1, \Delta x^2$; but this fact is readily given by the representation (B.7), since, as they prove in calculus, if a function ϕ possesses *continuous* partial derivatives of orders $\leq \ell$ in a neighbourhood of a point, then these derivatives in this neighbourhood are independent of the order in which they are taken; it follows that

$$\begin{aligned} D^2 f(x)[\Delta x^1, \Delta x^2] &= \frac{\partial^2}{\partial t_2 \partial t_1}\Big|_{t_1=t_2=0} \underbrace{f(x + t_1 \Delta x^1 + t_2 \Delta x^2)}_{\phi(t_1,t_2)} \\ &\qquad\qquad [(B.9)] \\ &= \frac{\partial^2}{\partial t_1 \partial t_2}\Big|_{t_1=t_2=0} \phi(t_1, t_2) \\ &= \frac{\partial^2}{\partial t_1 \partial t_2}\Big|_{t_1=t_2=0} f(x + t_2 \Delta x^2 + t_1 \Delta x^1) \\ &= D^2 f(x)[\Delta x^2, \Delta x^1] \\ &\qquad\qquad [\text{the same (B.9)}]. \end{aligned}$$

(3) Now, it is clear how to proceed: To define $D^3 f(x)[\Delta x^1, \Delta x^2, \Delta x^3]$, we fix in the second-order derivative $D^2 f(x)[\Delta x^1, \Delta x^2]$ the arguments $\Delta x^1, \Delta x^2$ and treat it as a function of x only, thus arriving at a mapping which maps U into $\mathbf{R}^m$ and depends on $\Delta x^1, \Delta x^2$ as on parameters (linearly in every one of them). Differentiating the resulting mapping in x, we arrive at a function $D^3 f(x)[\Delta x^1, \Delta x^2, \Delta x^3]$ which by construction is linear in every one of the arguments Δx^1, Δx^2, Δx^3 and satisfies (B.7); the latter relation, due to the Calculus result on the symmetry of partial derivatives, implies that $D^3 f(x)[\Delta x^1, \Delta x^2, \Delta x^3]$ is symmetric in $\Delta x^1, \Delta x^2, \Delta x^3$. After we have in our disposal the third derivative $D^3 f$, we can build from it in the already explained fashion the fourth derivative and so on until kth derivative is defined.

Remark B.2. Since $D^\ell f(x)[\Delta x^1, ..., \Delta x^\ell]$ is linear in every one of Δx^i, we can expand the derivative in a multiple sum:

$$\Delta x^i = \sum_{j=1}^{n} \Delta x^i_j e_j$$
$$\Downarrow$$
$$D^\ell f(x)[\Delta x^1, \ldots, \Delta x^\ell] = D^\ell f(x) \left[\sum_{j_1=1}^{n} \Delta x^1_{j_1} e_{j_1}, \ldots, \sum_{j_\ell=1}^{n} \Delta x^\ell_{j_\ell} e_{j_\ell} \right]$$
$$= \sum_{1 \le j_1, \ldots, j_\ell \le n} D^\ell f(x)[e_{j_1}, \ldots, e_{j_\ell}] \Delta x^1_{j_1} \ldots \Delta x^\ell_{j_\ell}. \tag{B.10}$$

What is the origin of the coefficients $D^\ell f(x)[e_{j_1}, ..., e_{j_\ell}]$? According to (B.7), one has

$$D^\ell f(x)[e_{j_1}, ..., e_{j_\ell}]$$
$$= \left. \frac{\partial^\ell}{\partial t_\ell \partial t_{\ell-1} ... \partial t_1} \right|_{t_1 = \cdots = t_\ell = 0} f(x + t_1 e_{j_1} + t_2 e_{j_2} + \cdots + t_\ell e_{j_\ell})$$
$$= \frac{\partial^\ell}{\partial x_{j_\ell} \partial x_{j_{\ell-1}} ... \partial x_{j_1}} f(x),$$

so that the coefficients in (B.10) are nothing but the partial derivatives, of order ℓ, of f.

Remark B.3. An important particular case of relation (B.7) is the one when $\Delta x^1 = \Delta x^2 = \cdots = \Delta x^\ell$; let us call the common value of these ℓ vectors d. According to (B.7), we have

$$D^\ell f(x)[d, d, ..., d] = \left. \frac{\partial^\ell}{\partial t_\ell \partial t_{\ell-1} ... \partial t_1} \right|_{t_1 = \cdots = t_\ell = 0} f(x + t_1 d + t_2 d + \cdots + t_\ell d).$$

This relation can be interpreted as follows: consider the function

$$\phi(t) = f(x + td)$$

of a real variable t. Then (check it!)

$$\phi^{(\ell)}(0) = \left. \frac{\partial^\ell}{\partial t_\ell \partial t_{\ell-1} \ldots \partial t_1} \right|_{t_1 = \ldots = t_\ell = 0}$$
$$\times f(x + t_1 d + t_2 d + \cdots + t_\ell d) = D^\ell f(x)[d, ..., d].$$

In other words, $D^\ell f(x)[d,\dots,d]$ is what is called *ℓth directional derivative of f taken at x along the direction d*; to define this quantity, we pass from function f of several variables to the univariate function $\phi(t) = f(x+td)$ — restrict f onto the line passing through x and directed by d — and then take the "usual" derivative of order ℓ of the resulting function of single real variable t at the point $t=0$ (which corresponds to the point x of our line).

Representation of higher-order derivatives: kth order derivative $D^k f(x)[\cdot,\dots,\cdot]$ of a C^k function $f : \mathbf{R}^n \to \mathbf{R}^m$ is what it is — it is a symmetric k-linear mapping on $\mathbf{R}^n$ taking values in $\mathbf{R}^m$ and depending on x as on a parameter. Choosing somehow coordinates in $\mathbf{R}^n$, we can represent such a mapping in the form

$$\begin{aligned}&D^k f(x)[\Delta x_1, ..., \Delta x_k]\\&\quad= \sum_{1\le i_1,\dots,i_k\le n} \frac{\partial^k f(x)}{\partial x_{i_k}\partial x_{i_{k-1}}\dots\partial x_{i_1}}(\Delta x_1)_{i_1}\dots(\Delta x_k)_{i_k}.\end{aligned}$$

We may say that the derivative can be represented by k-index collection of m-dimensional vectors $\frac{\partial^k f(x)}{\partial x_{i_k}\partial x_{i_{k-1}}\dots\partial x_{i_1}}$. This collection, however, is a difficult-to-handle entity, so such a representation does not help. There is, however, a case when the collection becomes an entity we know to handle; this is the case of the second-order derivative of a scalar function ($k=2, m=1$). In this case, the collection in question is just a symmetric matrix $H(x) = \left[\frac{\partial^2 f(x)}{\partial x_i \partial x_j}\right]_{1\le i,j\le n}$. This matrix (same as the linear map $h \mapsto H(x)h$) is called the *Hessian* of f at x. Note that

$$D^2 f(x)[\Delta x_1, \Delta x_2] = \Delta x_1^T H(x) \Delta x_2.$$

B.3.8 *Calculus of C^k mappings*

The calculus of C^k mappings can be summarized as follows.

Theorem B.7.

(i) *Let U be an open set in $\mathbf{R}^n$, $f_1(\cdot), f_2(\cdot) : \mathbf{R}^n \to \mathbf{R}^m$ be C^k in U and let real-valued functions $\lambda_1(\cdot), \lambda_2(\cdot)$ be C^k in U. Then the* *function*

$$f(x) = \lambda_1(x) f_1(x) + \lambda_2(x) f_2(x)$$

is C^k *in* U.

(ii) *Let* U *be an open set in* $\mathbf{R}^n$, V *be an open set in* $\mathbf{R}^m$, *let a mapping* $f : \mathbf{R}^n \to \mathbf{R}^m$ *be* C^k *in* U *and such that* $f(x) \in V$ *for* $x \in U$ *and, finally, let a mapping* $g : \mathbf{R}^m \to \mathbf{R}^p$ *be* C^k *in* V. *Then the superposition*

$$h(x) = g(f(x))$$

is C^k *in* U.

Remark B.4. For higher-order derivatives, in contrast to the first-order ones, there is no simple "chain rule" for computing the derivative of superposition. For example, the second-order derivative of the superposition $h(x) = g(f(x))$ of two C^2-mappings is given by the formula

$$\begin{aligned} Dh(x)[\Delta x^1, \Delta x^2] &= Dg(f(x))[D^2 f(x)[\Delta x^1, \Delta x^2]] \\ &\quad + D^2 g(x)[Df(x)[\Delta x^1], Df(x)[\Delta x^2]] \end{aligned}$$

(check it!). We see that both the first- and the second-order derivatives of f and g contribute to the second-order derivative of the superposition h.

The only case when there does exist a simple formula for high-order derivatives of a superposition is the case when the inner function is affine: if $f(x) = Ax + b$ and $h(x) = g(f(x)) = g(Ax + b)$ with a C^ℓ mapping g, then

$$D^\ell h(x)[\Delta x^1, ..., \Delta x^\ell] = D^\ell g(Ax + b)[A\Delta x^1, ..., A\Delta x^\ell]. \qquad \text{(B.11)}$$

B.3.9 *Examples of higher-order derivatives*

Example 1: Second-order derivative of an affine function $f(x) = a + b^T x$ is, of course, identically zero. Indeed, as we have seen,

$$Df(x)[\Delta x^1] = b^T \Delta x^1$$

is independent of x, and therefore, the derivative of $Df(x)[\Delta x^1]$ in x, which should give us the second derivative $D^2 f(x)[\Delta x^1, \Delta x^2]$, is zero. Clearly, the third, the fourth, etc., derivatives of an affine function are zero as well.

Example 2: Second-order derivative of a homogeneous quadratic form: $f(x) = x^T Ax$ (A is a symmetric $n \times n$ matrix). As we have seen,

$$Df(x)[\Delta x^1] = 2x^T A\Delta x^1.$$

Differentiating in x, we get

$$\begin{aligned} D^2 f(x)[\Delta x^1, \Delta x^2] &= \lim_{t\to+0} t^{-1}\left[2(x+t\Delta x^2)^T A\Delta x^1 - 2x^T A\Delta x^1\right] \\ &= 2(\Delta x^2)^T A\Delta x^1, \end{aligned}$$

so that

$$\boxed{D^2 f(x)[\Delta x^1, \Delta x^2] = 2(\Delta x^2)^T A\Delta x^1}.$$

Note that the second derivative of a quadratic form is independent of x; consequently, the third, the fourth, etc., derivatives of a quadratic form are identically zero.

Example 3: Second-order derivative of the log-det barrier: $F(X) = \ln \mathrm{Det}\,(X)$. As we have seen, this function of an $n \times n$ matrix is well defined and differentiable on the set U of matrices with positive determinant (which is an open set in the space $\mathbf{R}^{n\times n}$ of $n \times n$ matrices). In fact, this function is C^∞ in U. Let us compute its second-order derivative. As we remember,

$$DF(X)[\Delta X^1] = \mathrm{Tr}(X^{-1}\Delta X^1). \tag{B.12}$$

To differentiate the right-hand side in X, let us first find the derivative of the mapping $G(X) = X^{-1}$ which is defined on the open set of nondegenerate $n \times n$ matrices. We have

$$\begin{aligned} DG(X)[\Delta X] &= \lim_{t\to+0} t^{-1}\left[(X+t\Delta X)^{-1} - X^{-1}\right] \\ &= \lim_{t\to+0} t^{-1}\left[(X(I+tX^{-1}\Delta X))^{-1} - X^{-1}\right] \\ &= \lim_{t\to+0} t^{-1}[(I+t\underbrace{X^{-1}\Delta X}_{Y})^{-1}X^{-1} - X^{-1}] \\ &= \left[\lim_{t\to+0} t^{-1}\left[(I+tY)^{-1} - I\right]\right] X^{-1} \end{aligned}$$

$$= \left[\lim_{t \to +0} t^{-1} [I - (I + tY)] (I + tY)^{-1} \right] X^{-1}$$

$$= \left[\lim_{t \to +0} [-Y(I + tY)^{-1}] \right] X^{-1}$$

$$= -YX^{-1}$$

$$= -X^{-1} \Delta X X^{-1},$$

and we arrive at the important by its own right relation

$$\boxed{D(X^{-1})[\Delta X] = -X^{-1}\Delta X X^{-1}, \quad [X \in \mathbf{R}^{n \times n}, \mathrm{Det}\,(X) \neq 0]},$$

which is the "matrix extension" of the standard relation $(x^{-1})' = -x^{-2}$, $x \in \mathbf{R}$.

Now, we are ready to compute the second derivative of the log-det barrier:

$$F(X) = \ln \mathrm{Det}\,(X)$$
$$\Downarrow$$
$$DF(X)[\Delta X^1] = \mathrm{Tr}(X^{-1}\Delta X^1)$$
$$\Downarrow$$
$$\begin{aligned} D^2F(X)[\Delta X^1, \Delta X^2] &= \lim_{t \to +0} t^{-1} \left[\mathrm{Tr}((X + t\Delta X^2)^{-1}\Delta X^1) - \mathrm{Tr}(X^{-1}\Delta X^1) \right] \\ &= \lim_{t \to +0} \mathrm{Tr}\left(t^{-1} \left[(X + t\Delta X^2)^{-1}\Delta X^1 - X^{-1}\Delta X^1 \right] \right) \\ &= \lim_{t \to +0} \mathrm{Tr}\left(\left[t^{-1} \left[(X + t\Delta X^2)^{-1} - X^{-1} \right] \right] \Delta X^1 \right) \\ &\underbrace{=}_{(*)} \mathrm{Tr}\left([-X^{-1}\Delta X^2 X^{-1}] \Delta X^1 \right), \end{aligned}$$

where $(*)$ is due to the already known to us formula for $DG(X)[\Delta X]$, $G(X) = X^{-1}$, and we arrive at the formula

$$\boxed{\begin{aligned} D^2F(X)[\Delta X^1, \Delta X^2] = -\,\mathrm{Tr}(X^{-1}\Delta X^2 X^{-1}\Delta X^1) \\ [X \in \mathbf{R}^{n \times n}, \mathrm{Det}\,(X) > 0] \end{aligned}}.$$

Since $\mathrm{Tr}(AB) = \mathrm{Tr}(BA)$ for all matrices A, B such that the product AB makes sense and is square (check it or see C.4.B.4), the right-hand side in the above formula is symmetric in ΔX^1, ΔX^2, as it should be for the second derivative of a C^2 function.

B.3.10 *Taylor expansion*

Assume that $f : \mathbf{R}^n \to \mathbf{R}^m$ is C^k in a neighbourhood U of a point $\bar{x}$. The *Taylor expansion of order* k of f, built at the point $\bar{x}$, is the function

$$\begin{aligned} F_k(x) &= f(\bar{x}) + \frac{1}{1!}Df(\bar{x})[x-\bar{x}] + \frac{1}{2!}D^2 f(\bar{x})[x-\bar{x}, x-\bar{x}] \\ &\quad + \frac{1}{3!}D^2 f(\bar{x})[x-\bar{x}, x-\bar{x}, x-\bar{x}] \\ &\quad + \cdots + \frac{1}{k!}D^k f(\bar{x})[\underbrace{x-\bar{x}, ..., x-\bar{x}}_{k \text{ times}}]. \end{aligned} \tag{B.13}$$

We are already acquainted with the Taylor expansion of order 1

$$F_1(x) = f(\bar{x}) + Df(\bar{x})[x-\bar{x}].$$

This is the affine function of x which "very well" approximates $f(x)$ in a neighbourhood of $\bar{x}$, namely, within approximation error $\bar{o}(|x-\bar{x}|)$. Similar fact is true for Taylor expansions of higher order:

Theorem B.8. *Let $f : \mathbf{R}^n \to \mathbf{R}^m$ be C^k in a neighbourhood of $\bar{x}$, and let $F_k(x)$ be the Taylor expansion of f at $\bar{x}$ of degree k. Then*

(i) *$F_k(x)$ is a vector-valued polynomial of full degree $\leq k$ (i.e., every one of the coordinates of the vector $F_k(x)$ is a polynomial of $x_1, ..., x_n$, and the sum of powers of x_i's in every term of this polynomial does not exceed k);*
(ii) *$F_k(x)$ approximates $f(x)$ in a neighbourhood of $\bar{x}$ up to a remainder which is $\bar{o}(|x-\bar{x}|^k)$ as $x \to \bar{x}$:*

For every $\epsilon > 0$, there exists $\delta > 0$ such that

$$|x-\bar{x}| \leq \delta \quad \Rightarrow \quad |F_k(x) - f(x)| \leq \epsilon|x-\bar{x}|^k.$$

$F_k(\cdot)$ is the unique polynomial with components of full degree $\leq k$ which approximates f up to a remainder which is $\bar{o}(|x-\bar{x}|^k)$.
(iii) *The value and the derivatives of F_k of orders $1, 2, ..., k$, taken at $\bar{x}$, are the same as the value and the corresponding derivatives of f taken at the same point.*

As stated in theorem, $F_k(x)$ approximates $f(x)$ for x close to $\bar{x}$ up to a remainder which is $\bar{o}(|x-\bar{x}|^k)$. In many cases, it is not enough to know that the remainder is "$\bar{o}(|x-\bar{x}|^k)$" — we need an explicit bound on this remainder. The standard bound of this type is as follows.

Theorem B.9. *Let k be a positive integer, and let $f : \mathbf{R}^n \to \mathbf{R}^m$ be C^{k+1} in a ball $B_r = B_r(\bar{x}) = \{x \in \mathbf{R}^n : |x - \bar{x}| < r\}$ of a radius $r > 0$ centered at a point $\bar{x}$. Assume that the directional derivatives of order $k+1$, taken at every point of B_r along every unit direction, do not exceed certain $L < \infty$:*

$$|D^{k+1} f(x)[d, ..., d]| \leq L \quad \forall (x \in B_r) \forall (d, |d| = 1).$$

Then for the Taylor expansion F_k of order k of f taken at $\bar{x}$, one has

$$|f(x) - F_k(x)| \leq \frac{L|x - \bar{x}|^{k+1}}{(k+1)!} \quad \forall (x \in B_r).$$

Thus, in a neighbourhood of $\bar{x}$, the remainder of the kth-order Taylor expansion, taken at $\bar{x}$, is of order of $L|x - \bar{x}|^{k+1}$, where L is the maximal (over all unit directions and all points from the neighbourhood) magnitude of the directional derivatives of order $k+1$ of f.

Appendix C

Symmetric Matrices

C.1 Spaces of matrices

Let $\mathbf{S}^m$ be the space of symmetric $m \times m$ matrices, and $\mathbf{M}^{m,n}$ be the space of rectangular $m \times n$ matrices with real entries. From the viewpoint of their linear structure (i.e., the operations of addition and multiplication by reals), $\mathbf{S}^m$ is just the arithmetic linear space $\mathbf{R}^{m(m+1)/2}$ of dimension $\frac{m(m+1)}{2}$: By arranging the elements of a symmetric $m \times m$ matrix X in a single column, say, in the row-by-row order, you get a usual m^2-dimensional column vector; multiplication of a matrix by a real and addition of matrices correspond to the same operations with the "representing vector(s)." When X runs through $\mathbf{S}^m$, the vector representing X runs through $m(m+1)/2$-dimensional subspace of $\mathbf{R}^{m^2}$ consisting of vectors satisfying the "symmetry condition" — the coordinates coming from cells i,j and j,i in X are equal to each other. Similarly, $\mathbf{M}^{m,n}$ as a linear space is just $\mathbf{R}^{mn}$, and it is natural to equip $\mathbf{M}^{m,n}$ with the inner product defined as the usual inner product of the vectors representing the matrices:

$$\begin{aligned}\langle X, Y\rangle &= \sum_{i=1}^{m}\sum_{j=1}^{n} X_{ij}Y_{ij} = \mathrm{Tr}(X^TY) = \mathrm{Tr}(Y^TX)\\ &= \mathrm{Tr}(XY^T) = \mathrm{Tr}(YX^T),\end{aligned}$$

where Tr stands for the *trace* — the sum of diagonal elements of a (square) matrix; the above equalities are given by direct computation. With this inner product (called the *Frobenius inner product*),

$\mathbf{M}^{m,n}$ becomes a legitimate Euclidean space, and we may use in connection with this space all notions based upon the Euclidean structure, e.g., the (Frobenius) norm of a matrix

$$\|X\|_2 = \sqrt{\langle X, X\rangle} = \sqrt{\sum_{i=1}^{m}\sum_{j=1}^{n} X_{ij}^2} = \sqrt{\mathrm{Tr}(X^TX)},$$

and likewise the notions of orthogonality, orthogonal complement of a linear subspace, etc. The same applies to the space $\mathbf{S}^m$ equipped with the Frobenius inner product; of course, the Frobenius inner product of symmetric matrices can be written without the transposition sign:

$$\langle X, Y\rangle = \mathrm{Tr}(XY) = \mathrm{Tr}(YX), \quad X, Y \in \mathbf{S}^m.$$

C.2 Eigenvalue decomposition

Let us focus on the space $\mathbf{S}^m$ of symmetric matrices. The most important property of these matrices is as follows.

Theorem C.1 (Eigenvalue decomposition). *An $n \times n$ matrix A is symmetric if and only if it admits an orthonormal system of eigenvectors: There exist orthonormal basis $\{e_1, ..., e_n\}$ such that*

$$Ae_i = \lambda_i e_i, \quad i = 1, ..., n \tag{C.1}$$

for reals λ_i.

In connection with Theorem C.1, it is worthy to recall the following notions and facts.

C.2.A. Eigenvectors and eigenvalues: An *eigenvector* of an $n\times n$ matrix A is a *nonzero* vector e (real or complex) such that $Ae = \lambda e$ for (real or complex) scalar λ; this scalar is called the *eigenvalue* of *A corresponding to the eigenvector e.*

Eigenvalues of A are exactly the roots of the *characteristic polynomial*

$$\pi(z) = \det(zI - A) = z^n + b_1 z^{n-1} + b_2 z^{n-2} + \cdots + b_n$$

of A.

Theorem C.1 states, in particular, that for a symmetric matrix A, all eigenvalues are real, and the corresponding eigenvectors can be chosen to be real and form an orthonormal basis in $\mathbf{R}^n$.

C.2.B. Eigenvalue decomposition of a symmetric matrix: Theorem C.1 admits equivalent reformulation as follows (check the equivalence!).

Theorem C.2. *An $n \times n$ matrix A is symmetric if and only if it can be represented in the form*

$$A = U\Lambda U^T, \tag{C.2}$$

where

- *U is an orthogonal matrix: $U^{-1} = U^T$ (or, which is the same, $U^T U = I$ or, which is the same, $UU^T = I$ or, which is the same, the columns of U form an orthonormal basis in $\mathbf{R}^n$ or, which is the same, the transposes of rows of U form an orthonormal basis in $\mathbf{R}^n$).*
- *Λ is the diagonal matrix with the diagonal entries $\lambda_1, ..., \lambda_n$.*

Representation (C.2) with orthogonal U and diagonal Λ is called the *eigenvalue decomposition* of A. In such a representation,

- the columns of U form an orthonormal system of eigenvectors of A;
- the diagonal entries in Λ are the eigenvalues of A corresponding to these eigenvectors.

C.2.C. Vector of eigenvalues: When speaking about eigenvalues $\lambda_i(A)$ of a symmetric $n \times n$ matrix A, we always arrange them in the nonascending order:

$$\lambda_1(A) \geq \lambda_2(A) \geq \cdots \geq \lambda_n(A);$$

$\lambda(A) \in \mathbf{R}^n$ denotes the vector of eigenvalues of A taken in the above order.

C.2.D. Freedom in eigenvalue decomposition: Part of the data Λ, U in the eigenvalue decomposition (C.2) is uniquely defined by A, while the other data admit certain "freedom." Specifically, the sequence $\lambda_1, ..., \lambda_n$ of eigenvalues of A (i.e., diagonal entries of Λ) is exactly the sequence of roots of the characteristic polynomial of A (every root is repeated according to its multiplicity) and is thus uniquely defined by A (provided that we arrange the entries of the sequence in the nonascending order). The columns of U are not uniquely defined by A. What is uniquely defined are the *linear spans*

$E(\lambda)$ of the columns of U corresponding to all eigenvalues equal to certain λ; such a linear span is nothing but the *spectral subspace* $\{x : Ax = \lambda x\}$ of A corresponding to the eigenvalue λ. There are as many spectral subspaces as many different eigenvalues; spectral subspaces corresponding to different eigenvalues of symmetric matrix are orthogonal to each other, and their sum is the entire space. Denoting by $\mu_1 > \mu_2 > \cdots > \mu_k$ the distinct eigenvalues of A, when building an orthogonal matrix U in the spectral decomposition, one chooses an orthonormal eigenbasis in the spectral subspace corresponding to the largest eigenvalue μ_1 and makes the vectors of this basis the first columns in U, then chooses an orthonormal basis in the spectral subspace corresponding to the second largest eigenvalue μ_2 and makes the vector from this basis the next columns of U and so on.

C.2.E. "Simultaneous" decomposition of commuting symmetric matrices: Let $A_1, ..., A_k$ be $n \times n$ symmetric matrices. It turns out that *the matrices commute with each other* ($A_iA_j = A_jA_i$ *for all* i, j) *if and only if they can be "simultaneously diagonalized," i.e., there exist a single orthogonal matrix* U *and diagonal matrices* $\Lambda_1, \ldots, \Lambda_k$ *such that*

$$A_i = U\Lambda_i U^T, \quad i = 1, \ldots, k.$$

You are welcome to prove this statement by yourself; to simplify your task, here are two simple and statements, important by their own right, which help reach your target:

C.2.E.1: *Let* λ *be a real and* A, B *be two commuting* $n \times n$ *matrices. Then the spectral subspace* $E = \{x : Ax = \lambda x\}$ *of* A *corresponding to* λ *is invariant for* B *(i.e.,* $Be \in E$ *for every* $e \in E$*).*

C.2.E.2: *If* A *is an* $n \times n$ *matrix and* L *is an invariant subspace of* A *(i.e.,* L *is a linear subspace such that* $Ae \in L$ *whenever* $e \in L$*), then the orthogonal complement* $L^\perp$ *of* L *is invariant for the matrix* A^T*. In particular, if* A *is symmetric and* L *is invariant subspace of* A*, then* $L^\perp$ *is invariant subspace of* A *as well.*

C.3 Variational characterization of eigenvalues

Theorem C.3 (VCE — Variational Characterization of Eigenvalues, a.k.a. Courant–Fischer–Weyl min–max principle).

Let A be a symmetric matrix. Then

$$\lambda_\ell(A) = \min_{E\in\mathcal{E}_\ell} \max_{x\in E, x^Tx=1} x^TAx, \quad \ell = 1, ..., n, \tag{C.3}$$

where $\mathcal{E}_\ell$ is the family of all linear subspaces in $\mathbf{R}^n$ of the dimension $n - \ell + 1$.

VCE says that to get the largest eigenvalue $\lambda_1(A)$, you should maximize the quadratic form x^TAx over the unit sphere $S = \{x \in \mathbf{R}^n : x^Tx = 1\}$; the maximum is exactly $\lambda_1(A)$. To get the second largest eigenvalue $\lambda_2(A)$, you should act as follows: You choose a linear subspace E of dimension $n - 1$ and maximize the quadratic form x^TAx over the cross-section of S by this subspace; the maximum value of the form depends on E, and you minimize this maximum over linear subspaces E of the dimension $n - 1$; the result is exactly $\lambda_2(A)$. To get $\lambda_3(A)$, you replace in the latter construction subspaces of the dimension $n - 1$ by those of the dimension $n - 2$ and so on. In particular, the smallest eigenvalue $\lambda_n(A)$ is just the minimum, over all linear subspaces E of the dimension $n - n + 1 = 1$, i.e., over all lines passing through the origin, of the quantities x^TAx, where $x \in E$ is unit ($x^Tx = 1$); in other words, $\lambda_n(A)$ is just the minimum of the quadratic form x^TAx over the unit sphere S.

Proof of the VCE is pretty easy. Let $e_1, ..., e_n$ be an orthonormal eigenbasis of A: $Ae_\ell = \lambda_\ell(A)e_\ell$. For $1 \le \ell \le n$, let $F_\ell = \text{Lin}\{e_1, ..., e_\ell\}$, $G_\ell = \text{Lin}\{e_\ell, e_{\ell+1}, ..., e_n\}$. Finally, for $x \in \mathbf{R}^n$, let $\xi(x)$ be the vector of coordinates of x in the orthonormal basis $e_1, ..., e_n$. Note that

$$x^Tx = \xi^T(x)\xi(x),$$

since $\{e_1, ..., e_n\}$ is an orthonormal basis, and that

$$\begin{aligned} x^TAx &= x^TA\sum_i \xi_i(x)e_i = x^T\sum_i \lambda_i(A)\xi_i(x)e_i \\ &= \sum_i \lambda_i(A)\xi_i(x)\underbrace{(x^Te_i)}_{\xi_i(x)} \\ &= \sum_i \lambda_i(A)\xi_i^2(x). \end{aligned} \tag{C.4}$$

Now, given ℓ, $1 \le \ell \le n$, let us set $E = G_\ell$; note that E is a linear subspace of the dimension $n - \ell + 1$. In view of (C.4), the maximum of the quadratic form $x^T Ax$ over the intersection of our E with the unit sphere is

$$\max\left\{\sum_{i=\ell}^{n}\lambda_i(A)\xi_i^2 : \sum_{i=\ell}^{n}\xi_i^2 = 1\right\},$$

and the latter quantity clearly equals to $\max_{\ell\leq i\leq n}\lambda_i(A) = \lambda_\ell(A)$. Thus, for appropriately chosen $E \in \mathcal{E}_\ell$, the inner maximum in the right-hand side of (C.3) equals to $\lambda_\ell(A)$, whence the right-hand side of (C.3) is $\leq \lambda_\ell(A)$. It remains to prove the opposite inequality. To this end, consider a linear subspace E of the dimension $n-\ell+1$ and observe that it has nontrivial intersection with the linear subspace F_ℓ of the dimension ℓ (indeed, $\dim E + \dim F_\ell = (n-\ell+1)+\ell > n$, so that $\dim(E\cap F) > 0$ by the dimension formula). It follows that there exists a unit vector y belonging to both E and F_ℓ. Since y is a unit vector from F_ℓ, we have $y = \sum_{i=1}^{\ell}\eta_i e_i$ with $\sum_{i=1}^{\ell}\eta_i^2 = 1$, whence, by (C.4),

$$y^T Ay = \sum_{i=1}^{\ell}\lambda_i(A)\eta_i^2 \geq \min_{1\leq i\leq \ell}\lambda_i(A) = \lambda_\ell(A).$$

Since y is in E, we conclude that

$$\max_{x\in E: x^T x = 1} x^T Ax \geq y^T Ay \geq \lambda_\ell(A).$$

Since E is an arbitrary subspace form $\mathcal{E}_\ell$, we conclude that the right-hand side in (C.3) is $\geq \lambda_\ell(A)$. □

A simple and useful byproduct of our reasoning is the relation (C.4).

Corollary C.1. *For a symmetric matrix A, the quadratic form $x^T Ax$ is weighted sum of squares of the coordinates $\xi_i(x)$ of x taken with respect to an orthonormal eigenbasis of A; the weights in this sum are exactly the eigenvalues of A :*

$$x^T Ax = \sum_i \lambda_i(A)\xi_i^2(x).$$

C.3.1 *Corollaries of the VCE*

VCE admits a number of extremely important corollaries as follows.

C.3.A. Eigenvalue characterization of positive (semi)definite matrices: Recall that a matrix A is called positive definite (notation: $A \succ 0$) if it is symmetric and the quadratic form $x^T Ax$ is positive outside the origin; A is called positive semidefinite (notation:

$A \succeq 0$) if A is symmetric and the quadratic form $x^T Ax$ is nonnegative everywhere. VCE provides us with the following eigenvalue characterization of positive (semi)definite matrices.

Proposition C.1. *A symmetric matrix A is positive semidefinite if and only if its eigenvalues are nonnegative; A is positive definite if and only if all eigenvalues of A are positive.*

Indeed, A is positive definite if and only if the minimum value of $x^T Ax$ over the unit sphere is positive and is positive semidefinite if and only if this minimum value is nonnegative; it remains to be noted that by VCE, the minimum value of $x^T Ax$ over the unit sphere is exactly the minimum eigenvalue of A.

C.3.B. $\succeq$-Monotonicity of the vector of eigenvalues: Let us write $A \succeq B$ ($A \succ B$) to express that A, B are symmetric matrices of the same size such that $A - B$ is positive semidefinite (respectively, positive definite).

Proposition C.2. *If $A \succeq B$, then $\lambda(A) \geq \lambda(B)$ and if $A \succ B$, then $\lambda(A) > \lambda(B)$.*

Indeed, when $A \succeq B$, then, of course,

$$\max_{x \in E: x^T x = 1} x^T Ax \geq \max_{x \in E: x^T x = 1} x^T Bx$$

for every linear subspace E, whence

$$\begin{aligned} \lambda_\ell(A) &= \min_{E \in \mathcal{E}_\ell} \max_{x \in E: x^T x = 1} x^T Ax \\ &\geq \min_{E \in \mathcal{E}_\ell} \max_{x \in E: x^T x = 1} x^T Bx = \lambda_\ell(B), \quad \ell = 1, ..., n, \end{aligned}$$

i.e., $\lambda(A) \geq \lambda(B)$. The case of $A \succ B$ can be considered similarly.

C.3.C. Eigenvalue interlacement theorem: We shall formulate this extremely important theorem as follows.

Theorem C.4 (Eigenvalue interlacement theorem). *Let A be a symmetric $n \times n$ matrix and $\bar{A}$ be the angular $(n-k) \times (n-k)$ submatrix of A. Then, for every $\ell \leq n - k$, the ℓth eigenvalue of $\bar{A}$ separates the ℓth and the $(\ell + k)$th eigenvalues of A:*

$$\lambda_\ell(A) \succeq \lambda_\ell(\bar{A}) \succeq \lambda_{\ell+k}(A). \tag{C.5}$$

Indeed, by VCE, $\lambda_\ell(\bar{A}) = \min_{E\in\bar{\mathcal{E}}_\ell} \max_{x\in E:x^Tx=1} x^TAx$, where $\bar{\mathcal{E}}_\ell$ is the family of all linear subspaces of the dimension $n-k-\ell+1$ contained in the linear subspace $\{x \in \mathbf{R}^n : x_{n-k+1} = x_{n-k+2} = \cdots = x_n = 0\}$. Since $\bar{\mathcal{E}}_\ell \subset \mathcal{E}_{\ell+k}$, we have

$$\lambda_\ell(\bar{A}) = \min_{E\in\bar{\mathcal{E}}_\ell} \max_{x\in E:x^Tx=1} x^TAx \geq \min_{E\in\mathcal{E}_{\ell+k}} \max_{x\in E:x^Tx=1} x^TAx = \lambda_{\ell+k}(A).$$

We have proved the right inequality in (C.5). Applying this inequality to the matrix $-A$, we get

$$-\lambda_\ell(\bar{A}) = \lambda_{n-k-\ell}(-\bar{A}) \geq \lambda_{n-\ell}(-A) = -\lambda_\ell(A),$$

or, which is the same, $\lambda_\ell(\bar{A}) \leq \lambda_\ell(A)$, which is the first inequality in (C.5).

C.4 Positive semidefinite matrices and semidefinite cone

C.4.A. Positive semidefinite matrices. Recall that an $n \times n$ matrix A is called *positive semidefinite* (notation: $A \succeq 0$) if A is symmetric and produces nonnegative quadratic form:

$$A \succeq 0 \Leftrightarrow \{A = A^T \quad \text{and} \quad x^TAx \geq 0 \quad \forall x\}.$$

A is called positive *definite* (notation: $A \succ 0$) if it is positive semidefinite and the corresponding quadratic form is positive outside the origin:

$$A \succ 0 \Leftrightarrow \{A = A^T \quad \text{and} \quad x^TAx > 00 \quad \forall x \neq 0\}.$$

It makes sense to list a number of equivalent definitions of a positive semidefinite matrix.

Theorem C.5. *Let A be a symmetric $n \times n$ matrix. Then the following properties of A are equivalent to each other:*

(i) $A \succeq 0$,
(ii) $\lambda(A) \geq 0$,
(iii) $A = D^TD$ *for certain rectangular matrix* D,
(iv) $A = \Delta^T\Delta$ *for certain upper triangular* $n \times n$ *matrix* Δ,
(v) $A = B^2$ *for certain symmetric matrix* B,
(vi) $A = B^2$ *for certain* $B \succeq 0$.

The following properties of a symmetric matrix A also are equivalent to each other:

(i′) $A \succ 0$,
(ii′) $\lambda(A) > 0$,
(iii′) $A = D^T D$ *for certain rectangular matrix* D *of rank* n,
(iv′) $A = \Delta^T \Delta$ *for certain nondegenerate upper triangular* $n \times n$ *matrix* Δ,
(v′) $A = B^2$ *for certain nondegenerate symmetric matrix* B,
(vi′) $A = B^2$ *for certain* $B \succ 0$.

Proof. (i)⇔(ii): This equivalence is stated by Proposition C.1.

(ii)⇔(vi): Let $A = U\Lambda U^T$ be the eigenvalue decomposition of A, so that U is orthogonal and Λ is diagonal with nonnegative diagonal entries $\lambda_i(A)$ (we are in the situation of (ii) !). Let $\Lambda^{1/2}$ be the diagonal matrix with the diagonal entries $\lambda_i^{1/2}(A)$; note that $(\Lambda^{1/2})^2 = \Lambda$. The matrix $B = U\Lambda^{1/2}U^T$ is symmetric with nonnegative eigenvalues $\lambda_i^{1/2}(A)$, so that $B \succeq 0$ by Proposition C.1, and

$$B^2 = U\Lambda^{1/2}\underbrace{U^T U}_{I}\Lambda^{1/2}U^T = U(\Lambda^{1/2})^2 U^T = U\Lambda U^T = A,$$

as required in (vi).

(vi)⇒(v): Evident.

(v)⇒(iv): Let $A = B^2$ with certain symmetric B, and let b_i be ith column of B. Applying the Gram–Schmidt orthogonalization process (see proof of Theorem A.5(iii)), we can find an orthonormal system of vectors $u_1, \ldots, u_n$ and lower triangular matrix L such that $b_i = \sum_{j=1}^{i} L_{ij}u_j$ or, which is the same, $B^T = LU$, where U is the orthogonal matrix with the rows $u_1^T, \ldots, u_n^T$. We now have $A = B^2 = B^T(B^T)^T = LUU^TL^T = LL^T$. We see that $A = \Delta^T\Delta$, where the matrix $\Delta = L^T$ is upper triangular.

(iv)⇒(iii): evident.

(iii)⇒(i): If $A = D^T D$, then $x^T A x = (Dx)^T(Dx) \geq 0$ for all x. We have proved the equivalence of the properties (i)–(vi). Slightly modifying the reasoning (do it yourself!), one can prove the equivalence of the properties (i′)–(vi′). □

Remark C.1. (i) (Checking positive semidefiniteness): Given an $n \times n$ symmetric matrix A, one can check whether it is positive semidefinite by a purely algebraic finite algorithm (the so-called *Lagrange diagonalization of a quadratic form*), which requires at most $O(n^3)$ arithmetic operations. Positive definiteness of a matrix can also be checked by the *Choleski factorization algorithm* which finds the decomposition in (iv′) if it exists, in approximately $\frac{1}{6}n^3$ arithmetic operations.

There exists another useful algebraic criterion (Sylvester's criterion) for positive semidefiniteness of a matrix; according to this criterion, a symmetric matrix A is positive definite if and only if its angular minors are positive, and A is positive semidefinite if and only if all its principal minors are nonnegative. For example, a symmetric 2×2 matrix $A = \begin{bmatrix} a & b \\ b & c \end{bmatrix}$ is positive semidefinite if and only if $a \geq 0$, $c \geq 0$ and $\det(A) \equiv ac - b^2 \geq 0$.

(ii) [Square root of a positive semidefinite matrix]: By the first chain of equivalences in Theorem C.5, a symmetric matrix A is $\succeq 0$ if and only if A is the square of a positive semidefinite matrix B. The latter matrix is uniquely defined by $A \succeq 0$ and is called the *square root* of A (notation: $A^{1/2}$).

C.4.B. The semidefinite cone: When adding symmetric matrices and multiplying them by reals, we add, respectively multiply by reals, the corresponding quadratic forms. It follows that

> **C.4.B.1:** *the sum of positive semidefinite matrices and a product of a positive semidefinite matrix and a nonnegative real is positive semidefinite* or, which is the same (see Section 2.1.4);
>
> **C.4.B.2:** *$n \times n$ positive semidefinite matrices form a cone $\mathbf{S}^n_+$ in the Euclidean space $\mathbf{S}^n$ of symmetric $n \times n$ matrices, the Euclidean structure being given by the Frobenius inner product $\langle A, B \rangle = \mathrm{Tr}(AB) = \sum_{i,j} A_{ij} B_{ij}$.*

The cone $\mathbf{S}^n_+$ is called the *semidefinite* cone of size n. It is immediately seen that the semidefinite cone $\mathbf{S}^n_+$ is "good," specifically,

- $\mathbf{S}^n_+$ is closed: the limit of a converging sequence of positive semidefinite matrices is positive semidefinite;
- $\mathbf{S}^n_+$ is pointed: the only $n \times n$ matrix A such that both A and $-A$ are positive semidefinite is the zero $n \times n$ matrix;
- $\mathbf{S}^n_+$ possesses a nonempty interior composed of positive definite matrices.

Note that the relation $A \succeq B$ means exactly that $A - B \in \mathbf{S}^n_+$, while $A \succ B$ is equivalent to $A - B \in \operatorname{int} \mathbf{S}^n_+$. The "matrix inequalities" $A \succeq B$ ($A \succ B$) match the standard properties of the usual scalar inequalities:

$A \succeq A$ [reflexivity],
$A \succeq B,\ B \succeq A \Rightarrow A = B$ [antisymmetry],
$A \succeq B,\ B \succeq C \Rightarrow A \succeq C$ [transitivity],
$A \succeq B,\ C \succeq D \Rightarrow A + C \succeq B + D$ [compatibility with linear operations, I],
$A \succeq B, \lambda \geq 0 \Rightarrow \lambda A \succeq \lambda B$ [compatibility with linear operations, II],
$A_i \succeq B_i, A_i \to A, B_i \to B$ as $i \to \infty \Rightarrow A \succeq B$ [closedness],

with evident modifications when $\succeq$ is replaced with $\succ$, like

$$A \succeq B, C \succ D \quad \Rightarrow \quad A + C \succ B + D,$$

etc. Along with these standard properties of inequalities, the inequality $\succeq$ possesses a nice additional property:

C.4.B.3: *In a valid $\succeq$-inequality*

$$A \succeq B,$$

one can multiply both sides from the left and by the right by a (rectangular) matrix and its transpose:

$$\begin{array}{c} A, B \in \mathbf{S}^n, \quad A \succeq B, \quad V \in \mathbf{M}^{n,m}, \\ \Downarrow \\ V^T A V \succeq V^T B V. \end{array}$$

Indeed, we should prove that if $A - B \succeq 0$, then also $V^T(A - B)V \succeq 0$, which is immediate — the quadratic form $y^T[V^T(A - B)V]y = (Vy)^T(A - B)(Vy)$ of y is nonnegative along with the quadratic form $x^T(A - B)x$ of x.

An important additional property of the semidefinite cone is its *self-duality*.

Theorem C.6. *A symmetric matrix Y has nonnegative Frobenius inner products with all positive semidefinite matrices if and only if Y itself is positive semidefinite.*

Proof. *"If" part:* Assume that $Y \succeq 0$, and let us prove that then $\mathrm{Tr}(YX) \geq 0$ for every $X \succeq 0$. Indeed, the eigenvalue decomposition of Y can be written as

$$Y = \sum_{i=1}^{n} \lambda_i(Y) e_i e_i^T,$$

where e_i are the orthonormal eigenvectors of Y. We now have

$$\begin{aligned}\mathrm{Tr}(YX) &= \mathrm{Tr}\left(\left(\sum_{i=1}^{n} \lambda_i(Y) e_i e_i^T\right) X\right) = \sum_{i=1}^{n} \lambda_i(Y)\,\mathrm{Tr}(e_i e_i^T X)\\ &= \sum_{i=1}^{n} \lambda_i(Y)\,\mathrm{Tr}(e_i^T X e_i),\end{aligned} \tag{C.6}$$

where the concluding equality is given by the following well-known property of the trace:

C.4.B.4: *Whenever matrices A, B are such that the product AB makes sense and is a square matrix, one has*

$$\mathrm{Tr}(AB) = \mathrm{Tr}(BA).$$

Indeed, we should verify that if $A \in \mathbf{M}^{p,q}$ and $B \in \mathbf{M}^{q,p}$, then $\mathrm{Tr}(AB) = \mathrm{Tr}(BA)$. The left-hand side quantity in our hypothetic equality is $\sum_{i=1}^{p} \sum_{j=1}^{q} A_{ij} B_{ji}$, and the right-hand side quantity is $\sum_{j=1}^{q} \sum_{i=1}^{p} B_{ji} A_{ij}$; they are indeed equal.

Looking at the concluding quantity in (C.6), we see that it indeed is nonnegative whenever $X \succeq 0$ (since $Y \succeq 0$ and thus $\lambda_i(Y) \geq 0$ by Proposition C.1).

"only if" part: We are given Y such that $\mathrm{Tr}(YX) \geq 0$ for all matrices $X \succeq 0$, and we should prove that $Y \succeq 0$. This is immediate: for every vector x, the matrix $X = xx^T$ is positive semidefinite (Theorem C.5(iii)), so that $0 \leq \mathrm{Tr}(Yxx^T) = \mathrm{Tr}(x^T Y x) = x^T Y x$. Since the resulting inequality $x^T Y x \geq 0$ is valid for every x, we have $Y \succeq 0$. □

Bibliography

Ben-Tal, A. and Nemirovski, A. (2000). Robust solutions of linear programming problems contaminated with uncertain data, *Mathematical Programming, Series A* **88**, 411–424.

Ben-Tal, A. and Nemirovski, A. (2001). *Lectures on Modern Convex Optimization: Analysis, Algorithms and Engineering Applications* (SIAM, Philadelphia).

Ben-Tal, A. and Nemirovski, A. (2022). Lectures on modern convex optimization, https://www.isye.gatech.edu/~nemirovs/LMCOLN2023Spring.pdf.

Ben-Tal, A., El Ghaoui, L., and Nemirovski, A. (2009). *Robust Optmization* (Princeton University Press. Princeton, NJ, USA).

Bertsimas, D. and Den Hertog, D. (2022). *Robust and Adaptive Optmization* (Dynamic Ideas. Charlestown, MA, USA).

Bertsimas, D. and Tsitsiklis, J. N. (1997). *Introduction to Linear Optimization* (Athena Scientific, Nashua, NH, USA).

Bertsimas, D., Iancu, D., and Parrilo, P. (2010). Optimality of affine policies in multi-stage robust optimization, *Mathematics of Operations Research* **35**, 363–394.

Boyd, S., El Ghaoui, L., Feron, E., and Balakrishnan, V. (1994). *Linear Matrix Inequalities in System and Control Theory* (SIAM, Philadelphia).

Boyd, S. P. and Vandenberghe, L. (2004). *Convex Optimization* (Cambridge University Press, Cambridge, UK).

Frenk, H., Roos, C., Terlaky, T., and Zhang, S. (eds.) (1999). *High Performance Optimization* (Springer, Berlin/Heidelberg/Dordrecht/New York City).

Grotschel, M., Lovasz, L., and Schrijver, A. (1987). *Geometric Algorithms and Combinatorial Optimization* (Springer-Verlag, Berlin).

Khachiyan, L. G. (1979). A polynomial algorithm in linear programming, *Doklady Akademii Nauk SSSR* (in Russian) **244**, pp. 1093–1097, English translation in *Soviet Math Doklady* v. 20, pp. 191–104.

Monteiro, R. D. and Zhang, Y. (1998). A unified analysis for a class of long-step primal–dual path-following interior-point algorithms for semidefinite programming, *Mathematical Programming* **81**, 281–299.

Nemirovski, A. and Yudin, D. (1976). Informational complexity and efficient methods for the solution of convex extremal problems, *Ekonomika i Matematicheskie Metody* (in Russian) **12**, pp. 357–379, English translation in *Matekon* **13**(2), 3–25.

Nemirovski, A., Onn, S., and Rothblum, U. (2010). Accuracy certificates for computational problems with convex structure, *Mathematics of Operations Research* **35**, 1, 52–78.

Nesterov, Y. (2018). *Lectures on Convex Optimization*, Vol. 137 (Springer, Berlin/Heidelberg/Dordrecht/New York City).

Nesterov, Y. and Nemirovskii, A. (1994). *Interior-Point Polynomial Algorithms in Convex Programming* (SIAM, Philadelphia).

Nesterov, Y. and Todd, M. J. (1997). Self-scaled barriers and interior-point methods for convex programming, *Mathematics of Operations Research* **22**, 1–42.

Nesterov, Y. and Todd, M. J. (1998). Primal–dual interior-point methods for self-scaled cones, *SIAM Journal on Optimization* **8**, 324–364.

Nesterov, Y., Todd, M., and Ye, Y. (1996). Infeasible-start primal–dual methods and infeasibility detectors for nonlinear programming problems, Tech. Rep. 1156, School of Operations Research and Industrial Engineering, Cornell University, Ithaca, New York.

Renegar, J. (2001). *A Mathematical View of Interior-Point Methods in Convex Optimization* (SIAM, Philadelphia).

Roos, C., Terlaky, T., and Vial, J.-P. (2005). *Interior Point Methods for Linear Optimization* (Springer, New York, NY, USA).

Shor, N. Z. (1977). Cut-off method with space extension in convex programming problems, *Kibernetika* (in Russian) **13**, 1, pp. 94–95, English translation in *Cybernetics* **13**(1), 94–96.

Tardos, E. (1985). A strongly polynomial minimum cost circulation algorithm, *Combinatorica* **5**, 247–256.

Vavasis, S. A. and Ye, Y. (1996). A primal–dual interior point method whose running time depends only on the constraint matrix, *Mathematical Programming* **74**, 79–120.

Wright, S. J. (1997). *Primal-Dual Interior-Point Methods* (SIAM, Philadelphia).

Ye, Y. (1997). *Interior Point Algorithms: Theory and Analysis* (John Wiley & Sons, Hoboken, NJ, USA).

Solutions to Selected Exercises

Exercises for Chapter 1

Exercise 1.5. Run the experiment as follows:

(1) Pick at random in the segment $[0, 1]$ two "true" parameters θ_0^* and θ_1^* of the regression model

$$y = \theta_0 + \theta_1 x;$$

(2) Generate a sample of $N = 1000$ observation errors $\xi_i \sim P$, where P is a given distribution, and then generate observations y_i according to

$$y_i = \theta_0^* + \theta_1^* x_i + \xi_i, \; x_i = i/N;$$

(3) Estimate the parameters from the observations according to the following three estimation schemes:

$$\begin{array}{ll} \text{uniform fit}: & [\theta_{0,\infty}; \theta_{1,\infty}] = \text{argmin}_\theta \max\limits_{1 \le i \le N} |y_i - [\theta_0 + \theta_1 x_i]|, \\ \text{least squares fit}: & [\theta_{0,2}; \theta_{1,2}] = \text{argmin}_\theta \sum\limits_{1 \le i \le N} (y_i - [\theta_0 + \theta_1 x_i])^2, \\ \ell_1 \text{ fit}: & [\theta_{0,1}; \theta_{1,1}] = \text{argmin}_\theta \sum\limits_{1 \le i \le N} |y_i - [\theta_0 + \theta_1 x_i]|, \end{array}$$

and compare the estimates with the true values of the parameters.

Run 3 series of experiments:

- P is the uniform distribution on $[-1, 1]$;
- P is the standard Gaussian distribution with the density $\frac{1}{\sqrt{2\pi}} \exp\{-t^2/2\}$;
- P is the Cauchy distribution with the density $\frac{1}{\pi(1+t^2)}$.

Try to explain the results you get. When doing so, you can think about a simpler problem, where you are observing N times a scalar parameter θ^* according to $y_i = \theta^* + \xi_i$, $1 \leq i \leq N$, and then use the above techniques to estimate θ^*.

Solution: My results were as follows (in the table, I present average, over 10 experiments, recovering errors $\|\theta - \theta^*\|_2$):

	Uniform	Gauss	Cauchy
uniform fit	0.0079	0.7738	1729.0
least squares fit	0.0746	0.1177	18.848
ℓ_1 fit	0.1110	0.1128	0.1466

An explanation, in the case of estimating a parameter θ^* from N noisy observations $y_i = \theta^* + \xi_i$, is as follows. Observe that in this simple case, the estimates $\widehat{\theta}$ are as follows:

- uniform fit: $\widehat{\theta} = \frac{\max_i y_i + \min_i y_i}{2}$;
- least squares fit: $\widehat{\theta} = \frac{1}{N}\sum_i y_i$;
- ℓ_1 fit: $\widehat{\theta}$ is the empirical median of the sample of observations: A point such that $N/2$ of y_i are to the left, and $N/2$ of y_i are to the right of this point (think why).

Now,

- When the noises are uniformly distributed in $[-1, 1]$, the estimate based on the uniform fit is typically within $1/N$ of the true value, since in this case $\max_i y_i$ typically is within $O(1/N)$ of $\theta^* + 1$, and $\min_i y_i$ typically is within $O(1/N)$ of $\theta^* - 1$. Indeed, the probability of the event $\max_i y_i < 1 - \delta$ is $(1 - \delta/2)^N$, which is non-negligible only when $\delta \leq O(1/N)$. As a result, the estimate based on the uniform fit exhibits *superefficiency*: "normal" estimates, like the empirical mean or the empirical median, have typical accuracy $O(1/\sqrt{N})$ (to see it, compute the variance of $\theta^* - \frac{1}{N}\sum_i y_i$), and this is what we clearly see in the table. Note, however, that this phenomenon would disappear if instead of distribution "with sharp edges," like the uniform one, we would use distribution with smooth density vanishing outside of $[-1, 1]$.
- When the noises are Gaussian, the uniform fit does not make much sense — it is just inconsistent (i.e., the estimate does not converge

to the true value in probability when $N \to \infty$), which is intuitively clear: since the distribution has now infinite support, the larger the sample size N, the larger the variances in the maximum and the minimum of the observations become. Theoretically, the best under the circumstances, in a certain precise sense, is the empirical mean estimate, although the quality of the empirical median estimate is within a constant factor of the one of empirical mean.

- When the noises are distributed according to the Cauchy distribution, the situation becomes quite different. Now, the distribution has really "heavy tails" so heavy that even the expectation of the modulus of the noise, not speaking of the variance, is infinite. As a result, the uniform fit leads to a complete disaster, and the empirical mean — to nearly complete disaster, since this estimate is sensitive to the magnitude of noise. Formally speaking, the major argument in favor of empirical mean is that if the distribution of noise has zero mean and finite variance σ^2, then the expected squared error of recovering θ^* by the empirical mean estimate taken over N observations is σ^2/N, that is, it goes to 0 as $N \to \infty$. Since the Cauchy distribution has infinite variance, we have no reasons to expect the empirical mean to behave itself in this case. In contrast to this, the empirical median still works — what matters here is that the median of the noise distribution is 0, and that the density of this distribution is continuous and positive at 0. Indeed, in order for the empirical median of observations to be $> \theta^* + \epsilon$, at least half of the observations should be $> \theta^* + \epsilon$. For a single observation, the chances to be $> \theta^* + \epsilon$ are $1/2 - \delta$, with δ of order of $p(0)\epsilon$, where $p(0)$ is the density of the noise distribution at the origin. By the Law of Large Numbers, the fraction of observations which are $> \theta^* + \epsilon$ for large N should, with overwhelming probability, be close to $1/2 - \delta$, so that the probability of this fraction being $\geq 1/2$ goes to 0 as $N \to \infty$. Thus, the probability of the empirical median being $> \theta^* + \epsilon$ for every $\epsilon > 0$ goes to 0 as $N \to \infty$, and similarly — for the probability of the empirical median being $\leq \theta^* - \epsilon$. This being said, note that the applicability of the empirical median to estimating the shift parameter of a "heavy tail" distribution requires the shift parameter to be the median of the shifted distribution.

Exercise 1.10. (Computational study): Generate at random and solve 1000 LO problems

$$\max_{x\in\mathbf{R}^5}\left\{\sum_{j=1}^{5} c_j x_j : \sum_{j=1}^{5} a_{ij}x_j \le b_j, i \le 10\right\}$$

to get an impression of what are the chances to get solvable/infeasible/unbounded instances. Draw the entries in the data, independently of each other, from the uniform distribution on $\{-1, 1\}$.

Solution: My results are: 31.4% of solvable, 40.8% of infeasible, and 27.8% of unbounded instances.

Exercises for Chapter 2

Exercise 2.8. Let X be a nonempty convex set in $\mathbf{R}^n$ and $x \in X$. Prove that x is an extreme point of X

(1) if and only if the set $X\backslash\{x\}$ is convex,
(2) if and only if for every representation $x = \sum_{i=1}^{k} \lambda_i x_i$ of x as a convex combination of points from X with positive coefficients λ_i, $x_i = x$, $i = 1, \ldots, k$ holds.

Solution: Assume that x is not an extreme point of X, so that there exists $h \neq 0$ such that $x \pm h \in X$. Then the set $X\backslash\{x\}$ contains both $x_1 = x - h$ and $x_2 = x + h$, but does not contain $x = \frac{1}{2}[x_1 + x_2]$ and thus is not convex. Similarly, we have $x = \frac{1}{2}x_1 + \frac{1}{2}x_2$ with $x_1, x_2 \in X$ and x_1 distinct from x. Thus, both properties claimed to be equivalent to x being an extreme point of X are violated.

Now, assume that x is an extreme point of X. Let us prove that the set $X' = X\backslash\{x\}$ is convex. Indeed, assuming the opposite, there exist $x_1, x_2 \in X'$ and $\lambda \in [0, 1]$ such that $\lambda x_1 + (1-\lambda)x_2 \notin X'$; since X is convex and $X = X' \cup \{x\}$, it follows that $\lambda x_1 + (1-\lambda)x_2 = x$. Since x_1 and x_2 are distinct from x, we should have $\lambda > 0, 1 - \lambda > 0$ and $x_1 \neq x_2$. Setting $e = x_1 - x_2$, $\mu = \min[\lambda, 1-\lambda]$ and $h = \mu e$, we get $h \neq 0$ and $x - h = \lambda x_1 + (1-\lambda)x_2 - \mu(x_1 - x_2) = (\lambda - \mu)x_1 + (1 - \lambda + \mu)x_2$, which is a convex combination of x_1 and x_2, that is, $x - h \in X$. Similarly, $x + h \in X$, which is a desired contradiction with the geometric characterization of extreme points.

Still assuming that x is an extreme point of X, let us prove that if $x = \sum_{i=1}^k \lambda_i x_i$ with $x_i \in X$, $\lambda_i > 0$ and $\sum_i \lambda_i = 1$, then $x_1 = \cdots = x_k = x$. Assuming the opposite, one of x_i's, say, x_1, is different from x. Then $\lambda_1 < 1$, so that the point $\bar{x} = \frac{1}{1-\lambda_1}\sum_{i=2}^k \lambda_i x_i$ is well defined; this point is a convex combination of points from X and thus belongs to X, and we have $x = \lambda_1 x_1 + (1-\lambda_1)\bar{x}$. Since $x_1 \neq x$, we have $x_1 \neq \bar{x}$, whence, same as above, there exists $h \neq 0$ such that $x \pm h \in X$, which is a desired contradiction.

Exercise 2.9. Let $X = \{x \in \mathbf{R}^n : a_i^T x \leq b_i, 1 \leq i \leq m\}$ be a polyhedral set in $\mathbf{R}^n$. Prove that

(1) If X' is a face of X, then there exists a linear function $e^T x$ such that

$$X' = \operatorname*{Argmax}_{x \in X} e^T x := \left\{ x \in X : e^T x = \sup_{x' \in X} e^T x' \right\}.$$

(2) If v is a vertex of X, then there exists a linear function $e^T x$ such that v is the unique maximizer of this function on X.

Solution: (1) Since X' is a face of X, we have $X' = X_I := \{x \in \mathbf{R}^n : a_i^T x \leq b_i \,\forall i \leq m, a_i^T x = b_i, i \in I\}$ for some $I \subset \{1, \ldots, n\}$. Let us set $e = \sum_{i \in I} a_i$, and let us prove that e is as required. Indeed, let $w \in \operatorname{Argmax}_{x \in X} e^T x$. Choosing $u \in X_I$ (a face is nonempty by definition!), we have $e^T w \geq e^T u$, that is, $\sum_{i \in I} a_i^T [w - u] \geq 0$. For $i \in I$, we have $a_i^T u = b_i$, while $a_i^T w \leq b_i$ due to $w \in X$, that is, $a_i^T [w - u] \leq 0$. Taking into account that $\sum_{i \in I} a_i^T [w - u] \geq 0$, we conclude that $a_i^T [w - u] = 0$ for all $i \in I$, that is, $w \in X_I$. Thus, $\operatorname{Argmax}_{x \in X} e^T x \subset X_I$. Since $e^T x$ by construction is constant on X_I, we conclude that $\operatorname{Argmax}_{x \in X} e^T x = X_I = X'$. □

(2) This is a particular case of (1), since an extreme point of X by definition is a singleton face of X. □

Exercise 2.11. By Birkhoff Theorem, the extreme points of the polytope $\Pi_n = \{[x_{ij}] \in \mathbf{R}^{n \times n} : x_{ij} \geq 0, \sum_i x_{ij} = 1 \,\forall j, \sum_j x_{ij} = 1 \,\forall i\}$ are exactly the Boolean (with entries 0 and 1) matrices from this set. Prove that the same holds true for the "polytope of sub-double-stochastic" matrices $\Pi_{m,n} = \{[x_{ij}] \in \mathbf{R}^{m \times n} : x_{ij} \geq 0, \sum_i x_{ij} \leq 1 \,\forall j, \sum_j x_{ij} \leq 1 \,\forall i\}$.

Solution: First, every Boolean matrix $[x_{ij}]$ from $\Pi_{m,n}$ is an extreme point. Indeed, we know that every Boolean matrix is an extreme point of the box $B_{m,n} = \{[x_{ij}] \in \mathbf{R}^{m \times n} : 0 \le x_{ij} \le 1 \, \forall i, j\}$, and it remains to refer to the evident fact: *When $Y \subset X$ is a nested pair of convex sets, then every extreme point v of X which happens to be in Y is an extreme point of Y.* Indeed, were v the midpoint of a nontrivial segment in Y, it would be the midpoint of a nontrivial segment in X, which is not the case.

To prove that all extreme points of $\Pi_{m,n}$ are Boolean matrices from this polytope, it suffices to show that $\Pi_{m,n} = \text{Conv}(B)$, where B is the set of all Boolean matrices from $\Pi_{m,n}$ (see Corollary 2.9). Our plan is as follows: given a matrix $x \in \Pi_{m,n}$, we will show that x can be made a North-Western $m \times n$ submatrix of $k \times k$ double-stochastic matrix $\overline{x}$, with properly selected k. This is all we need: by Birkhoff Theorem, $\overline{x}$ is a convex combination of $k \times k$ permutation matrices, implying that x is a convex combination of the $m \times n$ North-Western submatrices of these permutation matrices, and these submatrices clearly are Boolean matrices from $\Pi_{m,n}$.

Thus, let a matrix $x \in \Pi_{m,n}$ be given; we want to extend it by adding several rows and columns to a larger double-stochastic matrix. First, by adding to x $n-m$ zero rows (if $n > m$) or $m-n$ zero columns (if $m > n$), we can reduce the situation to the one where $m = n$, which we assume from now on. Next, let $S = \sum_{i,j=1}^{n} x_{ij}$. Note that since the row sums in x are ≤ 1, we have $S \le n$, so that $\kappa := n - S$ is nonnegative; let d be the smallest integer which is $\ge \kappa$. This is how we can embed x, as the North-Western $n \times n$ submatrix, into $(n+d) \times (n+d)$ double-stochastic matrix. Denote by r_i, $i \le n$, the sum of entries in ith row of x, and by c_j, $j \le n$, the sum of entries in jth column of x. Note that $0 \le r_i \le 1$, $0 \le c_j \le 1$ and $\sum_i r_i = \sum_j c_j = S$. Let also $\rho_i = 1 - r_i$, $\sigma_j = 1 - c_j$, so that $\rho_i \ge 0$, $\sigma_j \ge 0$, and $\sum_i \rho_i = \sum_j \sigma_j = \kappa$. Now, let $\rho = [\rho_1/d; \rho_2/d; \ldots; \rho_n/d]$ and $\sigma = [\sigma_1/d; \sigma_2/d; \ldots; \sigma_n/d]$, so that ρ and σ are nonnegative vectors with sums of entries equal to $\kappa/d \le 1$. Setting $\theta = (1 - \kappa/d)/d$ and specifying $\overline{x}$ as the $(n+d) \times (n+d)$ matrix $\left[\begin{array}{c|c|c|c} x & \rho & \ldots & \rho \\ \hline \sigma^T & \theta & \ldots & \theta \\ \hline \vdots & \vdots & \ddots & \ldots \\ \hline \sigma^T & \theta & \ldots & \theta \end{array}\right]$, we, as is immediately seen, get a double-stochastic matrix, and this is the desired double-stochastic extension of x. □

Exercise 2.12. (Follow-up to Exercise 2.11): Let x be an $n \times n$ entrywise nonnegative matrix with all row and all column sums ≤ 1. Is it true that for some double-stochastic matrix $\overline{x}$, the matrix $\overline{x} - x$ is entrywise nonnegative?

Solution: Yes. By the result of Exercise 2.11, x is a convex combination of Boolean matrices with column and row sums ≤ 1. Every matrix with the latter property clearly is obtained from an appropriate permutation matrix by replacing some of the unit entries with zeros. Thus, every Boolean matrix with row and column sums ≤ 1 is entrywise $\leq$ a permutation matrix, and therefore a convex combination of the matrices of the former class is entrywise $\leq$ a convex combination of permutation matrices, which is a double-stochastic matrix. $\square$

Exercise 2.15. Prove that if K is a polyhedral cone, then the dual cone K_* is so, and $(K_*)_* = K$.

Solution: By the theorem on the structure of a polyhedral set, we have

$$K = \text{Conv}(\{v_1, \ldots, v_N\}) + \text{Cone}(\{r_1, \ldots, r_M\}),$$

whence, by the previous exercise,

$$K = \text{Rec}(K) = \text{Cone}(\{r_1, \ldots, r_M\}).$$

The fact that $(\text{Cone}(\{r_1, \ldots, r_M\}))_*$ is a polyhedral cone is evident: $K_* = \{x : x^T r_j \geq 0, 1 \leq j \leq M\}$. The fact that $((\text{Cone}(\{r_1, \ldots, r_M\}))_*)_* = \text{Cone}(\{r_1, \ldots, r_M\})$ was proved in Theorem 2.10 (this immediate consequence of the homogeneous Farkas lemma was the key instrument in our proof of the theorem on the structure of a polyhedral set.)

An alternative proof is as follows: K contains the origin and thus Polar (K) is polyhedral along with K, and $K = \text{Polar}(\text{Polar}(K))$ (items A and B in Section 2.4.2). It remains to note that for a cone K, $\text{Polar}(K) = -K_*$.

Exercise 2.16. Prove that if $K = \{x \in \mathbf{R}^n : a_i^T x \geq 0, i = 1, \ldots, m\}$, then $K_* = \text{Cone}(\{a_1, \ldots, a_m\})$.

Solution: By the theorem on the structure of polyhedral sets, the cone $\text{Cone}(\{a_1, \ldots, a_m\})$ is polyhedral, and we clearly have $(\text{Cone}(\{a_1, \ldots, a_m\})_* = \{x : a_i^T x \geq 0, 1 \leq i \leq m\}$. Taking the dual cones to the left- and the right-hand side and invoking the result of Exercise 2.15, we get $\text{Cone}(\{a_1, \ldots, a_m\}) = ((\text{Cone}(\{a_1, \ldots, a_m\}))_*)_* = (\{x : a_i^T x \geq 0, 1 \leq i \leq m\})_*$. $\square$

Exercise 2.17. Prove that if K is a polyhedral cone and d is a generator of an extreme ray of K, then in every representation

$$d = d_1 + \cdots + d_M, \; d_i \in K \; \forall i$$

d_i are nonnegative multiples of d.

Solution: Setting $d^i = d_1 + \cdots + d_{i-1} + d_{i+1} + \cdots + d_M$, observe that $d^i \in K$ and $d = d_i + d^i$; since d is a generator of an extreme ray in K, it follows that d_i is a nonnegative multiple of d, and this is so for every i. □

Exercise 2.18. Let,

$$K = \text{Cone}(\{r_1, \ldots, r_M\}).$$

Prove that if R is an extreme ray of K, then one of r_j can be chosen as a generator of R. What is the "extreme point" analogy of this statement?

Solution: Let d be a generator of R. Then

$$d = \sum_j \mu_j r_j$$

with appropriate $\mu_j \geq 0$. By the result of the previous exercise, all $\mu_j r_j$ are nonnegative multiples of d; since $d \neq 0$, there exists j_* such that $\mu_{j_*} r_{j_*} \neq 0$, that is, $\mu_{j_*} r_{j_*}$ is a positive multiple of d, whence d is a positive multiple of r_{j_*} (recall that $\mu_j \geq 0$). It follows that R is generated by r_{j_*}. □

The "extreme point" analogy of the statement is $\text{Ext}(\text{Conv}(V)) \subset V$.

Exercise 2.19. Let $K_1, \ldots, K_m$ be polyhedral cones in $\mathbf{R}^n$. Prove that

(1) *$K_1 + \cdots + K_m$ is a polyhedral cone in $\mathbf{R}^n$.*

Solution: One way to justify the claim is to note that by calculus of polyhedral representations the arithmetic sum of polyhedral sets is polyhedral; when the sets are cones, the sum is a cone as well. Another reasoning goes as follows: by the theorem on the structure of polyhedral sets, $K_i = \text{Cone}(R_i)$ for a finite set R_i, whence $K_1 + \cdots + K_m = \text{Cone}(R_1 \cup \cdots \cup R_m)$, and the right-hand side is a polyhedral cone by the same theorem.

(2) *$(K_1 \cap K_2 \cap \cdots \cap K_m)_* = (K_1)_* + \cdots + (K_m)_*$.*

Solution: For arbitrary cones $M_1, \ldots, M_m$, we clearly have $(M_1 + \cdots + M_m)_* = (M_1)_* \cap \cdots \cap (M_m)_*$. Setting $M_i = (K_i)_*$ we get $((K_1)_* + \cdots + (K_m)_*)_* = K_1 \cap \cdots \cap K_m$, whence

$$(K_1 \cap \cdots \cap K_m)_* = (((K_1)_* + \cdots + (K_m)_*)_*)_*.$$

Since the cone $(K_1)_* + \cdots + (K_m)_*$ is polyhedral by the previous item, the right-hand side, by the result of Exercise 2.15, is $(K_1)_* + \cdots + (K_m)_*$.

Exercise 2.22. Implement in software the approaches presented in the illustration in Section 2.4.4.3 and run the simulations. Recommended setup (for the notation, see Section 2.4.4.3):

- $n = 20$, $p = r = 10$,
- $X_j = \{x \in \mathbf{R}^p : 0 \leq x \leq [1;\ldots;1]\}$, $j = 1,\ldots,n$,
- $Y_j[x_j] = \{y \in \mathbf{R}^p : 0 \leq y \leq P^{i,j}x_j\}$, where $P^{i,j}$, $1 \leq i \leq p$, $1 \leq j \leq n$, are generated at random permutation matrices,
- R, P: generated at random with entries P_i, R_i uniformly distributed on $[0, n]$.

Quantify the inaccuracy of implementable solution $\{[x_j; y_j], j \leq n\}$ to (P) by the quantity

$$\epsilon = \frac{[\sum_j x_j - R]_+ + [P - \sum_j y_j]_+}{\sum_i P_i + \sum_i R_i},$$

where $[z]_+$ is the sum, over i, of positive parts $\max[z_i, 0]$ of entries z_i in vector z.

Apply both deterministic and randomized techniques for building implementable solutions to problem (P) described in the illustration, select the result with smaller inaccuracies and register the resulting inaccuracies in a series of simulations.

Solution: In my experiments, the randomized correction utilized $q = 2$. The observed inaccuracies in a series of 20 simulations were as follows:

	Deterministic correction (%)	Randomized correction (%)	Best correction (%)
Mean	3.89	4.69	2.32
Median	1.56	3.78	1.56
Max	32.5	12.4	10.1

Note: Inaccuracies of implementable solutions to (P).

Exercise 2.23. (Semi-computational study): Let n and m be positive integers with $n \geq 2$. Let us call an (m, n)-*bundle* a collection $\mathcal{E}$ of m $(n-1)$-dimensional linear subspaces $E_1, \ldots, E_m$ in $\mathbf{R}^n$. Such a collection can be represented by m nonzero vectors $e_i \in \mathbf{R}^n$, $i \leq n$, according to $E_i = \{x \in \mathbf{R}^n : e_i^T x = 0\}$. Let us call an (m, n)-bundle $\mathcal{E} = \{E_i, i \leq m\}$ *regular* if for every $k \leq \min[m, n]$, the intersection of every k of the linear subspaces from the collection $\mathcal{E}$ is of dimension

$n-k$. Equivalently: (m,n)-bundle $\mathcal{E} = \{E_i = \{x \in \mathbf{R}^n : e_i^T x = 0\}, i \leq m\}$ is regular if and only if for every $k \leq \min[m,n]$, every k of vectors $e_1, \ldots, e_m$ are linearly independent.

Given (m,n)-bundle, we can partition $\mathbf{R}^n$ as follows: at every point x outside of $\cup_{i \leq m} E_i$ the m reals $e_1^T x, e_2^T x, \ldots, e_m^T x$ are nonzero; let us call the sequence of their signs the *signature* of x. The set of points $x \notin \cup_{i \leq m} E_i$ with a given signature is clearly convex and open; let us call a nonempty set of this type the *cell* of the partition given by $\mathcal{E}$ *with the signature in question.* Thus, given $\mathcal{E}$, we can partition E_n into the union of hyperplanes E_i and a finite number of cells — nonempty open convex sets; their closures are cones bounded by hyperplanes E_i.

As a simple example: let $1 \leq m \leq n$ with $n > 1$ and E_i, $1 \leq i \leq m$, be coordinate hyperplanes: $E_i = \{x \in \mathbf{R}^m : x_i = 0\}$. The cells of the partition associated with the just described regular (m,n)-bundle are the 2^m sets $\{x \in \mathbf{R}^n : \epsilon_i x_i > 0, i \leq m\}$ associated with 2^m sequences $\vec{\epsilon} = (\epsilon_1, \ldots, \epsilon_m)$ with $\epsilon_i = \pm 1$.

The goal of the exercise is to understand how many cells there are in a partition associated with an (m,n)-bundle.

(1) Prove the following fact:

> (!) The number of cells in the partition associated with a regular (m,n)-bundle, $m \geq 1, n \geq 2$, depends solely on m, n, but not on the specific (m,n)-bundle in question. This number, let it be denoted $N(m,n)$, satisfies the relation
>
> $$\begin{aligned} n \geq 2, m = 1 &\Rightarrow N(m,n) = 2, \\ n \geq 2, m \geq 2 &\Rightarrow N(m,n) = N(m-1,n) + N(m-1,n-1), \end{aligned} \tag{2.37}$$
>
> where by definition $N(\mu, 1) = 2$ when $\mu \geq 1$.

Solution: To save words, let us call the partition stemming from an (m,n)-bundle, *an* (m,n)*-partition.* It is clear that the number of cells in a regular $(1,n)$-partition, $n \geq 2$, is 2. Now, let us prove by induction in $m \geq 2$ that the number of cells in an (m,n)-partition with $n \geq 2$ is independent of the partition and satisfies the recurrence in (2.37).

Base $m = 2$: Given a regular $(2,n)$-partition — that is, two distinct hyperplanes E_1,E_2 passing through the origin in $\mathbf{R}^n$ — we "see by eyes" that the number of cells is 4 (as it clearly is 2^m for every regular partition with $m \leq n$), independently of what is the partition, and the recurrence in (2.37) does take place.

Inductive step: Assume that the claim in question holds true for $m = \overline{m} \geq 2$, and let us prove that it holds true for $m = \overline{m} + 1$. Thus, let $\mathcal{E} = \{E_1, \ldots, E_m\}$ be an (m,n)-bundle. The number of cells in the $(m-1,n)$-partition $\{E_1, \ldots, E_{m-1}\}$ by inductive hypothesis is independent of what exactly is the bundle with which we are operating and is $N(m-1,n)$. Now, certain number K of cells of this partition intersect with E_m, and each of these cells yields exactly two cells, in the partition associated with the bundle $\{E_1, \ldots, E_m\}$; that is, the number of cells in the latter partition is $N(m-1,n) + K$. It is immediately clear what K is: this is the number of cells which one gets in the partition of E_m (that is, $\mathbf{R}^{n-1}$) by bundle $E_1, \ldots, E_{m-1}$; it indeed is a partition when $n > 2$, and, as is immediately seen, is regular along with the partition $E_1, \ldots, E_m$, that is, $K = N(m-1, n-1)$. When $n = 2$, $E_1, \ldots, E_m$ are distinct lines in the 2D plane passing through the origin, and therefore $K = 2 = N(m-1, 1)$. We have established the recurrence in (!), implying, in particular, that the number of cells in the partition stemming from the regular (m,n) bundle $E_1, \ldots, E_m$ with $m = \overline{m} + 1$ depends on m, n only. The inductive step is completed. □

(2) Given an (m,n)-bundle with $n \geq 2$, we can subject the vectors $e_1, \ldots, e_m$ to whatever small perturbations to make the perturbed bundle regular; on the other hand, small enough perturbations of the data of an (m,n)-bundle cannot reduce the number of cells in the associated partition. Derive from these observations that

> (!!) Specifying the function $N(m,n)$ of $m \geq 1, n \geq 2$ by the recurrence
>
> $$N(m+1, n+1) = N(m, n+1) + N(m,n), \quad m \geq 1, n \geq 2,$$
>
> and "initial conditions" $N(m,1) \equiv 2$, $m \geq 1$ and $N(1,n) = 2$, $n \geq 1$, we get an upper bound on the number of cells in a partition associated with the (m,n)-bundle; this bound coincides with the actual number of cells whenever the bundle is regular.

Use the recurrence to compute $N(5,10)$ and $N(10,20)$.

Solution: (!!) is readily given by the results of item (1) and the reasoning preceding (!!). Direct computation via the recurrence yields $N(5,10) = 512 = 2^9$, $N(10,20) = 524288 = 2^{19}$.

(3) Imagine you draw at random, in a symmetric fashion with respect to orthogonal rotations, 11-dimensional linear subspace of $\mathbf{R}^{22}$.[1] What are the chances to get a linear subspace containing a strictly positive vector? Answer the same question when you draw your 11-dimensional subspace at random from $\mathbf{R}^{21}$.

Solution: The 22 coordinate hyperplanes of $\mathbf{R}^{22}$ intersected with a random 11-dimensional subspace E of $\mathbf{R}^{22}$ with probability 1 form a regular $(22, 11)$-bundle in E, and the number of cells in the associated partition is $N(22, 11) = 2097152 = 2^{21}$ (computation according to (!)). By symmetry, the probabilities of each of the 2^{22} possible signatures being a signature of some cell are the same, implying that the chances to get a particular signature, e.g., the all-pluses one, is $1/2$. When E is drawn from $\mathbf{R}^{21}$, computation says that the number of cells in a similar partition drops down to 1233332, so that the probability to get every particular signature grows to $1233332/2^{21} \approx 0.5881$.

(4) Consider m hyperplanes $F_i = \{x \in \mathbf{R}^n : a_i^T x = b_i\}$, $a_i \neq 0$, $b_i \neq 0$; let us call such a collection an *inhomogeneous* (m, n)-*bundle*. For a point $x \in \mathbf{R}^n \backslash (\cup_i F_i)$, the reals $a_i^T x - b_i$, $i = 1, \ldots, m$, are nonzero; let us call the sequence of their signs *the signature* of x. Same as in the case when the hyperplanes pass through the origin, the entire $\mathbf{R}^n$ is partitioned into the union of $\cup_i F_i$ and a number of cells — convex sets with nonempty interior composed of points with common signature. Prove that with $N(m, n)$ given by (!), the function

$$\frac{1}{2} N(m + 1, n + 1)$$

is a tight upper bound on the number of cells in the partition of $\mathbf{R}^n$ associated with an inhomogeneous (m, n)-bundle.

Solution: Given an inhomogeneous (m, n)-bundle, let us associate with it a homogeneous $(m + 1, n + 1)$-bundle

$$E_i = \{[x; t] \in \mathbf{R}^{n+1} : a_i^T x - b_i t = 0\},\ i \leq m, E_{m+1} = \{[x; t] : t = 0\}.$$

[1]To get a random k-dimensional subspace E in $\mathbf{R}^m$ in a fashion invariant w.r.t. rotations of the space, it suffices to generate an $(n - k) \times n$ *Gaussian matrix* G — random matrix with entries drawn at random, independently of each other, from the standard Gaussian distribution, and to take $E = \operatorname{Ker} G$. Indeed, when U is an $n \times n$ orthogonal matrix, GU has exactly the same distribution as G, so that with our construction the distribution of $U^{-1}E$ is the same as the distribution of E. Alternatively, you can generate an $n \times k$ Gaussian matrix A and take its image space as E.

Cells of the partition of $\mathbf{R}^{n+1}$ associated with this homogeneous bundle are of two types — "upper" — those living in the half-space $t > 0$, and "lower" — those living in the half-space $t < 0$. If C is an upper cell with signature $\vec{\epsilon} = (\epsilon_1, \ldots, \epsilon_m, 1)$, then $-C$ is a lower cell with signature $-\vec{\epsilon}$, and vice versa, so that the numbers of upper and lower cells are upper-bounded by $\frac{1}{2}N(m+1, n+1)$ (and equal to the latter quantity, when every $k \leq \min[m+1, n+1]$ vector in the collection $\{[a_i; -b_i], i \leq m, [0; \ldots; 0; -1]\}$ is linearly independent). Clearly, intersections of the upper cells with the hyperplane $t = 1$ in $\mathbf{R}^{n+1}$ are exactly the cells of the partition of $\mathbf{R}^n$ associated with our initial inhomogeneous bundle. □

(5) Given positive integers m, n, consider a random polyhedral set

$$P = \{x \in \mathbf{R}^n : Ax \geq b\},$$

with $m \times n$ matrix A and entries in A, b drawn, independently of each other, from the standard Gaussian distribution. What is the probability $p(m, n)$ for this set to have a nonempty interior? Fill the following table:

$p(10,10)$	$p(20,10)$	$p(30,10)$	$p(40,10)$	$p(50,10)$	$p(60,10)$	$p(70,10)$

Solution: We are asking what is the probability of the partition of $\mathbf{R}^n$ induced by an inhomogeneous (m, n)-bundle $F_i = \{x \in \mathbf{R}^n : a_i^T x = b_i\}$ including a cell with all-ones signature. By symmetry, this is the same as the probability of having a cell with any other fixed signature. By (4), the number of cells in the partition in question with probability 1 is $\frac{1}{2}N(m+1, n+1)$, so that

$$p(m, n) = N(m+1, n+1)2^{-m-1},$$

resulting in

$p(10,10)$	$p(20,10)$	$p(30,10)$	$p(40,10)$	$p(50,10)$	$p(60,10)$	$p(70,10)$
1.0000	0.5881	0.0494	1.111e−3	1.193e−5	8.082e−8	4.002e−10

(6) Consider the standard form LO program with n variables and $m \leq n$ equality constraints:

$$\max \left\{ \sum_{j=1}^{n} c_j x_i : \sum_{j=1}^{n} A_{ij} x_j = b_i, i \leq m, \ x \geq 0. \right\}$$

Assuming A_{ij} and b_i drawn at random, independently of each other, from the standard Gaussian distribution, what is the probability $q(m,n)$ of the problem having a strictly positive feasible solution? Fill the following table:

$q(10,10)$	$q(10,20)$	$q(10,30)$	$q(10,40)$	$q(10,50)$
$q(100,100)$	$q(100,150)$	$q(100,200)$	$q(100,250)$	$q(100,300)$

Solution: Let us look at the kernel $E \subset \mathbf{R}^{n+1}$ of the linear mapping $[x;t] \to [Ax - tb] : \mathbf{R}^{n+1} \to \mathbf{R}^m$; we are asking whether E (which with probability 1 has dimension $n+1-m$) contains a strictly positive vector. Since E is drawn at random from a rotationally invariant distribution, the same as in (3) reasoning demonstrates that the probability in question is

$$q(m,n) = N(n+1, n+1-m)2^{-n-1},$$

resulting in

$q(10,10)$	$q(10,20)$	$q(10,30)$	$q(10,40)$	$q(10,50)$
$2^{-10} \approx 9.766\text{e}-4$	0.5881	0.9786	0.9997	1.0000
$q(100,100)$	$q(100,150)$	$q(100,200)$	$q(100,250)$	$q(100,300)$
$2^{-100} \approx 7.889\text{e}-31$	$2.724\text{e}-5$	0.5282	0.9994	1.0000

Exercise 2.24. (Computational study): Find the number of extreme points in the polyhedral set

$$Q = \left\{ x \in \mathbf{R}^5 : \sum_{j=1}^{5} \cos(i \cdot j) x_j \leq 10,\ i = 1, \ldots, 10 \right\}$$

and compare it with the upper bound $\binom{10}{5} = 252$ given by Algebraic characterization of extreme points.

Solution: My computation, based on brute search listing of extreme points via their algebraic characterization, gives the answer 42.

Exercise 2.25. (Computational study): Generate 1000 polyhedral sets

$$\left\{ x \in \mathbf{R}^5 : \sum_{j=1}^{5} a_{ij} x_j \leq b_i,\ i \leq 10 \right\}$$

with the data entries drawn at random, independently of each other, from the uniform distribution on $\{-1, 0, 1\}$, and find percentage of instances which are

- empty,
- contain lines,
- bounded,
- unbounded,
- nonempty and bounded.

Solution: My computations yielded percentages as follows:

- empty – 40.8%,
- contain lines – 31.2%,
- bounded – 45.9%,
- unbounded – 54.1%,
- nonempty and bounded – 5.1%.

Exercises for Chapter 3

Exercise 3.2. (Do we need strict inequalities?: Given the system

$$\begin{aligned} Ax &< p, \\ Bx &\leq q, \end{aligned} \tag{S}$$

of finitely many strict and nonstrict linear inequalities in variables $x \in \mathbf{R}^n$, build a system of nonstrict linear inequalities (S') which is solvable if and only if (S) is, with feasible solutions to (S) easily convertible into feasible solutions to (S'), and vice versa.

Solution: It suffices to define (S') as

$$\begin{aligned} Ay - tp &\leq -[1; \ldots; 1] \\ By - tq &\leq 0 \\ t &\geq 1 \end{aligned} \tag{S'}$$

the variables being y, t. If $[y; t]$ solves (S'), then $t > 0$, so that $x = y/t$ is well defined; clearly, x solves (S). Vice versa, if x solves (S), then $Ax - p \leq -[\epsilon; \ldots; \epsilon]$ for some $\epsilon > 0$, and $Bx \leq q$. Assuming w.l.o.g. that $\epsilon \leq 1$ and setting $y = \epsilon^{-1}x$, $t = \epsilon^{-1}$, we get a solution to (S'). □

Exercise 3.9. Consider a primal–dual pair of LO programs

$$\max_x \left\{ c^T x : \begin{array}{c} Px \le p \\ Qx \ge q \\ Rx = r \end{array} \right\} \tag{P}$$

$$\min_{\lambda=[\lambda_\ell,\lambda_g,\lambda_e]} \left\{ p^T\lambda_\ell + q^T\lambda_g + r^T\lambda_e : \begin{array}{c} \lambda_\ell \ge 0 \\ \lambda_g \le 0 \\ P^T\lambda_\ell + Q^T\lambda_g + R^T\lambda_e = c \end{array} \right\}. \tag{D}$$

Assume that both problems are feasible, and that the primal problem does contain inequality constraints. Prove that the feasible set of at least one of these problems is unbounded.

Solution: A nonempty polyhedral set is bounded iff its recessive cone is trivial. Assuming both problems feasible, their recessive cones are

$$\begin{aligned} \mathrm{Rec}_P &= \{x : Px \le 0, Qx \ge 0, Rx = 0\}, \\ \mathrm{Rec}_D &= \left\{ [\lambda_\ell; \lambda_g; \lambda_e] : \begin{array}{c} \lambda_\ell \ge 0, \lambda_g \le 0 \\ P^T\lambda_\ell + Q^T\lambda_g + R^T\lambda_e = 0 \end{array} \right\}, \end{aligned}$$

and we should lead to contradiction the assumption that both Rec_P and Rec_D are trivial.

Now, denoting $n = \dim x$ and assuming $\mathrm{Rec}_P = \{0\}$, the cone dual to Rec_P is the entire R^n, while by the homogeneous Farkas lemma this dual cone is $\{y = P^T\lambda_P + Q^T\lambda_Q + R^T\lambda_R : \lambda_P \le 0, \lambda_Q \ge 0\}$. Assume first that both P and Q are present, let us choose somehow vectors $\bar\lambda_P > 0$ and $\bar\lambda_Q < 0$ and set $\bar y = P^T\bar\lambda_P + Q^T\bar\lambda_Q$. By what was said about the cone dual to Rec_P, there exist $\lambda_P \le 0$, $\lambda_Q \ge 0$ and λ_R such that $\bar y = P^T\lambda_P + Q^T\lambda_Q + R^T\lambda_R$, that is, $P^T[\bar\lambda_P - \lambda_P] + Q^T[\bar\lambda_Q - \lambda_Q] + R^T[-\lambda_R] = 0$. Setting

$$\lambda_\ell = \bar\lambda_P - \lambda_P, \lambda_g = \bar\lambda_Q - \lambda_Q, \lambda_e = -\lambda_R,$$

and taking into account that $\bar\lambda_P > 0$, $\lambda_P \le 0$, $\bar\lambda_Q < 0$, $\lambda_Q \ge 0$, we get $\lambda_\ell > 0, \lambda_g < 0$, $P^T\lambda_\ell + Q^T\lambda_g + R^T\lambda_e = 0$, that is, Rec_D contains a nonzero vector $[\lambda_\ell; \lambda_g; \lambda_e]$.

Now, assume that exactly one of P and Q is present (we know that both P and Q cannot be absent!), for the sake of definiteness, let the present matrix be P (the case when P is absent and Q is present is completely similar). We are in the case when

$\mathbf{R}^n = \{y = P^T\lambda_P + R^T\lambda_R : \lambda_P \leq 0\}$. Selecting, same as above, $\bar{\lambda}_P > 0$ and setting $\bar{y} = P^T\bar{\lambda}_P$, we have $\bar{y} = P^T\lambda_P + R^T\lambda_R$ with some $\lambda_P \leq 0$, so that

$$P^T \underbrace{[\bar{\lambda}_P - \lambda_P]}_{=:\lambda_\ell > 0} + R^T \underbrace{[-\lambda_R]}_{=:\lambda_e} = 0.$$

We see that $[\lambda_e; \lambda_e]$ is a nontrivial recessive direction of the dual feasible set, which is a desired contradiction. □

Exercise 3.10. For positive integers $k \leq n$, let $s_k(x)$ be the sum of the k largest entries in a vector $x \in \mathbf{R}^n$, e.g., $s_2([1;1;1]) = 1 + 1 = 2$, $s_2([1;2;3]) = 2 + 3 = 5$. Find a polyhedral representation of $s_k(x)$.

Hint: Take into account that the extreme points of the set $\{x \in \mathbf{R}^n : 0 \leq x_i \leq 1, \sum_i x_i = k\}$ are exactly the 0/1 vectors from this set, and derive from this that

$$s_k(x) = \max_y \left\{ y^T x : 0 \leq y_i \leq 1 \quad \forall i, \sum_i y_i = k \right\}.$$

Solution: By the Hint, we have

$$s_k(x) = \max_y \left\{ y^T x : 0 \leq y_i \leq 1 \quad \forall i, \sum_i y_i = k \right\}.$$

Passing to the dual problem, we have

$$s_k(x) = \min_{\lambda = [\lambda_\ell; \lambda_g; \lambda_e]} \left\{ \sum_{i=1}^n [\lambda_\ell]_i + k\lambda_e : \begin{array}{l} \lambda_\ell \geq 0, \lambda_g \leq 0 \\ \lambda_\ell + \lambda_g + \lambda_e[1; \ldots; 1] = x \end{array} \right\},$$

which gives a polyhedral representation of $s_k(\cdot)$ as follows:

$$\tau \geq s_k(x) \quad \Leftrightarrow \quad \exists \lambda_\ell, \lambda_g, \lambda_e : \begin{cases} \lambda_\ell \geq 0, \lambda_g \leq 0 \\ \sum_i [\lambda_\ell]_i + k\lambda_e \leq \tau \\ x_i = [\lambda_\ell]_i + [\lambda_g]_i + \lambda_\ell, \quad 1 \leq i \leq n \end{cases},$$

which can be simplified to (we set $z = \lambda_\ell$, $t = \lambda_e$) the following:

$$\tau \geq s_k(x) \quad \Leftrightarrow \quad \exists z, t : \begin{cases} z \geq 0 \\ x \leq z + t[1; \ldots; 1]. \\ \sum_i z_i + kt \leq \tau \end{cases}$$

Exercise 3.11. Consider the scalar linear constraint

$$a^T x \leq b, \tag{1}$$

with uncertain data $a \in \mathbf{R}^n$ (b is certain) varying in the set

$$\mathcal{U} = \left\{ a : |a_i - a_i^*|/\delta_i \leq 1, 1 \leq i \leq n, \sum_{i=1}^{n} |a_i - a_i^*|/\delta_i \leq k \right\}, \tag{2}$$

where a_i^* are given "nominal data," $\delta_i > 0$ are given quantities and $k \leq n$ is an integer (in the literature, this is called "budgeted uncertainty"). Rewrite the robust counterpart

$$a^T x \leq b \,\forall a \in \mathcal{U} \tag{RC}$$

in a tractable LO form (that is, write down an explicit system (S) of linear inequalities in variables x and additional variables such that x satisfies (RC) if and only if x can be extended to a feasible solution of (S)).

Solution: Let D be diagonal $n \times n$ matrix with diagonal entries δ_i, and let $a_i - a_i^* = \delta_i \epsilon_i$, so that

$$\begin{aligned} \mathcal{U} &= \{a = a^* + D\epsilon : -1 \leq \epsilon_i \leq 1 \,\forall i, \textstyle\sum_i |\epsilon_i| \leq k\} \\ &= \{a = a^* + D\epsilon : -u \leq \epsilon \leq u, u_i \leq 1 \,\forall i, \textstyle\sum_i u_i \leq k\}. \end{aligned}$$

x is robust feasible iff

$$\begin{aligned} b \geq \max_a \{x^T a : a \in \mathcal{U}\} &= \max_{\epsilon,u} \left\{ x^T[a^* + D\epsilon] : -u \leq \epsilon \leq u, \right. \\ &\qquad \left. u \leq [1;\ldots;1], \sum_i u_i \leq k \right\}, \\ &= x^T a^* + \max_{\epsilon,u} \left\{ [Dx]^T \epsilon : -u \leq \epsilon \leq u, u \leq [1;\ldots;1], \sum_i u_i \leq k \right\}, \\ &= x^T a^* + \min_{\lambda_{\ell,u},\lambda_{g,u},\lambda_{\ell,1},\lambda_{\ell,k}} \left\{ [1;\ldots;1]^T \lambda_{\ell,1} + k\lambda_{\ell,k} : \right. \\ &\qquad \left. \begin{array}{l} \lambda_{\ell,u} \geq 0, \lambda_{\ell,1} \geq 0, \lambda_{\ell,k} \geq 0, \lambda_{g,u} \leq 0 \\ \lambda_{\ell,u} + \lambda_{g,u} = Dx \\ -\lambda_{\ell,u} + \lambda_{g,u} + \lambda_{\ell,1} + \lambda_{\ell,k}[1;\ldots;1] = 0 \end{array} \right\} . \end{aligned}$$

[LO duality],

$$= x^T a^* + \min_{\lambda_{\ell,1},\lambda_{\ell,k}} \left\{ [1;\ldots;1]^T \lambda_{\ell,1} + k\lambda_{\ell,k} : \begin{array}{l} \lambda_{\ell,1} \geq 0, \lambda_{\ell,k} \geq 0 \\ Dx \leq \lambda_{\ell,1} + \lambda_{\ell,k}[1;\ldots;1] \end{array} \right\}$$

[eliminating $\lambda_{\ell,u}, \lambda_{g,u}$]

Thus, (RC) can be represented as the system of linear constraints

$$\lambda_{\ell,1} \geq 0, \lambda_{\ell,k} \geq 0, Dx \leq \lambda_{\ell,1} + \lambda_{\ell,k}[1;\ldots;1], [a^*]^T x \\ + [1;\ldots;1]^T \lambda_{\ell,1} + k\lambda_{\ell,k} \leq b,$$

in variables $x, \lambda_{\ell,1}, \lambda_{\ell,k}$.

Exercises for Chapter 4

Exercise 4.4. (1) Consider a parametric LO program

$$\max_{x,s} \left\{ c^T x - M \sum_{i=1}^{m} s_i : Ax + s = b, x \geq 0, s \geq 0 \right\}, \qquad (P_M),$$

with $b \geq 0$, along with the LO program

$$\max_{x} \left\{ c^T x : Ax = b, x \geq 0 \right\}, \qquad (P),$$

with $m \times n$ matrix A of rank m. Prove that if (P) is solvable, then there exists $M_* > 0$ such that whenever $M > M_*$, problem (P_M) is solvable, the optimal values of (P_M) and (P) are the same, and all optimal solutions (x, s) to (P_M) are of the form $(x, 0)$, where x is an optimal solution to (P).

Solution: Indeed, when (P) is solvable, its feasible set X is a face in the feasible set of (P_M). Now, the feasible set X^+ of (P_M) is independent of M and does not contain lines (since (P_M) is in the standard form). Let $v_1 = (x_1, s_1), \ldots, v_I = (x_I, s_I)$ be the vertices of X^+, and $d^1 = (x^1, s^1), \ldots, d^J = (x^J, s^J)$ be the directions of the extreme rays of the recessive cone of X^+, so that

$$X^+ = \mathrm{Conv}(\{v_1, \ldots, v_I\}) + \mathrm{Cone}\left(\{d^1, \ldots, d^J\}\right).$$

Observe that the set of extreme points of X is nonempty (since X is nonempty and does not contain lines); since X is a face in X^+, the extreme points of X are extreme points of X^+ as well, that is, some of s_i are zeros. Without loss of generality, assume that s_i are zeros for $1 \leq i \leq K$, where $1 \leq K \leq I$. It may also happen that some of s^j are zeros; without loss of generality assume that this is the case when $1 \leq j \leq L$, where $0 \leq L \leq J$. Finally, let e be the all-one m-dimensional vector, so that the objective in (P_M) is $c_M := [c; -M\,\mathrm{e}]$.

When $K < i \leq I$, we have $s_i \geq 0$ and $s_i \neq 0$, whence $c_M^T v_i = c^T x_i - M\sigma_i$ with $\sigma_i = \mathrm{e}^T s_i > 0$. Similarly, when $L < j \leq J$, we have $s^j \geq 0$ and $s^j \neq 0$, whence $c_M^T d^j = c^T x^j - M\sigma^j$, where $\sigma^j = \mathrm{e}^T s^j > 0$. We can clearly find $M_* \geq 0$ such that $c^T v_i - M\sigma_i < c^T x_1$ for all i, $K < i \leq I$, and $c^T x^j - M\sigma^j < 0$ for all j, $L < j \leq M$. We claim that when $M \geq M_*$, (P_M) is solvable, and all optimal solutions to (P_M) are of the form $(x, 0)$, where x is an optimal solution to (P). Indeed, we have $c^T x^j \leq 0$ when $j \leq L$ (since otherwise (P) would be unbounded, which is not the case); with this in mind and taking into account that $M \geq M_*$, we get $c_M^T d^j \leq 0$ for all j, so that (P_M) is solvable. Now, let

$$[x; s] = \sum_{i=1}^{I} \lambda_i [x_i; s_i] + \sum_{j=1}^{M} \mu_j [x^j; s^j] \qquad [\lambda \geq 0, \textstyle\sum_i \lambda_i = 1,\ \mu \geq 0],$$

be an optimal solution to (P_M). From optimality, μ_j are nonzero only for those j for which $c_M^T d^j = 0$, and this by our analysis can happen only for $j \leq L$ (specifically, for those $j \leq L$ for which $c^T x^j = 0$). Similarly, λ_i can be positive only for those i for which $c_M^T v_i = \min_{1 \leq i' \leq I} c_M^T v_{i'}$, and this by construction can happen only when $i \leq K$. The bottom line is that positive λ_i correspond to $i \leq K$, and positive μ_j correspond to $j \leq L$, meaning that every optimal solution (x, s) to (P_M) has $s = 0$ and thus is optimal for (P). □

(2) The result of (1) suggests the following single-phase "Big M"-implementation of the PSM: Given a standard form LO program (P) with $b \geq 0$ (the latter can always be achieved by multiplying appropriate equality constraints by -1), we associate with it the program (P_M) for which we can easily point out an initial basis composed of (indexes of) the slack variables s along with the associated basic feasible solution $(x = 0, s = b)$. We solve the resulting problem (P_M) considering M as a large constant. From the description of the PSM, it follows that M will affect only the subsequent vectors of reduced costs (and the decisions we make based on these costs) and will appear linearly in the reduced costs. Now, making decisions based on reduced costs requires to compare them with zero, and here we think about M as a large constant, meaning that reduced cost of the form $a + pM$ (a and p are known reals) should be treated as positive when $p > 0$ and as negative when $p < 0$; when $p = 0$, the decision is made based on what a is. Assuming that (P) is solvable,

by (1) the problem (P_M) for all large enough values of M is solvable, and every optimal solution to this problem is $(x, 0)$, where x is an optimal solution to (P). In other words, when (P) is solvable, running the Big M method, modulo highly unlikely cycling, will result in arriving at a basic feasible solution with the s-components equal to 0 and the reduced costs of the form $a_i + p_i M$ where for every i either $a_i \leq 0$, either $p_i < 0$, meaning that the current basic solution is feasible for (P) and optimal for all problems (P_M) with large enough values of M and therefore is optimal for (P).

Use the Big M version of the PSM to solve the problem

$$\begin{array}{l} \max \ -2x_1 + x_2 + x_3 - 6x_4 \\ x_1 + x_2 \qquad\qquad = 2 \\ \qquad\qquad x_3 + x_4 = 2 \\ x_1 \qquad + x_3 \qquad = 2 \\ \qquad x \geq 0 \end{array}$$

Solution: The initial tableau is

	x_1	x_2	x_3	x_4	s_1	s_2	s_3
$6M$	$-2+2M$	$1+M$	$1+2M$	$-6+M$	0	0	0
$s_1 = 2$	$\underline{1}$	1	0	0	1	0	0
$s_2 = 2$	0	0	1	1	0	1	0
$s_3 = 2$	1	0	1	0	0	0	1

The reduced costs are obtained from the objective $[-2; 1; 1; -6; -M; -M; -M]$ of (P_M) by adding a linear combination of rows of A chosen from the requirement to zero out the reduced costs in the basic columns of s_1, s_2, s_3.

For M large, there are positive costs, e.g., the one of x_1. Let x_1 enter the basis, s_1 leave it. The pivoting element is underlined. The new tableau is

	x_1	x_2	x_3	x_4	s_1	s_2	s_3
$4+2M$	0	$3-M$	$1+2M$	$-6+M$	$2-2M$	0	0
$x_1 = 2$	1	1	0	0	1	0	0
$s_2 = 2$	0	0	1	1	0	1	0
$s_3 = 0$	0	-1	$\underline{1}$	0	-1	0	1

For large M, there are positive reduced costs, e.g., the one of x_3. x_3 enters the basis, s_3 leaves it; pivoting element is underlined. The new tableau is

	x_1	x_2	x_3	x_4	s_1	s_2	s_3
$4+2M$	0	$4+M$	0	$-6+M$	3	0	$-1-2M$
$x_1=2$	1	1	0	0	1	0	0
$s_2=2$	0	1	0	1	$\underline{1}$	1	-1
$x_3=0$	0	-1	1	0	-1	0	1

The reduced cost of s_1 is positive. s_1 enters the basis, s_2 leaves the basis, the pivoting element is underlined. The new tableau is

	x_1	x_2	x_3	x_4	s_1	s_2	s_3
$-2+2M$	0	$1+M$	0	$-9+M$	0	-3	$2-2M$
$x_1=0$	1	0	0	-1	0	-1	1
$s_1=2$	0	1	0	$\underline{1}$	1	1	-1
$x_3=2$	0	0	1	1	0	1	0

When M is large enough, the reduced cost of x_4 is positive. x_4 enters the basis, s_1 leaves the basis, pivoting element is underlined. The new tableau is

	x_1	x_2	x_3	x_4	s_1	s_2	s_3
16	0	10	0	0	$9-M$	$6-M$	$-7-M$
$x_1=2$	1	1	0	0	1	0	0
$x_4=2$	0	$\underline{1}$	0	1	1	1	-1
$x_3=0$	0	-1	1	0	-1	0	1

The reduced cost of x_2 is positive; x_2 enters the basis, and, say, x_4 leaves it. The pivoting element is underlined. The new tableau is

	x_1	x_2	x_3	x_4	s_1	s_2	s_3
-4	0	0	0	-10	$-1-M$	$-4-M$	$3-M$
$x_1=0$	1	0	0	-1	0	-1	1
$x_2=2$	0	1	0	1	1	1	-1
$x_3=2$	0	0	1	1	0	1	0

Now, for all large enough M, all the reduced costs are nonpositive; we have found the optimal solution to (P_M) for large M. In this solution, all s's are zero, that is, we have found the optimal solution to the problem of interest (P); this solution is $x = [0;2;2;0]$, the optimal value being 4.

Exercises for Chapter 5

Exercise 5.3. Consider a transportation problem

$$\mathrm{Opt} = \min_{x_{ij}} \left\{ \sum_{i=1}^{p}\sum_{j=1}^{q} c_{ij}x_{ij} : \begin{array}{l} \sum_{j=1}^{q} x_{ij} = a_i,\ 1 \le i \le p \\ \sum_{i=1}^{p} x_{ij} = b_j,\ 1 \le j \le q \\ x_{ij} \ge 0 \end{array} \right\},$$

where $a_i \ge 0$, $b_i \ge 0$ and $\sum_i a_i = \sum_j b_j$.

Assume that supplies a_i and the demands b_j are somehow decreased (but remain nonnegative), while the new total supply is equal to the new total demand. Is it true that the optimal value in the problem cannot increase?

Solution: Not necessarily. Indeed, let $p = q = 2$, $a_1 = a_2 = b_1 = b_2 = 1$ and $c_{1,1} = c_{2,2} = 1$, $c_{1,2} = c_{2,1} = 100$. Now, let the new supplies and demands be $a_1' = 1/2$, $a_2' = 1$, $b_1' = 1$, $b_2' = 1/2$. The initial problem has a trivial solution $x_{11} = x_{22} = 1$, $x_{12} = x_{21} = 0$, and the initial optimal value is 2. In the new problem, for every feasible solution x we should have $x_{11} \le a_1' = 1/2$ and $x_{11} + x_{21} = b_1' = 1$, whence $x_{21} = 1 - x_{11} \ge 1/2$. But then the cost of the shipment x is at least $c_{21}x_{21} \ge 50$, whence the optimal value in the new problem is at least 50!

Exercise 5.4. The Maximal Flow problem is as follows:

> *Given an oriented graph $G = (\mathcal{N}, \mathcal{E})$ with arc capacities $\{u_\gamma \in (0,\infty) : \gamma \in \mathcal{E}\}$ and two selected nodes distinct from each other: source $\underline{i}$ and sink $\bar{i}$, find the largest flow from source to sink, that is, find the largest s such that the vector of external supplies "s at the source, $-s$ at the sink, zero at all remaining nodes" fits a flow f satisfying the conservation law and obeying the bounds $0 \le f_\gamma \le u_\gamma$ for all $\gamma \in \mathcal{E}$.*

(1) Write down the Maximal Flow problem as an LO program in variables s (external supply at the source) and f (flow in the network) and write down the dual problem.

Solution: The LO program is

$$\mathrm{Opt}(P) = \max_{s,f} \left\{ s : 0 \leq f \leq u, (Pf)_i = \left\{ \begin{array}{l} s,\ i = \underline{i} \\ -s,\ i = \overline{i} \\ 0,\ i \notin \{\underline{i}, \overline{i}\} \end{array} \right. , \forall i \right\}, \tag{P}$$

where P is the incidence matrix of G. The dual program is

$$\mathrm{Opt}(D) = \min_{\lambda_e, \lambda_\ell, \lambda_g} \left\{ u^T \lambda_\ell : \left\{ \begin{array}{l} \lambda_\ell \geq 0, \lambda_g \leq 0 \\ P^T \lambda_e + \lambda_\ell + \lambda_g = 0 \\ (\lambda_e)_{\overline{i}} - (\lambda_e)_{\underline{i}} = 1 \end{array} \right. \right\}. \tag{D}$$

(2) Let us define a *cut* as a partition of the set $\mathcal{N}$ of nodes of G into two nonoverlapping subsets $\underline{I}$ and $\overline{I} = \mathcal{N} \backslash \underline{I}$ such that the source is in $\underline{I}$, and the sink is in $\overline{I}$. For a cut $(\underline{I}, \overline{I})$, let the capacity $U(\underline{I}, \overline{I})$ of the cut be defined as

$$U(\underline{I}, \overline{I}) = \sum_{\gamma = (i,j) \in \mathcal{E} : i \in \underline{I}, j \in \overline{I}} u_\gamma.$$

Prove that if (s, f) is a feasible solution to the LO reformulation of the Maximal Flow problem, then for every cut $(\underline{I}, \overline{I})$ one has

$$s \leq U(\underline{I}, \overline{I}).$$

Prove that the inequality here is an equality if and only if the flow f_γ in every arc γ which starts in $\underline{I}$ and ends in $\overline{I}$ ("forward arc of the cut") is equal to u_γ, and the flow f_γ in every arc γ which starts in $\overline{I}$ and ends in $\underline{I}$ ("backward arc of the cut") is zero; if this is the case, s is the optimal value in the Maximal Flow problem.

Solution: We have, denoting $s(\gamma)$ and $e(\gamma)$ the start and the end nodes of an arc γ, respectively,

$$\begin{aligned} s &= \sum_{i \in \underline{I}} (Pf)_i = \sum_{i \in \underline{I}} \left[\sum_{p:(i,p) \in \mathcal{E}} f_{ip} - \sum_{q:(q,i) \in \mathcal{E}} f_{qi} \right] \\ &= \sum_{\gamma \in \mathcal{E} : s(\gamma) \in \underline{I}} f_\gamma - \sum_{\gamma \in \mathcal{E} : e(\gamma) \in \underline{I}} f_\gamma \\ &= \sum_{\gamma \in \mathcal{E} : s(\gamma) \in \underline{I}, e(\gamma) \in \overline{I}} f_\gamma - \sum_{\gamma \in \mathcal{E} : s(\gamma) \in \overline{I}, e(\gamma) \in \underline{I}} f_\gamma. \end{aligned},$$

the concluding difference clearly is $\leq U(\underline{I}, \overline{I})$, and the equality is achieved iff $f_\gamma = u_\gamma$ for all forward arcs of the cut, and $f_\gamma = 0$ for all backward arcs. In this case, s achieves its upper bound $U(\underline{I}, \overline{I})$ (which is valid for all feasible solutions to the Maximal Flow problem) and thus is the optimal value in the problem.

(3) *Invoking the LP Duality Theorem, prove the famous *Max Flow–Min Cut Theorem: If the Max Flow problem is solvable, then its optimal value is equal to the minimum, over all cats, capacity of a cut.*

Solution: We already know that the capacity of a cut is an upper bound on the optimal value in the Max Flow problem; thus, all we need to prove is that if the Max Flow problem is solvable, then there exists a cut with the capacity equal to the magnitude Opt of the maximal flow. Let $s = \text{Opt}$, f^* be the optimal value and the optimal flow vector in the Max Flow problem. By the previous item, it suffices to find a cut such that the flow f^*_γ is equal to u_γ in every forward arc of the cut and is equal to 0 in every backward arc of the cut.

By the LO Duality Theorem, we can augment $(s = \text{Opt}, f^*)$ with a feasible dual solution $\lambda^*_e, \lambda^*_\ell, \lambda^*_g$ such that the complementary slackness holds, namely, $(\lambda^*_\ell)_\gamma > 0 \Rightarrow f^*_\gamma = u_\gamma$ and $(\lambda^*_g)_\gamma < 0 \Rightarrow f^*_\gamma = 0$. Setting $\mu_i = (\lambda^*_e)_i$ and calling μ_i the potential of node i, the constraints of the dual problem read

$$\begin{aligned} &\forall \gamma \in \mathcal{E} : \mu_{s(\gamma)} - \mu_{e(\gamma)} = -(\lambda^*_\ell)_\gamma - (\lambda^*_g)_\gamma, && \text{(a)} \\ &\mu_{\overline{i}} - \mu_{\underline{i}} = 1. && \text{(b)} \end{aligned}$$

Besides this, for every γ at least one of the quantities $(\lambda^*_\ell)_\gamma$, $(\lambda^*_g)_\gamma$ is zero (indeed, otherwise, due to sign constraints on λ^*_ℓ and λ^*_g and complementary slackness, we would have simultaneously $f^*_\gamma = 0$ and $f^*_\gamma = u_\gamma$, which is impossible, since arc capacities are > 0).

Now, let $\underline{I} = \{i : \mu_i \leq \mu_{\underline{i}}\}$, $\overline{I} = \{i : \mu_i > \mu_{\underline{i}}\}$, which clearly is a partition of the set of nodes. By (b), we have $\overline{i} \in \overline{I}$, and clearly $\underline{i} \in \underline{I}$, that is, the partition is a cut. Let γ be a forward arc of this partition; then

$$0 > \mu_{s(\gamma)} - \mu_{e(\gamma)} = -(\lambda^*_\ell)_\gamma - (\lambda^*_g)_\gamma \qquad (!)$$

Since $(\lambda^*_\ell)_\gamma \geq 0$, $(\lambda^*_g)_\gamma \leq 0$ and both these quantities cannot be nonzero simultaneously, (!) says that $(\lambda^*_\ell)_\gamma > 0$, whence, by complementary slackness, $f^*_\gamma = u_\gamma$. Now, let γ be a backward arc of the cut. Then

$$0 < \mu_{s(\gamma)} - \mu_{e(\gamma)} = -(\lambda^*_\ell)_\gamma - (\lambda^*_g)_\gamma,$$

whence, by the same argument as above, $(\lambda^*_g) < 0$ and therefore $f^*_\gamma = 0$. Thus, the flow f^* fully fills all forward arcs of the cut and is zero in all backward arcs, as required. □

Exercise 5.5. Consider the *Assignment problem* as follows:

> *There are p jobs and p workers. When job j is carried out by worker i, we get profit c_{ij}. We want to associate jobs with workers in such a way that every work is assigned to exactly one worker, and no worker is assigned two or more jobs. Under these restrictions, we seek to maximize the total profit.*

(1) Let G be a graph with $2p$ nodes, p of them representing the workers, and remaining p representing the jobs. Every worker node i, $1 \leq i \leq p$, is linked by an arc $(i, p+j)$ with every job node $p+j$, $1 \leq j \leq p$, the capacity of the arc is $+\infty$, and the transportation cost is $-c_{i,p+j}$. The vector of external supplies has entries indexed by the worker nodes equal to 1 and entries indexed by the job nodes equal to -1.

Prove that the basic feasible solutions $f = \{f_{i,p+j}\}_{1\leq i,j\leq p}$ of the resulting Network Flow problem are exactly the assignments, that is, flows given by permutations $i \mapsto \sigma(i)$ of the index set $\{1, 2, \ldots, p\}$ according to $f_{i,p+\sigma(i)} = 1$ and $f_{i,p+j} = 0$ when $j \neq \sigma(i)$.

Solution: We know that every basic feasible solution to a Network Flow problem with integral vector of external supplies is integral. Thus, if f is a basic feasible solution, then $f_{i,p+j}$ are nonnegative integers. From the Conservation law,

- for every $i \leq p$, $\sum_{j=1}^p f_{i,p+j} = 1$, which combines with the fact that $f_{i,p+j}$ are nonnegative integers to imply that there exists exactly one $j = j(i)$ such that $f_{i,p+j(i)} = 1$, and $f_{i,p+j} = 0$ when $j \neq j(i)$;
- for every $j \leq p$, $\sum_{i=1}^p f_{i,p+j} = 1$, meaning that the mapping $i \mapsto j(i)$ is a mapping of $\{1, \ldots, p\}$ onto $\{1, \ldots, p\}$, and thus $i \mapsto j(i)$ is a permutation.

Thus, every basic feasible solution is an assignment.

Vice versa, given an assignment f associated with a permutation σ, consider the corresponding arcs $(i, p + \sigma(i))$. After their directions are ignored, these p arcs of G form a graph with no cycles on the nodes $1, \ldots, 2p$; as such, this graph can be extended by adding arcs of G (G is clearly connected!) to a spanning tree; clearly, f is the basic feasible solution associated with this spanning tree.

(2) Extract from the previous item that the Assignment problem can be reduced to a Network Flow problem (more exactly, to finding an optimal basic feasible solution in a Network Flow problem).

Solution: By the previous item, assignments are exactly the basic feasible solutions to the Network Flow problem from the previous item; the objective of this problem, as evaluated at such a basic feasible solution, is the minus profit of the assignment in question. Finally, the Network Flow problem in question is clearly feasible with a bounded feasible set; consequently, the feasible set is the convex hull of assignments, and every vertex optimal solution to the Network Flow problem is an optimal solution to the Assignment problem.

Pay attention to the following two facts:

- Every spanning tree of the graph in question has $2p-1$ arcs, while every basic feasible solution has just p nonzero entries, meaning that when $p > 1$, all basic feasible solutions are degenerate.
- The flows $\{f_{i,p+j}\}_{i,j=1}^{p}$ on G can be associated with $p \times p$ matrices $F_{ij} = f_{i,p+j}$, and with this interpretation, the feasible set of the above Network Flow problem is nothing but the set of all double-stochastic $p \times p$ matrices. Thus, the fact established in the first item: "the extreme points of the feasible set of the problem (i.e., the vertices of its feasible set) are exactly the assignments" recovers anew the Birkhoff theorem: "The extreme points of the set of double-stochastic matrices are exactly the permutation matrices."

Exercise 5.6. Consider the following modification of assignment problem:

> (!) *There are q jobs and $p \geq q$ workers; assigning worker i to job j, we get profit c_{ij}. We want to assign every job to exactly one worker in such a way that no worker is assigned to two or more jobs (although some of the workers can be unassigned to any jobs at all) and want to maximize the total profit under this assignment.*

(1) Reformulate the problem as an Assignment problem.

Solution: when $p > q$, let us add $p-q$ fictitious jobs and set $c_{ij} = 0$ for all $i \leq p$ and every fictitious job j. The resulting Assignment problem with p workers and p jobs is equivalent to (!).

(2) Assume that all c_{ij} are either 0 or 1; let us say that $c_{ij} = 1$ means that worker i knows how to do job j, and $c_{ij} = 0$ means that worker i does not know how to do job j, and that an assignment ("exactly one worker for every job, no worker with more than one job assigned") is good if every job is assigned to a worker who knows how to do this job. Assume that at least one worker knows how to do at least one job, that is, not all c_{ij} are zeros. Consider the network with $p+q+2$ nodes: the source (node 0), p worker nodes $(1, 2, \ldots, p)$, q job nodes $(p+1, \ldots, p+q)$ and the sink (node $p+q+1$). The arcs, every one of capacity 1, are as follows:

- p arcs "source $\mapsto$ worker node" (i.e., arcs $(0, i)$, $1 \leq i \leq p$);
- the arcs "worker $i \mapsto$ job j which the worker i knows how to do" (i.e., arcs $(i, p+j)$, $1 \leq i \leq p$, $1 \leq j \leq q$, corresponding to the pairs (i, j) with $c_{ij} = 1$);
- q arcs "job node $\mapsto$ sink" (i.e., arcs $(p+j, p+q+1)$, $1 \leq j \leq q$) along with the Maximal Flow problem on this network.

(a) Prove that the existence of a good assignment is equivalent to the fact that the magnitude of the maximal flow in the above Max Flow problem is equal to q

Solution: Clearly, a good assignment (one where every job is assigned to a worker who knows how to do the job) exists if and only if the maximal flow in the just defined network is $\geq q$. Indeed, a good assignment clearly defines a feasible flow of magnitude q. Now, assume that a feasible flow of magnitude $\geq q$ exists. Since there exists a cut of capacity q (all nodes but the sink one form one set, and the sink forms the other set), feasible flow of capacity $> q$ does not exist, and the flow in question is of capacity q. We are in the situation when the Network Flow problem with external supply equal to q in the source, $-q$ at the sink and 0 at all remaining nodes is feasible; adding to this problem a cost vector and minimizing the cost, we get a solvable Network Flow problem with integral vector of external supplies; since the capacities of arcs are integral, a vertex of the (clearly bounded) nonempty feasible set of the problem (such a vertex does exist!) is an integral feasible flow f, and since the arc capacities are 1, this flow is a 0/1 one. Since the magnitude of this flow is equal to q and the outgoing flow at every one of the q job nodes is ≤ 1, the outgoing (and thus, the incoming) flow at every job node is exactly 1. And since the flow in every worker–job arc is either zero or 1, and the incoming flow at every job node j is exactly 1, for every $j \leq q$ there is exactly one worker node $i(j)$ such that $f_{i(j),p+j}$ differs from zero and thus is equal to 1. Since the incoming flow at every worker node is at most 1, different j's correspond to different $i(j)$'s. Thus, flow f associates with every job j exactly one worker $i(j)$ such that $(i(j), j)$ is an arc in the network, that is, $c_{i(j),j} = 1$. Besides this, different values of j correspond to different values of $i(j)$, that is, no worker is associated with more than one job, and we get a good assignment. The bottom line is that *a good assignment exists if and only if the maximal flow in the network we have built is exactly* q. □

(b) Use the Max Flow–Min Cut Theorem to prove the following fact:

> (!!) *In* (!) *with zero/one* c_{ij}, *an assignment with profit* q *(i.e., an assignment where every job* j *is assigned to a worker who knows how to do this job) exists if and only if for every subset* S *of the set* $\{1, \ldots, q\}$ *of jobs the total number of workers who know how to do a job from* S *is at least the cardinality of* S.

Note: (!!) is called the *Marriage Lemma*, according to the following interpretation: There are q young ladies and p young men, some of the ladies being acquainted with some of the men. When is it possible to select for every one of the ladies a bridegroom from the above group of men in such a way that different ladies get different bridegrooms, and every lady is acquainted with her bridegroom? The answer is: It is possible if and only if for every set S of the ladies, the total number of men acquainted each with at least one of the ladies from S is at least the cardinality of S.

Solution: If there exists a set S of jobs such that the total number of workers knowing each how to do at least one job from S is $< \mathrm{Card}\, S$, then a good assignment clearly does not exist.[2] It remains to prove that if the magnitude of the maximal flow is $< q$, then there exists a set of job nodes S such that the cardinality of the set W of workers knowing how to do at least one job from S is $< \mathrm{Card}\, S$. Indeed, by Max Flow–Min Cut Theorem there exists a cut $(\underline{I}, \overline{I})$ with capacity $< q$. We have $\underline{I} = \{0\} \cup \bar{U} \cup \bar{S}$, $\overline{I} = \{p+q+1\} \cup U \cup S$, where $\bar{U}$ is some set of worker nodes, $\bar{S}$ is some set of job nodes, U is the complement of $\bar{U}$ in the set of worker nodes, and S is the complement of $\bar{S}$ in the set of job nodes. The arcs which start in $\underline{I}$ and end in $\overline{I}$ are the $\mathrm{Card}\, U$ arcs which start at the source 0 and end at a node from U, $\mathrm{Card}\, \bar{S}$ arcs which start at the nodes from $\bar{S}$ and end at the sink node $p+q+1$, and M arcs which start in $\bar{U}$ and end in S. Since all arc capacities are 1, the capacity of the cut is $\mathrm{Card}\, U + \mathrm{Card}\, \bar{S} + M < q$. We see that $\mathrm{Card}\, U + M < q - \mathrm{Card}\, \bar{S} = \mathrm{Card}\, S$. Now, M is the number of arcs from $\bar{U}$ to S, and it is *at least* the number of workers in $\bar{U}$ knowing how to do at least one job from S. Therefore, the total number of workers knowing how to do at least one job from S clearly is $\leq \mathrm{Card}\, U + M < \mathrm{Card}\, S$. Thus, there exists a set of jobs — namely, S — such that the total number of workers knowing how to do at least one job from S is $< \mathrm{Card}\, S$, and we are done. $\square$

Exercise for Chapters 6 and 7

Exercise 7.1. (Computational study: stabilizing LTIS): *LTIS Background*: Dynamics of *Linear Time-Invariant System* (LTIS) is

[2]A formal reasoning could be as follows: With S as above, let W be the set of all workers knowing how to do at least one job from S, and let $\mathrm{Card}\, W < \mathrm{Card}\, S$. Let $\bar{S}$ be the set of job nodes not in S, and $\bar{W}$ be the set of worker nodes not in W. Let us set $\underline{I} = \{0\} \cup \bar{W} \cup \bar{S}$ and $\overline{I} = \{p+q+1\} \cup W \cup S$; $(\underline{I}, \overline{I})$ clearly is a cut. The arcs which start in $\underline{I}$ and end in $\overline{I}$ clearly are the $\mathrm{Card}\, W$ arcs from the source to the worker nodes in W and $q - \mathrm{Card}\, S$ arcs from job nodes in $\bar{S}$ to the sink; since all arcs are of capacity 1, the capacity of the cut is $q - \mathrm{Card}\, S + \mathrm{Card}\, W < q$, and thus the maximal flow is $< q$.

given by the recurrence

$$\begin{array}{c} x_0 = z \\ x_{t+1} = Ax_t + Bu_t + d_t,\ t = 0, 1, \ldots \\ \left[\begin{array}{l} \bullet\ \ x_t \in \mathbf{R}^{n_x}\text{: state at time } t \\ \bullet\ \ u_t \in \mathbf{R}^{n_u}\text{: control at time } t \\ \bullet\ \ d_t \in \mathbf{R}^{n_x}\text{: external disturbance at time } t \end{array}\right]. \end{array}$$

The conceptually simplest control policy is the *state-based linear feedback* given by

$$\begin{array}{c} u_t = Kx_t \\ {[K \in \mathbf{R}^{n_u \times n_x}\text{: controller}]} \end{array}.$$

Equipped with feedback control policy, the system exhibits *closed-loop* dynamics given by

$$\begin{array}{c} x_0 = z, \\ x_{t+1} = [A + BK]x_t + d_t,\ t = 0, 1, \ldots \end{array}$$

The closed-loop dynamics is called *stable* if in the absence of external disturbances — in the case when $d_t \equiv 0$ — all trajectories go to 0 as $t \to \infty$. It is known that closed-loop dynamics is stable iff the moduli of all eigenvalues of the matrix $A+BK$ are < 1, same as iff this matrix admits *Lyapunov Stability Certificate* (LSC) X — positive definite $n \times n$ matrix X such that for some *decay rate* $\gamma \in [0, 1)$, it holds

$$[A + BK]^T X^{-1}[A + BK] \preceq \gamma X^{-1}. \tag{!}$$

As for the necessity of this condition, we refer the reader to a whatever standard control textbook. Sufficiency is immediate: Assuming (!), and setting $Z = X^{-1/2}[A + BK]X^{1/2}$, (!) says that

$$\|Zy\|_2 \le \sqrt{\gamma}\|y\|_2 \quad \forall y.$$

On the other hand, passing from states x to states $y = X^{-1/2}x$, closed-loop dynamics becomes

$$y_{t+1} = Zy_t + X^{-1/2}d_t, \quad t = 1, 2, \ldots$$

implying that when $d_t \equiv 0$, $\|y_t\|_2 \leq \gamma^{t/2}\|y_0\|_2$ holds, whence $\|x_t\|_2 \leq \|X^{1/2}\|_{2,2}\|X^{-1/2}x_0\|_2\gamma^{t/2}$; here

$$\|Q\|_{2,2} = \max_{\xi:\|\xi\|_2\leq 1} \|Q\xi\|_2,$$

is the spectral norm of matrix Q. It follows that

$$\|x_t\|_2 \leq \sqrt{\lambda_{\max}(X)/\lambda_{\min}(X)}\gamma^{t/2}\|x_0\|_2, \quad t = 0, 1, \ldots \qquad (*)$$

where $\lambda_{\max}(X)$ and $\lambda_{\min}(X)$ are the largest and the smallest eigenvalues of $X \succ 0$.

Building stabilizing controller: One of the standard problems related to LTIS is to specify, given A and B, a feedback matrix K resulting in stable closed-loop dynamics. We intend to consider this problem in a slightly simplified setting, where the decay rate $\gamma \in [0, 1)$ is given in advance. Observe that (!) is homogeneous in X, so the constraint $X \succ 0$ can be without loss of generality replaced with $X \succeq I_{n_x}$. Passing from variables X and K to X and $F = KX$ and multiplying both sides in (!) by X from the left and from the right, (!) reduces to solving the system

$$X \succeq I_{n_x}, \; [AX + BF]^T X^{-1}[AX + BF] \preceq \gamma X$$

of matrix inequalities in variables $X \in \mathbf{S}^{n_x}$ and $F \in \mathbf{R}^{n_u \times n_x}$. Invoking the Schur complement lemma (Lemma 7.2) this system is equivalent to the system of *linear* matrix inequalities

$$X \succeq I_{n_x}, \; \left[\begin{array}{c|c} \gamma X & AX + BF \\ \hline [AX + BF]^T & X \end{array}\right] \succeq 0 \qquad (\#)$$

in variables X, F, which is an SDO feasibility problem.

Under the circumstances, the natural objective to be minimized is the maximal eigenvalue of X, resulting in the optimization problem

$$\mathrm{Opt} = \min_{X,\lambda}\left\{\lambda : \lambda I \succeq X \succeq I_{n_x}, \; \left[\begin{array}{c|c} \gamma X & AX + BF \\ \hline [AX + BF]^T & X \end{array}\right] \succeq 0\right\}, \qquad (D)$$

the rationale being the desire to minimize the right-hand side in $(*)$.

Now, for your task:

(1) Implement in software the design of stabilizing the controller via solving (D) and carry out a numerical experimentation.

Recommended platform: `MATLAB`,
Recommended setup:

$$A = \left[\begin{array}{c|c} \cos(\theta) & \sin(\theta) \\ \hline -\sin(\theta) & \cos(\theta) \end{array}\right] \text{ with } \theta = \tfrac{\pi}{32}, \quad B = \begin{bmatrix} 0 \\ 1 \end{bmatrix}.$$

You could use as a solver, either the Ellipsoid Algorithm, or an IPM like `SDPT3` or `Mosek` invoked via `cvx`.[3]

Recommended experimentation is as follows. With no control (i.e., with zero K), the closed-loop dynamics given by the above A is what is called "neutral" — with zero external disturbances, the subsequent 2D states are clockwise rotations of each other by angle $\pi/32$, so that the trajectory neither goes to 0, nor runs to ∞. However, it is easily seen that nonzero disturbances, even approaching zero as time grows, with zero controls may push the trajectory to ∞. In contrast, with stable dynamics, a bounded sequence of external disturbances results in bounded trajectories, and converging to zero as $t \to \infty$ disturbances result in converging to 0 as $t \to \infty$ trajectories.

You are recommended to observe these phenomena in numerical simulations, by making the sequence of disturbances a linear combination of harmonic oscillations: $d_t = \sum_{i=1}^k \cos(\omega_i t) f^i$, $t = 1, 2, \ldots$.

Solution: In my experiments, the initial state z was selected at random, and the external disturbances were selected as

$$d_t = \cos(\theta t)e + \cos(2\theta t)f,$$

with randomly selected unit 2D vectors e and f. The summary of my experiments is as follows:

[3] "CVX: Matlab Software for Disciplined Convex Programming," Michael Grant and Stephen Boyd, http://cvxr.com/cvx/.

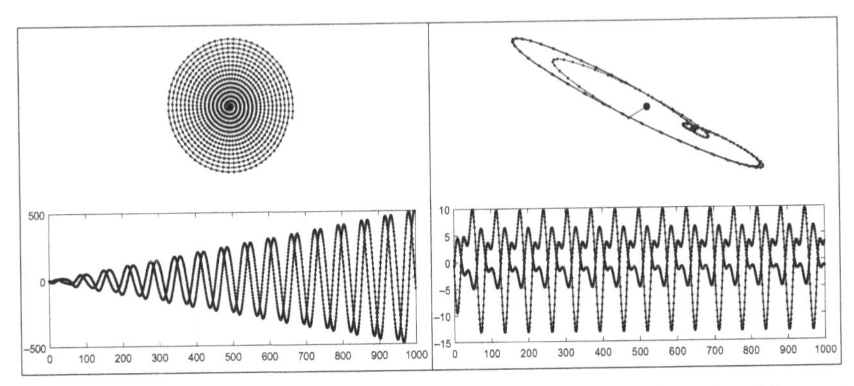

Fig. C.1. • Top: The curve $[x_1(t); x_2(t)]$ • Bottom: $x_1(t)$ and $x_2(t)$ vs. t • Left: no control ($K = 0$) • Right: Optimized static feedback • Bold black dot: Initial state.

- EA and IPM results:

	Opt	K	CPU, sec
IPM	2.7993	[−0.4316,−1.0423]	0.8250
EA	2.7993	[−0.6221,−1.5023]	0.1320

- sample trajectories with and without control are depicted in Fig. C.1.

The solution set to all exercises is available here:

Index

D

E

F

G

H

I

L

M

N

O

P

R

S

T

U

V

Printed in the United States
by Baker & Taylor Publisher Services